Self, Culture and Consciousness

Sangeetha Menon · Nithin Nagaraj
V.V. Binoy
Editors

Self, Culture and Consciousness

Interdisciplinary Convergences on Knowing and Being

Springer

Editors
Sangeetha Menon
NIAS Consciousness Studies Programme
National Institute of Advanced Studies
Bengaluru, Karnataka
India

V.V. Binoy
NIAS Consciousness Studies Programme
National Institute of Advanced Studies
Bengaluru, Karnataka
India

Nithin Nagaraj
NIAS Consciousness Studies Programme
National Institute of Advanced Studies
Bengaluru, Karnataka
India

ISBN 978-981-10-5776-2 ISBN 978-981-10-5777-9 (eBook)
DOI 10.1007/978-981-10-5777-9

Library of Congress Control Number: 2017946664

Printed on acid-free paper

This Springer imprint is published by Springer Nature
The registered company is Springer Nature Singapore Pte Ltd.
The registered company address is: 152 Beach Road, #21-01/04 Gateway East, Singapore 189721, Singapore

Contents

Editors and Contributors

About the Editors

Sangeetha Menon is Professor at the National Institute of Advanced Studies (NIAS), Bengaluru, India, and heads the NIAS Consciousness Studies Programme of NIAS. She is a nominated member of the International Society for Science and Religion (Cambridge), a Board Member of the International Association for Transpersonal Psychology, and a Council Member of the Indian Council of Philosophical Research, Ministry of Human Resources Development, Government of India. Professor Menon has authored *Brain, Self and Consciousness* (Springer, 2014) and *The Beyond Experience: Consciousness in Bhagavad Gita* (2007); coauthored *Dialogues: Philosopher Meets Seer* (2003); and coedited *Interdisciplinary Perspectives on Consciousness and the Self* (Springer, 2014); *Consciousness, Experience and Ways of Knowing: Perspectives from Science, Philosophy and the Arts* (2006); *Science and Beyond: Cosmology, Consciousness and Technology in Indic Traditions* (2004); *Consciousness and Genetics* (2002); and *Scientific and Philosophical Studies on Consciousness* (1999). She has visited and spoken at many universities in India, the United States, England, Australia, Germany, France, Italy, Spain, Japan, Taiwan, Singapore and Moscow. She has been visiting professor at the Oxford Centre for Hindu Studies, Oxford University, and at the Nanzan Institute of Religion and Culture, Nanzan University, Japan. She was invited to be a panellist at the World Parliament of Religions, Melbourne, in 2009. Apart from her academic interests, she writes poetry and fiction and is an avid photographer, artist and web designer. She also engages in charity programmes, being a trustee of the Sambodh Foundation, India. Her website is http://niasconsciousnesscentre.com/sm-cv.html.

Nithin Nagaraj is Assistant Professor with the NIAS Consciousness Studies Programme, National Institute of Advanced Studies, Indian Institute of Science Campus, Bengaluru. He was a visiting faculty member in Mathematics at IISER Pune for a semester before joining as Assistant Professor at the Department of Electronics and Communication Engineering, Amrita University for the period 2009–13. He also has several years of industry research experience. He worked as a Research Scientist (2001–04) and Lead Scientist (2013–15) at GE Global Research (Bengaluru) in the area of biomedical signal and image analysis. At GE Global Research, he conducted innovative work on medical image compression algorithms, medical image segmentation and registration, lossless data embedding and ultrasound liver tissue characterisation. During his stint in the industry, he was a co-inventor of eight US patent applications (two patents granted, all patents owned by GE). He joined the Consciousness Studies Programme at NIAS in October 2015. His current research interests include Complexity Theories of Consciousness, Neural Signal

Multiplexing, Causality Measures, Nonlinear Signal Processing and Chaos Theory. He has published 14 papers in international journals, and over 40 national and international conference presentations with a total of 700+ citations and an h-index of 10 (source: Google Scholar). He has an Erdös number of 3. He is an invited reviewer for the following international journals—*Chaos: An Interdisciplinary Journal of Nonlinear Science*, *European Physics Journal*, *Communications in Nonlinear Science and Numerical Simulation* (Elsevier), *IEEE Transactions on Image Processing*, *Acta Applicandae Mathematicae*, *International Journal of Imaging*, *EURASIP Journal of Information Security*, *Journal of Information Sciences* (Elsevier), *International Journal of Bifurcations and Chaos*, *IEEE Transactions on Information Forensics & Security*, *Computers and Mathematics with Applications* (Elsevier), *The Journal of the Franklin Institute* (Elsevier), *Mathematical Problems in Engineering*, *Journal of Theoretical Biology*. His personal interests include popularising mathematical thinking, the study of Indian scriptures (Vedanta) and the practice of Atma Vichara.

V.V. Binoy is Assistant Professor, School of Natural Sciences and Engineering, and a professor at the National Institute of Advanced Studies (NIAS), Indian Institute of Science Campus, Bengaluru. Binoy is interested in understanding the biological and environmental basis of social cognition in both animals and human beings. He explores the determinants of social decision-making and personality traits (also referred to as individual variation in the behaviour, coping style or behavioural syndrome) in vertebrates using fish and amphibian model systems. His research also focuses on the development of attitudes towards biotic and abiotic natural resources and environmental decision-making in schoolchildren from various cultures across India. Cultural variation in the autobiographical memory, cognitive style and fluid intelligence in children is another topic of his research. Binoy leads the biology education team of the Connected Learning Initiative (CLIx), a joint venture of the Massachusetts Institute of Technology, USA, and the Tata Institute of Social Science (TISS), Mumbai. Binoy is a research affiliate in the Center for the Study of Neuroeconomics, George Mason University, USA and Krasnow Institute for Advanced Study, USA. He has a master's degree and doctorate in Zoology and has been a recipient of the 'Cognitive Science Research Initiative Postdoctoral Fellowship' and 'Young Scientist' Start-up Research Grant from the Department of Science and Technology, Government of India. He is also passionate about yoga and Asian martial arts and has been a keen practitioner. For further details see: http://social-cognition.weebly.com/. He is interested in science education and communication. He hosts a citizen science initiative named Student-Network (http://www.nias.res.in/wash/), which aims to enhance student-scientist interaction and joint knowledge production.

Contributors

Michel Bitbol Archives Husserl, CNRS/ENS, Paris, France

Owen Flanagan The Center for Comparative Philosophy, Duke University, Durham, NC, USA

Harsha N. Halahalli Multimodal Brain Image Analysis Laboratory (MBIAL), National Institute of Mental Health and Neurosciences (NIMHANS), Bengaluru, India; Department of Neurophysiology, National Institute of Mental Health and Neurosciences (NIMHANS), Bengaluru, India; Department of Physiology, KS Hegde Medical Academy, Nitte University, Mangalore, India

Morris B. Hoffman University of Denver, Denver, CO, USA

Soumya Iyengar National Brain Research Centre, Manesar, Gurgaon, Haryana, India

V.N. Jha Centre of Advanced Study in Sanskrit, University of Pune, Pune, India

John P. John Multimodal Brain Image Analysis Laboratory (MBIAL), National Institute of Mental Health and Neurosciences (NIMHANS), Bengaluru, India; Department of Psychiatry, National Institute of Mental Health and Neurosciences (NIMHANS), Bengaluru, India; Department of Clinical Neurosciences, National Institute of Mental Health and Neurosciences (NIMHANS), Bengaluru, India; NIAS Consciousness Studies Programme, National Institute of Advanced Studies, Bengaluru, India

Jagadish S. Kallimani Department of Computer Science and Engineering, Ramaiah Institute of Technology, Bengaluru, India

Prathibha Karanth The Com DEALL Trust, Bengaluru, India

Samir Karmakar Jadavpur University, Kolkata, India

Rajesh Kasturirangan Mind and Society Initiative, Azim Premji University, Bengaluru, India

Frank Krueger George Mason University, Fairfax, VA, USA

Bindu M. Kutty Department of Neurophysiology, National Institute of Mental Health and Neurosciences (NIMHANS), Bengaluru, India; NIAS Consciousness Studies Programme, National Institute of Advanced Studies, Bengaluru, India

Victoria Lysenko Department for Oriental Philosophies, Institute of Philosophy, Russian Academy of Sciences, Moscow, Russian Federation

Ramesh Kumar Mishra Centre for Neural and Cognitive Sciences, University of Hyderabad, Hyderabad, India

Alok Nath Mohapatra National Brain Research Centre, Manesar, Gurgaon, Haryana, India

Ajay Kumar Nair Department of Neurophysiology, National Institute of Mental Health and Neuro Sciences (NIMHANS), Bengaluru, India

Deepti Navaratna Indira Gandhi National Centre for the Arts—Southern Regional Centre - Bengaluru, Bengaluru, India

Georg Northoff Brain Imaging and Neuroethics Research Unit, Institute of Mental Health Research, Royal Ottawa Mental Health Centre, Ottawa, Canada

Pravesh Parekh Multimodal Brain Image Analysis Laboratory (MBIAL), National Institute of Mental Health and Neurosciences (NIMHANS), Bengaluru, India; Department of Psychiatry, National Institute of Mental Health and Neurosciences (NIMHANS), Bengaluru, India

Pooja Parishar National Brain Research Centre, Manesar, Gurgaon, Haryana, India

L.M. Patnaik National Institute of Advanced Studies, Indian Institute of Science Campus, Bengaluru, India

Stephen G. Post School of Medicine, Center for Medical Humanities, Compassionate Care, and Bioethics, Stony Brook University, Stony Brook, USA

Seema Prasad Centre for Neural and Cognitive Sciences, University of Hyderabad, Hyderabad, India

Shankar Rajaraman NIAS Consciousness Studies Programme, National Institute of Advanced Studies, Bengaluru, India

Ambika Rathore Centre for Converging Technologies, University of Rajasthan, Jaipur, India

S. Siddharth NIAS Consciousness Studies Programme, National Institute of Advanced Studies, Bengaluru, India

Ravindra M. Singh Department of Philosophy, University of Delhi, New Delhi, Delhi, India

Ishan Vashishta NIAS Consciousness Studies Programme, National Institute of Advanced Studies, Bengaluru, India

Mohit Virmani NIAS Consciousness Studies Programme, National Institute of Advanced Studies, Bengaluru, India

Marieke van Vugt Institute of Artificial Intelligence & Cognitive Engineering, University of Groningen, Groningen, The Netherlands

Wenqing Zhao The Center for Comparative Philosophy, Duke University, Durham, NC, USA

List of Figures

Promises and Limitations of Conscious Machines

Auditory Verbal Hallucinations in Schizophrenia: A Model for Aberrant Self-consciousness

The Neuroscience of Blame and Punishment

Brain and Self: A Neurophilosophical Account

List of Tables

Bridging Self, Culture and Consciousness

Sangeetha Menon, Nithin Nagaraj and V.V. Binoy

We live in a time when the emotions we possess, the identities we carry, the memories we retain, the decisions we take, the unconscious influences to which we are beholden, and the free will we exercise all have an impact upon the fundamental nature of our consciousness, and determine how the self and identity will express themselves or evolve in a multicultural and pluralistic world. Such evolutionary changes are to be seen in the context of variations in cultural practices, decision-making, social constructions and our self-identity. We are at a critical point in history when never before have the intersections between culture, self and consciousness been so vital that the human species itself is redefining its distinctiveness as an evolving primate as well as a thinking and transforming person for the better. This book raises questions, reflects upon the most exciting possibilities, and debates the fundamental aspects of consciousness and self in the context of cultural, philosophical and multidisciplinary divergences. We believe that such an understanding lies at the cusp of philosophy, neurosciences, psychiatry and the medical humanities.

S. Menon (✉) · N. Nagaraj · V.V. Binoy
NIAS Consciousness Studies Programme, National Institute of Advanced Studies, Bengaluru 560012, India
e-mail: prajnanata@gmail.com

N. Nagaraj
e-mail: nithin.nagaraj@gmail.com

V.V. Binoy
e-mail: vvbinoy@gmail.com

S. Menon et al. (eds.), *Self, Culture and Consciousness*,
DOI 10.1007/978-981-10-5777-9_1

The Larger Questions

Who are we really? What drives us with insatiable appetite towards an accumulation of knowledge, experience, beliefs, values and attitudes? What can we know for sure—individually and collectively? In pursuit of answering these questions, human endeavours have led to incredible progress in the last 200 years in philosophy, the arts, humanities, sciences, engineering and technology. The first of these questions pertains to the nature and identity of 'self'. Is the self identical with a deeper core being which is ontologically significant, or is it culturally created, or is at an illusion produced by the brain in collaboration with the frail fringes of the body maps in our heads?

The arts, philosophy, sciences and every knowledge system that we have today make social contributions, and the role played by society in shaping our collective beliefs and attitudes has become inalienable. To be human is to share our being with a larger coexisting world as well as to adapt to changing tools and ways of knowing and creating knowledge. We have rich inner lives and we all have direct and immediate access to our inner self and the organic consciousness that surrounds it. Yet, we are unable to articulate our experiences accurately in third-person language. The search for the neural correlates of consciousness is continuing, with an emphasis on theories and methodologies of the physical and biological sciences. But the pertinent and unavoidable question is: will the understanding of consciousness ever reach a resolution and consensus regarding its fundamental features? The challenge is to have exhaustive third-person data that can be integrated with knowledge from alternative sources and methods which determine the specific nature of experiences of people from across cultures and who possess divergent cognitive and emotional traits. The impact on the individual's self of acculturation and migration to a different culture is vital, since there are also the subliminal threads connecting cultural expressions with the individual's wellbeing. In other words, the persisting question is whether we need to review the monolithic approaches that rely entirely on single disciplinary points of view? Can complex components of culture and mind be explained without the overarching force of consciousness which in a way is a unitary force as well as across species and life expressions? The nature and character of consciousness have remained elusive to exclusive objective analyses which science has perfected over the centuries. Yet, we are quite sure that we are conscious beings and that consciousness has a deeper connection with the continuing identity that each of us carries in ease most of the time!

While we do not understand the nature and function of consciousness per se in species other than humans and higher mammals, serious initiatives are being taken to understand the nature of intelligence and basic cognitive capabilities across the animal world and plant species. Are life, intelligence, cognition and consciousness interrelated phenomena? Can we talk about culture in the context of the life and behaviour of other species? What is it that demarcates the cultural traits of non-human species from those of humans? Alongside these questions exists another debate which goes to the root of activity such as intentionality and agency.

Who is an agent? And is consciousness always intentional? Is consciousness crucial to understand and even 'generate' reality? And, if it is, can we say that consciousness is bereft of the influences of the cultural contexts and societal practices within which one lives? These questions provoke significant discussion on considering consciousness in the realm of content and information, and raising the minimal features for an intelligent system to have agency and authorship. Further, important issues in this context are the existence of free will, decision-making, biological determinism and moral agency. How can these be placed within human cultural frameworks?

There is general agreement that consciousness is a complex phenomenon that should be looked upon as an entity constitutive of more than cognitive and rational capabilities, and thus beyond the Baconian and Cartesian models of the mind. One way of perceiving the complexity of consciousness is to consider it as a unitary phenomenon which integrates beneath its 'umbrella' deeper realms of mind, such as imagination and emotion. What is the role of emotions in understanding consciousness? Are emotions of cultural origin? If there are emotions that are pan-cultural and transcultural, what is the implication for understanding consciousness within the context of emotions and feelings? What is the neurological basis of meditative states? How do we understand concepts of beauty and the sublime, and their relation to mind? How are these related to cultural, spiritual, philosophical and aesthetic notions?

Experiential primacy and the subjective quality of experience are the hallmarks of consciousness. A dominant discussion in consciousness studies has been the question of 'qualia', or the subjective nature of experience, and whether or how it is produced. Phenomenology and phenomenologically inspired methods and approaches focus on first-person data and narratives to understand consciousness. In this context, how can the divide between subject and object, subjectivity and objectivity, experience and representation be understood? Will an emphasis on the subjective and experiential nature of consciousness bring in the importance of cultural embeddedness of human experiences? What are the philosophical implications of these questions in the context of Eastern or Western philosophical and literary traditions? Are there experiences, or features of experiences, that can be conceptually generalised beyond individual cultures? Is culture itself a unitary framework for understanding the human mind? Where can spiritual experiences be placed?

Though one might argue that consciousness is a unitary concept, or an entity with cognitive and linguistic overload, it is impossible to place consciousness outside the larger living space of human culture. The repertoire of behaviour, attitudes and values that we embrace has its origins in the cradle of culture which is both biological and societal. The emergence of dispositions such as empathy and altruism, to make decisions in varied ways, or our abilities to cope with challenges, and to move on—are these networked neurologically? What are the implications of cultural neuroscience for understanding the human self?

Consciousness and self are like two sides of a coin. It is almost impossible to talk about consciousness without a self who is the experiencer, and who possesses self-identity accrued upon and influenced by one's experiences and interactions with the environment and culture. But then how do we understand the altered self, the self that is on the fringes due to neurological and neuropsychiatric challenges? How are mind, consciousness and self interrelated? What are the different models of self? Do they have a biological or a cultural basis? What are the role and influence of culture in understanding the cause of neurological and psychiatric challenges? What is the role of family, society and workplace values and practices towards interventional measures for alleviation? What is the role of the medical humanities in the context of understanding the deeper realms of mind and consciousness?

Through the twenty-three essays in this volume, and the five parts into which the volume is divided, our attempt is to raise some of the above questions in order to present the overarching influence of divergent methodologies, interdisciplinary engagements and conceptual frameworks in projecting the three major puzzles for the coming century that are complex as well as leading towards evolutionary and humanistic connectors: self, culture and consciousness. We wish to speculate that the convergences in knowing and being will happen in the interstitial spaces between these three fundamental forces.

Bridging Self, Culture and Consciousness

The larger goal of this volume is to connect and contrast the essential features of the concepts of self, culture and consciousness and demonstrate that these three fundamental forces are both subliminally and explicitly tied with our ways of knowing and being in the contemporary world. In other words, the metaphysics of the self and the epistemology of culture are bound in the ways in which we theorise and project consciousness. The chapters in Part I present a succinct picture of the scenario of the emergence of consciousness, and intelligence as a product of evolution, which can also be contextualised in the machine world. Part II highlights the fragile and hence very serious subject matter of self-transformation and self-reflection, and how alternative ways of care and conceptions of identity help us to understand those communities of people who struggle to create and experience meaning in their daily lives, due to various challenges to their sense of self. The authors of these works strongly argue for the connection between healing, agency and being.

Often, our records and methods of understanding consider the knowing process as being exclusively, or at least heavily, rational. Rational entities and rational choices rule our worlds. But is that really true in our lives? Of what are our lives actually constituted? These two questions can be addressed to a great extent with the help of situating the person and the self in culture, language and a world of emotional meanings. The individualistic nature of cognitive functions and capabilities that we present aside, our ways of knowing are also socially graded and gauged in a continuous manner. And this is why a field like cultural neuroscience can contribute much in approaching a subject like social cognition. Part III provides six perspectives from which to understand memory, emotions, meaning-making and the other self.

With this background, Part IV presents four alternative epistemologies on the self from the point of view of its subjective nature. These four epistemologies consider, on the one side, the intricate connectedness between the experiential and the neural nature of the self. Both the brain and the self seem to narrate the stories of our lives but in different languages, while creating a unitary background of meaning through the hallmark context of experience. Discussion on the qualitative nature of conscious experiences and the cognitive frameworks where the meaning-making happens demands the interdisciplinary engagement of not just methods but of fundamental concepts that we draw upon for knowing. How do we know the experience? In other words, is knowledge possible without the overarching element of being brought into our epistemological process? The four chapters in Part IV provide different responses to such questions.

Part V, which comprises five chapters, is an attempt to bring in both the scientific and philosophical significance gained by the field of consciousness studies. The dialogues that existed a hundred years ago were not able to move across epistemologies of a strict scientific tradition and a speculative philosophical discourse. Today, however, it is possible to discuss physicalism, intelligence, the experiential primacy of consciousness and the future of the self itself in a language that passes through the hard doors of empirical methods while retaining that essence of our ways of knowing and being in an interdisciplinary world which is both pluralistic culturally and individualistic experientially.

Emergence of Consciousness

While we talk about consciousness mostly from a humanistic platform, it becomes imperative that a fair amount of discussion be devoted to the emergence of consciousness from an evolutionary and artificial intelligence perspective. Part I of this book begins with a discussion on the emergence of consciousness, from both the biological and machine points of view, along with an investigation on the varied nature of the cognitive system in the deaf, and how meditation inspires the neural wirings to change through the mechanism of neuroplasticity. Soumya Iyengar, Pooja Parishar and Alok Nath Mohapatra, in their chapter (Chapter "Avian Cognition and Consciousness—From the Perspective of Neuroscience and Behaviour"), give a review of avian cognition and consciousness, from the perspective of neuroscience and behaviour. The answer to the question of what is likely to be 'x animal' is not possible until we find a mechanism to decode the 'animal mind' into a language accessible to human beings. However, indirect approaches exploring the behavioural and neural analogies for human cognitive capacities have given some results. Among species, after mammals it is the species of birds that has been studied widely to understand cognitive abilities. Significant interspecific variation exists in the cognitive abilities of birds. Yet, there is growing evidence to prove that popular model systems, such as pigeon, parrot, etc., exhibit several cognitive capacities once thought to be unique to human beings. Interestingly, the bird brain is strikingly different from the mammalian brain and does not possess a neopallium, the seat of higher cognitive abilities present in human beings, which make the studies on bird cognition significant. Soumya Iyengar and her collaborators offer discussions on a topic that is yet to take centre-stage in Indian academia. This chapter helps the reader to bridge the function and structural basis of cognitive abilities in a brain that is strikingly different from but functionally similar to the mammal's.

According to modern neuroscience, the behaviour of an organism, including that of human beings, is represented in the brain by neural circuits and any ensuing modifications lead to an alteration in the specific behaviour. Though meditation has been widely accepted as a tool for self-regulation and self-transformation across cultures, the scientific study of how meditation practices affect behaviour has gained popularity only in recent times. The chapter on meditation, cognitive reserve and the neural basis of consciousness by Ajay Kumar Nair and Bindu M. Kutty (Chapter "Meditation, Cognitive Reserve and the Neural Basis of Consciousness") elucidates how the continuous practice of meditation techniques transforms the brain and behaviour by utilising a unique capacity of the brain, called 'plasticity'. The study points out that long-term practitioners of meditation acquire higher levels of attention and awareness.

The exploration of the impact of developmental abnormality in any of the components of the cognitive system, such as attention, perception, processing of information, memory and decision-making, and the behavioural and neurological adaptation exhibited by an individual to compensate for its negative impact can provide a window to understanding the biological bases of consciousness. The chapter on the 'plasticity of the brain' by Seema Prasad and Ramesh Kumar Mishra (Chapter "Attention and Perception in the Deaf: A Case for Plasticity in Consciousness") details the processes of attention and perception in deaf people. Due to the lack of availability of auditory stimulus, the brains of deaf people depend more on visual stimuli for cognising the world and, due to the plastic nature of the nervous system, their system of visual information-processing becomes more sophisticated in comparison with people who can hear as well as see. This study on cognitive divergence in deaf people may offer insights to understand how attention, awareness and consciousness are connected. This chapter states that, due to the adaptational divergence in the brain circuitry, deaf people differ from others in the conscious and unconscious processing of visual stimuli. The authors emphasise the need for analysing cognitive systems with modification at different levels, such as perception and cognition, to understand the biological bases of awareness and consciousness.

The future of machines and artificial intelligence (AI) holds exciting prospects in the context of consciousness, but also presents some tough challenges. We, humans, experience our surrounding environments with five senses, and we respond with sentiments, feelings and emotions, features that are not yet present in machines. L.M. Patnaik and Jagadish S. Kallimani (Chapter "Promises and Limitations of Conscious Machines") survey the challenging issues facing AI, the promise inherent in these challenges, and the limitations of machines in the context of consciousness, cognition, intelligence and ethics. Self-awareness is a puzzle for machines, while unawareness does not play any role in machine consciousness. Machine cognition could play an important role in handling hazardous and critical environments that are beyond the experience of humans. In a futuristic society where machines will interact with humans, it is imperative that machines are endowed with emotional intelligence. Another important aspect is ethics and the need to protect human dignity and confidentiality while designing such intelligent machines. Building ethical values in machines is a complex topic which first requires a deeper understanding of ethical values in humans.

Healing, Agency and Being

What is the role of personal agency in healing and understanding the altered self of the person with neural and psychological challenges? In Part II of the book, three significant areas are discussed by the first three authors—specifically, dementia, autism spectrum disorders (ASD) and schizophrenia. The last chapter of this part offers a detailed discussion on the classical text of the *Yoga Sutra* in the context of defining the place of self-reflection in addressing psychological challenges and the ensuing self-alteration. This part as a whole focuses on the change in self-perception caused by the challenges received by the mind and brain, and on whether the same can be approached with newer perspectives on self, self-reflection and ways of healing. The chapter by Stephen G. Post (Chapter "'Is Grandma Still There?' A Pastoral and Ethical Reflection on the Soul and Continuing Self-identity in Deeply Forgetful People") is a critical account of the exclusive methods of scientific reductionism and how it fails to understand the subjective experiences of people with dementia, and how the importance of an enduring self is dismissed, ignoring the personhood of people with Alzheimer's disease. The author narrates several accounts of care movements for people with dementia that involve art and music, etc. In such movements, what is endorsed and respected is an enduring self of the person. In the scientific account of self, several contemporary philosophers place emphasis on 'hypercognitive values', though the other dimensions of the human self, such as emotional, relational, aesthetic, creative and spiritual values, contribute to building meaning in the lives of the deeply forgetful. Post remarks that, according to those who endorse 'hypercognitive values', to be a 'person' requires the ability to project rational plans as moral agents into the future. Since people with Alzheimer's disease do not generally have this capacity, they are deemed to be something less than 'persons'. This chapter raises the important question of whether by labelling a deeply forgetful individual a 'non-person', we are ignoring the moral place in the human community of equal regard.

Autism spectrum disorders (ASD) represent the most studied developmental abnormality where malformed vital cognitive abilities, especially those required to live in a complex social system (social cognition), is the focus. The chapter by Prathibha Karanth (Chapter "Children with Autism Spectrum Disorders: A Case for 'Alternative Selves'?") presents an alternative view of the approach to understanding the self through the discussion of children with ASD. The experiences and difficulties of these individuals raise a whole range of questions on the role of the brain in our notion of self, including the body and mind of the self vis-à-vis the external world and the others in it. The author argues that autistic children, being the victims of divergence in the development of sensory and motor systems and higher cognitive abilities, fail to make a 'normal mental image' of the world and hence to adapt by choosing the socially right responses. In the recent past, a number of autobiographical narratives from those diagnosed with ASD have increasingly sensitised the public to the many issues faced by individuals diagnosed with autism, that result in the many 'symptoms of ASD' which are listed as diagnostic characteristics of autism spectrum disorders. By studying the experiences of the subjects of ASD in their own words, we can change the perception of this disease and suggest rehabilitation measures.

Agency is the sense of ownership of one's actions, a failure of which has been associated with various symptoms of schizophrenia. A failure of self-agency is characterised by symptoms such as hallucinations, delusions and disorganisation symptoms, including formal thought disorder, inappropriate affect, bizarre behaviour, reduced or absent verbal output, reduced initiative to perform goal-directed actions, inability to derive pleasure from otherwise pleasurable activities, general social withdrawal, attention and memory deficits and impaired executive functions, emotional dysregulation (e.g. depression, anxiety), a sense of hopelessness, demoralisation, suicidal tendencies, and lack of insight. Such symptoms, seen in varying extents of severity, have a significant impact on the social and occupational wellbeing of the patients, affecting work, interpersonal relationships and self-care. John et al., in their chapter (Chapter "Auditory Verbal Hallucinations in Schizophrenia: A Model for Aberrant Self-consciousness"), investigate auditory verbal hallucinations (AVH)—a hallmark symptom of schizophrenia, as a model for understanding the impaired self. Using advanced neuroimaging methods like functional magnetic resonance imaging (fMRI), one can probe the underlying mechanism of self-agency and its possible failure in patients experiencing AVH. This in turn can lead to insights into the neural correlates of self. To this end, the researchers introduce a novel fMRI paradigm—Hallucination Attention Modulation Task (HAMT)—to study the neural correlates of AVH and discuss some preliminary results from a pilot study which link AVH to an underlying disorder of self-agency. The preliminary results are promising and link AVH to an impaired self-agency in patients with schizophrenia, hinting at the superior parietal lobule being the neural substrate for the same.

The chapter by Sangeetha Menon (Chapter "Body and Self-reflection: The Crux of Yoga Philosophy and Practice") suggests another perspective on agency, offering a theoretical framework to understand self-reflection from the point of view of yoga philosophy and practice. Self-reflection is awareness about one's mental states in the context of collective awareness of the continuing personal identity frilled by memories, beliefs and worldviews. Reflecting upon another's mind entails complex processes that are facilitated by prior self-awareness. The author writes that Patanjali's system of Ashtanga yoga or the eight-fold path that consists of *yama, niyama, āsana, prānāyāma, pratyāhāra, dhārana, dhyāna* and *samādhi* is an integral model covering the social, emotional and personal world of the individual. The lack of complexity in the construal of agency is also the result of the lack of phenomenological and ontological richness in the conceptualisation of the self and the agent. The phenomenal aspects of being an agent include an array of both apparent and subliminal experiential content that pertains to beliefs, worldviews, perceptions and understanding about one's sociocultural context. The *Yoga Sutra* approach to consciousness, according to this study, is based on a model that integrates various emotional, cognitive and social aspects of the agency with a purpose of transforming the agent for positive experiences. The yoga theory of mind is not focused on the awareness of the other, but gives focus to self-awareness, and to the ability to self-reflect. Yoga phenomenology focuses on the inner awareness of the self, rather than on the awareness of information received from an external environment. In this chapter, Sangeetha Menon presents four important elements of self-reflection derived from yoga phenomenology based on its model of the mental levels (*citta bhumi*), pain (*kleśa*), and the techniques (*viveka khāyti, prati prasavam, kriya yoga, samyama* and *dhyānam*) for the restraining of mental distractions. These are self-certainty, body-centrality, cognitive centrality and contented experience. It is hoped that a complex framework of agency, helped by yoga phenomenology and psychology, will contribute to an understanding of the nature of self-reflection, and its role in addressing psychiatric and psychological challenges.

The Social Self, Culture and Cognition

Our experiences and cognitions are not only situated in but also influenced by the culture we possess, and the society in which we live. The self which is the central force behind human experiences is constantly changed or shaped by the values we hold, the emotions we experience, and the complex ways in which our knowing and experiencing are interspersed. The intersubjective modes of communication with the community in which we share our lived values create new sources of information and knowledge. Though in our daily living we take the perceptions we have and the meanings that accrue to be based on true facts, we rarely question the reality of the notions and concepts that are subliminal to thinking and lie unexamined. Focusing on the influences of the self, such as memory about oneself, influences from cultural practices, how concepts and meanings are attributed by users of language, and how emotions and mental states can be understood from a classical dualistic system, Part III highlights alternative epistemologies in the context of cultural and pluralistic diversity. This part delves into these convergences.

How is it that we live in a coherent, complete and stable world despite such poor access to it from our senses? This is one of the central questions that Rajesh Kasturirangan addresses in his chapter (Chapter "Fullness, Trust and the Self"). Our perceptual experience of the world is woefully incomplete, yet we persist in believing and 'trusting' that our world is 'full'. Kasturirangan argues for the existence of a background which is the substratum of the 'full' world, as well as a background organiser of our percepts—the self. Trust comes from the world we live in and is mediated by the self, and the self's trustworthiness is public and verifiable. This self is prior to sensory consciousness and mind-body and is embedded in a trustworthy world, contrary to the Cartesian view which is predicated on the mistrust of our senses, of our conceptual capabilities and our bodily needs. However, can we rest in the security of this newfound trustworthiness of the world, and the unique authenticator of trust and trust-bearer, the self? Rajesh ends with the possibility that reality could be more diabolical than we can imagine—what if the world is being created and destroyed every moment by a capricious divinity whose only purpose is to turn our dreams and illusions into reality? The self plays an important role in this new scepticism, since it is through the manipulation of this self that we end up losing the ability to distinguish between human life, the world we inherit through evolution, and the mechanical world we have created in the last 2000 years.

Although the capacity to make, store and retrieve a replica of stimuli perceived, ranging from objects to complex episodes and events witnessed, is present in various animal species, memories where mental time-travel and subjective experience of an event are involved, known as autobiographical memory, is unique to human beings. The chapter by Binoy et al. (Chapter "Autobiographical Memory: Where Self, Wellbeing and Culture Congregate") focuses on autobiographical memory as 'a subjective perspective on specific events experienced at particular time points linked together on a personal timeline'. With semantic memory we recall the what, when and where aspects of an event. Autobiographical memory is the remembrance of the subjective experience of an event, and the personal history reliving a specific event, without losing the sense of self and experience. Understanding the formation and utilisation of autobiographical memory is important, since this cognitive capacity is vital in determining the formation of identity, maintaining social relationships, solving problems and emotion regulation. Studies suggest that people from Eastern and Western cultures are significantly different in the nature and usage of autobiographical memory. There is ample evidence to indicate that individuals who use autobiographical memories have higher levels of positive relationships and wellbeing.

Traditionally, the influence of music on the brain, cognition and consciousness has been studied as a subsystem of auditory cognition where understanding the link between stimulus and response is the paradigm and musical memory is considered only a trace of that reaction. The chapter on musical memory by Deepti Navaratna (Chapter "The Alchemy of Musical Memory: Connecting Culture to Cognition") states that current research reveals learning, practising, generating and experiencing music to involve both conscious and unconscious processes. Musicians and listeners go through a multimodal possessing of information and emotions. These studies also point to the need for integrating the theories of embodiment and the self, involving both the creator and receiver of music, while studying the mechanism of music-processing and the formation of musical memory. Increasingly, evidence is available to prove that people from divergent cultural backgrounds exhibit noticeable variations in their ability to understand and enjoy various elements of musical stimuli. Focusing on musical memory, the author introduces a relatively young branch of cognitive science—namely, musical cognition.

Morris B. Hoffman and Frank Krueger in their chapter (Chapter "The Neuroscience of Blame and Punishment") on the neuroscience of blame and punishment discuss 'third-party punishment' where a punitive action is 'taken by one individual against a second individual in response to a harm or threat inflicted by the second on a *third* individual'. This chapter discusses the link between 'self' and 'intentionality' by drawing examples from the contemporary literature on third-party punishment. The intentionality of the person who has committed a crime is considered and more stringent action is recommended against those who break the law deliberately than against someone who did the same violation inadvertently. The authors argue that self-related aversive experiences are associated with the distress experienced by the victim while penalising a third person. The brain networks activated while taking a decision to punish an individual who has caused suffering to another person are also involved in processing the information associated with self-monitoring and personal distress. Further studies in this line of work are expected to shed more light on our understanding of how the brain comes up with blame and punishment, and the ensuing insights could even change the way in which the judiciary perceives the criminal states of mind (e.g. purposeful, knowing, reckless and negligent) as well as the intentionality behind the violation when determining a punishment.

Questions on what is real, what is experienced and what is known are central to discussing meaning. What is the nature of existence common to fictitious entities, for instance, the different characters in children's literature, such as a horse with wings, giants, dwarves and talking beasts? These are characters construed by adults to engage young minds in a transaction which creates a world other than the one ingrained in the collective (un)conscious of people. Samir Karmakar in his chapter (Chapter "Becoming Conscious About the Existence of the Non-existents: Logic, Language and Speech Acts") on becoming conscious about the existence of the non-existents presents a topic that is linguistically important and metaphysically time-tested. The study on how meaning is attributed by users of a language while using non-existential naming words helps establish the centrality of seemingly irrational, non-existent expressions in developing the initial years of young minds. The author believes that the development of rationality relates to mental exercises which may otherwise be considered *irrational*, and that the existence of an entity is qualified by the set of conventions that are associated with our thinking. Speech has a crucial role in the formation of the world of conventions. It is crucial also in maintaining as well as transferring conventions. While expressing our thought through speech, we perform certain kinds of intentional acts. These acts are crucial not only in expressing our thoughts but also in recreating the world again and again.

Indian philosophical systems offer rich and thorough theories of consciousness, self and materiality. The Samkhya system, one of the six *darsana* systems of Indian philosophy, has a rigorous approach to understanding matter and the self. Matter is thought of as continuously evolving, owing to the dynamic interplay between the three elements or *guna* that constitute it—namely, *sattva, rajas and tamas.* The psychophysical apparatus of *jiva,* the individual selves, are also a product of matter and are therefore constantly changing. The predominance of one of the three *guna*s determines the kind of change and allows beings to be classified into distinct personalities. In his chapter, Shankar Rajaraman (Chapter "Bhoja's Model for Analysing the Mental States of Literary Characters Based on Samkhya Metaphysics") delves deeply into the eleventh-century Sanskrit aesthetician Bhoja's model for analysing the mental states of literary characters based on Samkhya metaphysics. The key to understanding the mental state of a literary character, according to Bhoja, is to figure out the personality type from the knowledge of character traits. Bhoja conceptualises literary characters as possessing a personality and classifies them into four categories as *santa, udatta, lalita,* and *uddhata*, corresponding to *sattvodrikta* (excess of *sattva guna*), sāttvika (predominantly *sattva, rajasa* (predominantly *rajas*) and *tamasa* (predominantly *tamas*) categories. Further, he provides 24 traits that can (but need not necessarily) form the basis of a character's self-identity. Characters that possess all 24 traits are classified as superior (*uttama*) and those with 18 and 12 traits as intermediate (*madhyama*) and inferior (*kaniṣṭha*) respectively. He gives a further taxonomy of unique traits and the different life-goals pursued by these four personality types. The author argues for the richness of Bhoja's model in going beyond literature, and claims that it has implications for understanding the mental states of real-world individuals, their traits and personality types, self-identity and life-goals. For example, it is tempting to explore the utility of Bhoja's model for understanding the mental states of real-world individuals since they feel, think and behave like different literary character types at different times and thus possess a dynamic, continuously evolving personality. Another application of Bhoja's model which is proposed in this chapter is a tool for self-reflection and personal growth.

The Self and Alternative Epistemologies

How are brain processes and self-processes related? Do we need to use alternative epistemologies to delineate the intricate ways in which the brain and the self influence each other? These two concerns further inspire us to look at the ways in which self-perceptions can be carved by different traditions and ways of living. The underlying challenge for describing the self in an exhaustive manner is to explain the problem of qualia. The four chapters in Part III help us to re-examine the notion of qualia and the primacy of experience from the point of view of both the discussions on the self and on consciousness. Conscious experience arises from and refers to a subjective self. Georg Northoff, in his chapter (Chapter "Brain and Self: A Neurophilosophical Account"), explores different concepts of the self and how they are related to recent findings about neural mechanisms related to the self-reference of stimuli. Neural activity in the middle regions of the brain, the cortical midline structures, is increased whenever self-specific stimuli are presented (the self-reference effect in neuroscience). Moreover, increased neuronal synchronisation in the gamma frequency domain can be observed. In order to relate empirical neuroscientific findings of self-reference to the philosophical concepts of the self, neurophilosophical discussion is necessary. This is addressed by investigating the degree of specificity for the self of these neuroscientific findings. Neuronal specificity, psychological specificity, experimental specificity, phenomenal specificity and conceptual specificity of the self are discussed in detail in this chapter. Each of these concepts attempts to determine how specifically they relate to the self. For example, neuronal specificity describes whether the spatial and temporal patterns of neural activity observed in studies about self-specificity are really specific to the self. It turns out that in each of these domains (neural, experimental/psychological, conceptual, phenomenal) current imaging studies on self-reference suffer from non-specificity. In other words, these are unspecific with regard to the concept of the self itself. They tell us about self-reference as the relationship of particular contents to the self, but not about the self directly. The idea of a 'minimal self'—a basic sense of self that occurs immediately and is always already part of our experience of the world—is discussed and its relation to neuroscientific results is explored. Neuroscience is yet to demonstrate which neuronal mechanisms underlie the experience of 'mine'-ness and belongingness.

Owen Flanagan and Wenqing Zhao (Chapter "The Self and Its Good Vary Cross-Culturally: A Dozen Self-variations and Chinese Familial Selves"), with a focus on Chinese philosophy, demonstrate in their chapter that self and its conception of good vary cross-culturally, through the concept they term as 'self-variations'. This chapter gives an extended analysis of a distinctive conception of the self, its nature, boundaries and fate, which both explains and justifies certain characteristic ways of thinking about goodness and wellbeing in Chinese philosophy. According to Chinese philosophy, the metaphysical aspects of the self are connected to a family lineage. It is proposed that the culturally situated Chinese notion of *wellbeing* can provide important dimensions of the self in the Chinese tradition. The Chinese family-oriented self is different from the interdependent self (such as the Buddhist all-sentient-beings-oriented self) and the independent self (such as the neo-liberal self). The individual's nature is relational, and individual wellbeing is understood in terms of the wellbeing of the family.

The 'sensation of pain', 'the taste of tea', 'what it feels like to see the colour red' are familiar examples of our subjective experiences and these are referred to as 'qualia'. The problem of qualia is indeed a hard one—being purely subjective and unique, it resists all attempts to be described by a reductionist point of view—the hallmark of 'objective' scientific inquiry. Victoria Lysenko (Chapter "The Problem of Qualia: Perspectives on the Buddhist Theories of Experience") investigates this problem of qualia from a Buddhist perspective. In particular, she draws several parallels between qualia and the notion of 'dharmas' peculiar to Abhidharma schools. Arguments of Buddhist epistemologists such as Dignāga and Dharmakīrti may, according to the author, help to explain the appropriation of experience as 'mine' with no reference to the metaphysical subject (*atman* or self). The Buddhist approach is experientialist, and to know the world means to experience, feel and live it through our body and not only to think about it. Buddhism emphasises three important principles (*trilakṣaṇas*): impermanence (*anityata*); selflessness —'non-self' (*anātman, anatta*); and unsatisfactoriness (*duḥkha*). According to the Buddhists, the experience of 'I' is wrongly taken as evidence for the existence of the eternal and unchanging self (*atman*). Thus, for the Buddha and Buddhist thinkers, the word *ātman* is a purely conventional designation of the impermanent mind. The word *adhyatmika* refers, in fact, not to the *atman*, as its etymology may suggest, but to certain cognitive phenomena which are 'close to mind', like the six *indriya*s (the faculties of sight, hearing, taste, smell, touch and mind) and the six *vijnana*s, or the six modalities of cognitive discernment (visual, auditory, gustatory, olfactory, tactile and mental).

Self-related thoughts can be persistent and lead to self-induced, self-inflicted suffering. Such spontaneous thought activity, commonly known as mind-wandering and termed 'perseverative cognition', can sometimes be beneficial for skills such as creativity, but it can turn into impairing rumination and becomes a risk factor for depression and other forms of psychopathology. What are the cognitive and neural processes that build up this type of sticky, self-related thinking and can we start to measure it in the laboratory? Marieke van Vugt, in her chapter (Chapter "Getting Stuck on Myself: The Cognitive Processes Underlying Mental Suffering"), reviews the basic cognitive science of spontaneous thought, together with the clinical psychology literature on different types of self-related thinking. The theories in the literature to explain perseverative cognition include a bias towards attending to negative information, a defect in inhibition, and habits of thought that make one get stuck in rehearsing negative patterns of thinking. Computational modelling can help us to arrive at a fine-grained understanding of a cognitive process and enable quantitative predictions about the consequences of different mental strategies. Van Vugt goes on to propose a simple computational model of mind-wandering implemented in the adaptive control of thought-rational (ACT-R) cognitive architecture. Her model satisfactorily reproduces some hallmark phenomena in the mind-wandering literature. Analytical meditation practices, inspired by the age-old contemplative practices of Tibetan monks, may hold promise as interventions for reducing perseverative cognition. Such practices help in deconstructing self-referential thought, one of the key ingredients of perseverative cognition. Some of these methods hold the key to liberate people to be less of a victim of their own self-referential thinking, thereby unlocking the potential of their minds.

Consciousness, Experiential Primacy and Knowing

The two traditions that bring in scientific concepts as central to solving the problem of subjectivity in consciousness studies are phenomenology and panpsychism. Both traditions carry an acute sense of subjectivity in conceptualising the place of experience and its interiority. The last part of this book, Part V, gives a bird's-eye view of one of the current excitements in consciousness studies—namely, bringing in information theory to understand experiential primacy, and its prospects to explain the qualitative nature of all experiences. The greater challenge in the dialogue between these two opposing camps of phenomenology and panpsychism is that we once again have to revisit the division and demarcated spaces between what we mean by that which is material and that which is conscious. Hence this section ends with a review of the conceptual differences held between the classical Indian traditions in defining matter and consciousness. This deeply philosophical and theoretical approach might also add richness to the ways by which the phenomenologist and the panpsychist shape the notions of experience and the self.

Scientific inquiry is highly successful at objectifying reality—but when it comes to the study of the subjective self it leaves us with a blind spot. Michel Bitbol (Chapter "Beyond Panpsychism: The Radicality of Phenomenology") reminds us of this blind spot and compares and contrasts two approaches that have been adopted by the West to address it, in his chapter on the radicality of phenomenology going beyond panpsychism. The method of phenomenology starts off by suspending all judgements about the alleged objective world, together with the very activity of objectification, on the basis of lived experience. The other approach is panpsychism, or rather pan-experientialist metaphysics, which attempts to put lived experience back into the very domain that was deprived of it by the act of objectifying. Both these approaches are dramatic or radical from the perspective of science but have diametrically opposite trajectories. Panpsychism takes the 'progressive' direction of scientific research and tries to identify a missing ingredient which is construed as a further property of things. Phenomenology, on the other hand, adopts the 'regressive' direction and tries to identify the experiential background out of which the world of manifest objects is picked out and constituted. Bitbol documents both directions in succession, starting with a critical exposition of the first, 'progressive' one (panpsychist metaphysics), and then defending the second, 'regressive' one (phenomenology), as the most promising in terms of making global sense of our being-in-a-world.

We live in the information age today, thanks to rapid developments in communication systems, networking and computer technology. The birth of the information age can be traced back to Claude Shannon's paper in 1948 (Shannon 1948), where he developed a mathematical theory of communication and gave a probabilistic definition to the notion of 'information'. The act of choosing 'one' unique message from 'many' possible messages gives rise to the generation of information which can be precisely quantified by the celebrated Shannon entropy equation. Having convincingly argued against physicalism or reductionism to explain consciousness, David Chalmers, in his path-breaking paper (Chalmers 1995), proposed a new programme for building a non-reductive theory of consciousness. Chalmers argued for treating consciousness as a fundamental entity (like mass, charge or spin) rather than attempting to reduce consciousness to brain function or explained purely in terms of computational or neural mechanisms. Chalmers observed the isomorphism between physically embodied information spaces with phenomenal or experiential information spaces, and thus postulated the notion of 'double-aspect information' (Chalmers 1996). In his proposed non-reductive theory of consciousness, information is truly fundamental and comprises dual aspects—corresponding to the physical and the phenomenal features of the world. Chalmers' speculative proposition of the existence of a fundamental link between 'information' and 'experience' did not receive much attention in the science of the mind, until recently, when Tononi (and others) proposed a theory of consciousness borrowing similar ideas from information theory (Tononi 2004a; Oizumi et al. 2014). Tononi's Integrated Information Theory, or IIT, is an axiomatic non-reductive scientific approach which takes the phenomenology of experience and its characteristics as primary and builds a theory to explain consciousness as the capacity of a system of mechanisms (constituted of elements which can either be neurons or logic gates) to integrate intrinsic information. Intrinsic integrated information, though not in the sense of Shannon, plays a central role in this new theory to measure the quantity of experience and to represent its quality. In their chapter, Nithin Nagaraj and Mohit Virmani (Chapter "Is 'Information' Fundamental for a Scientific Theory of Consciousness?") trace the fundamental ideas behind IIT and also highlight its limitations. They discuss another important clinical measure of consciousness known as Perturbational Complexity Index, or PCI (Casali et al. 2013). Both IIT and PCI belong to the new genre of scientific measures which are rooted in what are called 'complexity theories of consciousness' (Seth et al. 2008). Nagaraj and Virmani then discuss the link between integrated information, consciousness and data compression. They also explain their latest approach to bridging the gap between IIT and PCI through a novel approach known as compression complexity (Virmani and Nagaraj 2016). Thus, the latest trend in scientific theories of consciousness has been on the problem of 'measurement'. To this end, recent scientific literature in the field has witnessed the proposal of a medley of scientific measures which attempt to capture brain complexity and its link to consciousness. Whether *information* and *complexity* are fundamental notions necessary to explain and *measure consciousness* remains to be seen.

The chapter by Ravindra M. Singh (Chapter "Encircling the Consciousness Conundrum") on encircling the consciousness conundrum is another study to demonstrate that the current state of theorising the IIT as developed by Edelman and Tononi (Edelman and Tononi 2000; Tononi 2004b, 2008, 2012, 2015) and its subsequent supplementing through the work of Koch and his research team (Tononi and Koch 2015; Koch et al. 2016) represent the most rigorously articulated and empirically testable hypothesis available on the topic. The author starts with a review of the progress in consciousness studies, from the philosophical point of view, in the last few decades. While acknowledging that not everything is yet settled as far as consciousness is concerned, and distancing himself from the sceptics about consciousness, the author argues that researchers have recently made definite progress through novel methods for isolating neural correlates of consciousness, such as the no-report paradigm, the perceptual awareness scale, attentional blink and other online functional imaging techniques. The author believes that IIT offers a new paradigm that offers promise for science and a big challenge for philosophical exercise.

The following chapter in this part by S. Siddharth and Sangeetha Menon (Chapter "What Does It Mean for Qualia to be Intrinsic?") examines the nature of intrinsic property. Debates in consciousness studies consider the intrinsic property of an object to be one of the formidable challenges to address. The experiential quality or the qualia is mostly considered as intrinsic except for the physicalist schools of its extrinsic and relational nature. An intrinsic property of an object is intuitively understood as the property which the object can have independently of all other objects. What does it mean for a property to be intrinsic, and why do physical properties not qualify as intrinsic? The distinction between intrinsic and extrinsic properties is based on the notion that any relations (extrinsic properties) require *relata* (or substances) of which they are relations. This chapter details the distinction between being intrinsic and extrinsic, by examining two approaches to defining an intrinsic property. (i) Combination of experiences is not possible. If one further assumes ontological monism, this thesis leads us to the conclusion that all entities are *simples*—such that they cannot be divided. (ii) Relational properties are not reducible to intrinsic properties. According to the hard problem of consciousness, there is a fundamental difference between qualia and physical properties, because of which the former cannot be reduced to the latter. In this study, the authors argue that this difference can be captured using the intrinsic/extrinsic distinction between properties. The two implications of the *qualia-as-intrinsic* thesis that are emphasised in this paper are the impossibility of combination, and the irreducibility of extrinsic to intrinsic properties.

The six orthodox systems and the three heterodox systems of Indian philosophy together form *darsana* (*astika* and *nastika darsana*s respectively). The orthodox systems which considered the Vedas as an authority, and the heterodox systems (which did not consider the Vedas to be authoritative), were further developed by a very rich tradition of commentaries and sub-commentaries with intense dialogues within and across these systems. These dialogues added depth to the Indian philosophical systems. V.N. Jha, in his chapter (Chapter "Matter and Consciousness: The Classical Indian Philosophical Approach"), expounds on these systems, highlighting the holistic approach of enquiry into matter and consciousness in the traditional Indian philosophical systems. He classifies these systems into four types based on their distinctive worldviews. The validity of the fundamental classification of the universe into *cetana* (conscious) and *acetana* or *jada* (non-conscious or inert matter) is debated in the Indian philosophical system with a divergence of views. The nature and identity of the self, the locus of consciousness, and the role of the mind, body and senses in acquiring knowledge are discussed in detail. Looking closely at our experience, Indian logicians have postulated the existence of a 'self' as distinct from the body, senses and mind, and thus distinct from matter. Jha then delves into the nature of the self or *atman* and how this is treated by the six *darsanas*. Based on the treatments of various Indian philosophical schools, seven models of understanding consciousness are highlighted. These are the Carvaka, Samkhya, Nyaya-Vaisesika, the Advaita Vedanta model of Sankaracharya, Bhakti Vedanta, the Buddhist and the Jaina models.

When a muscle twitches in our body, or the tongue tastes the fine flavour of an Indian coffee brew, have we thought about the origin of the intention, will and initiation to feel the twitch, or of tasting the flavour? What is the origin of the qualitative and refined nature of human experience? Moreover, what is the purpose of our capacity to self-reflect and ask a question such as the above, even while having a distinct and nuanced experience? Why does neural functioning correspond to experiences of a personal self? And how do we know the interrelations between the physiological brain and the subjective self? In this book, we propose that the self, culture and consciousness are three significant spaces of convergence of the thinking human mind and the experiencing human self. The goal of this book is to present and initiate a new set of approaches to objectively, as well as passionately, respond to those convergences which define our life and living in subliminal ways.

References

Casali, A. G., Gosseries, O., Rosanova, M., Boly, M., Sarasso, S., Casali, K. R., et al. (2013). A theoretically based index of consciousness independent of sensory processing and behavior. *Science Translational Medicine, 5*(198), 198ra105.

Chalmers, D. J. (1995). Facing up to the problem of consciousness. *Journal of Consciousness Studies, 2*(3), 200–219.

Chalmers, D. J. (1996). *The conscious mind: In search of a fundamental theory*. Oxford: Oxford University Press.

Edelman, G., & Tononi, G. (2000). *A universe of consciousness: How matter becomes imagination*. New York: Basic Books.

Koch, C., Massimini, M., Boly, M., & Tononi, G. (2016). The neural correlates of consciousness: Progress and problems. *Nature Review Neuroscience, 17*, 307–321.

Oizumi, M., Albantakis, L., & Tononi, G. (2014). From the phenomenology to the mechanisms of consciousness: Integrated information theory 3.0. *PLoS Computational Biology, 10*(5), e1003588.

Seth, A. K., Dienes, Z., Cleeremans, A., Overgaard, M., & Pessoa, L. (2008). Measuring consciousness: Relating behavioural and neurophysiological approaches. *Trends in Cognitive Sciences, 12*(8), 314–321.

Shannon, C. E. (1948). A mathematical theory of communication. *Bell System Technical Journal, 27*, 379–423 and 623–656.

Tononi, G. (2004a). An information integration theory of consciousness. *BMC Neuroscience, 5*(1), 42.

Tononi, G. (2004b). An information integration theory of consciousness. *BMC Neuroscience, 5*(42), 1–22. doi:10.1186/1471-2202-5-42

Tononi, G. (2008). Consciousness as integrated information: A provisional manifesto. *Biological Bulletin, 215*, 216–242. doi:10.2307/25470707

Tononi, G. (2012). *Phi: A voyage from the brain to the soul*. New York, NY: Pantheon Books.

Tononi, G. (2015). Integrated information theory. *Scholarpedia*, 10, 4164. doi:10.4249/scholarpedia.4164

Tononi, G., & Koch, C. (2015). Consciousness: Here, there and everywhere? *Philosophical Transactions of the Royal Society B*, *370*, 20140167. http://dx.doi.org/10.1098/rstb.2014.0167.

Virmani, M., & Nagaraj, N. (2016). A compression-complexity measure of integrated information. arXiv preprint arXiv:1608.08450v2

Part I
Emergence of Consciousness

Avian Cognition and Consciousness—From the Perspective of Neuroscience and Behaviour

Soumya Iyengar, Pooja Parishar and Alok Nath Mohapatra

Are birds conscious? Birds and primates occupy similar ecological niches, face similar challenges in foraging for food, and live in large social groups. Despite the fact that brain evolution is divergent in birds and mammals, the evolution of brain function is convergent. A number of studies on neural networks, structure, function and behaviour have demonstrated striking similarities between the overall organisation of the brain in humans, other primates and birds. Taken together, this data suggests that both cognition and consciousness may have evolved independently and in parallel across different species of birds and mammals. The present review focuses on the remarkable cognitive abilities of different species of birds such as problem-solving, tool use, mathematical abilities and self-awareness, the neural circuits underlying these behaviours and attempts to link the avian brain and behaviour to consciousness.

Cognition and Consciousness in Animals

Cognition

The myriad forms of life on earth encompassing humans and various species of vertebrates and invertebrates live in a complex environment and face similar challenges in foraging for food, finding mates, protecting and rearing their young

S. Iyengar (✉) · P. Parishar · A.N. Mohapatra
National Brain Research Centre, Manesar, Gurgaon, Haryana, India
e-mail: soumya@nbrc.ac.in

P. Parishar
e-mail: poojaparishar@nbrc.ac.in

A.N. Mohapatra
e-mail: thinkalok@gmail.com

S. Menon et al. (eds.), *Self, Culture and Consciousness*,
DOI 10.1007/978-981-10-5777-9_2

and having complex social interactions with others. Organisms other than humans, including non-human primates, birds, insects (for example, bees) and even invertebrates such as octopi display a number of highly complex behaviours in response to a given situation. However, the idea that life forms other than humans had a mind at all or could think has been debated by a number of philosophers over the years. On a positive note, David Hume (Hume 2014; cf. Lurz 2009) suggested that animals could think and reason since many of the behaviours that they exhibited were akin to those demonstrated by humans. According to Hume, animals could associate responses to sensory stimuli with various facts or propositions (such as colours, for example, the colour blue representing the sky) as humans do. However, his idea of 'thoughts' in animals was equivalent to 'beliefs' or 'ideas', which is certainly untestable and cannot be proven.

Amongst the naysayers, René Descartes averred that animals have neither thought nor reason, based on his language-test and action-test arguments (Radner and Radner 1989). Descartes based his language-test arguments on the premise that declarative speech, or the ability to state facts or beliefs, is present only in humans. Declarative speech is also used to express occurrent thought, which one 'entertains, brings to mind or is suddenly struck by' (Malcolm 1972). Although Descartes agreed that animals (both birds and mammals) have a number of vocalisations, he suggested that these were only used to express 'passions'. He also proposed that, since animals are incapable of declarative speech, they were also unable to think, since speech expresses thought. Descartes also proposed an action-test argument to prove that animals lack reasoning where he defined 'reason' as 'a universal instrument which can be used in all kinds of situations' (Descartes 1988). In this case, although Descartes agreed that animals did occasionally follow certain general principles of reasoning to solve a problem in a certain situation, these behaviours were not transferred to novel situations. Most of the other philosophers who have postulated that animals are incapable of thinking or reasoning have also based their theories on the fact that the ability for speech and language is a pre-requisite for thought and reasoning (Davidson 1984; Bennett 1976, 2003; Bermúdez 2003; Searle 1994, 2003).

Over the years, many researchers have countered the language-test argument proposed by Descartes. Malcolm (1972) has suggested that, rather than occurrent thought, animals are likely to have inclinations which may reflect motivation and the production of behaviour specific to an occasion (dispositional thinking). Additionally, Hauser et al. (2002) have suggested that, while animals have occurrent thoughts, they are simply unable to express them. Other researchers have shown that some species of animals, such as African grey parrots (Pepperberg 1999), chimpanzees (Savage-Rumbaugh et al. 1998) and honeybees (Tetzlaff and Rey 2009) can use vocalisations as well as specific signs (gestures) for communicating about their environment (such as warning calls or the presence of food) with other members of their species.

Further, some scientists believe that insight learning may be taken as evidence for occurrent thought in animals. Insight learning can be defined as the abrupt realisation of a problem's solution and does not depend on trial and error, responding to environmental cues or by observing and imitating another subject attempting to solve the task. Animals such as chimpanzees use insight learning to solve problems, such as being able to use multiple boxes stacked one on top of the other to obtain an otherwise out of reach reward, without trial and error. It is a completely cognitive experience which requires internal visualisation, and can then be repeated in future to solve similar problems (Köhler 1925; Heinrich 2002). Others have also suggested that, rather than having a very broad or universal power of reasoning to respond to every novel situation, the intelligence and powers of reasoning in animals have evolved to solve problems specific to their niche (Hauser 2000; 'massive modularity thesis', Carruthers 2006).

Consciousness

If the arguments for and against animal cognition were not long and daunting enough, the idea that animals are conscious has been fraught with equal dissension. What is consciousness? The word itself is derived from the Latin words *con* (with) and *scire* (to know), giving rise to the most basic definition of consciousness, that is, the state of being aware of oneself or of an external stimulus (http://www.dictionary.com). However, in reality, there is no single or universal definition of consciousness which applies to all life forms on earth, making it difficult to study and compare across different species. The study of consciousness has fascinated mankind and has been extensively examined by ancient cultures in the East and by Greek and Roman philosophers. Western philosophies have generally held that humans are exceptional in their abilities to use language to communicate and express their feelings and emotions. According to this view, since humans are also exceptional in their ability to convey to others that they are introspective or 'self-conscious', they are considered 'conscious', whereas animals which lack these abilities are not (Koch et al. 2016; Tononi et al. 2016).

Amongst the different kinds of consciousness initially identified in humans are access consciousness and phenomenal consciousness (Nagel 1974; Rosenthal 1986; Dennett 1991; Block 1995b; Carruthers 2000). Access consciousness includes the abilities of humans and other animals to discriminate between various stimuli, focus attention and produce specific behaviours in response to different situations (Long and Kelley 2010). In contrast, phenomenal consciousness refers to how a particular experience 'feels', that is, the subjective nature of the experience (Nagel 1974). Whereas access consciousness can be more readily studied across different species (referred to as the 'soft problem' of consciousness), phenomenal consciousness is, perhaps, impossible to test in the absence of language ('the hard problem', Block 1995a; Block et al. 1997).

Lurz (2009) has suggested that both conscious and unconscious mental states can have effects on the behaviour of different organisms (Marcel 1983; Tye 1995). Earlier philosophers have proposed first-order and higher-order theories to explain consciousness in animals. Higher-order theories suggest that there is a specific perception or higher-order thought or belief linked to conscious states. Some studies (Armstrong 1997; Lycan 1996) suggest that higher-order awareness implies that one is perceptually aware of different stimuli and this awareness is directed inwards, towards the mind (inner sense theories). Other studies (Rosenthal 1986; Carruthers 2000) suggest that conscious mental states exist if subjects have higher-order thoughts about being in such mental states, that is, if one is aware that he or she is aware. Several pros and cons exist for each case (reviewed in Lurz 2009), with studies on self-awareness in animals suggesting that they are capable of higher-order awareness (Gennaro 2004, 2009) and others suggesting that animals are incapable of such awareness since it requires language (Bennett and Bennett 1966, 1988; Bennett 1964; Davidson 1984, 1985; Dennett 1991).

In the case of first-order theories, subjects become conscious or aware of changes in their environment as a result of conscious mental states, as opposed to having higher-order thoughts about being in such mental states. For example, if an animal is looking at a tree, it is currently conscious of the tree, because according to first-order theories, its perception causes it to believe that there is a tree in front of it. Whereas unconscious mental states exist, they do not make subjects aware of their surroundings (Dretske 1995; Evans 1982; Tye 1995). Thus, the perception of various stimuli by different animals makes them aware of their environment. Additionally, higher-order thoughts make animals conscious of various beliefs and desires (Lurz 2004, 2006).

Whereas the question of the existence of cognition and consciousness in animals had earlier focused mainly on mammals including non-human primates, the idea that birds could have any such ability was not considered at all. Early neuroanatomists such as Edinger (Edinger and Rand 1908) and Ariëns-Kapper (1909) found that the overall appearance of the avian brain was different from that of mammals (discussed in detail below) and the idea that the evolution of different species and their brain structure is progressive [fish to amphibians, followed by reptiles, birds, 'lower' mammals, primates and culminating in humans (Darwin and Barrett 1988)] unfortunately led to the belief that birds could only produce stereotyped responses to changes in their environments.

However, more recently, Carroll (1988) and Evans (2000) have demonstrated that evolution is not linear or progressive. They showed that the common ancestor of birds and mammals, known as stem amniotes, existed 320 million years ago (MYA). Stem amniotes diverged to give rise to sauropsids (the ancestors of birds and reptiles) and therapsids (the mammalian ancestor) in the Triassic period, approximately 250 MYA. However, modern-day mammals arose from therapsids 203 MYA, *before* modern-day birds evolved from sauropsids (135 MYA). Further, concepts about the cognitive abilities of birds and their brain structure (discussed below) began to change with the implementation of a number of modern techniques in neuroanatomy and molecular biology as well as behavioural observations in the field and in laboratory settings to study avian brain structure and function.

Cognitive Abilities of Birds

A number of studies have shown that, far from merely producing stereotyped behaviours, different species of birds are capable of learning and producing highly complex behaviours in response to cognitively challenging situations. The cognitive abilities of birds of the genus Corvidacea (including species such as crows, jays, ravens and jackdaws) and passerine songbirds, parrots and hummingbirds have been studied in the wild and under lab conditions and are reviewed here.

Problem-Solving

'The Thirsty Crow', a well-loved and oft repeated fable from the Panchatantra (3rd century, BCE) and Aesop (620–520 BCE), suggests that observations on avian cognition started a long time ago. The basic premise was that a thirsty crow (a corvid) came upon a partially filled pitcher of water which it could not drink. The crow then placed stones in the pitcher and raised the level of water to quench its thirst. Bird and Emery (2009) replicated these findings in the lab by posing a similar problem to rooks (another species of corvid). Birds were faced with raising the level of water in a narrow tube to obtain a food reward (a worm) floating on top, which they could not reach directly with their beaks. Rooks soon learned to use stones provided for the purpose and got the worm—a different take on the old saying 'the early bird gets the worm'! In other modifications of the experiment, when the rooks were provided with a choice between smaller and larger stones, they chose larger ones, suggesting that they had analysed the situation in order to achieve their target sooner. Replacing water in the container with sawdust led to an interesting result—rooks did not even attempt the experiment, suggesting that they knew or had learned that the level of sawdust could not be raised with stones, as in the case of water.

Tool Use

Besides using stones, a number of studies have shown that New Caledonian crows are specialists in tool use and manufacture. They use their beaks to craft tools of different sizes and shapes out of pandanus leaves to extract insect larvae from crevices and holes in trees and under leaves (Chappell and Kacelnik 2002, 2004). These findings were replicated in the lab by Weir et al. (2002) who provided different lengths of wire to Betty, a New Caledonian crow, in the lab. The task at hand was to extract a food reward placed in a small bucket inside a narrow container, which could not be reached directly. Betty was able to choose a piece of wire of the correct length and then bent it to the correct angle, hooking it around the bucket handle to retrieve the food reward from within the container. Recently, a remarkable video recording demonstrated how an eight-step puzzle was solved by a New Caledonian crow called 007 (Taylor 2014) to reach a food reward placed in a box with bars. The food could only be reached through the bars with a long stick. The crow was provided with a much smaller stick tied to a string, hanging from a perch, while the longer stick of the length required to get the food reward was placed in a specially designed apparatus containing a swinging door. After initially surveying the scenario, in the first trial itself, the crow retrieved the small stick, then used it to retrieve two stones from different boxes. In the next step, it placed the stones inside the container with the swinging lid and retrieved the longer stick to finally obtain the food reward. These findings suggest that corvids understand the problem at hand and adapt to it, by modifying and making suitable tools and using specific tools to match specific situations. However, despite being able to solve complex problems spontaneously, New Caledonian crows have been shown to use perceptual feedback rather than insight to do so (Taylor et al. 2012).

Theory of Mind (ToM) and Prospective Thinking Abilities

In addition to these problem-solving abilities, corvids have also been shown to possess a 'theory of mind'—that is, the ability of an individual to think about another's mental state, earlier thought to be restricted to primates. These results were based on experiments performed on corvids which store food ('cache' their food) for lean periods in winter. Since species such as Clarke's nutcrackers can have as many as 30,000 caches to be recovered over a period of 6 months, food caching is indeed a daunting task (Balda and Kamil 1992). Food-caching birds have to remember not only the location of the caches but also what was cached—that is, whether the stores consisted of perishable or non-perishable food, since perishable food would need to be consumed first. Western scrub-jays (another food-caching species of corvid) have been shown to consume perishable food before starting on their non-perishable stores, demonstrating that birds can think and plan for the future (reviewed in Emery and Clayton 2004). Further, Clayton et al. (2001) and Emery and Clayton (2004) have demonstrated that birds who pilfer others' stores are much more careful about protecting their own, both in nature and in lab settings. 'Thief' scrub-jays used various strategies, such as avoiding being seen by other birds, hiding food at sites which cannot be seen by others, re-caching stores several times and hiding inedible material in place of caches when they are observed by others (Bugnyar and Kotrschal 2002, 2004; Dally et al. 2004, 2006a; Emery and Clayton 2001; Emery et al. 2004). However, the same birds will not re-cache their food in the presence of their mates (Dally et al. 2006b). These results suggest that Western scrub-jays possessed a 'theory of mind', since these birds were aware that, if they could pilfer others' stores, their own could also be stolen.

In another variant of these experiments, a raven (a food-caching bird) was provided with food in an experimental arena, next to a compartment which contained another raven (Bugnyar et al. 2016). The two compartments were separated by a solid wall or a window through which the subject could be seen by the other bird. When the subject knew that it was being observed through the window, it would take longer and use more evasive methods of caching food than in the other condition, when it could not be observed. The experiment was modified by replacing the window with a peephole, after which subjects were placed in this compartment and allowed to watch experimenters cache in the experimental arena. Subjects were then replaced in the arena and the window in the adjoining compartment was provided with a peephole. Even when there was no observer and the sounds of a raven moving around were played to the subject from the next compartment, subjects resorted to evasive manoeuvres to cache their food. These findings suggest that ravens were capable of generalising from their experiences, even when they were observing humans, not other ravens, and strengthens the idea that they possess a theory of mind.

Vocalisation, Vocal Learning and Mimicry

One of the basic tenets of human exceptionalism rests on the fact that humans can produce speech and learn languages from role models during early development in order to communicate with other members of their species. Behavioural observations on various species of animals have shown that, besides humans, cetaceans (such as whales and dolphins) and bats have the 'ability to acquire vocalisations through imitation rather than instinct' (Jarvis 2004) and are classified as vocal learners. In addition to these species, three distantly related species of birds (parrots, hummingbirds and songbirds) also learn their vocalisations from a tutor (typically their parent/s) during development and then perfect these songs to acquire and produce a unique (although species-typical) song in adulthood. Birds of these species are diverse in the way that they learn songs—some are close-ended learners, learning their song during a sensitive period in their development and producing a highly stereotyped song in adulthood throughout their lives (for example, Australian zebra finches), others are open-ended learners and change their songs every year (canaries, Brainard and Doupe 2002). There is also a rich diversity in the size of songbirds' repertoires. For example, zebra finches learn and produce a single song which does not change appreciably over different iterations, whereas birds like nightingales can imitate and produce thousands of songs (Bartsch et al. 2015). The functions of singing may also vary, with some species of songbirds using their songs mainly for attracting mates whereas others use them for attracting mates as well as fending off rivals (Catchpole and Slater 2008). Jarvis (2004) has speculated upon the evolution of vocalisation and vocal learning, and suggested that neural pathways underlying vocal behaviour may have evolved in both birds and mammals from common stem amniote ancestors which existed 320 million years ago, and vocal learning itself may have been influenced by a combination of genetic and epigenetic factors (Jarvis 2004). Other possibilities include independent evolution of neural circuits for vocalisation and vocal learning in birds and mammals and/or the fact that all birds and mammals are capable of vocal learning to some extent and this ability and the underlying neural circuits have become highly specialised in some species of birds and mammals. Jarvis (2004) has also proposed that perhaps some degree of each of these hypotheses is true for the evolution of vocal learning and the neural circuitry underlying this behaviour.

Another group of birds which have fascinated mankind with their abilities at vocal mimicry are parrots. Being open-ended vocal learners, parrots can learn to use human speech (Pepperberg 1981, 1999) to label different objects as in the well-documented case of the African grey parrot, Alex. Just as in humans, parrots learn well in enriched environments when the training paradigms are referential and functional (Marler and Slabbekoorn 2004; Pepperberg 1997). Alex could produce at least 20,000 distinct vocalisations and had the remarkable ability to name or vocally label 100 objects of different colours, shapes and sizes and made from different materials (Pepperberg et al. 1991). Not only did he categorise different objects and learn to differentiate between them, he was also capable of combining the names of different objects in novel ways to express his need for or refuse certain objects (Pepperberg 2007). Alex was also observed to practise his vocalisations when he was alone, just humans do during skill learning (Pepperberg et al. 1991). Interestingly, Alex had also developed the concept of the personal pronoun 'I' versus 'you', suggesting some familiarity with grammar (Pepperberg et al. 2000).

Besides parrots, amongst some of the most skilled and amazing mimics are lyrebirds (superb lyrebirds, *Menura superba* and Albert's lyrebirds, *Menura alberti*, both native to Australia). Lyrebirds perform a courtship dance and also use their songs to attract mates (Putland et al. 2006). In addition to the ability to mimic perfectly all other species of songbirds and mammals like koalas and dingoes in their environment, lyrebirds increase their vocal repertoire, especially during the breeding season, to include car alarms, ringing phones, crying babies, human voices, chainsaws, camera shutters, rifle-shots and laser guns (Lee 2014).

Taken together, these findings suggest that, just as humans learn speech from their parents, young songbirds learn their vocalisations from tutors (generally their parents) during a sensitive period in early development. Interestingly, songbirds and mimics can choose what to learn from their tutors' repertoires and even those of conspecific males and their siblings (Brainard and Doupe 2002; Derégnaucourt and Gahr 2013) in order to enrich their song for attracting mates and competing with rival males. A specialist like Alex was able to communicate his needs to humans by using different words and combinations of words for novel objects, showing almost human-like abilities in communication, which contrasts with Descartes' language-test argument.

Object Permanence

Piaget (1952) found that children learn to track their immediate surroundings (other individuals and objects) at different stages during early development, even when they are out of sight. Children initially track the movements of objects that they can see (Stages 1 and 2), followed by tracking partly hidden objects (Stage 3). This is followed by being able to form mental representations of objects which are completely hidden (Stage 4), and understanding the displacement of hidden objects whether performed visibly or invisibly (Stages 5 and 6 respectively) (Uzgiris and Hunt 1975). Whereas Alex demonstrated object permanence up to Stage 6 (Pepperberg et al. 1997; Piaget 1952), magpies displayed object permanence up to Stage 4 (Pollok et al. 2000), suggesting that they are able to cognise changes in their immediate environment.

Mathematical Abilities

Koehler (1950, reviewed in Thorpe 1958) was the first to test the numerical abilities of different species of birds. He provided birds with cards on which a specific number of dots (from 2 to 6) was printed and two boxes, one with the same number of dots as the card and the other with either one more or one less than this number. Koehler found that, on simultaneous presentation of the card and the boxes, a raven and a grey parrot could learn to pick the box with the same number of dots as those on the card. Ravens could also add numbers successively, which was demonstrated by experiments where they were given a reward for eating a specific number of items from boxes (for example, 6) provided one after the other. The number of food items provided in five boxes presented one after the other were 1, 2, 1, 2 and 1 and the raven had to stop at 6 to get the reward, even if more items were available. Koehler (1950) demonstrated that a jackdaw (*Corvus monedula*) could be trained to remove specific numbers of food items from boxes of different colours presented successively or simultaneously. The performance of ravens and jackdaws in these experiments was similar to those obtained from chimpanzees when tested on similar paradigms (Boysen and Berntson 1989).

Another example was the African grey parrot Alex, who was able to count up to six, (Pepperberg et al. 2000) since he could report the number of familiar and novel objects provided to him in trays (Pepperberg 1987). Alex was also capable of discriminating between and counting specific objects provided to him in a group, for example, orange versus purple sticks. Pepperberg (2006) reported that Alex had developed the concept of zero or 'none' and could also add numbers, with his skills being comparable to those of non-human primates and young children.

Visual Discrimination and Memory

The visual system of birds is very well-developed and they are excellent at visual discrimination. They are capable of categorising images presented to them based on perceptual similarities and also whether they are the same or different. Different studies have shown that pigeons can discriminate between photographs of humans, conspecifics, aerial photographs, scenes from nature like water and trees, inanimate objects such as chairs and cars and letters of the alphabet (reviewed in Emery 2006). A particularly interesting set of studies also demonstrated that pigeons were able to discriminate between the paintings of famous impressionist artists such as Van Gogh, Chagall, Monet and Picasso (Watanabe 2001; Watanabe et al. 1995). Pigeons can also match objects when presented with a group of two similar and one different object (Macphail 1982). Of course, one of the best examples of visual discrimination was provided by Alex (mentioned above), who was able to name or label more than 100 different kinds of object on the basis of their colour, size, shape and material (Pepperberg 1999).

Amongst the corvids, our own experiments have shown that Indian house crows (*Corvus splendens*) can be trained easily on tasks based on visually discriminating between numbers, shapes (Parishar and Iyengar, unpublished) and simple and complex images (Sharma and Iyengar, unpublished). Other examples of crows using visual discrimination to good effect and also long-term memory is the famous example of mobbing, wherein crows recognise humans who have harmed them or their young and the entire flock can attack the perpetrator. This was tested by Marzluff and his group (Cornell et al. 2012) on the University of Seattle campus by one of the members donning a mask, capturing American crows and banding them, which was a traumatic experience for the birds. Once a couple of birds had been banded, each time anyone don the mask, the entire flock would be alerted, whether or not they had been banded and would scold and fly after the masked offender until out of sight. Members of the flock would not pay attention to people wearing other masks but would even recognise the mask if worn upside down or in a different orientation, suggesting that crows do pay attention to and remember individual features of human faces. This was also confirmed in a separate study in which jungle crows recognised photographs of human faces. In this study, Bogale et al. (2011) demonstrated that specific features (eyes and mouth) were necessary for the crows to discriminate between different faces.

Can the Cognitive Abilities of Birds Be Linked to Behavioural Correlates of Consciousness?

These results suggest that birds are capable of solving complex problems including those linked to visual discrimination, manufacturing and using tools, that they possess a theory of mind and the ability to think and plan for the future, that they are able to perform tasks requiring short- and long-term memory, that they can learn and produce patterned vocalisations (song), mimic and communicate with humans (in the case of parrots such as Alex) and that they possess mathematical abilities. Koch et al. (2016) have argued that any organism (such as humans) which possess such abilities would definitely be deemed conscious. According to Koch, humans can be distinguished from other animals on the basis of introspection—that is, self-consciousness or the ability to reflect upon one's self, and the ability to produce language. As explained above, birds—for example, Alex—are capable of communicating with conspecifics and even with humans. Even though it is not possible to test whether birds or other animals reflect upon themselves, another test of consciousness—that is, self-awareness—can be tested using the mirror self-recognition test developed by Gallup (1970).

Mirror Self-awareness

One of the most important hallmarks of consciousness is self-awareness or the sense of self, defined as 'an awareness of being aware—a unified subjective experience of being that extends in space and time' (Toda and Platt 2015). To test self-awareness, (Gallup 1970) developed an elegantly simple test called the mirror self-recognition or 'mark test', wherein the face or head of the human or animal being tested is marked in such a way that the mark can be observed only in a reflection. Mirror self-awareness develops in children between 18 and 24 months of age (Amsterdam 1972) and is also seen in great apes (gorillas, chimpanzees and orangutans) (Anderson and Gallup 2011; Suddendorf and Collier-Baker 2009). Gallup initially carried out his experiments on captive chimpanzees and found that after initial aggression towards the reflection (since it was thought to be a novel conspecific), animals engaged in exploratory and social behaviours to test the mirror image. This was followed by contingency testing, where the animal would keep making unusual gestures such as touching its face, grimacing or turning from side to side to additionally test the presence of its mirror image. Once it was convinced that the image in the mirror was its own reflection, a chimpanzee would initiate self-directed behaviours to explore or remove the mark using the mirror (Gallup 1970). However, there is huge variation in the ability of different species in their performance of the mark test. Whereas elephants (Plotnik et al. 2006) and dolphins (Reiss and Marino 2001) could recognise themselves in mirrors, dogs, cats and harbour seals and even Old World and New World monkeys did not pass the 'mark test' (Suddendorf and Butler 2013), unless they were intensively trained to pay attention to a particular point on their face which was rewarded or stimulated by a somatosensory stimulus (a red laser, which was slightly irritating) (Chang et al. 2015). Most species of birds tested for mirror self-recognition pay attention to their reflections but tend to regard them as conspecifics, generally attacking and pecking at their reflections. We decided to test Indian house crows in our lab on the mark test and found that, despite their abilities at visual discrimination, they failed the mirror self-recognition task (Parishar, Mohapatra and Iyengar, unpublished). Whereas the mark was ignored, birds initially attacked and pecked their reflections, as if they were novel conspecifics. However, after the initial training trials, crows mostly ignored their reflections, except for occasionally directing vocalisations towards the mirror. This was also true of New Caledonian crows, who fail the mark test, despite the fact that they have learned to use mirrors to retrieve a food reward in a task where the food could not be seen except in the reflection (Medina et al. 2011). However, magpies (Prior et al. 2008) and zebra finches (Parishar, Mohapatra and Iyengar, unpublished) do pass the mark test and a study on pigeons has demonstrated that they can learn to recognise themselves in mirrors with intensive training (Epstein et al. 1981), as can macaques (Chang et al. 2015). Taken together, these results suggest that, just as in different species of mammals, various levels of self-awareness exist in different species of birds, which does not correlate with their ability to solve complex cognitive tasks. In fact, different degrees of self-awareness are demonstrated by different individuals of the same species, even when tested under the same conditions on the mark test seen in magpies, wherein four out of a total of six subjects tested could recognise their reflections (Prior et al. 2008).

Neural Basis of Consciousness

Currently, there is general agreement that the brain is the seat of consciousness and the search is on, in humans and different species of animals, for neuronal correlates of consciousness (NCC) or 'the minimal neural mechanisms that are jointly sufficient for any one conscious percept, thought or memory, under constant background conditions' (Koch et al. 2016; Tonini et al. 2016; Oizumi et al. 2014). Studies on NCC in humans and animals (mostly mammalian) included both bottom-up and top-down approaches to the problem (reviewed in Butler et al. 2005). Whereas bottom-up theories of consciousness suggest that specific groups of neurons in the brain generate activity leading to consciousness, top-down theories posit that the neural activity of a number of brain regions collectively give rise to consciousness.

Top-down theories suggest that motor and homeostatic functions are important for animals to be aware of themselves. Tononi and Edelman (1998) and Edelman (2003) focus on general features of consciousness such as complexity and unity which are associated with neural activity in the temporal and frontal associative and motor regions of the cortex and the thalamus. These regions interact with the septal region, amygdala, hippocampus, dorsal thalamus, hypothalamus and reticular activating system (related to emotions and learning). Another top-down theory suggests that consciousness is associated with activity in circuits consisting of sensory and motor, cortical and thalamic structures. Muscular activity activates muscle spindles (specialised sensory receptors which sense the degree of stretch in muscle fibres) and are critical for the generation of consciousness (Cotterill 2001). Cotterill (2001) also includes the amygdala, hippocampus, dorsal thalamus, subthalamus, hypothalamus, caudate, putamen and globus pallidus (forebrain structures) as well as brain stem structures such as the superior colliculus, cerebellum, substantia nigra, pontine nucleus, red nucleus and inferior olive, and parts of the autonomic nervous system as generators for consciousness.

Bottom-up theories proposed by Crick and Koch (1995), Rees et al. (2002) suggest that visual awareness is very important for consciousness and that connections between the secondary or association visual cortices and prefrontal cortex are needed to give rise to consciousness. They have further suggested that neurons in Layer V of the temporal, parietal and prefrontal cortex which fire in burst patterns generate consciousness. Besides these regions, they also suggest that neural activity is seen in other parts of the cortex, dorsal thalamus, claustrum, dorsal components of the basal ganglia (caudate, putamen and the globus pallidus), cerebellum and brain stem nuclei related to motor control and may be important for consciousness. Another bottom-up theory (Eccles 1982, 1990; Beck and Eccles 1992) suggest that consciousness is generated in cortical columns of pre- or supplementary motor areas by pyramidal neurons.

Recently, Koch et al. (2016) defined neuronal correlates of consciousness (NCC) as the 'minimum neuronal mechanisms jointly sufficient for any one conscious percept' (Crick and Koch 1990). NCC include not only specific neurons but the neuronal mechanisms whose activity can lead to a specific experience. Studies on REM and NREM sleep in human subjects suggest that temporal, parietal and occipital cortices are associated with perception, whereas the frontal cortex is important for generating thought-like experiences. Interestingly, different types of perceptions which are experienced during dreams in REM sleep (such as faces, locations, movements and speech) lead to a predictable increase in high-frequency neural activity in posterior cortical regions (Siclari et al. 2014), suggesting that the posterior cortical region may represent NCC. The posterior cortex is also activated when matching performance with expectations, when subjects are asked to compare awareness and the relevance of the task, suggesting a role for this region in content-specific consciousness (Koivisto and Revonsuo 2010; Pitts et al. 2014; Melloni et al. 2011; Sandberg et al. 2014; Andersen et al. 2016).

Besides the reticular activating system in the brain stem, small bilateral lesions of the intralaminar nuclei of the thalamus, which are connected to widespread regions of the cortex, led to coma, whereas stimulating this region led to recovery from unconscious states (Van der Werf et al. 2002; Schiff et al. 2007). However, lesions of regions of the thalamus other than the intralaminar nuclei and EEG studies in rats as well as patients with thalamic atrophy did not affect consciousness, suggesting that, other than decreased motor function and the ability to communicate, the thalamus was not important for generating consciousness (Fuller et al. 2011; Lutkenhoff et al. 2015).

In addition to the involvement of neural activity in specific brain regions, Von Economo neurons (VENS), large spindle-shaped bipolar neurons located in Layer V of the frontoinsular (FI) and anterior cingulate cortex (ACC) in humans and in other large-brained mammals such as great apes, elephants, whales and bottle-nosed dolphins, have been implicated in generating consciousness (Hakeem et al. 2009; Von Economo and Koskinas 1925; Allman et al. 2010; Cauda et al. 2014, reviewed in Koch et al. 2016). The ACC is involved in anticipation and modulation of arousal and both FI and ACC are actively involved in modulating emotions, subjective pain, embarrassment, guilt and decision-making (Allman et al. 2005), all of which deal with self-awareness. VENs express receptors for the neurotransmitters vasopressin, dopamine and serotonin, which are known to be important for cognition (Insel and Young 2001; Daw et al. 2002; Nakajimaa et al. 2013). Structurally, VENs appear to be specialised to integrate information between local circuits and subcortical regions, through their distal dendrites (Larkum 2013). Other studies showed that there is a profound disruption in social functioning and awareness as a result of frontotemporal dementia wherein the number of VENs is markedly reduced (Kaufman et al. 2008) or in autism wherein there is a decrease in the number of VENs in the ACC (Barnea-Goraly et al. 2004). Other studies have suggested that supragranular neurons (in Layers II and III) are better poised to generate consciousness since they receive a number of feed-forward and feed-back projections from within the cortex itself as well as the thalamus. Additional support comes from the fact that activity in the supragranular neurons can be correlated with conscious sensation, which ceases in case of anaesthesia or sleep (reviewed in Koch et al. 2016).

As mentioned earlier, if problems of linking brain structure, cognition and consciousness were not difficult enough in humans and mammals other than primates, the results from studies on avian brain structure and their interpretations from earlier literature made it even more difficult to even conceive of the idea that birds could perform tasks requiring complex cognition. This stemmed from the fact that birds diverged from mammals about 250 MYA and, at first glance, avian brain structure is very different from that of mammals. However, as detailed in the next section, there is a great degree of homology between different brain regions and neural circuits in birds and mammals.

Avian Brain Structure

Pallium Versus Cortex

A popular misconception in older literature which led to the belief that birds are incapable of cognition and led to the derogatory term ‘bird brains’ was the lack of a laminated cortex in birds, which was believed to endow humans and other primates with their ‘exceptional’ cognitive abilities. Edinger and Rand (1908) presented the unified theory of brain evolution wherein different parts of the mammalian cortex classified as paleocortex (oldest cortex, three-layered), archicortex (archaic cortex with three to four layers) and neocortex (new or six-layered) (Edinger and Rand 1908; Arïens-Kapper 1909) were added around the basal ganglia or the core of the brain across evolution. According to Edinger, the exceptional cognitive abilities of humans arose from the neocortex, which formed a significantly larger part of the brain compared to other mammals. Since neurons in the avian brain are not arranged in the form of lamina, Edinger suggested that the avian brain is mainly composed of the striatum (a part of the basal ganglia). According to his schema, the outer part of the avian telencephalon was divided into the hyperstriatum (hypertrophied striatum), paleostriatum (older striatum which was the precursor of the mammalian amygdala), neostriatum (new striatum which would give rise to the caudate and putamen in mammals) and archistriatum. These findings fuelled the (wrong) belief that birds were capable of reacting in a stereotyped manner to their environment and were incapable of solving problems, as seen in mammals.

Recently, scientific research on the avian brain, using tools related to neuroanatomy, molecular biology and magnetic resonance imaging, has led to an enormous body of connectional, morphological and genetic data relating to the avian brain. Far from the cortex being absent in birds, several studies have shown that, although not laminated as in mammals, birds certainly possess a cortical homologue (the outer part of the telencephalon or the pallium (Jarvis et al. 2005; Reiner et al. 2004 for review). Results from Karten's group (Karten 1969; Nauta and Karten 1970; Karten and Shimizu 1989) have demonstrated that the pallium can constitute a significant part (as much as 75%) of birds' brains. Based on his studies, Karten has proposed the 'nuclear-layered hypothesis' based on the arrangement of neurons scattered throughout the brain but organised into specialised groups called nuclei in some parts of the brain. This hypothesis suggests that the brain of the common ancestor of birds, reptiles and mammals (the stem amniote) was arranged in the nuclear pattern and evolved into the laminar pattern, when mammalian ancestors (therapsids) diverged from avian ancestors (sauropsids) 250 MYA.

The Organisation of Neural Circuits in the Avian Brain

Neuroanatomical tract-tracing studies have shown that most of the neural circuits and brain regions thought to be involved in generating mammalian consciousness are also present in birds (reviewed in Butler and Cotterill 2006; Iyengar 2012). The pallium can be divided into the hyperpallium (or Wulst) which receives sensory input from the lemnisci and from the midbrain colliculi (mesopallium and nidopallium; Fig. 1a). Besides sensory areas, the hyperpallium also contains motor areas. Further, the avian cortex also possesses two possible candidates which may be homologous to the mammalian prefrontal cortex (Fig. 1b). One of these regions is the NCl (nidopallium caudolaterale) which receives information from the midbrain colliculi (collopallial projections) and also has reciprocal connections with secondary sensory areas of the pallium for all modalities (Güntürkün et al. 1993; Waldmann and Gu 1993; Leutgeb et al. 1996; Hartmann and Güntürkün 1998; Metzger et al. 1998; Durstewitz et al. 1999; Diekamp et al. 2002; Lissek et al. 2002; Rose and Colombo 2005; see Emery 2006 for review, Fig. 1). These findings are

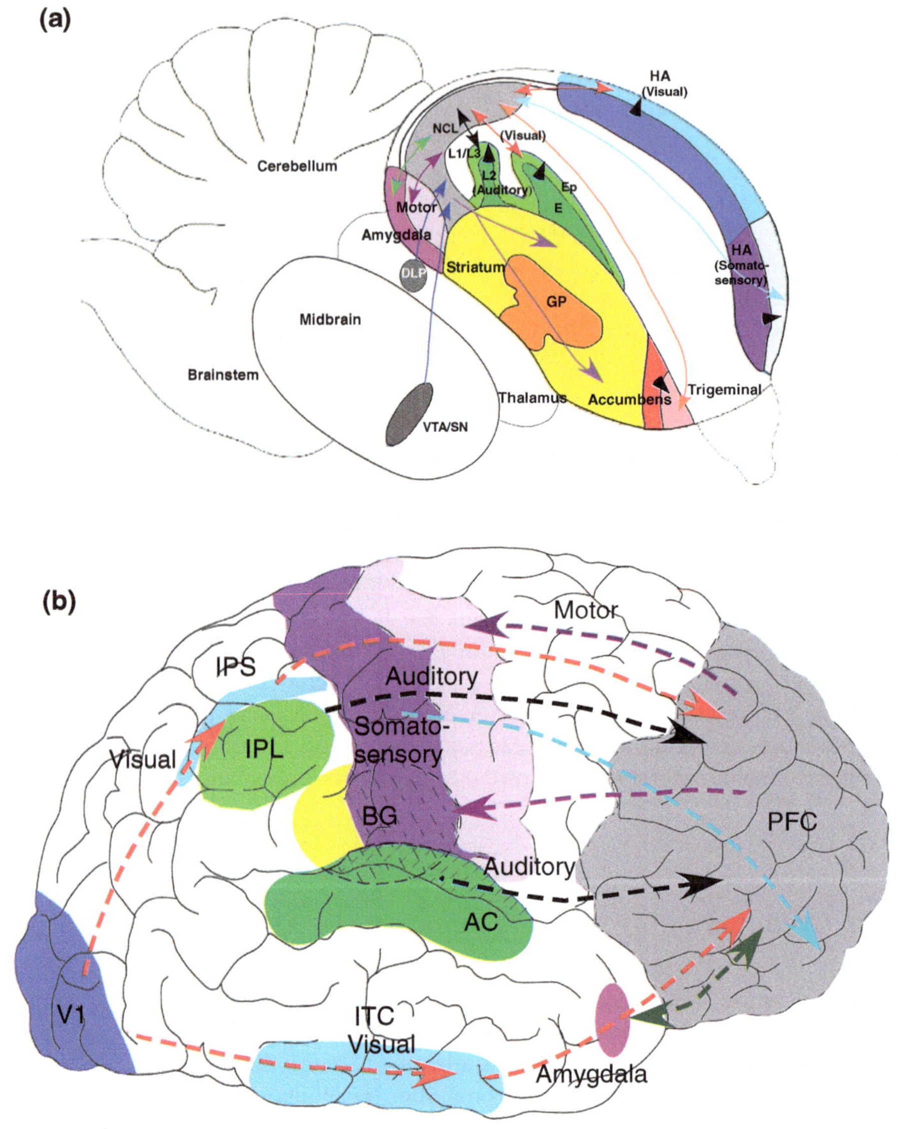

Fig. 1 A comparison of major connections of the caudal nidopallium in birds and prefrontal cortex in mammals. **a** Avian brain, Abbr.: *DLP* nucleus dorsolateralis posterior thalami, *E* entopallium, *Ep* entopallial belt, *GP* globus pallidus, *HA* hyperpallium, *NCL* caudal nidopallium, *VTA/SN* ventral tegmental area/substantia nigra. **b** Mammalian brain, Abbr.: *AC* auditory cortex, *BG* basal ganglia, *IPL* inferior parietal lobule, *IPS* intraparietal sulcus, *ITC* inferotemporal cortex, *PFC* prefrontal cortex, *V1* primary visual cortex. *Colours* represent homologous areas in the avian pallium and mammalian cortex (Auditory, *green*, Visual, *blue*, Somatosensory, *purple*, Motor, *light purple*, Basal ganglia, *yellow*, Prefrontal cortex, *grey*, Amygdala, *pink*) and *arrows* indicate major connections. **a** Adapted from Kröner and Güntürkün (1999) and Rattenborg et al. (2011), with permission for both from John Wiley and Sons and **b** adapted from Klemen and Chambers (2012) with permission from Elsevier

based on lesion studies or blocking D1 (dopamine Type 1) receptors present in this region which give rise to deficits in tasks involving visual discrimination, working memory, reversal learning and delayed alternation (Reiner 1986; Mogensen and Divac 1993; Hartmann and Güntürkün 1998; Diekamp et al. 2002; Aldavert-Vera et al. 1999). The other region is the dorsolateral corticoid area (CDL; Montagnese et al. 2003) which receives lemniscal input. However, rather than the prefrontal cortex, this region has been suggested by Atoji and Wild (2005) to represent the mammalian cingulate cortex which is a part of the limbic system. This theory is based on the fact that both the mammalian cingulate cortex and the pigeon CDL have connections with the hippocampal complex, amygdala, basal ganglia, midline nuclei of the thalamus, and visual and motor cortices (Van Hoesen et al. 1993; Atoji and Wild 2005; Vogt et al. 1997).

Besides the pallium itself, thalamo-cortical-basal ganglia loops which are present in mammalian brains and underlie not only motor functions but also various cognitive functions (Alexander et al. 1986; Alexander and Crutcher 1990; Parent and Hazrati 1995) are also present in avian brains. Homologues of the mammalian thalamic intralaminar nuclei are present in the avian dorsal thalamic zone (Veenman et al. 1997) which receive projections from the globus pallidus and project to the striatum, forming parts of a loop including the thalamus, striatum and pallidum which is similar to that in mammals (Jarvis et al. 2005). Further, relay nuclei for various sensory modalities (visual, auditory, somatosensory) within the thalamus project to specific parts of the nidopallium and, as in mammals, sensory parts of the pallium project back onto their respective sensory thalamic nuclei (Karten et al. 1973; Wild and Williams 2000). Interestingly, the song-control system of songbirds provides an example of a thalamo-cortical-basal ganglia loop which is specialised to learn and produce song. Homologous nuclei are arranged in the form of an anterior forebrain pathway in the brains of avian vocal learners (songbirds, parrots and hummingbirds) (Johnson et al. 1995; Vates and Nottebohm 1995; Iyengar et al. 1999; Luo et al. 2001; Reiner et al. 2004a, b). A group of neurons in the pallial nucleus HVC projects to Area X, a nucleus of the avian basal ganglia (Nottebohm et al. 1976, 1982). Area X contains both striatal and pallidal neurons, unlike the mammalian basal ganglia (in which striatum and pallidum are distinct entities; (Reiner et al. 2004a, b; Carrillo and Doupe 2004; Farries and Perkel 2002; Farries et al. 2005) of which pallidal neurons project to the thalamic nucleus DLM (nucleus dorsolateralis anterior, pars medialis). In turn, DLM projects to LMAN (lateral magnocellular nucleus of the anterior nucleus) (Carrillo and Doupe 2004; Reiner et al. 2004b) located in the pallium. Since LMAN projects to Area X as well as RA (Robust nucleus of the arcopallium), a motor cortical region of the pallium, loops are formed within the song-control system. Another component of the song-control system is the posterior pathway, which consists of premotor neurons within the pallial nucleus HVC (higher vocal centre) (McCasland 1987), which project to the premotor region RA (Nottebohm et al. 1976, 1982). Neurons in RA then project to the nXIIts (12th nerve nucleus) consisting of motor neurons which innervate the musculature of the syrinx or the vocal organ in songbirds (Vicario 1994). A homologue of the posterior pathway in humans is a neural circuit which is important for speech. It connects the motor cortical representation of the face to the nucleus ambiguus which controls the larynx and the 12th nerve nucleus which innervates the musculature of the tongue (Nauta and Kuypers 1958; Kuypers 1958; Jürgens 1998; Zhang et al. 1995).

Another homology between avian and mammalian brains is the mesolimbic dopaminergic pathway which is important for the motivation to perform goal-directed behaviours (Berridge and Robinson 1998; Robinson et al. 2005). Dopamine is important for the anticipation of a reward and dopaminergic neurons present in the ventral tegmental area—subtantia nigra complex (VTA-SNc)—innervate the ventral striatum (nucleus accumbens) in mammals (Andén et al. 1964; Lynd-Balta and Haber 1994). GABAergic interneurons in VTA-SNc inhibit the dopaminergic neurons in this region and prevent them from releasing dopamine. Endogenous opioids bind to μ-opioid receptors (a family of inhibitory G-protein couple receptors) present on GABA interneurons, leading to the disinhibition of VTA-SNc neurons, allowing them to secrete dopamine (De Vries and Shippenberg 2002; Tomkins and Sellers 2001; Xi and Stein 2002). The basic organisation of the mesolimbic dopaminergic system is similar in birds including passerine songbirds (Lewis et al. 1981; Gale and Perkel 2006; Khurshid et al. 2010) which produce a highly motivated behaviour (song) as are the effects of the opioid system (Khurshid et al. 2010) and dopamine release in the avian striatum (Sasaki et al. 2006).

In terms of similarities at the level of single neurons, Jarvis (2004), Reiner et al. (2004a) has suggested that projection neurons within the cortical nucleus LMAN in songbirds are similar to neurons of Layer III and upper Layer V of the mammalian premotor cortex. These neurons project to spiny neurons in Area X (a part of the avian basal ganglia) as well as RA (homologous to the motor cortex). The neural circuit connecting pallidal neurons in Area X which project to DLM (a part of the dorsal thalamus) are comparable to neurons in the globus pallidus of mammals which project to the ventral lateral and ventral anterior nuclei in the dorsal thalamus. Neurons in these thalamic nuclei project to Layer III of the premotor cortex which is homologous to LMAN (Jacobson and Trojanowski 1975; Alexander et al. 1986; Alexander and Crutcher 1990). Whereas it is tempting to speculate that the neurons in LMAN may be avian counterparts of VENs (present in Layer V of the mammalian cortex), there is as yet no proof that species of birds like corvids and magpies which have been shown to live in complex social environments and are capable of exhibiting higher cognitive behaviours possess these neurons (Cauda et al. 2014).

Conclusions

The huge diversity of behaviour, both innate and learned, produced by birds in response to their environment as well as the neural circuits underlying these behaviours is truly remarkable. Despite the fact that avian brain structure has diverged from that of mammals during evolution, the functions of various neural circuits as well as different behaviours that emerge from these circuits are similar across mammals and birds. Rather than Descartes' original views on the fact that animals are not capable of thought, present-day cognitive neuroscientists, philosophers, neuroanatomists and neurophysiologists have suggested that consciousness is a product or an 'emergent property' of the rich complexity in neural circuits present in different organisms. It has also been suggested that consciousness has evolved along with the brain and behaviour, from 'non-conscious precursors' to that observed in humans (Seager and Bourget 2007; Sinha 1999; Dehaene and Naccache 2001; Gupta and Sinha 2014). The results of painstaking research on brain structure, cognition and consciousness in various species have led to the Cambridge Declaration on Consciousness (Lowe et al. 2012) in which it was suggested that consciousness was not unique to humans. Consciousness emerged early in evolution based on the homologies in neural circuits and their functions, including cognition and emotions across different species, including birds. In particular, birds provide an example of the parallel evolution of the brain, its functions and consciousness.

Gupta and Sinha (2014) suggest that access consciousness which is present in all species and important for reasoning, focusing attention and the conscious control of action and speech (Block 1995a; Long and Kelley 2010) is linked to executive functions of the brain. Executive functions include the performance of novel tasks, decision-making, working memory, detection of errors, the regulation of mental states and, subsequently, the performance of actions (Posner and Rothbart 1998; Stuss and Alexander 2000; Hughes and Graham 2002), which stem from access consciousness and are linked to frontoparietal networks in the brain (Sergent et al. 2005). It would be important to analyse and understand the emergence of consciousness in birds by comparing the links between access consciousness, executive functions and the underlying neural circuits. Comparing homologous brain regions and neural circuits and their activity, while birds perform different cognitively demanding tasks, may provide answers as to how consciousness has evolved. At the same time, it appears at present almost impossible to look through the eyes of a bird (phenomenal consciousness, Nagel 1974), to understand whether they can introspect or whether they sometimes wonder: 'Are humans conscious?'

References

Aldavert-Vera, L., Costa-Miserachs, D., Divac, I., & Delius, J. D. (1999). Presumed 'prefrontal cortex' lesions in pigeons: Effects on visual discrimination performance. *Behavioural Brain Research, 102*(1), 165–170.

Alexander, G. E., & Crutcher, M. D. (1990). Functional architecture of basal ganglia circuits: Neural substrates of parallel processing. *Trends in Neurosciences, 13*(7), 266–271.

Alexander, G. E., DeLong, M. R., & Strick, P. L. (1986). Parallel organization of functionally segregated circuits linking basal ganglia and cortex. *Annual Review of Neuroscience, 9*(1), 357–381.

Allman, J. M., Tetreault, N. A., Hakeem, A. Y., Manaye, K. F., Semendeferi, K., Erwin, J. M., et al. (2010). The von Economo neurons in frontoinsular and anterior cingulate cortex in great apes and humans. *Brain Structure and Function, 214*(5–6), 495–517.

Allman, J. M., Watson, K. K., Tetreault, N. A., & Hakeem, A. Y. (2005). Intuition and autism: A possible role for Von Economo neurons. *Trends in Cognitive Science, 9*(8), 367–373. doi:10.1016/j.tics.2005.06.008.

Amsterdam, B. (1972). Mirror self image reactions before age two. *Developmental Psychobiology, 5*(4), 297–305.

Andén, N.-E., Carlsson, A., Dahlström, A., Fuxe, K., Hillarp, N.-Å., & Larsson, K. (1964). Demonstration and mapping out of nigro-neostriatal dopamine neurons. *Life Sciences, 3*(6), 523–530.

Anderson, J. R., & Gallup, G. G., Jr. (2011). Which primates recognize themselves in mirrors? *PLoS Biology, 9*(3), e1001024.

Andersen, L. M., Pedersen, M. N., Sandberg, K., & Overgaard, M. (2016). Occipital MEG activity in the early time range (<300 ms) predicts graded changes in perceptual consciousness. *Cerebral Cortex, 26*(6), 2677–2688.

Ariëns Kappers, C. U. (1909). The phylogenesis of the paleocortex and archicortex compared with the evolution of the visual neocortex. *Archives of Neurology, 4,* 161–272.

Armstrong, I. (1997). Technology and text: Glass consciousness and nineteenth-century culture. *Culture, Landscape and Environment: The Linacre Lectures*, 149–75.

Atoji, Y., & Wild, J. M. (2005). Afferent and efferent connections of the dorsolateral corticoid area and a comparison with connections of the temporoparietooccipital area in the pigeon (*Columba livia*). *Journal of Comparative Neurology, 485*(2), 165–182.

Balda, R. P., & Kamil, A. C. (1992). Long-term spatial memory in clark's nutcracker. *Nucifraga columbiana. Animal Behaviour, 44*(4), 761–769. doi:10.1016/S0003-3472(05)80302-1.

Barnea-Goraly, N., Kwon, H., Menon, V., Eliez, S., Lotspeich, L., & Reiss, A. L. (2004). White matter structure in autism: Preliminary evidence from diffusion tensor imaging. *Biological Psychiatry, 55*(3), 323–326.

Bartsch, C., Weiss, M., & Kipper, S. (2015). Multiple song features are related to paternal effort in common nightingales. *BMC Evolutionary Biology, 15*(1), 1–8. doi:10.1186/s12862-015-0390-5.

Beck, F., & Eccles, J. (1992). Quantum aspects of consciousness and the role of consciousness. *Proceedings of the National Academy of Science, 89,* 11357.

Bennett, J. (1976). *Linguistic behaviour.* Hackett Publishing Company.

Bennett, J. (2003). *Learning from six philosophers: Descartes, Spinoza, Leibniz, Locke, Berkeley.* Hume: Clarendon Press.

Bennett, J., & Bennett, J. F. (1966). *Kant's analytic.* Cambridge University Press.

Bennett, J., & Bennett, J. F. (1988). *Events and their names.* Hackett Publishing Company.

Bennett, J. F. (1964). *Rationality: An essay towards an analysis.* Routledge and Kegan Paul.

Bermúdez, J. L. (2003). *Thinking without words* (vol. 113). Oxford University Press.

Berridge, K. C., & Robinson, T. E. (1998). What is the role of dopamine in reward: Hedonic impact, reward learning, or incentive salience? *Brain Research Reviews, 28*(3), 309–369.

Bird, C. D., & Emery, N. J. (2009). Insightful problem solving and creative tool modification by captive nontool-using rooks. *Proceedings of the National Academy of Science USA, 106*(25), 10370–10375. doi:10.1073/pnas.0901008106.

Block, N. (1995a). How many concepts of consciousness? *Behavioral and Brain Sciences, 18*(02), 272–277.

Block, N. (1995b). On a confusion about a function of consciousness. *Brain and Behavioral Sciences, 18*(2), 227–247.

Block, N. J., Flanagan, O. J., & Güzeldere, G. (1997). *The nature of consciousness: Philosophical debates.* Bradford Books.

Bogale, B. A., Aoyama, M., & Sugita, S. (2011). Categorical learning between 'male' and 'female' photographic human faces in jungle crows (*Corvus macrorhynchos*). *Behavioural Processes, 86*(1), 109–118. doi:10.1016/j.beproc.2010.10.002.

Boysen, S. T., & Berntson, G. G. (1989). Numerical competence in a chimpanzee (*Pan troglodytes*). *Journal of Comparative Psychology, 103*(1), 23–31.
Brainard, M. S., & Doupe, A. J. (2002). What songbirds teach us about learning. *Nature, 417* (6886), 351–358.
Bugnyar, T., & Kotrschal, K. (2002). Observational learning and the raiding of food caches in ravens, *Corvus corax*: Is it 'tactical' deception? *Animal Behaviour, 64*(2), 185–195. doi:10.1006/anbe.2002.3056.
Bugnyar, T., & Kotrschal, K. (2004). Leading a conspecific away from food in ravens (*Corvus corax*)? *Animal Cognition, 7*(2), 69–76. doi:10.1007/s10071-003-0189-4.
Bugnyar, T., Reber, S. A., & Buckner, C. (2016). Ravens attribute visual access to unseen competetors. *Nature Communications, 7,* 10506. doi:10.1038/ncomms10506.
Butler, A. B., & Cotterill, R. M. (2006). Mammalian and avian neuroanatomy and the question of consciousness in birds. *The Biological Bulletin, 211*(2), 106–127.
Butler, A. B., Manger, P. R., Lindahl, B., & Århem, P. (2005). Evolution of the neural basis of consciousness: A bird–mammal comparison. *BioEssays, 27*(9), 923–936.
Carrillo, G. D., & Doupe, A. J. (2004). Is the songbird area X striatal, pallidal, or both? An anatomical study. *Journal of Comparative Neurology, 473*(3), 415–437.
Carroll, R. L. (1988). *Vertebrate palaeontology and evolution.* Spektrum Akademischer Verlag.
Carruthers, P. (2000). *Phenomenal consciousness: A naturalistic theory* (vol. 207). Cambridge University Press.
Carruthers, P. (2006). *The architecture of the mind: Massive modularity and the flexibility of thought.* Clarendon Press.
Catchpole, C. K., & Slater, P. J. B. (2008). *Bird song: Biological themes and variations.* New York: Cambridge University Press.
Cauda, F., Geminiani, G. C., & Vercelli, A. (2014). Evolutionary appearance of von Economo's neurons in the mammalian cerebral cortex. *Frontiers Human Neuroscience, 8,* 104. doi:10.3389/fnhum.2014.00104.
Chang, L., Fang, Q., Zhang, S., Poo, M.-M., & Gong, N. (2015). Mirror-induced self-directed behaviors in rhesus monkeys after visual-somatosensory training. *Current Biology, 25*(2), 212–217.
Chappell, J., & Kacelnik, A. (2002). Tool selectivity in a non-primate, the New Caledonian crow (*Corvus moneduloides*). *Anim Cognition, 5*(2), 71–78. doi:10.1007/s10071-002-0130-2.
Chappell, J., & Kacelnik, A. (2004). Selection of tool diameter by New Caledonian crows *Corvus moneduloides*. *Animal Cognition, 7*(2), 121–127. doi:10.1007/s10071-003-0202-y.
Clayton, N. S., Griffiths, D. P., Emery, N. J., & Dickinson, A. (2001). Elements of episodic-like memory in animals. *Philosophical Transactions of the Royal Society of London B Biological Science, 356*(1413), 1483–1491. doi:10.1098/rstb.2001.0947.
Cornell, H. N., Marzluff, J. M., & Pecoraro, S. (2012). Social learning spreads knowledge about dangerous humans among American crows. *Proceedings of the Royal Society of London B: Biological Sciences, 279*(1728), 499–508.
Cotterill, R. M. (2001). Cooperation of the basal ganglia, cerebellum, sensory cerebrum and hippocampus: Possible implications for cognition, consciousness, intelligence and creativity. *Progress in Neurobiology, 64*(1), 1–33.
Crick, F., & Koch, C. (1990). *Towards a neurobiological theory of consciousness.* Paper presented at the Seminars in the Neurosciences.
Crick, F., & Koch, C. (1995). Why neuroscience may be able to explain consciousness. *Scientific American, 273*(6), 84–85.
Dally, J. M., Emery, N. J., & Clayton, N. S. (2004). Cache protection strategies by western scrub-jays (*Aphelocoma californica*): Hiding food in the shade. *Proceedings of the Royal Society of London B: Biological Sciences, 271*(Suppl 6), S387–S390. doi:10.1098/rsbl.2004.0190.
Dally, J. M., Clayton, N. S., & Emery, N. J. (2006a). The behaviour and evolution of cache protection and pilferage. *Animal Behaviour, 72*(1), 13–23. doi:10.1016/j.anbehav.2005.08.020.

Dally, J. M., Emery, N. J., & Clayton, N. S. (2006b). Food-caching western scrub-jays keep track of who was watching when. *Science, 312*(5780), 1662–1665. doi:10.1126/science.1126539.
Darwin, C., & Barrett, P. H. (1988). *On the origin of species, 1859*. W. Pickering.
Davidson, D. (1984). *Inquiries into truth and interpretation*. Oxford University Press.
Davidson, D. (1985). Adverbs of action. In B. Vermazen & M. B. Hintikka (Eds.), *Essays on Davidson: Actions and events* (pp. 230–241). Oxford University Press.
Daw, N. D., Kakade, S., & Dayan, P. (2002). Opponent interactions between serotonin and dopamine. *Neural Networks, 15*(4), 603–616.
De Vries, T. J., & Shippenberg, T. S. (2002). Neural systems underlying opiate addiction. *The Journal of Neuroscience, 22*(9), 3321–3325.
Dehaene, S., & Naccache, L. (2001). Towards a cognitive neuroscience of consciousness: Basic evidence and a workspace framework. *Cognition, 79*(1), 1–37.
Dennett, D. C. (1991). *Consciousness explained*. Penguin.
Derégnaucourt, S., & Gahr, M. (2013). Horizontal transmission of the father's song in the zebra finch (*Taeniopygia guttata*). *Biology Letters, 9*, 20130247.
Descartes, R. (1988). Descartes: Selected philosophical writings. Cambridge University Press.
Diekamp, B., Gagliardo, A., & Güntürkün, O. (2002). Nonspatial and subdivision-specific working memory deficits after selective lesions of the avian prefrontal cortex. *The Journal of neuroscience, 22*(21), 9573–9580.
Dretske, F. (1995). *Naturalizing the mind*. MIT Press.
Durstewitz, D., Kröner, S., & Güntürkün, O. (1999). The dopaminergic innervation of the avian telencephalon. *Progress in Neurobiology, 59*(2), 161–195.
Eccles, J. (1990). A unitary hypothesis of mind-brain interaction in the cerebral cortex. *Proceedings of the Royal Society of London B: Biological Sciences, 240*(1299), 433–451.
Eccles, J. C. (1982). Animal consciousness and human self-consciousness. *Cellular and Molecular Life Sciences, 38*(12), 1384–1391.
Edelman, G. M. (2003). Naturalizing consciousness: A theoretical framework. *Proceedings of the National Academy of Sciences, 100*(9), 5520–5524.
Edinger, L., & Rand, H. W. (1908). The relations of comparative anatomy to comparative psychology. *Journal of Comparative Neurology and Psychology, 18*(5), 437–457. doi:10.1002/cne.920180502.
Emery, N. J. (2006). Cognitive ornithology: The evolution of avian intelligence. *Philosophical Transactions of the Royal Society of London B: Biological Sciences, 361*(1465), 23–43.
Emery, N. J., & Clayton, N. S. (2001). Effects of experience and social context on prospective caching strategies by scrub jays. *Nature, 414*(6862), 443–446. doi:10.1038/35106560.
Emery, N. J., & Clayton, N. S. (2004). The mentality of crows: Convergent evolution of intelligence in corvids and apes. *Science, 306*(5703), 1903–1907. doi:10.1126/science.1098410.
Emery, N. J., Dally, J. M., & Clayton, N. S. (2004). Western scrub-jays (*Aphelocoma californica*) use cognitive strategies to protect their caches from thieving conspecifics. *Animal Cognition, 7* (1), 37–43. doi:10.1007/s10071-003-0178-7.
Epstein, R., Lanza, R. P., & Skinner, B. F. (1981). Self-awareness in the pigeon. *Science, 212* (4495), 695–696.
Evans, G. (1982). *The varieties of reference* (vol. 137). Oxford University Press.
Evans, S. E. (2000). Amniote evolution. In G. Bock & G. Cardew (Eds.), *Evolutionary developmental biology of the cerebral cortex* (pp. 109–113). Wiley.
Farries, M. A., Ding, L., & Perkel, D. J. (2005). Evidence for "direct" and "indirect" pathways through the song system basal ganglia. *Journal of Comparative Neurology, 484*(1), 93–104.
Farries, M. A., & Perkel, D. J. (2002). A telencephalic nucleus essential for song learning contains neurons with physiological characteristics of both striatum and globus pallidus. *The Journal of Neuroscience, 22*(9), 3776–3787.
Fuller, P., Sherman, D., Pedersen, N. P., Saper, C. B., & Lu, J. (2011). Reassessment of the structural basis of the ascending arousal system. *Journal of Comparative Neurology, 519*(5), 933–956.

Gale, S. D., & Perkel, D. J. (2006). Physiological properties of zebra finch ventral tegmental area and substantia nigra pars compacta neurons. *Journal of Neurophysiology, 96*(5), 2295–2306.

Gallup, G. G. (1970). Chimpanzees: Self-recognition. *Science, 167*(3914), 86–87.

Gennaro, R. J. (2004). Higher-order theories of consciousness: An overview. In R. J. Gennaro (ed.), *Higher-order theories of consciousness*. John Benjamin.

Gennaro, R. J. (2009). Animals, consciousness, and I-thoughts. In R. W. Lurz (ed.), *The philosophy of animal minds* (pp. 184–200). Cambridge University Press.

Güntürkün, O., Miceli, D., & Watanabe, M. (1993). Anatomy of the avian thalamofugal pathway. *Vision, brain and behavior in birds* (pp. 115–135). Cambridge: MIT.

Gupta, S., & Sinha, A. (2014). Executive functions as a path to understanding nonhuman consciousness: Looking under the light. In S. Menon, A. Sinha, & B. V. Sreekantan (Eds.), *Interdisciplinary perspectives on consciousness and the self* (pp. 101–116). Springer.

Hakeem, A. Y., Sherwood, C. C., Bonar, C. J., Butti, C., Hof, P. R., & Allman, J. M. (2009). Von Economo neurons in the elephant brain. *The Anatomical Record, 292*(2), 242–248.

Hartmann, B., & Güntürkün, O. (1998). Selective deficits in reversal learning after neostriatum caudolaterale lesions in pigeons: Possible behavioral equivalencies to the mammalian prefrontal system. *Behavioural Brain Research, 96*(1), 125–133.

Hauser, M. (2000). *Wild minds: What animals really think*. Henry Holt and Company.

Hauser, M. D., Chomsky, N., & Fitch, W. T. (2002). The faculty of language: What is it, who has it, and how did it evolve? *Science, 298*(5598), 1569–1579. doi:10.1126/science.298.5598.1569.

Heinrich, B. (2002). Raven consciousness. In M. Bekoff, C. Allen, & G. M. Burghardt (Eds.), *The cognitive animal: Empirical and theoretical perspectives on animal cognition* (pp. 47–52). MIT Press.

Hughes, C., & Graham, A. (2002). Measuring executive functions in childhood: Problems and solutions? *Child and Adolescent Mental Health, 7*(3), 131–142.

Hume, D. (2014). *A treatise of human nature*. Start Classics.

Insel, T. R., & Young, L. J. (2001). The neurobiology of attachment. *Nature Review of Neuroscience, 2,* 129–136.

Iyengar, S. (2012). Comparative and evolutionary aspects of cognition. In P. N. Tandon, R. C. Tripathi, & N. Srinivasan (Eds.), *Expanding horizons of the mind science* (pp. 249–266). New York: Nova Science.

Iyengar, S., Viswanathan, S. S., & Bottjer, S. W. (1999). Development of topography within song control circuitry of zebra finches during the sensitive period for song learning. *The Journal of Neuroscience, 19*(14), 6037–6057.

Jacobson, S., & Trojanowski, J. Q. (1975). Corticothalamic neurons and thalamocortical terminal fields: An investigation in rat using horseradish peroxidase and autoradiography. *Brain Research, 85*(3), 385–401.

Jarvis, E. D. (2004). Learned birdsong and the neurobiology of human language. *Annal of New York Academy of Science, 1016,* 749–777. doi:10.1196/annals.1298.038.

Jarvis, E. D., Güntürkün, O., Bruce, L., Csillag, A., Karten, H., & Kuenzel, W. (2005). Avian brains and a new understanding of vertebrate brain evolution. *Nature Review of Neuroscience, 6*(2), 151–159. doi:10.1038/nrn1606.

Johnson, F., Sablan, M. M., & Bottjer, S. W. (1995). Topographic organization of a forebrain pathway involved with vocal learning in zebra finches. *Journal of Comparative Neurology, 358*(2), 260–278.

Jürgens, U. (1998). Neuronal control of mammalian vocalization, with special reference to the squirrel monkey. *Naturwissenschaften, 85*(8), 376–388.

Karten, H. J. (1969). The organization of the avian telencephalon and some speculations on the phylogeny of the amniote telencephalon. *Annals of the New York Academy of Sciences, 167*(1), 164–179.

Karten, H. J., Hodos, W., Nauta, W. J., & Revzin, A. M. (1973). Neural connections of the "visual wulst" of the avian telencephalon. Experimental studies in the pigeon (*Columba livia*) and owl (*Speotyto cunicularia*). *Journal of Comparative Neurology, 150*(3), 253–277.

Karten, H. J., & Shimizu, T. (1989). The origins of neocortex: Connections and lamination as distinct events in evolution. *Journal of Cognitive Neuroscience, 1*(4), 291–301.
Kaufman, J. A., Paul, L. K., Manaye, K. F., Granstedt, A. E., Hof, P. R., Hakeem, A. Y., et al. (2008). Selective reduction of Von Economo neuron number in agenesis of the corpus callosum. *Acta Neuropathologica, 116*(5), 479–489.
Khurshid, N., Jayaprakash, N., Hameed, L. S., Mohanasundaram, S., & Iyengar, S. (2010). Opioid modulation of song in male zebra finches (*Taenopygia guttata*). *Behavioural Brain Research, 208*(2), 359–370.
Klemen, J., & Chambers, C. D. (2012). Current perspectives and methods in studying neural mechanisms of multisensory interactions. *Neuroscience Biobehavioral Reviews, 36*(1), 111–133.
Koch, C., Massimini, M., Boly, M., & Tononi, G. (2016). Neural correlates of consciousness: Progress and problems. *Nature Reviews Neuroscience, 17*(5), 307–321.
Köhler, W. (1925). *The mentality of apes*. London: Routledge and Kegan Paul.
Koehler, O. (1950). The ability of birds to 'count'. *Bulletin of the Animal Behavioral Society, 9*, 41–45.
Koivisto, M., & Revonsuo, A. (2010). Event-related brain potential correlates of visual awareness. *Neuroscience and Biobehavioral Reviews, 34*(6), 922–934.
Kröner, S., & Güntürkün, O. (1999). Afferent and efferent connections of the Caudolateral Neostriatum in the Pigeon (*Columba livia*): A Retro- and Anterograde pathway tracing study. *The Journal of Comparative Neurology, 407*, 228–260.
Kuypers, H. G. J. M. (1958). Corticobulbar connexions to the pons and lower brain stem in man. *Brain, 81*(3), 364–388.
Larkum, M. (2013). A cellular mechanism for cortical associations: An organizing principle for the cerebral cortex. *Trends in Neurosciences, 36*(3), 141–151.
Lee, W. H. (2014, 11 November). Lyrebird: The best songbird ever! [video file]. Retrieved from http://www.sciencealert.com/watch-lyrebird-perfectly-imitates-laser-gun-sounds.
Leutgeb, S., Husband, S., Riters, L. V., Shimizu, T., & Bingman, V. P. (1996). Telencephalic afferents to the caudolateral neostriatum of the pigeon. *Brain Research, 730*(1), 173–181.
Lewis, J. W., Ryan, S. M., Arnold, A. P., & Butcher, L. L. (1981). Evidence for a catecholaminergic projection to area X in the zebra finch. *Journal of Comparative Neurology, 196*(2), 347–354.
Lissek, S., Diekamp, B., & Güntürkün, O. (2002). Impaired learning of a color reversal task after NMDA receptor blockade in the pigeon (*Columbia livia*) associative forebrain (*Neostriatum caudolaterale*). *Behavioral Neuroscience, 116*(4), 523.
Long, L. N., & Kelley, T. D. (2010). Review of consciousness and the possibility of conscious robots. *Journal of Aerospace Computing, Information, and Communication, 7*(2), 68–84. doi:10.2514/1.46188.
Lowe, P., Panksepp, J., Reiss, D., Edelman, D., Swinderen, B. V., & Koch, C. (2012, 7 July). The Cambridge declaration on consciousness. http://fcmconference.org/img/CambridgeDeclarationOnConsciousness.pdf. Accessed December 18, 2016.
Luo, M., Ding, L., & Perkel, D. J. (2001). An avian basal ganglia pathway essential for vocal learning forms a closed topographic loop. *The Journal of Neuroscience, 21*(17), 6836–6845.
Lurz, R. W. (2004). In search of the metaphor of the mind: A critical review of Baars' in the theater of consciousness. *Philosophical Psychology, 17*(2), 297–307.
Lurz, R. W. (2006). Conscious beliefs and desires: A same-order approach. In U. Kriegel & K. Williford (Eds.), *Self-representational approaches to consciousness*. MIT Press.
Lurz, R. W. (2009). *Animal minds*. Internet Encyclopedia of Philosophy.
Lutkenhoff, E. S., Chiang, J., Tshibanda, L., Kamau, E., Kirsch, M., Pickard, J. D., et al. (2015). Thalamic and extrathalamic mechanisms of consciousness after severe brain injury. *Annals of Neurology, 78*(1), 68–76.
Lycan, W. G. (1996). *Consciousness and experience* (vol. 72). MIT Press.
Lynd-Balta, E., & Haber, S. (1994). The organization of midbrain projections to the striatum in the primate: Sensorimotor-related striatum versus ventral striatum. *Neuroscience, 59*(3), 625–640.

Macphail, E. M. (1982). *Brain and intelligence in vertebrates*. Clarendon Press.
Malcolm, N. (1972). Thoughtless Brutes. *Proceedings and Addresses of the American Philosophical Association, 46,* 5–20. doi:10.2307/3129585.
Marcel, A. J. (1983). Conscious and unconscious perception: Experiments on visual masking and word recognition. *Cognitive Psychology, 15,* 197–237.
Marler, P., & Slabbekoorn, H. W. (2004). *Nature's music: The science of birdsong*. Elsevier Academic.
McCasland, J. S. (1987). Neuronal control of bird song production. *The Journal of Neuroscience, 7*(1), 23–39.
Medina, F., Taylor, A., Hunt, G., & Gray, R. D. (2011). New Caledonian crows' responses to mirrors. *Animal Behaviour, 82*(5), 981–993.
Melloni, L., Schwiedrzik, C. M., Müller, N., Rodriguez, E., & Singer, W. (2011). Expectations change the signatures and timing of electrophysiological correlates of perceptual awareness. *The Journal of Neuroscience, 31*(4), 1386–1396.
Metzger, M., Jiang, S., & Braun, K. (1998). Organization of the dorsocaudal neostriatal complex: A retrograde and anterograde tracing study in the domestic chick with special emphasis on pathways relevant to imprinting. *Journal of Comparative Neurology, 395*(3), 380–404.
Mogensen, J., & Divac, I. (1993). Behavioural effects of ablation of the pigeon-equivalent of the mammalian prefrontal cortex. *Behavioural Brain Research, 55*(1), 101–107.
Montagnese, C. M., Mezey, S. E., & Csillag, A. (2003). Efferent connections of the dorsomedial thalamic nuclei of the domestic chick (*Gallus domesticus*). *Journal of Comparative Neurology, 459*(3), 301–326.
Nagel, T. (1974). What is it like to be a bat? *The Philosophical Review, 83*(4), 435–450.
Nakajimaa, S., Gerretsena, P., Takeuchic, H., Caravaggioa, F., Chowb, T., Follc, B. L., et al. (2013). The potential role of dopamine D3 receptor neurotransmission in cognition. *European Neuropsychopharmacology, 23*(8), 799–813. doi:10.1016/j.euroneuro.2013.05.006.
Nauta, W. J., & Karten, H. J. (1970). A general profile of the vertebrate brain, with sidelights on the ancestry of cerebral cortex. In *Neuroanatomy* (pp. 520–539). Springer.
Nauta, W. J., & Kuypers, H. G. M. (1958). Some ascending pathways in the brain stem reticular formation. In J. H. Herbert, L. D. Proctor, R. S. Knighton, W. C. Noshay, & R. T. Costello (Eds.), *Reticular formation of the brain* (pp. 3–30). Oxford: Little Brown.
Nottebohm, F., Paton, J. A., & Kelley, D. B. (1982). Connections of vocal control nuclei in the canary telencephalon. *Journal of Comparative Neurology, 207*(4), 344–357.
Nottebohm, F., Stokes, T. M., & Leonard, C. M. (1976). Central control of song in the Canary, Serinus Canarius. *Journal of Comparative Neurology, 165*(4), 457–486.
Oizumi, M., Albantakis, L., & Tononi, G. (2014). From the phenomenology to the mechanisms of consciousness: Integrated information theory 3.0. *PLoS Computational Biology, 10*(5), e1003588.
Parent, A., & Hazrati, L.-N. (1995). Functional anatomy of the basal ganglia. I. The cortico-basal ganglia-thalamo-cortical loop. *Brain Research Reviews, 20*(1), 91–127.
Pepperberg, I. M. (1981). Functional vocalizations by an African Grey Parrot (*Psittacus erithacus*). *Zeitschrift für Tierpsychologie, 55*(2), 139–160. doi:10.1111/j.1439-0310.1981.tb01265.x.
Pepperberg, I. M. (1987). Acquisition of the same/different concept by an African Grey parrot (*Psittacus erithacus*): Learning with respect to categories of color, shape, and material. *Animal Learning and Behavior, 15*(4), 423–432. doi:10.3758/bf03205051.
Pepperberg, I. M. (1997). Social influences on the acquisition of human-based codes in parrots and nonhuman primates. In C. T. Snowdon & M. Hausberger (Eds.), *Social influences on vocal development* (pp. 157–177). Cambridge University Press.
Pepperberg, I. M. (1999). *The Alex studies*. Harvard University Press.
Pepperberg, I. M. (2006). Grey Parrot (*Psittacus erithacus*) numerical abilities: Addition and further experiments on a zero-like concept. *Journal of Comparative Psychology, 120*(1), 1–11. doi:10.1037/0735-7036.120.1.1.

Pepperberg, I. M. (2007). Grey parrots do not always 'parrot': The roles of imitation and phonological awareness in the creation of new labels from existing vocalizations. *Language Sciences, 29*(1), 1–13. doi:10.1016/j.langsci.2005.12.002.

Pepperberg, I. M., Brese, K. J., & Harris, B. J. (1991). Solitary sound play during acquisition of English vocalizations by an African Grey parrot (*Psittacus erithacus*): Possible parallels with children's monologue speech. *Applied Psycholinguistics,* 12(02), 151–178. doi:10.1017/S0142716400009127.

Pepperberg, I. M., Sandefer, R. M., Noel, D. A., & Ellsworth, C. P. (2000). Vocal learning in the Grey parrot (*Psittacus erithacus*): Effects of species identity and number of trainers. *Journal of Comparative Psychology, 114*(4), 371–380.

Pepperberg, I. M., Willner, M. R., & Gravitz, L. B. (1997). Development of Piagetian object permanence in a grey parrot (*Psittacus erithacus*). *Journal of Comparative Psychology, 111*(1), 63–75.

Piaget, J. (1952). *La naissance de l'intelligence chez l'enfant*. International Universities Press.

Pitts, M. A., Metzler, S., & Hillyard, S. A. (2014). Isolating neural correlates of conscious perception from neural correlates of reporting one's perception. *Frontiers in Psychology, 5*, 1078. doi:10.3389/fpsyg.2014.01078.

Plotnik, J. M., De Waal, F. B., & Reiss, D. (2006). Self-recognition in an Asian elephant. *Proceedings of the National Academy of Sciences, 103*(45), 17053–17057.

Pollok, B., Prior, H., & Güntürkün, O. (2000). Development of object permanence in food-storing magpies (*Pica pica*). *Journal of Comparative Psychology, 114*(2), 148–157. doi:10.1037/0735-7036.114.2.148.

Posner, M. I., & Rothbart, M. K. (1998). Attention, self–regulation and consciousness. *Philosophical Transactions of the Royal Society of London B: Biological Sciences, 353* (1377), 1915–1927.

Prior, H., Schwarz, A., & Güntürkün, O. (2008). Mirror-induced behavior in the magpie (Pica pica): Evidence of self-recognition. *PLoS Biology, 6*(8), e202.

Putland, D. A., Nicholls, J. A., Noad, M. J., & Goldizen, A. W. (2006). Imitating the neighbours: Vocal dialect matching in a mimic-model system. *Biology Letters, 2*(3), 367–370. doi:10.1098/rsbl.2006.0502.

Radner, D., & Radner, M. (1989). *Animal consciousness*. Prometheus Books.

Rattenborg, N., Martinez, D., Roth, T. C., & Pravosudov, V. V. (2011). Hippocampal memory consolidation during sleep: A comparison of mammals and birds. *Biological Reviews Cambridge Philosophical Society, 86*(3), 658–691.

Rees, G., Kreiman, G., & Koch, C. (2002). Neural correlates of consciousness in humans. *Nature Reviews Neuroscience, 3*(4), 261–270.

Reiner, A. (1986). The co-occurrence of substance P-like immunoreactivity and dynorphin-like immunoreactivity in striatopallidal and striatonigral projection neurons in birds and reptiles. *Brain Research, 371*(1), 155–161.

Reiner, A., Laverghetta, A. V., Meade, C. A., Cuthbertson, S. L., & Bottjer, S. W. (2004a). An immunohistochemical and pathway tracing study of the striatopallidal organization of area X in the male zebra finch. *Journal of Comparative Neurology, 469*(2), 239–261.

Reiner, A., Perkel, D. J., Bruce, L. L., Butler, A. B., Csillag, A., Kuenzel, W., et al. (2004b). Revised nomenclature for avian telencephalon and some related brainstem nuclei. *Journal of Comparative Neurology, 473*(3), 377–414.

Reiss, D., & Marino, L. (2001). Mirror self-recognition in the bottlenose dolphin: A case of cognitive convergence. *Proceedings of the National Academy of Sciences USA, 98*(10), 5937–5942.

Robinson, S., Sandstrom, S. M., Denenberg, V. H., & Palmiter, R. D. (2005). Distinguishing whether dopamine regulates liking, wanting, and/or learning about rewards. *Behavioral Neuroscience, 119*(1), 5.

Rose, J., & Colombo, M. (2005). Neural correlates of executive control in the avian brain. *PLoS Biology, 3*(6), e190.

Rosenthal, D. M. (1986). Two concepts of consciousness. *Philosophical Studies, 49,* 329–359.

Sandberg, K., Del Pin, S. H., Bibby, B. M., & Overgaard, M. (2014). Evidence of weak conscious experiences in the exclusion task. *Frontiers in Psychology, 23*(5), 1080. doi:10.3389/fpsyg.2014.01080. eCollection 2014.

Sasaki, A., Sotnikova, T. D., Gainetdinov, R. R., & Jarvis, E. D. (2006). Social context-dependent singing-regulated dopamine. *The Journal of Neuroscience, 26*(35), 9010–9014.

Savage-Rumbaugh, S., Shanker, S. G., & Taylor, T. J. (1998). *Apes, language, and the human mind.* Oxford University Press.

Schiff, N. D., Giacino, J., Kalmar, K., Victor, J., Baker, K., Gerber, M., et al. (2007). Behavioural improvements with thalamic stimulation after severe traumatic brain injury. *Nature, 448*(7153), 600–603.

Seager, W., & Bourget, D. (2007). Representationalism about consciousness. In *The Blackwell companion to consciousness* (pp. 261–276).

Searle, J. R. (1994). Animal Minds. *Midwest Studies IN Philosophy, 19*(1), 206–219. doi:10.1111/j.1475-4975.1994.tb00286.x.

Searle, J. R. (2003). *Rationality in action.* MIT Press.

Sergent, C., Baillet, S., & Dehaene, S. (2005). Timing of the brain events underlying access to consciousness during the attentional blink. *Nature Neuroscience, 8*(10), 1391–1400.

Siclari, F., Bernardi, G., Riedner, B. A., LaRocque, J. J., Benca, R. M., & Tononi, G. (2014). Two distinct synchronization processes in the transition to sleep: A high-density electroencephalographic study. *Sleep, 37*(10), 1621.

Sinha, A. (1999). Almost minds? The search for consciousness in nonhuman primates. *Scientific and Philosophical Studies on Consciousness*, 149–158.

Stuss, D. T., & Alexander, M. P. (2000). Executive functions and the frontal lobes: A conceptual view. *Psychological Research, 63*(3–4), 289–298.

Suddendorf, T., & Butler, D. L. (2013). The nature of visual self-recognition. *Trends in Cognitive Science, 17*(3), 121–127.

Suddendorf, T., & Collier-Baker, E. (2009). The evolution of primate visual self-recognition: Evidence of absence in lesser apes. *Proceedings of the Royal Society of London B: Biological Sciences, 276*(1662), 1671–1677.

Taylor, A. H. [BBC]. (2014, 4 February). *Are crows the ultimate problem solvers?—Inside the animal mind: episode 2. BBC2* [video file]. https://www.youtube.com/watch?v=AValTA7eBZE. Accessed December 12, 2016.

Taylor, A. H., Knaebe, B., & Gray, R. D. (2012). An end to insight? New Caledonian crows can spontaneously solve problems without planning their actions. *Proceedings of the Royal Society of London B: Biological Sciences, 279*(1749), 4977–4981.

Tetzlaff, M., & Rey, G. (2009). Systematicity and intentional realism in honeybee navigation. In R. W. Lurz. *The philosophy of animal minds.* Cambridge University Press.

Thorpe, W. H. (1958). *Learning and instinct in animals.* Methuen.

Toda, K., & Platt, M. L. (2015). Animal cognition: Monkeys pass the mirror test. *Current Biology, 25*(2), R64–R66.

Tomkins, D. M., & Sellers, E. M. (2001). Addiction and the brain: The role of neurotransmitters in the cause and treatment of drug dependence. *Canadian Medical Association Journal, 164*(6), 817–821.

Tononi, G., Boly, M., Massimini, M., & Koch, C. (2016). Integrated information theory: From consciousness to its physical substrate. *Nature Review of Neuroscience, 17*(7), 450–461. doi:10.1038/nrn.2016.44.

Tononi, G., & Edelman, G. M. (1998). Consciousness and complexity. *Science, 282*(5395), 1846–1851.

Tye, M. (1995). *Ten problems of consciousness: A representational theory of the phenomenal mind* (vol. 282). MIT Press.

Uzgiris, I. C., & Hunt, J. M. (1975). *Assessment in infancy: Ordinal scales of psychological development.* Champaign, IL: University of Illinois Press.

Van der Werf, Y. D., Witter, M. P., & Groenewegen, H. J. (2002). The intralaminar and midline nuclei of the thalamus. Anatomical and functional evidence for participation in processes of arousal and awareness. *Brain Research Reviews, 39*(2), 107–140.
Van Hoesen, G. W., Morecraft, R. J., & Vogt, B. A. (1993). Connections of the monkey cingulate cortex. In B. A. Vogt & M. Gabriel (Eds.), *Neurobiology of cingulate cortex and limbic thalamus* (pp. 249–284). Boston: Birkhäuser.
Vates, G. E., & Nottebohm, F. (1995). Feedback circuitry within a song-learning pathway. *Proceedings of the National Academy of Sciences USA, 92*(11), 5139–5143.
Veenman, M. V., Elshout, J. J., & Meijer, J. (1997). The generality vs domain-specificity of metacognitive skills in novice learning across domains. *Learning and Instruction, 7*(2), 187–209.
Vicario, D. S. (1994). Motor mechanisms relevant to auditory-vocal interactions in songbirds. *Brain, Behavior and Evolution, 44*(4–5), 265–278.
Vogt, B., Vogt, L., Nimchinsky, E., & Hof, P. (1997). Chapter VIII Primate cingulate cortex chemoarchitecture and its disruption in Alzheimer's disease. *Handbook of Chemical Neuroanatomy, 13,* 455–528.
Von Economo, C., & Koskinas, G. (1925). *The cytoarchitectonics of the adult human cortex.* Vienna and Berlin: Julius Springer Verlag.
Waldmann, C., & Gu, O. (1993). The dopaminergic innervation of the pigeon caudolateral forebrain: Immunocytochemical evidence for a 'prefrontal cortex' in birds? *Brain Research, 600*(2), 225–234.
Watanabe, S. (2001). Discrimination of cartoons and photographs in pigeons: Effects of scrambling of elements. *Behavioural Processes, 53*(1–2), 3–9.
Watanabe, S., Sakamoto, J., & Wakita, M. (1995). Pigeons' discrimination of paintings by Monet and Picasso. *Journal of the Experimental Analysis of Behavior, 63*(2), 165–174. doi:10.1901/jeab.1995.63-165.
Weir, A. A., Chappell, J., & Kacelnik, A. (2002). Shaping of hooks in New Caledonian crows. *Science, 297*(5583), 981.
Wild, J., & Williams, M. (2000). Rostral wulst in passerine birds. I. Origin, course, and terminations of an avian pyramidal tract. *Journal of Comparative Neurology, 416*(4), 429–450.
Xi, Z.-X., & Stein, E. A. (2002). GABAergic mechanisms of opiate reinforcement. *Alcohol and Alcoholism, 37*(5), 485–494.
Zhang, S., Bandler, R., & Davis, P. J. (1995). Brain stem integration of vocalization: Role of the nucleus retroambigualis. *Journal of Neurophysiology, 74*(6), 2500–2512.

Meditation, Cognitive Reserve and the Neural Basis of Consciousness

Ajay Kumar Nair and Bindu M. Kutty

Introduction

The quest for self-development is perhaps an integral part of being human. The quest to understand ourselves and our place in nature is another such endeavour that has stood the test of time. As such, the study of consciousness is weaved into the history and narratives of cultures around the world. Religious and spiritual traditions as well as several schools of philosophy have dwelled at length about the nature of consciousness, developed theories and systematised approaches to improve our wellbeing and finally reach a state of consciousness that is free and unperturbed by life's challenges. Theories vary and so do the practices recommended. Some traditions envisioned that prayer and penance might invoke divine grace and promotion to a state of emancipation where one could enjoy sustained wellbeing for eternity, while others have focused on systematic efforts for self-development without divine intervention. In spite of the diversity of views and approaches, a common tool that these traditions have, is meditation or some form of contemplative practice.

While meditative practices have a long history over several millennia, the scientific study of meditation has a short past. Perhaps the earliest clinical study of the effects of yoga/meditation was carried out at the yoga institute in 1918 at Versova, Mumbai followed by studies at Kaivalyadhama yoga institute at Lonavala near Pune (Khalsa 2004). The earliest electroencephalography (EEG) studies were carried out in the 1950s (Das and Gastaut 1955) and 1960s by BK Anand and others who reported the high prevalence of alpha waves during meditation which did not attenuate even when challenged with distractor stimuli (Anand et al. 1961). These were the times when the impact of meditation on brain functions was unknown and

A.K. Nair · B.M. Kutty (✉)
Department of Neurophysiology, National Institute of Mental Health and Neuro Sciences (NIMHANS), Bengaluru, India
e-mail: bindu.nimhans@gmail.com

S. Menon et al. (eds.), *Self, Culture and Consciousness*,
DOI 10.1007/978-981-10-5777-9_3

the existence of adult brain plasticity—that adult brains have the capacity to re-organise its circuitry and even its strategies—was yet to be discovered. The next wave of meditation studies came in the 1970s and 1980s from Robert Keith Wallace, Fred Travis and their colleagues who catalogued a number of effects of transcendental meditation practice. Their studies helped to understand the importance of meditation enhanced brain synchrony and coherence help to integrate the brain functions (Wallace et al. 1971; Wallace 1970). In particular, they reported the occurrence of enhanced EEG theta and alpha power as well as enhanced EEG synchrony and coherence in during meditation (Jevning et al. 1992; Shapiro and Walsh 1984). The last couple of decades have seen an explosive growth in meditation studies using many modalities including EEG, event related potentials (ERPs), structural and functional magnetic resonance imaging (MRI, fMRI) and so on (Boccia et al. 2015; Cahn and Polich 2006; Gonçalves et al. 2015; Lomas et al. 2015; Singh and Telles 2015; Tang et al. 2015).

These developments were not in isolation. The scientific study of consciousness also became popular during the last few decades and the development of cognitive neuroscience and associated enabling technologies made it possible to start a movement from philosophical concepts to empirical testing and validation. In this chapter, we review some of the recent scientific studies on short-term and long-term meditative practices on brain plasticity and cognitive reserve capacities and consider how they might help elucidate the development of higher states of consciousness.

Meditation and the Brain

The brain is a highly complex network consisting of billions of densely interconnected neurons and glia that interact using a wide variety of electrochemical mechanisms. Individual neurons convey very little information; they act in a binary mode—either they 'fire' or they stay silent. EEG oscillatory rhythms are emergent properties of neural networks that enable unified cognitive operations by making neuronal assemblies in different parts of the brain work in concert (Buzsáki and Freeman 2015). Sensory perception, attention, perceptual binding, memory, sensory-motor integration and indeed consciousness, are all enabled by these brain oscillations and have been substantially preserved in mammalian evolution (Buzsáki and Freeman 2015). Different EEG rhythms are observed during the performance of different tasks. To simplify greatly, theta waves are associated with memory encoding and executive control; alpha waves are associated with relaxation, behavioural inhibition and memory formation; beta waves are involved in polymodal sensory processing and motor preparation; gamma waves have been implicated in sensory processing and perceptual binding (Uhlhaas et al. 2008). The functional significance of these rhythms is still a subject of active research.

Meditation is a term that bundles up a set of highly diverse mind-body practices with the common thread of conscious mental training leading to a positive experience (Fontana 2012). Some meditation techniques involve concentration on breath or a *mantra,* while others require detached observation with suspension of judgement, while still others may involve directed thinking, experiencing and enactment. Typically, meditation involves internalised attention, heightened awareness and better cognitive control (Dahl et al. 2015; Tang et al. 2015). Thus the neurophysiological effects of meditative practices can be studied through EEG, ERP, fMRI and other tools that are used in cognitive neuroscience. As is to be expected, differences in meditative techniques are accompanied by differences in neurophysiological effects. Nevertheless, resting state theta and alpha activity are often altered amongst meditation practitioners as compared to non-meditators, illustrating trait changes due to brain plasticity mechanisms (Cahn and Polich 2006). Changes in these rhythms are also typically seen within meditator groups (and not within non-meditators) between rest and meditative states demonstrating the interaction of state and trait effects of the meditative practice (Davidson and Kaszniak 2015).

A number of studies have shown that long-term meditative practices bring about changes in the physical structure of the brain. An early study of insight meditation practitioners found that, compared to age- and gender-matched controls, meditators had less age-associated cortical thinning in regions related to interoception, sensory and cognitive processing (Lazar et al. 2005). The insula, which is a region associated with the interoceptive processes and breath awareness techniques, had the largest between-group difference. These structural changes may be due to several factors, such as greater dendritic arborisation (more branches per neuron), enhanced synaptic restructuring (related to the connections between neurons), increase in the glial volume (which contribute to the scaffolding between neurons and also actively participate in the communication process) and accompanied by increased regional vasculature (necessary for the enhanced oxygen consumption). Similarly, a study of a heterogeneous group of 50 long-term meditators found differences in the folding of cortical gyri in specific areas (especially right anterior dorsal insula) that were positively correlated with years of meditation practice (Luders et al. 2012). Even very short-term meditation interventions (11 h) can bring about functional and structural changes. A study of Integrative Body-Mind Training (IBMT) training showed increased activity in the anterior cingulate cortex leading to better self-regulation. This was accompanied by improved fractional anisotropy (an index of white matter integrity and density) in the corona radiata which connects the anterior cingulate to other regions (Tang et al. 2010).

Meditative practices typically provide behavioural benefits in terms of relaxation, development of positive affect (feelings), reduction of negative affect and improvement in various cognitive abilities. These behavioural changes are associated with neurophysiological changes seen in terms of enhanced brain synchrony and coherence as well as autonomic modulation in the short term as well as trait changes in the long term. The trait changes can be attributable to the neural plasticity mechanisms that manifest as changes in grey matter volume and white matter integrity mentioned earlier. These changes are likely to be the basis of the cognitive reserve that provides the long-term meditation practitioners enhanced neuroprotection against aging and resilience in the face of situational challenges leading the cultivation of higher mental attributes such as heightened wellbeing and altruistic behaviour. Perhaps through the enhancement of cognitive reserve capacities, meditative practices are able to bring about a total change in brain network properties.

Meditation and Cognitive Reserve

An intriguing finding in neurology patients has been that the severity of cognitive deficits is not fully predicted by brain pathology or age-associated changes. Several factors such as education, occupation and engagement in social activities built up over a lifetime have been found to be protective against cognitive decline with trauma and brain degeneration. This capacity for effective cognitive functioning in spite of pathophysiological challenges has been termed 'cognitive reserve' and is attributed to possible compensatory or neuroprotective mechanisms. In addition, cognitive reserve also manifests as improved decision-making and strategy selection abilities in aged populations that do not manifest clinical symptoms (Barulli and Stern 2013). The underlying neural mechanisms are yet to be delineated but are partially explained by greater brain network efficiency or compensatory use of less advantageous but alternate neuronal circuits. Thus the concepts of brain plasticity, brain reserve, neural reserve, neural compensation and brain maintenance are related to, but slightly different from, the concept of cognitive reserve and may contribute to the development of resilience to brain insults and aging processes (Barulli and Stern 2013).

As discussed, a variety of long-term meditative practices have been found to bring about structural and functional changes in the brain that defy the degeneration that accompanies advancing age. Meditation could thus be an easily learnable tool for the development of cognitive reserve that could be of great utility for a larger segment of the population that does not have a background or resources for protective factors such as higher education, intellectually challenging work or an extensive social life. It is therefore important to study the meditative brain in closer detail to examine neural changes that accompany long-term meditative practices.

We have carried out a number of studies from a sleeping brain perspective as well as studies of cognitive capacities and wellbeing in the wake state in long-term practitioners of different schools of meditation such as Vipassana, Sudarshan Kriya and Rāja yoga. Sleep is a well-orchestrated sequence of brain-state changes that we all undergo every night, although we are not consciously aware of these changes. There are several sleep cycles in a typical night's sleep. Each sleep cycle goes progressively into deeper sleep with slower EEG rhythms and culminates with a period of paradoxical sleep called rapid eye movement (REM) sleep which is strongly correlated with dreaming episodes. We have found that senior Vipassana meditators have distinct sleep architecture with longer slow wave and REM durations as compared to healthy controls and novice meditators (Nagendra et al. 2012;

Pattanashetty et al. 2010; Sulekha et al. 2006). Scalp topography maps also show enhanced theta-alpha (6–10 Hz) power in senior meditators as compared to controls and novices. Additionally, senior meditators also show higher REM density with significantly more phasic and tonic events as compared to other groups (Maruthai et al. 2016). It was Mason and colleagues who reported the changes in sleep variables associated with meditation practice. They found that long-term transcendental meditation practitioners had higher theta-alpha (along with delta) during deep sleep than short-term practitioners (Mason et al. 1997). These long-term meditators had self-reported 'higher states of consciousness' during sleep and the neurophysiological changes during sleep were consistent with the reported periods of 'transcendental consciousness' during meditation which also had increased theta-alpha activity. The above study also found increased REM density in the long-term group of TM meditators, similar to our findings in long-term Vipassana meditators (Maruthai et al. 2016). REM activity dynamics can be linked with reactivation of neural circuitry and synaptic potentiation (Diekelmann and Born 2010) and that enhanced REM could be indicative of lucid dreaming (Hobson and McCarley 1977; Mota-Rolim et al. 2010). Lucid dreaming (where one is aware of the content of the dreams and is able to control them) is in itself a different state of consciousness with features of both waking and normal dreaming (Voss et al. 2009).

Meditation and Higher States of Consciousness

Overall, long-term meditative practices are accompanied by neuroplastic changes. While these meditation techniques are very different in content and process, there seem to be some common mechanisms that underlie the phenomenon of neural plasticity. These changes could be due to the development of higher states of consciousness as per the theories guiding the different meditation techniques.

In the waking brain, too, a number of electrophysiological correlates of higher mental functions have been found. Enhanced theta-alpha activity, especially in the frontal regions has been associated with sustained internalised attention and positive emotion processing (Aftanas and Golocheikine 2001). During meditation, apart from enhanced changes in the theta and alpha bands (found in practitioners from a number of meditation techniques), gamma band activity has also been prominently found (Cahn et al. 2010). Meditation has been considered to be a hypometabolic state with enhanced parasympathetic activity while staying responsive and alert (Young and Taylor 1998). The key difference is that, during most hypometabolic states such as sleep and hibernation, awareness levels are almost missing while, in most forms of meditation, awareness is maintained or heightened. This sense of awareness may be open to external stimuli (such as in mindfulness forms of meditation) or may remain unaffected by external stimuli (such as in the yogic meditation studied by Anand et al. 1961). Meditators have increased self-regulation, enhanced subjective wellbeing and psychological wellbeing as seen in a number of studies. Urry et al. (2004) found prefrontal asymmetry with greater left activation in those having higher wellbeing and in our own work. An important requirement that has been highlighted in the recent literature is that wellbeing considerations in meditators need to be studied in the context of their practice (Sedlmeier et al. 2012). We have found this to be very valuable in our studies and have found that proficiency in meditation (not just long-term practice) plays a major role in wellbeing enhancement.

A word of caution is in order. Not all meditative techniques are equally efficacious for all people and in a few situations there may be adverse incidents (Fingelkurts et al. 2015). To avoid adverse outcomes, especially in those with existing psychopathologies, it is important to consider psychological dispositions and personality traits as well as pre-existing health issues while deciding on the type of meditation practice to be adopted. While a personalised approach is useful when making a decision about which technique is suitable, after commencing the journey it might be better to apply best practices that may have been refined over (thousands of) years of that meditative tradition.

It is useful to consider how meditation might help in moving up the hierarchy of conscious experiences. Schooler has suggested that while people typically have mental processes that are non-conscious (unexperienced) or conscious (experienced), they also intermittently experience meta-conscious (re-represented) mental processes (Schooler 2002). Since meditation results in heightened awareness and since many meditation forms encourage a contemplative and non-judgemental stance; meta-cognitive processes are routinely employed by meditators. Since traditional practices encourage higher moral standards and self-regulation, it is reasonable to infer that sincere meditation practice could facilitate experiences of higher states of consciousness.

In most traditional approaches, meditation has been a tool; and wellbeing, the pathway; to the final goal of liberation or self-transcendence (Sedlmeier et al. 2012). Scientific evidence thus far supports the claim that meditation does lead to enhanced wellbeing, building up of cognitive reserve capacities, development of hidden potentialities and positive changes in states of consciousness. Whether the final goal of liberation and/or self-transcendence is valid, realistic and achievable may remain unverifiable until the puzzle of consciousness is solved.

References

Aftanas, L. I., & Golocheikine, S. A. (2001). Human anterior and frontal midline theta and lower alpha reflect emotionally positive state and internalized attention: High-resolution EEG investigation of meditation. *Neuroscience Letters, 310,* 57–60.

Anand, B. K., Chhina, G. S., & Singh, B. (1961). Some aspects of electroencephalographic studies in yogis. *Electroencephalography and Clinical Neurophysiology, 13*(3), 452–456.

Barulli, D., & Stern, Y. (2013). Efficiency, capacity, compensation, maintenance, plasticity: Emerging concepts in cognitive reserve. *Trends in Cognitive Science, 17,* 502–509. doi:10.1016/j.tics.2013.08.012.

Boccia, M., Piccardi, L., & Guariglia, P. (2015). The meditative mind: A comprehensive meta-analysis of MRI studies. *BioMed Research International, 2015*. doi:10.1155/2015/419808

Buzsáki, G., & Freeman, W. (2015). Editorial overview: Brain rhythms and dynamic coordination. *Current Opinion in Neurobiology*, 31, v. doi:10.1016/j.conb.2015.01.016

Cahn, B. R., & Polich, J. (2006). Meditation states and traits: EEG, ERP, and neuroimaging studies. *Psychology Bullettin, 132,* 180–211. doi:10.1037/0033-2909.132.2.180.

Cahn, B. R., Delorme, A., & Polich, J. (2010). Occipital gamma activation during Vipassana meditation. *Cognitive Process, 11,* 39–56. doi:10.1007/s10339-009-0352-1.

Dahl, C. J., Lutz, A., & Davidson, R. J. (2015). Reconstructing and deconstructing the self: Cognitive mechanisms in meditation practice. *Trends in Cognitive Sciences, 19*(9), 515–523. doi:10.1016/j.tics.2015.07.001.

Das, N., & Gastaut, H. (1955). Variations in the electrical activity of the brain, heart, and skeletal muscles during yogic meditation and trance. *Electroencephalography and Clinical Neurophysiology, 6,* 211–219.

Davidson, R. J., & Kaszniak, A. W. (2015). Conceptual and methodological issues in research on mindfulness and meditation. *American Psychologist, 70*(7), 581.

Diekelmann, S., & Born, J. (2010). The memory function of sleep. *Nature Review of Neuroscience, 11,* 114–126. doi:10.1038/nrn2762.

Fingelkurts, A. A., & Kallio-Tamminen, T. (2015). EEG-guided meditation: A personalized approach. *Journal of Physiology-Paris, 109*(4), 180–190. doi:10.1016/j.jphysparis.2015.03.001.

Fontana, D. (2012). *The meditation handbook: The practical guide to eastern and western meditation techniques*. Duncan Baird Publishers.

Gonçalves, J. P. B., Lucchetti, G., Menezes, P. R., & Vallada, H. (2015). Religious and spiritual interventions in mental health care: A systematic review and meta-analysis of randomized controlled clinical trials. *Psychological Medicine, 45*(14), 2937–2949. doi:10.1017/S0033291715001166.

Hobson, J. A., & McCarley, R. W. (1977). The brain as a dream state generator: An activation-synthesis hypothesis of the dream process. *American Journal of Psychiatry, 134,* 1335–1348. doi:10.1176/ajp.134.12.1335.

Jevning, R., Wallace, R. K., & Beidebach, M. (1992). The physiology of meditation: A review. A wakeful hypometabolic integrated response. *Neuroscience and Biobehaviour Reviews, 16,* 415–424. doi:10.1016/S0149-7634(05)80210-6.

Khalsa, S. B. S. (2004). Yoga as a therapeutic intervention: A Bibliometric analysis of published research studies. *Indian Journal of Physiology and Pharmacology, 48,* 269–285.

Lazar, S. W., Kerr, C. E., Wasserman, R. H., Gray, J. R., Douglas, N., Treadway, M. T., et al. (2005). Meditation experience is associated with increased cortical thickness. *NeuroReport, 16,* 1893–1897.

Lomas, T., Ivtzan, I., & Fu, C. H. (2015). A systematic review of the neurophysiology of mindfulness on EEG oscillations. *Neuroscience and Biobehavioral Reviews, 57,* 401–410.

Luders, E., Kurth, F., Mayer, E. A., Toga, A. W., Narr, K. L., & Gaser, C. (2012). The unique brain anatomy of meditation practitioners: Alterations in cortical gyrification. *Frontiers Human Neuroscience, 6,* 1–9. doi:10.3389/fnhum.2012.00034.

Maruthai, N., Nagendra, R. P., Sasidharan, A., Srikumar, S., Datta, K., Uchida, S., et al. (2016). Senior Vipassana Meditation practitioners exhibit distinct REM sleep organization from that of novice meditators and healthy controls. *International Review of Psychiatry, 28*(3), 279–287.

Mason, L. I., Alexander, C. N., Travis, F. T., Marsh, G., Orme-Johnson, D. W., Gackenbach, J., et al. (1997). Electrophysiological correlates of higher states of consciousness during sleep in longterm practitioners of the Transcendental Meditation program. *Sleep, 20*(2), 102–110.

Mota-Rolim, S. A., Erlacher, D., Tort, A. B., Araujo, J. F., & Ribeiro, S. (2010). Different kinds of subjective experience during lucid dreaming may have different neural sub-strates. *Journal of Neuroscience, 25*(3), 550–557.

Nagendra, R. P., Maruthai, N., & Kutty, B. M. (2012). Meditation and its regulatory role on sleep. *Frontiers in Neurology, 3,* 54. doi:10.3389/fneur.2012.00054.
Pattanashetty, R., Sathiamma, S., Talakkad, S., Nityananda, P., Trichur, R., & Kutty, B. M. (2010). Practitioners of Vipassana meditation exhibit enhanced slow wave sleep and REM sleep states across different age groups. *Sleep and Biological Rhythms, 8*(1), 34–41. doi:10.1111/j.1479-8425.2009.00416.x.
Schooler, J. W. (2002). Re-representing consciousness: Dissociations between experience and meta-consciousness. *Trends in Cognitive Sciences, 6,* 339–344.
Sedlmeier, P., Eberth, J., Schwarz, M., Zimmermann, D., Haarig, F., Jaeger, S., et al. (2012). The psychological effects of meditation: A meta-analysis. *Psychology Bullettin, 138,* 1139–1171. doi:10.1037/a0028168.
Shapiro, D. H., & Walsh, R. N. (Eds.). (1984). *Meditation: Classic and contemporary perspectives.* AldineTransaction.
Singh, N., & Telles, S. (2015). Neurophysiological effects of meditation based on evoked and event related potential recordings. *BioMed Research International*, 2015.
Sulekha, S., Thennarasu, K., Vedamurthachar, A., Raju, T. R., & Kutty, B. M. (2006). Evaluation of sleep architecture in practitioners of Sudarshan Kriya yoga and Vipassana meditation. *Sleep and Biological Rhythms, 4,* 207–214. doi:10.1111/j.1479-8425.2006.00233.x.
Tang, Y.-Y., Lu, Q., Geng, X., Stein, E. A., Yang, Y., & Posner, M. I. (2010). Short-term meditation induces white matter changes in the anterior cingulate. *PNAS, 107,* 15649–15652. doi:10.1073/pnas.1011043107.
Tang, Y.-Y., Holzel, B., & Posner, M. I. (2015). The neuroscience of mindfulness. *Nature Review of Neuroscience, 16,* 213–225. doi:10.1038/nrn3916.
Uhlhaas, P. J., Haenschel, C., Nikolić, D., & Singer, W. (2008). The role of oscillations and synchrony in cortical networks and their putative relevance for the pathophysiology of schizophrenia. *Schizophrenia Bulletin, 34,* 927–943. doi:10.1093/schbul/sbn062.
Urry, H. L., Nitschke, J. B., Dolski, I., Jackson, D. C., Dalton, K. M., Mueller, C. J., et al. (2004). Making a life worth living: Neural correlates of well-being. *Psychological Science, 15,* 367–372. doi:10.1111/j.0956-7976.2004.00686.x.
Voss, U., Holzmann, R., Tuin, I., & Hobson, J. A. (2009). Lucid dreaming: A state of consciousness with features of both waking and non-lucid dreaming. *Sleep, 32,* 1191–1200.
Wallace, R. K. (1970). Physiological effects of transcendental meditation. *Science, 167*(3926), 1751–1754.
Wallace, R., Benson, H., & Wilson, A. (1971). A wakeful hypometabolic physiologic state. *American Journal of Physiology,* 221, 795–799. doi: NLM.19711112
Young, J. D. E., & Taylor, E. (1998). Meditation as a voluntary hypometabolic state of biological estivation. *Physiology, 13*(3), 149–153.

Attention and Perception in the Deaf: A Case for Plasticity in Consciousness

Seema Prasad and Ramesh Kumar Mishra

Introduction

Although philosophers of mind have not yet settled if mental states and functions are to be tied to neural activity per se, much of the current research into brain plasticity suggests that every experience that we go through modifies cortical connections and functionality. Cognitive functions related to attention, memory, perception, language and decision-making evolve as experiences sculpt brain networks (Mahncke et al. 2006; Merzenich et al. 1996). At times philosophers of mind look into and discuss data from experimental literature in order to understand conceptual issues better. One good example is the recent works of Ned Block (Block 2014, 2015a, b). Block takes experimental data very seriously and uses them to argue on the nature of consciousness or attention or other related things. Data from cognitive neurosciences have been used abundantly on such matters (Bor and Seth 2012). Whether consciousness arises from neural computations carried out by designated neuronal ensembles has been an open issue. There also have been many debates on the nature of attention, awareness and their relationship to consciousness (Montemayor and Haladjian 2015; Srinivasan 2007). Different cognitive psychologists and philosophers of minds have taken radically opposite positions (Koch and Tsuchiya 2007; Cohen et al. 2012; Lamme 2003; Kentridge 2011). In this chapter, we look at cognitive enhancement as a consequence of sensory deprivation as in the case of the deaf and try to see what we can infer about critical mechanisms such as attention and visual awareness.

Reorganisation of brain structures can have a far-reaching impact on both neural functioning and behaviour. For instance, continual practice of a skill such as meditation could influence awareness and attention (Manna et al. 2010; Raffone and

S. Prasad · R.K. Mishra (✉)
Centre for Neural and Cognitive Sciences, University of Hyderabad,
Hyderabad 500046, India
e-mail: rkmishra@uohyd.ac.in

S. Menon et al. (eds.), *Self, Culture and Consciousness*,
DOI 10.1007/978-981-10-5777-9_4

Srinivasan 2009, 2010). These data suggest that one's neural apparatus can be altered through long-term practice. Other instances of brain reorganisation are seen in the case of sensory deprivation. Sensory deprivation since birth often leads to a complete and dramatic reorganisation of brain networks that assume spectacular power. This reorganisation helps the individual survive and dwell in the face of severe limitations that arise because of such deprivation. It is a similar case with deaf individuals where hearing impairment has been shown to promote superior visual awareness and visuospatial attentional abilities. Considering that the deaf participants have higher selective attention and accepting the theory that attention and awareness are in some form of causal relationship, then one must expect a fundamentally different kind of awareness in the deaf. Therefore, data from such participants are not only informative about their conditions; they can be used to answer deep questions about the mind. In this context, we present in this chapter data from hearing impaired people where they show enhancement in both perceptual and attentional mechanisms. Current research suggests that the dorsal visual processing stream undergoes considerable modification in the deaf (Bavelier et al. 2006). What does this imply for visual perception and visual awareness in deaf? Although our current understanding of visual awareness and attentional mechanisms comes from studies of normal participants, there is much to learn from the data on deaf. It is time we consider individual difference related factors while talking about awareness, perception and consciousness, at least while referring to experimental data.

Neuroplasticity of Cognitive Systems

Deaf signers need to pay constant attention to another signer's dynamically changing hand and lip movements for communication. Cognitive psychological studies with both children and adult deaf individuals have produced a range of data that testify to such excellent example of plasticity (Bavelier and Neville 2002; Dye et al. 2008). Deprivation in hearing strengthens vision. However, the mechanism of vision and its relations to such complex processes as attention and awareness has been a matter of debate for several decades. While we still do not know if our consciousness and awareness are dependent on our ability to attend, findings from the deaf have the power to throw new light on these questions. In this chapter, we will survey data that show the subtle nature of plasticity-induced cognitive abilities in the deaf in the area of visual attention and awareness. Research shows that deaf just do not see better but have developed a better sense of awareness of the visual world, at times extending into unconscious processing. Their heightened ability to respond to visual stimuli in the periphery is also seen with their superior ability to process unconscious stimuli presented below the thresholds of awareness. One of the hotly debated questions in cognitive science is whether attention and consciousness are one and the same thing? Or are they fully dissociated? Although reviewing all the differing viewpoints on this issue is beyond the scope of this chapter, we will discuss some of the major theoretical frameworks and later examine how changes induced by deafness can be connected to them.

Is attention necessary for conscious perception? Attention, in this case, is mostly referred to the mechanism which selects an object out of many for further selective processing. Visual-spatial attention to a location leads to higher processing of objects that appear in that location. Attention to one object or location leads to lesser acuity in another location. Evidence from paradigms such as change blindness and attentional blink suggests that, if attention is not sufficiently engaged, then the perceiver does not extract objective knowledge of things (Dux and Marois 2009). Therefore, based on available evidence one can say that attention, being a limited resource, only illuminates certain aspects of the world for our consciousness. Subjective knowledge about a broader visual field is at best gross and people do not remember any useful detailed information. Selective attention is required for detailed task specific knowledge. Therefore, this line of argument makes attention mandatory for conscious perception and suggests that attention and consciousness cannot be fully dissociated (Cohen et al. 2012). Opposite to this view is that attention and consciousness are fully distinct processes (Koch and Tsuchiya 2007). A commonly cited example of consciousness in the near absence of attention is gist perception. When participants are shown a photograph or a scene for a brief duration, they are still able to correctly perceive the gist of the stimulus. As the duration of presentation is too low for the attentional system to be engaged, it has been claimed that conscious awareness can exist without top-down attention. Another example is that of pop-out in visual search. When participants are asked to identify a red circle among a set of green circles, the red circle 'pops out' instantly. The time required to identify such a salient stimulus is independent of the number of other distracting stimuli. It is said that attention is not engaged in such situations, but individuals are nevertheless conscious of the stimulus. In sum, the necessity of attention for conscious perception is a heavily debated issue in the scientific study of consciousness and there is no final consensus on the topic.

Is consciousness necessary for attentional engagement? In contrast to the question discussed above, most researchers agree to the point that attention can be directed to stimuli that never reach conscious awareness (Dehaene et al. 2006; Kiefer and Martens 2010). For example, interocular grouping (combining different parts of an image into a holistic representation) can happen without awareness (Lin and Yeh 2016). In this study, continuous flash suppression was used as a method to make stimuli presented to each eye invisible. Similarly, binocular rivalry can arise without awareness (Klink and Roelfsema 2016). Thus, even if the participant is not aware of a certain stimulus, attention can be directed to it. Studies from masked priming have shown that directing attention to a masked prime can increase priming effects. (Naccache et al. 2002; Shin et al. 2009; van den Bussche et al. 2010) Thus, deployment of attention does not necessarily result in conscious perception. In the context of deafness, if attention and awareness are two different mechanisms, then plasticity-induced changes should be seen in them differently. If they are causally related, then plasticity should influence them similarly.

Given these interrelationships between attention and awareness, it is then important to examine how to ground this in an experimental paradigm. Unfortunately, much theorisation in cognitive psychology on these constructs has been an outcome of the paradigms used to study them and the constraints that they come with. For example, the attentional blink or the inattentional blindness tasks exploit only specific aspects of attention. As is now widely known, attention is not a unitary mechanism but a collection of different processes. It has various attributes, such as local versus global; focused versus dispersed; selective versus non-selective. A single paradigm may not capture all these attributes. Therefore, a causal explanation regarding awareness and attention is also limited by the particular task or paradigm used to infer about such processes. As different tasks have been used to understand visual perception and attention in the deaf, they have given rise to multiple explanations in the field. It is not clear what exactly the deaf individuals have which in turn enhances the brain mechanisms related to awareness and attention. Further, plasticity-related changes in core brain functions leading to noticeable outcomes in behaviour is also a matter of skill and habit. Therefore, many researchers have also tried to link sign language use and its proficiency while understanding behavioural task outcomes in the deaf.

In the next two sections, we will discuss the issue of sensory deprivation leading to brain plasticity which in turn modifies attentional and perceptual skills. We will then discuss if brain plasticity changes both? People can be aware of many things in their visual field while they can attend only to some. The limited capacity system does not explain the extensive awareness which most observers seem to experience and believe strongly (Cohen et al. 2016). Although objective knowledge is acquired from very few objects, participants show subjective awareness of many. Constraints on visual perception and eye movements also suggest that much of our visual awareness comes from parafoveal and peripheral awareness. Cohen et al. (2016) question this and ask how come our visual awareness and our intuitive knowledge about things are so rich when our attentional and working memory capacities are justifiably so less? This question is of central importance to the deaf issue since we need to know what exactly brain plasticity does to the deaf brain. Does it increase their attentional capability or it enhances their scope of awareness without changing their attentional or working memory capacity? The evidence is sparse since only a few studies have examined these issues.

Enhanced Visual Perception and Attention in the Deaf

What exactly does plasticity mean? It refers to the neural changes because of acquisition and practice of some skills. This is seen, for example, in bilingual speakers whose brains show plasticity because of the practice of bilingualism (Abutalebi et al. 2011; Costa and Sebastián-Gallés 2014). Higher levels of language use lead to changes in the frontoparietal attention network in bilinguals. Another understanding of plasticity is related to transfer of neural selectivity because of sensory deprivation. No cortical area is hard wired to process and be engaged in only one function. In the deaf, absence of audition leads to better vision. Therefore, these two types of plasticity are fundamentally different. Plasticity then refers to the possible changes in the brain as a function of adaptation to new skills. Even very short-term practice of a difficult task can improve plasticity in the neural system, in the form of new functional connectivity among areas or even structural changes. For example, a recent study (Rodrigue et al. 2016) shows that the practice of anti-saccade task in Schizophrenia leads to better connectivity between the pre-frontal region with the insular and temporal region. Playing video games like Super Mario changes structural plasticity in the grey matter (Kühn et al. 2014). Grey matter increase has long been thought to enhance cognitive skills. Children, if trained with music and second-language skills, show improvements in executive control (Janus et al. 2016). Training-induced plasticity in older adults has been shown to improve inhibitory control (Spierer et al. 2013). Often these studies measure changes in brain area volumes pre- and post-training to see the induced changes. Many more examples can be found in current research where some form of training in challenging skills leads to structural and functional changes in neural networks as well as behavioural improvements.

The deaf person's plasticity-induced changes in cognitive functions is a different issue. Here, sensory deprivation leads to one modality becoming stronger than what is normally seen. The deaf individuals may learn a sign language while they are children and many who do not learn a sign language may use just gestures. Therefore, enhancement of cognitive processes in the deaf are not induced by training. Congenital deafness leads to very profound modification of the brain, most importantly the primary visual cortex. Much has been learned about compensatory plasticity-related changes in animal models (Rauschecker 1995) and also in blind people (Cohen et al. 1997). Neural data from the congenitally deaf show that their primary visual cortex has a greater and more expansive receptive field (Neville et al. 1982). The deaf occipital cortex also shows higher cortical thickness. In the deaf, it is not just the visual cortex which shows enhanced efficiency of processing but the auditory cortex which has taken over visual processing (Finney et al. 2001). Therefore, the deaf can attend to peripheral stimuli better and discriminate using superior spatial attention. More recently it has been suggested that such cross-modal reorganisation of the auditory and visual cortex in the deaf is linked to the age of onset of sign language use. Therefore, not all deaf develop a higher visual ability since it is linked to their practice of sign language. Some minimal structural modifications may happen during development, but more prominent changes are probably linked to the acquisition of sign language skills.

Why should neuroplasticity alter fundamental mechanisms related to attention, awareness or even consciousness? Several neurological conditions have been used to understand issues related to consciousness. Blindsight is perhaps the most common condition used specifically as evidence for unconscious processing (Ajina and Bridge 2016; Block 2015a, b; Danckert and Goodale 2000). Blindsight is a condition where there is damage to certain parts of the visual cortex (Weiskrantz 1986). Individuals with this condition are 'blind' in certain regions of their visual field. Thus, they are not consciously aware of the objects presented in their blind regions. But they can nonetheless perform above chance level on tasks that require responding to such stimuli. The common explanation for blindsight is that, although the patients are unaware of the objects in their blind regions, they are nonetheless processing the features of the visual stimulus. This has provided convincing evidence that people can be influenced by unconscious stimuli. But not all researchers agree that blindsight is a convincing case of unconscious vision. This is because—as in most studies on unconscious processing—it is difficult to establish whether the patients were entirely unaware of the stimulus. Some researchers have argued that blindsight is a case of severely degraded conscious vision, but not unconscious vision (Overgaard 2012). It has been argued that participants fail to detect the stimulus during identification tests because they adopt a conservative criterion—responding that they have not seen the stimulus only because they are not sure. Studies with more sophisticated awareness tests, such as the perceptual awareness scale, have found a correlation between the reported visibility of the stimulus and the accuracy in responding to it (Overgaard et al. 2008). Thus, these researchers claim that traditional methods of testing awareness used in many studies underestimate visibility and so blindsight is a result of degraded form of conscious processing. This issue is still being heavily debated (Brogaard 2011) Interested readers can refer to the debates between Ned Block and Ian Phillips to know about the controversies regarding unconscious perception and the role of blindsight studies in examining this issue ('Block/Phillips debate on unconscious perception', NYU website). Controversies apart, the studies with blindsight and other conditions like hemineglect highlight the fact that naturally occurring conditions in human participants can shed new light on critical issues. The changes observed in the deaf due to plasticity is one such unexplored domain which has the potential to inform us more on conscious and unconscious visual processing in humans.

Visual Processing Advantages in the Deaf

Over the last few decades, several studies have provided evidence that other senses in the deaf—mostly sight—is modified due to the lack of auditory input. Most of these studies point towards a theory of compensation—that is, the deficit in a particular modality leads to advantages in the other. In contrast, there have also been deficit theories (Mitchell 1996) which are based on the division of labour hypothesis. They suggest that since the visual system has to perform the functions of both visual and auditory systems, certain deficits in visual processing are observed in the deaf population. This chapter will focus on the compensation aspect of the cross-modal plasticity in the deaf. Several factors contribute to the cross-modal changes observed in the deaf. Some of them are shown in Fig. 1. The deaf population can be extremely heterogeneous (A). Individuals can differ on the causes of deafness (congenital, neurological diseases, accidents), degree of deafness (moderate to profound) and the hearing status of their parents (deaf/non-deaf).

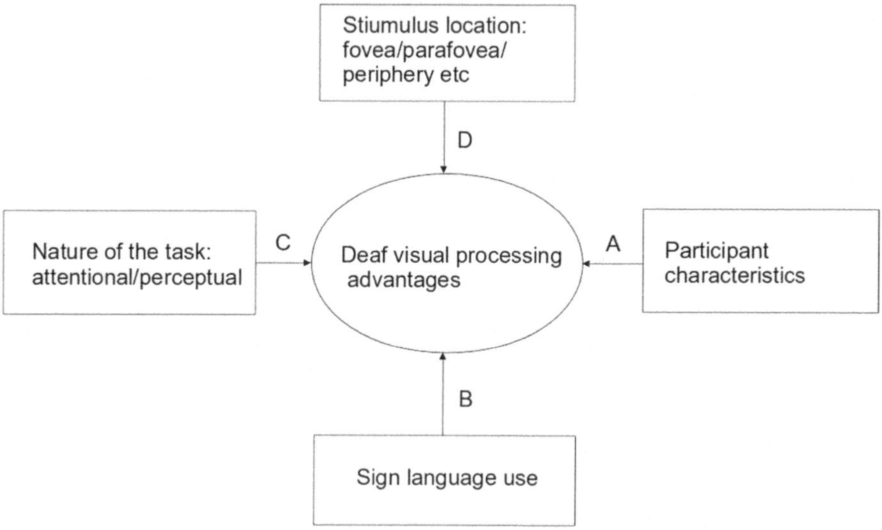

Fig. 1 Factors contributing to visual processing advantages in the deaf. The heterogeneity in the deaf population (*A*), proficiency in sign language use (*B*), whether the demands of the task are perceptual or attentional in nature (*C*) and whether the task requires responding to central or peripheral locations (*D*) are among the most important factors which decide whether processing advantages are seen for the deaf

Further differences can arise due to the age of acquisition of sign language and the proficiency in sign language use (B). Deaf individuals have to effectively monitor the sign language of interlocutors to communicate. Dynamically shifting attention between hands and features of the face (such as eyes or lips) contributes to changes in the visual processing system. Normal-hearing individuals do not have to indulge in such wide spanning of their visual field. Thus, sign language is an important contributing factor in the plasticity observed in the deaf. The nature of the task and its demands play an important role in determining if deaf show an advantage in performance (C). Most studies have found evidence of superior performance in the deaf only when the task had an attentional component in it (Bosworth and Dobkins 2002; Bottari et al. 2010). One of the most important factors that determine visual advantage in the deaf is the stimulus location (D). Several studies have shown enhanced peripheral processing in the deaf. But whether this enhancement is at the cost of processing at the fovea or whether enhancement is observed at all locations in the visual field is still an open question (Dye 2016). As is evident, the locus of the changes as a result of cross-modal plasticity is still debatable and several factors contribute to it. We will, however, limit our discussion in this chapter to factors C and D and more importantly examine how they throw light on the issues of attention and consciousness.

Several theoretical frameworks exist to account for the various findings observed in deaf literature (Dye and Bavelier 2013). Codina et al. (2011) examined retinal structures in deaf using optical coherence tomography (OCT) and observed significantly larger neural rim areas—suggesting a larger number of ganglion cells, compared to normal-hearing individuals. These results suggest that plasticity leads to changes in visual processing as early as the retinal level (perceptual enhancement hypothesis). In Bottari et al. (2010) deaf participants were found to be faster at detecting targets (at both centre and periphery) in a simple detection task compared to normal-hearing participants. In contrast, the two groups did not differ on a shape discrimination task. Since a discrimination task involves orienting of spatial attention to the targets, the authors conclude that enhanced attentional orienting is not a necessary component of deaf visual processing. Rather, deaf individuals are faster at reacting to visual stimuli. Further evidence for this theory comes from an ERP study by Bottari et al. (2010). In this study, deaf and normal-hearing were tested on a simple detection task while EEG recording was done. There were differences between the two groups in early visual components of ERP (C1/P1) suggesting that the changes in deaf are more sensory in nature rather than attentional. Similar results were found for central and peripheral targets. Thus, these studies suggest that the 'enhanced reactivity' is present across the visual field and not necessarily just at the periphery. Further, they suggest enhancement of perceptual representations of stimuli in the deaf.

In contrast, the dorsal route hypothesis predicts changes that are attentional in nature. These changes are said to arise due to modifications to dorsal processing stream in the deaf as a result of plasticity. Few studies have examined attentional orienting (Parasnis and Samar 1985; Colmenero et al. 2004; Prasad et al. 2015) and selective attention (Dye et al. 2007; Proksch and Bavelier 2002) in the deaf. The earliest study to examine orienting in deaf people was Parasnis and Samar (1985). They compared deaf and normal-hearing participants using a Posner cueing paradigm (Posner 1980). This paradigm is typically used to measure the orienting responses to targets. Cues are presented before target onset either at the same location as the target (valid trials) or at the opposite location (invalid trials). The valid cues are supposed to attract attention to the target location thereby speeding up responses to such targets. Invalid cues on the other hand orient attention away from the targets, thus, slowing down the responses on such trials. Parasnis and Samar (1985) found that deaf could disengage faster from an invalidly cued location leading to lesser costs for the deaf on invalid trials. There was no group difference in the ability to orient towards validly cued locations. Similarly, in two experiments, Colmenero et al. (2004) found that deaf subjects can disengage attention faster than normal-hearing participants. However, all of these studies examined attentional orienting through manual responses. Whereas, attention and eye movements are closely linked (Hoffman and Subramaniam 1995). So then, does the visual processing advantages in the deaf modulate spatial orienting with respect to eye movements? And does target eccentricity play a role?

To examine this, Prasad et al. (2015) compared deaf and normal-hearing participants on an oculomotor version of the Posner cueing task (Fig. 2a). Targets (white discs) appeared either at parafovea (7°) or at the periphery (21°). Spatial cues in the form of brief (100 m) flickers were presented just before the targets. Targets

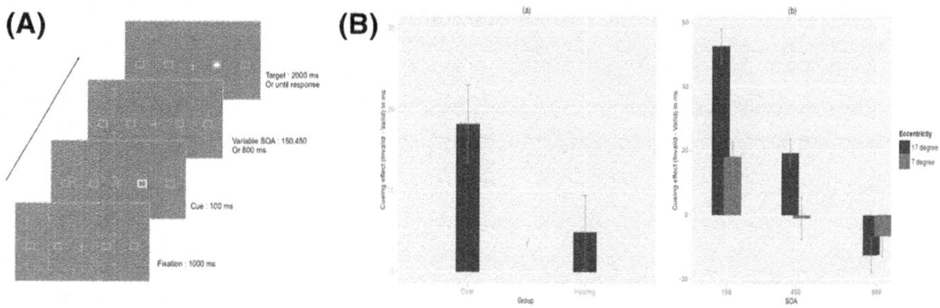

Fig. 2 **A** Sequence of events in a sample trial. A brief cue was presented for 100 m after the fixation cross. Target was presented either at the cued location or at an uncued location. The figure shows a sample sequence on a valid trial where there is match in the cue and target location. **B** Oculomotor responses (*a*) Cueing effect for deaf and hearing groups ($p < 0.05$) (*b*) cueing effect at 150, 450 and 800 m SOA at both perifovea and periphery (*right panel*), Prasad et al. (2015)

and the cues were either presented at the same location (valid trials) or at the opposite location (invalid trials). The stimulus onset asynchrony (SOA) between the cue and the target was varied: 150, 450 or 800 m. Participants were asked to make a saccade to the target as soon as it appeared on the screen. Facilitation (faster responses on valid trials) was observed at shorter SOA (150 m) and inhibition of return (faster responses on invalid trials) at long SOA (800 m). Inhibition of return (IOR) refers to delayed responding to already attended locations. This arises due to the reluctance of the attentional system to orient towards a previously attended location. In this study, the two groups did not differ with respect to facilitation or IOR. However, the overall cueing effect (RT difference between invalid and valid trials) was higher for the deaf compared to normal-hearing participants (Fig. 2B). This difference was not modulated by target eccentricity. In sum, the findings of this study suggest that deaf have a higher ability to utilise spatial cues to orient towards target locations.

In another study, Jayaraman et al. (2016) examined attentional disengagement without relying on ignored cues to generate IOR. Jayaraman et al. (2016) replicated the experiment by Vaughan (1984) in which participants were asked to track with their gaze a fixation stimulus that moved to different locations on the screen resulting in a sequence of saccades (Fig. 3a). The last saccade was designed to be at a previously fixated location or at the opposite location. The time delay between the third and the fourth fixation was also manipulated to examine the time course of IOR. IOR was observed as a slower saccade latency to the last fixation when it appeared at a previously viewed location. This effect was also modulated by the distance from the location that was previously visited. As the distance increased, the magnitude of IOR decreased. Importantly, the authors found that the deaf and normal-hearing participants did not differ with respect to the magnitude or time course of IOR (Fig. 3b). The findings of the two above studies suggest that deaf individuals are not necessarily better at disengaging attention. This is in line with

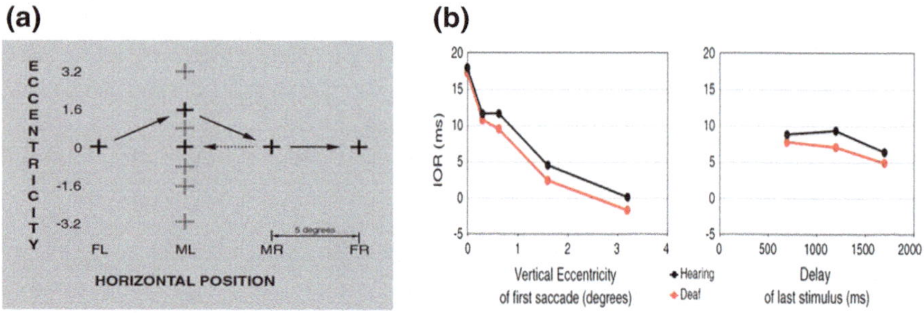

Fig. 3 **a** Illustration of the possible events on a trial. **b** IOR scores for the two groups as a function of vertical eccentricity of the first saccade and the time delay between the third and the fourth fixation (Jayaraman et al. 2016)

previous studies which have found comparable results in deaf and normal-hearing in terms of inhibition (Chen et al. 2006).

It is difficult to say which theoretical perspective explains the majority of findings observed in the deaf literature. However, as Dye and Bavelier (2013) note in their comprehensive review, 'a majority of the literature on visual adaptation to deafness in deaf native signers reveals compensatory changes that favor the dorsal route hypothesis and variants thereof (p. 244)'. It is important to first understand the role of dorsal stream in visual processing in general. The dual stream hypothesis has been an influential model of visual processing (Goodale and Milner 1992). According to this model, visual processing is mediated by two distinct streams or pathways in the brain. One is the ventral pathways concerned with 'vision for perception'. The other is the dorsal pathways which mediate 'vision for action'. Ventral pathways are said to be responsible for processing features of the stimuli that result in awareness, such as shape, colour or size of an object. On the other hand, dorsal pathways are responsible for guiding actions, such as grasping. Thus, aspects of the visual object processed in the dorsal stream do not reach conscious awareness. As Goodale and Milner put it, 'The dorsal stream is not in the business of providing any kind of a visual representation of the world: it just converts visual information directly into action (2004: 114)'. Several studies using continuous flash suppression have given evidence for the role of dorsal stream in unconscious processing. Fang et al. (2005) showed in a fMRI study that activation in the ventral stream, but not the dorsal stream depended on the visibility of the stimulus. Neuroimaging data showed substantial activity in the dorsal stream for invisible objects suggesting that dorsal stream is largely responsible for processing unconscious information. The dual stream hypothesis is important for a discussion on consciousness as it suggests that any processing carried out by the dorsal stream is essentially unconscious. Brogaard (2011) argues for this point and suggests that functioning of the dorsal stream has also been evolutionary beneficial to us. She takes the example of reaching for a coffee mug where the person performing the action is aware of the colour, size and location of the coffee mug. But reaching for the mug requires complex mathematical calculations, such as estimating the speed at which the arm must be moved and the trajectory to be followed in order to reach the mug precisely. We are never aware of these computations and we perform them on-the-fly. Brogaard argues that such computations which are truly unconscious are taken care of by the dorsal stream. Again, there have been oppositions to these ideas. Specifically, researchers have argued that the distinction between the two pathways is not so rigid and that there may be intersections (Hebart and Hesselmann 2012). But given what we know about the function of dorsal pathways and the changes to them due to deafness, what are the implications for unconscious processing in the deaf?

Unconscious Processing in the Deaf

The question is how does the deaf visual system process a stimulus that is below the threshold of awareness? It seems logical that there would be differences between deaf and normal-hearing in the processing of unconscious stimuli because of plasticity-induced changes in the dorsal stream in the deaf. But what is the exact nature of this influence? Specifically, if the changes in the deaf are largely attentional in nature as suggested by many researchers, then what does higher attention in the deaf particularly do to unconscious stimuli?

Dehaene et al. (2006) proposed the following taxonomy to explain the relationship between attention and (un)conscious processing based on the global workspace model (Dehaene and Naccache 2001). They refer to a 2 by 2 matrix resulting from an interaction between top-down attention and stimulus strength (Fig. 4). A sufficiently strong stimulus coupled with top-down attention is required

Bottom-up stimulus strength	**Top down attention** Absent	Present
Weak or interrupted	Subliminal(unattended)	Subliminal (attended)
Sufficiently strong	Preconscious	Conscious

Fig. 4 Taxonomy proposed by Dehaene et al. (2006) for the possible relationships between bottom-up strength of a stimulus and the top-down attention directed on it (adapted from Dehaene et al. 2006)

for conscious access to a stimulus. Consequently, lack of conscious access can occur due to failure at one of these levels resulting in two types of non-conscious processing: subliminal (as a result of weak bottom-up the strength of the stimulus) and preconscious (as a result of lack of top-down attention to an otherwise strong potentially accessible stimulus). Paradigms such as attentional blink are given as examples of preconscious processing. More important to our discussion, they discuss two types of subliminal processing as a result of weak stimulus strength—subliminal-attended and subliminal-unattended. The subliminal-unattended stimuli produce no or weak priming effects. A majority of studies in subliminal priming thus fall under the category of examining subliminal-attended stimulus. These are the stimuli that have weak bottom-up stimulus strength which is insufficient to enable conscious access in spite of top-down attention. Thus, the 'subliminal' nature or, in other words, the lack of conscious awareness of such stimuli is independent of the presence or absence of attention. The presence of attention can, however, amplify the depth of processing of such stimuli leading to greater priming effects. It is important to maintain the distinction between the perception of the subliminal stimulus and the processing of that stimulus. The former refers to the reportability of the stimulus which is related to the perceptual representation. The latter refers to the influence of a subliminal stimulus on a subsequently presented target. Thus, according to the taxonomy proposed by Dehaene et al. (2006), presence of attention amplifies subliminal processing leading to priming effects.

In the context of deafness, we refer to Fig. 5 to examine which aspect of subliminal processing is affected by the plasticity-induced changes in the deaf. There are two ways of processing a subliminal stimulus, referred to here as A and B. Process A refers to the perceptual processing of the stimulus which enables conscious reportability or the detection of the stimulus. Process B refers to the sensorimotor activations triggered by a subliminal prime, which is measured as the priming effect. Thus, A represents how well a participant can 'see' the prime stimulus whereas B represents what the primes do to a subsequent action. Which of these processes, if any, is modulated by the deaf visual system? According to the

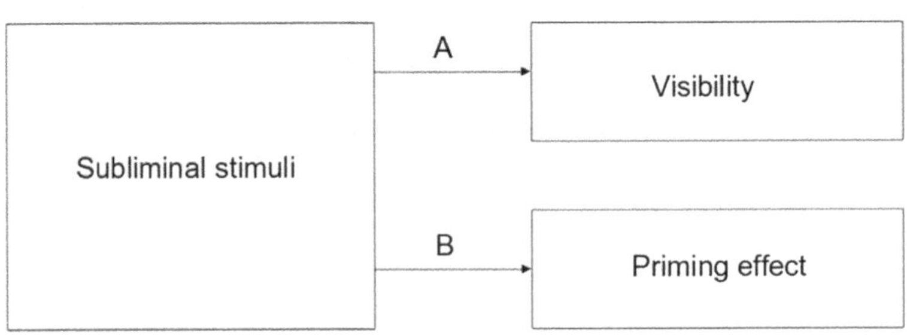

Fig. 5 Two distinct effects arising from the processing of a subliminal stimulus. The direct processing of a subliminal stimulus is measured by its visibility (*B*)—for example, on a prime identification test. Priming effects are an indirect measure of subliminal processing where the prime influences a subsequent conscious behaviour (*A*)

model based on the global workspace hypothesis, depth of processing of a subliminal stimulus can be modulated by attention and thus, enhanced attention should lead to enhanced priming effects. For instance, in Sumner et al. (2006), Participants were asked to respond to left or right arrows through a manual response. Masked primes were presented prior to the targets. The prime duration was selected such that they produced distinct positive (with 'visible' primes) and negative priming (with 'invisible' primes) effects. Attention was modulated using exogenous cues that were either presented at the location of the prime ('cued') or in the opposite ('uncued') location. Negative effects of the prime were found to be higher when primes appeared on cued locations as a result of attentional cueing. Several studies have similarly provided evidence regarding the role of attention on subliminal priming (Marzouki et al. 2007; Kiefer 2012).

What happens to the visibility of a subliminal stimulus under attention is less clear. Kouider and Dehaene (2007) note that 'subliminal information is information that cannot be brought into consciousness, in spite of all efforts of focused attention'. Does this mean that attention has no effect on the visibility of a subliminal stimulus? However, there is previous evidence that attention can modulate perceptual threshold (Müller and Humphreys 1991). Additionally, it is widely agreed that attention enhances the perceptual saliency of a stimulus. As Carrasco et al. (2004) have noted, 'attention changes the strength of a stimulus by increasing its "effective contrast" or salience'. In that case, can top-down attention turn an invisible stimulus into a visible one? But this seems contradictory to the very definition of 'subliminal' under the classification proposed by Dehaene et al. (2006). A possible explanation is that attention can enhance visibility in a graded manner but only till the conscious threshold is reached. Thus, although a subliminal stimulus can never cross the threshold and become visible, attention can nonetheless enhance the perceptual strength.

Prasad et al. (2017) compared deaf and normal-hearing participants on a subliminal priming paradigm to examine these issues. This paradigm has been used to understand the influence of invisible stimuli on a conscious action. Initially used in lexical priming studies, the paradigm is now extensively used to understand unconscious influences on any conscious behaviour. Typically, a prime is presented for a brief duration followed or preceded by a mask. The mask is supposed to suppress the retinal afterimage of the prime thus ensuring its invisibility. Following this, a target is presented to which the participants are asked to respond. The relationship between the prime and the target is manipulated between different trials. A prime similar (visually/semantically) to the target results in faster responses whereas a dissimilar prime leads to slower responses. This effect is known as priming.

In this study, deaf and normal-hearing participants completed 'fixed' and 'free' trials. On fixed trials, numbers 1 or 2 were used as target cues to which the participants had to respond based on the instruction (Fig. 6). On free trials, 0 was used as the free choice target and participants were asked to freely choose between two response alternatives. The free and fixed choice targets were presented only at

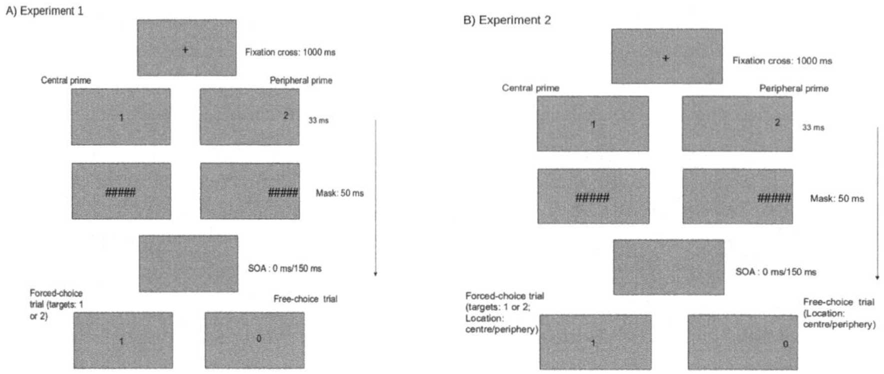

Fig. 6 Sequence of events on a sample trial—experiment 1 and experiment 2. In experiment 1, primes (1 or 2) were presented at centre or periphery followed by a mask (#####). Participants then responded to the target cue (fixed trials) or chose between two alternative responses (free trials). The target numbers (1, 2 or 0) were always presented at the centre. In experiment 2, the design was similar to experiment 1. The only difference was that the targets were also presented at periphery, always matching the prime location (Prasad et al. 2017)

centre in Experiment 1 and at both centre/periphery in Experiment 2. Both fixed and free trials were preceded by number primes (1 or 2) presented at centre or periphery.

Results of Experiment 1 showed that deaf participants chose the response congruent with the prime more often than hearing participants. Similarly, faster responses on free and fixed trials were observed on congruent trials—that is, trials on which the prime and the target matched. Importantly, this effect was enhanced in the deaf. Thus, higher priming effects were observed in the deaf compared to normal-hearing, in terms of both choices and response times. In Experiment 2, when the prime-target location matched on all trials, higher priming effects in the deaf were observed only at centre, but not at the periphery.

Could the deaf 'see' the primes better? No, and yes. The answer is ambiguous because although we largely did not find differences in prime visibility between the two groups, in Experiment 1 the prime visibility index was significantly higher for the deaf for peripheral primes. This means that deaf were more consciously aware of the primes at the periphery in Experiment 1 compared to normal-hearing. This suggests that there was also a perceptual enhancement of the stimulus representation when presented at the periphery. In sum, we found evidence of modulation of sensorimotor activations (process B) due to the plasticity-related changes in the deaf, in the form of enhanced priming effects in the deaf. The effect on prime visibility (process A) was observed only in a particular case—for peripheral primes in Experiment 1.

It is difficult to draw definitive conclusions based on Prasad et al. (2017) since the primes were not exactly at the perceptual threshold. The sensitivity index of the primes was above chance level. Nonetheless, the findings largely suggest that the visibility of the primes was not greater in the deaf. Thus, deaf were not better at 'seeing' the primes but we still observed that higher priming effects were observed.

Is it possible that the threshold for perception itself is lower for the deaf? Answering this question would require measuring the threshold at which a stimulus becomes invisible to deaf and normal participants and see if there is a difference. Few studies in the past have examined what happens when deaf are presented with stimuli just at the threshold of awareness (Bosworth and Dobkins 1999; Finny and Dobkins 2001). More studies are required to clearly examine which aspect of visual processing is modified in the deaf.

Conclusion

It is important to note that philosophers of mind interested in cognition allow only some type of data into their discussion. However, many long-held paradigms inside psychology are fast changing. Examples are the replication crises and also the inclusion of individual and cultural diversity in experimentation. In such times, it is then important to consider data from different areas to see whether they help us refine higher theories. The rich data from the deaf cast light on critical issues that include perception, awareness, attention and ultimately consciousness. Therefore, we have presented data from experiments that show a very different nature of cognitive organisation in the deaf. Although the deaf live among us and share our world, they may be altogether having a different consciousness?

What are the larger implications of these diverse experimental effects seen in the deaf on visual and attentional tasks? The experimental results that we have described for the deaf have far-reaching consequences for our conceptualisation of important cognitive mechanisms like visual perception, attention and visual awareness. One finds philosophers of mind debating on issues like attention and consciousness when they draw data from patients with visual neglect or from tasks administered with normal college going population. Of course, these data do have a lot to suggest on such controversial issues like the role of attention in consciousness. However, we also need data from neural plasticity on such issues. The deaf data indicates reorganisation of core neural networks that then function differently and aid in behaviour. We have shown that neuroplasticity in the deaf leads to substantial enhancement in several attentional and perceptual capabilities. This enhanced control over attention is central to contemporary theorising about executive control (Posner 2016). The data from subliminal priming experiment suggest that the deaf seem to strongly process stimuli that they cannot even perceive. It is not that neural reorganisation leads the deaf to see better. They do, however, seem to have a broader attentional and visual span. Invisible stimuli in the visual periphery seem to influence how they take decisions for visual actions. However, the different paradigms we have mentioned do not refer to similar mechanisms. One can find also a discrepancy in the experimental literature where the deaf have shown superior performance on some tasks and not on others. It is not of course clear what exactly is altered and modified in the deaf neutrally, the effects of which one can see behaviourally.

The connection to larger issues like awareness attention and consciousness are straight forward in the deaf studies. That is because all these mechanisms are altered differently due to neuroplasticity in the deaf. We know that there is no general agreement in the field if, for example, attention and awareness are related or unrelated processes (Lamme 2003). Some think that awareness and attention diverge as distinct mechanisms from the earliest stages of visual processing (Koivisto et al. 2005). It has also been suggested that one can be attentive without awareness as in the case of blindsight (Kentridge et al. 1999). Evidence from EEG studies has shown that participants can become subjectively aware of stimuli much before they have deployed attention to it. At the beginning of the chapter, we set out to examine whether one expects changes in the interrelationships between attention and consciousness because of neuroplasticity? From the evidence provided in this chapter, the answer to this question is yes. Neuroplasticity in the deaf leads to changes in the dorsal stream functioning. This influences the mechanisms of conscious and unconscious processing in the deaf. Because of extensive sign language practice, these mechanisms further develop in the deaf compared to normal-hearing. Therefore, it is important to consider data from such areas like the deaf when trying to propose theories about the causal nature attention and awareness.

The scientific study of consciousness has progressed to a large extent over the last few decades. Although still controversial, it has developed into a serious topic of empirical research. It has also captured public imagination because understanding what consciousness is and how and why we are conscious is a fundamental question for human beings. One of the foundational issues in consciousness science is the relationship between attention and consciousness. As current debates in the field are fought on hard experimental evidence, the data from hearing impaired individuals will offer a new perspective and contribute to this debate.

Acknowledgements We thank Dr. Gouri Shanker Patil for his invaluable support in all our studies on the deaf population.

References

Abutalebi, J., Della Rosa, P. A., Green, D. W., Hernandez, M., Scifo, P., Keim, R., et al. (2011). Bilingualism tunes the anterior cingulate cortex for conflict monitoring. *Cerebral Cortex*, bhr287.

Ajina, S., & Bridge, H. (2016). Blindsight and unconscious vision: What they teach us about the human visual system. *The Neuroscientist*, 1073858416673817.

Bavelier, D., Dye, M. W., & Hauser, P. C. (2006). Do deaf individuals see better? *Trends in Cognitive Sciences, 10*(11), 512–518.

Bavelier, D., & Neville, H. J. (2002). Cross-modal plasticity: Where and how? *Nature Reviews Neuroscience, 3*(6), 443–452.

Block, N. (2014). Rich conscious perception outside focal attention. *Trends in Cognitive Sciences, 18*(9), 445–447.

Block, N. (2015a). Consciousness, big science, and conceptual clarity. In *The Future of the Brain* (pp. 161–176). Princeton University Press.

Block, N. (2015b). The Anna Karenina principle and skepticism about unconscious perception. *Philosophy and Phenomenological Research*. doi:10.1111/phpr.12258.
Bor, D., & Seth, A. K. (2012). Consciousness and the prefrontal parietal network: Insights from attention, working memory, and chunking. *Frontiers in Psychology, 3,* 63.
Bosworth, R. G., & Dobkins, K. R. (1999). Left-hemisphere dominance for motion processing in deaf signers. *Psychological Science, 10*(3), 256–262.
Bosworth, R. G., & Dobkins, K. R. (2002). The effects of spatial attention on motion processing in deaf signers, hearing signers, and hearing nonsigners. *Brain and Cognition, 49*(1), 152–169.
Bottari, D., Nava, E., Ley, P., & Pavani, F. (2010). Enhanced reactivity to visual stimuli in deaf individuals. *Restorative Neurology and Neuroscience, 28*(2), 167–179.
Brogaard, B. (2011). Are there unconscious perceptual processes? *Consciousness and Cognition, 20*(2), 449–463.
Carrasco, M., Ling, S., & Read, S. (2004). Attention alters appearance. *Nature Neuroscience, 7*(3), 308–313.
Chen, Q., Zhang, M., & Zhou, X. (2006). Effects of spatial distribution of attention during inhibition of return (IOR) on flanker interference in hearing and congenitally deaf people. *Brain Research, 1109*(1), 117–127.
Codina, C., Pascalis, O., Mody, C., Toomey, P., Rose, J., Gummer, L., et al. (2011). Visual advantage in deaf adults linked to retinal changes. *PLoS ONE, 6*(6), e20417.
Cohen, L. G., Celnik, P., Pascual-Leone, A., Corwell, B., Faiz, L., Dambrosia, J., et al. (1997). Functional relevance of cross-modal plasticity in blind humans. *Nature, 389*(6647), 180–183.
Cohen, M. A., Cavanagh, P., Chun, M. M., & Nakayama, K. (2012). The attentional requirements of consciousness. *Trends in Cognitive Sciences, 16*(8), 411–417.
Cohen, M. A., Dennett, D. C., & Kanwisher, N. (2016). What is the bandwidth of perceptual experience? *Trends in Cognitive Sciences, 20*(5), 324–335.
Colmenero, J. M., Catena, A., Fuentes, L. J., & Ramos, M. M. (2004). Mechanisms of visuospatial orienting in deafness. *European Journal of Cognitive Psychology, 16*(6), 791–805.
Costa, A., & Sebastián-Gallés, N. (2014). How does the bilingual experience sculpt the brain? *Nature Reviews Neuroscience, 15*(5), 336–345.
Danckert, J., & Goodale, M. A. (2000). Blindsight: A conscious route to unconscious vision. *Current Biology, 10*(2), R64–R67.
Dehaene, S., & Naccache, L. (2001). Towards a cognitive neuroscience of consciousness: Basic evidence and a workspace framework. *Cognition, 79*(1), 1–37.
Dehaene, S., Changeux, J. P., Naccache, L., Sackur, J., & Sergent, C. (2006). Conscious, preconscious, and subliminal processing: A testable taxonomy. *Trends in Cognitive Sciences, 10*(5), 204–211.
Dux, P. E., & Marois, R. (2009). The attentional blink: A review of data and theory. *Attention, Perception, & Psychophysics, 71*(8), 1683–1700.
Dye, M. W. (2016). Foveal processing under concurrent peripheral load in profoundly deaf adults. *Journal of Deaf Studies and Deaf Education, 21*(2), 122–128.
Dye, M. W., & Bavelier, D. (2013). Visual attention in deaf humans: A neuroplasticity perspective. In A. Kral, A. N. Popper, & R. R. Fay (Eds.), *Deafness* (pp. 237–263). New York: Springer.
Dye, M. W., Baril, D. E., & Bavelier, D. (2007). Which aspects of visual attention are changed by deafness? The case of the Attentional Network Test. *Neuropsychologia, 45*(8), 1801–1811.
Dye, M. W., Hauser, P. C., & Bavelier, D. (2008). Visual attention in deaf children and adults. In M. Marschark & P. C. Hauser (Eds.), Deaf cognition: Foundations and outcomes (pp. 250–263). Oxford University Press.
Fang, F., Murray, S. O., Kersten, D., & He, S. (2005). Orientation-tuned fMRI adaptation in human visual cortex. *Journal of Neurophysiology, 94*(6), 4188–4195.
Finney, E. M., & Dobkins, K. R. (2001). Visual contrast sensitivity in deaf versus hearing populations: Exploring the perceptual consequences of auditory deprivation and experience with a visual language. *Cognitive Brain Research, 11*(1), 171–183.

Finney, E. M., Fine, I., & Dobkins, K. R. (2001). Visual stimuli activate auditory cortex in the deaf. *Nature Neuroscience, 4*(12), 1171–1173.
Goodale, M. A., & Milner, A. D. (1992). Separate visual pathways for perception and action. *Trends in Neurosciences, 15*(1), 20–25.
Goodale, M., & Milner, D. (2004). *Sight unseen*. Oxford: Oxford University Press.
Hebart, M. N., & Hesselmann, G. (2012). What visual information is processed in the human dorsal stream? *The Journal of Neuroscience, 32*(24), 8107–8109.
Hoffman, J. E., & Subramaniam, B. (1995). The role of visual attention in saccadic eye movements. *Perception and Psychophysics, 57*(6), 787–795.
Janus, M., Lee, Y., Moreno, S., & Bialystok, E. (2016). Effects of short-term music and second-language training on executive control. *Journal of Experimental Child Psychology, 144,* 84–97.
Jayaraman, S., Klein, R. M., Hilchey, M. D., Patil, G. S., & Mishra, R. K. (2016). Spatial gradients of oculomotor inhibition of return in deaf and normal adults. *Experimental Brain Research, 234*(1), 323–330.
Kentridge, R. W. (2011). Attention without awareness. In C. Mole, D. Smithies, & W. Wu (Eds.), *Attention: Philosophical and psychological essays* (pp. 228–246). USA: Oxford University Press.
Kentridge, R. W., Heywood, C. A., & Weiskrantz, L. (1999). Effects of temporal cueing on residual visual discrimination in blindsight. *Neuropsychologia, 37*(4), 479–483.
Kiefer, M. (2012). Executive control over unconscious cognition: Attentional sensitization of unconscious information processing. *Frontiers in Human Neuroscience, 6,* 61.
Kiefer, M., & Martens, U. (2010). Attentional sensitization of unconscious cognition: Task sets modulate subsequent masked semantic priming. *Journal of Experimental Psychology: General, 139*(3), 464.
Klink, P. C., & Roelfsema, P. R. (2016). Binocular rivalry outside the scope of awareness. *Proceedings of the National Academy of Sciences, 113*(30), 8352–8354.
Koch, C., & Tsuchiya, N. (2007). Attention and consciousness: Two distinct brain processes. *Trends in Cognitive Sciences, 11*(1), 16–22.
Koivisto, M., Revonsuo, A., & Salminen, N. (2005). Independence of visual awareness from attention at early processing stages. *NeuroReport, 16*(8), 817–821.
Kouider, S., & Dehaene, S. (2007). Levels of processing during non-conscious perception: A critical review of visual masking. *Philosophical Transactions of the Royal Society B: Biological Sciences, 362*(1481), 857–875.
Kühn, S., Gleich, T., Lorenz, R. C., Lindenberger, U., & Gallinat, J. (2014). Playing Super Mario induces structural brain plasticity: Gray matter changes resulting from training with a commercial video game. *Molecular Psychiatry, 19*(2), 265–271.
Lamme, V. A. (2003). Why visual attention and awareness are different. *Trends in Cognitive Sciences, 7*(1), 12–18.
Lin, S. Y., & Yeh, S. L. (2016). Interocular grouping without awareness. *Vision Research, 121,* 23–30.
Mahncke, H. W., Connor, B. B., Appelman, J., Ahsanuddin, O. N., Hardy, J. L., Wood, R. A., et al. (2006). Memory enhancement in healthy older adults using a brain plasticity-based training program: A randomized, controlled study. *Proceedings of the National Academy of Sciences USA, 103*(33), 12523–12528.
Manna, A., Raffone, A., Perrucci, M. G., Nardo, D., Ferretti, A., Tartaro, A., et al. (2010). Neural correlates of focused attention and cognitive monitoring in meditation. *Brain Research Bulletin, 82*(1), 46–56.
Marzouki, Y., Grainger, J., & Theeuwes, J. (2007). Exogenous spatial cueing modulates subliminal masked priming. *Acta Psychologica, 126*(1), 34–45.
Merzenich, M. M., Jenkins, W. M., Johnston, P., & Schreiner, C. (1996). Temporal processing deficits of language-learning impaired children ameliorated by training. *Science, 271*(5245), 77.
Mitchell, T. V. (1996). How audition shapes visual attention. Ph.D. dissertation. Bloomington, IN: Indiana University.

Montemayor, C., & Haladjian, H. H. (2015). *Consciousness, attention, and conscious attention*. Cambridge: MIT Press.

Müller, H. J., & Humphreys, G. W. (1991). Luminance-increment detection: Capacity-limited or not? *Journal of Experimental Psychology: Human Perception and Performance, 17*(1), 107.

Naccache, L., Blandin, E., & Dehaene, S. (2002). Unconscious masked priming depends on temporal attention. *Psychological Science, 13*(5), 416–424.

Neville, H. J., Kutas, M., & Schmidt, A. (1982). Event-related potential studies of cerebral specialization during reading: II. Studies of congenitally deaf adults. *Brain and Language, 16* (2), 316–337.

Overgaard, M. (2012). Blindsight: Recent and historical controversies on the blindness of blindsight. *Wiley Interdisciplinary Reviews: Cognitive Science, 3*(6), 607–614.

Overgaard, M., Fehl, K., Mouridsen, K., Bergholt, B., & Cleeremans, A. (2008). Seeing without seeing? Degraded conscious vision in a blindsight patient. *PLoS ONE, 3*(8), e3028.

Parasnis, I., & Samar, V. J. (1985). Parafoveal attention in congenitally deaf and hearing young adults. *Brain and Cognition, 4*(3), 313–327.

Posner, M. I. (1980). Orienting of attention. *Quarterly Journal of Experimental Psychology, 32*(1), 3–25.

Posner, M. I. (2016). Orienting of attention: Then and now. *Quarterly Journal of Experimental Psychology, 69*(10), 1864–1875.

Prasad, S. G., Patil, G. S., & Mishra, R. K. (2015). Effect of exogenous cues on covert spatial orienting in deaf and normal hearing individuals. *PLoS ONE, 10*(10), e0141324.

Prasad, S., Patil G. S., & Mishra, R. K. (2017). *Cross-modal plasticity in the deaf enhances processing of masked stimuli in the visual modality*, *7*(1), Scientific Reports.

Proksch, J., & Bavelier, D. (2002). Changes in the spatial distribution of visual attention after early deafness. *Journal of Cognitive Neuroscience, 14*(5), 687–701.

Raffone, A., & Srinivasan, N. (2009). An adaptive workspace hypothesis about the neural correlates of consciousness: insights from neuroscience and meditation studies. *Progress in Brain Research, 176,* 161–180.

Raffone, A., & Srinivasan, N. (2010). The exploration of meditation in the neuroscience of attention and consciousness. *Cognitive Processing, 11*(1), 1–7.

Rauschecker, J. P. (1995). Compensatory plasticity and sensory substitution in the cerebral cortex. *Trends in Neurosciences, 18*(1), 36–43.

Rodrigue, A. L., Austin, B. P., Dyckman, K. A., & McDowell, J. E. (2016). Brain activation differences in schizophrenia during context-dependent processing of saccade tasks. *Behavioral and Brain Functions, 12*(1), 19.

Shin, K., Stolte, M., & Chong, S. C. (2009). The effect of spatial attention on invisible stimuli. *Attention, Perception, & Psychophysics, 71*(7), 1507–1513.

Spierer, L., Chavan, C., & Manuel, A. L. (2013). Training-induced behavioral and brain plasticity in inhibitory control. *Frontiers in Human Neuroscience, 7,* 427.

Srinivasan, N. (2007). Interdependence of attention and consciousness. *Progress in Brain Research, 168,* 65–75.

Sumner, P., Tsai, P. C., Yu, K., & Nachev, P. (2006). Attentional modulation of sensorimotor processes in the absence of perceptual awareness. *Proceedings of the National Academy of Sciences USA, 103*(27), 10520–10525.

Van den Bussche, E., Hughes, G., Van Humbeeck, N., & Reynvoet, B. (2010). The relation between consciousness and attention: An empirical study using the priming paradigm. *Consciousness and Cognition, 19*(1), 86–97.

Vaughan, J. (1984). Saccades directed at previously attended locations in space. *Advances in Psychology, 22,* 143–150.

Weiskrantz, L. (1986). *Blindsight: A case study and implications*. Oxford: Oxford University Press.

Promises and Limitations of Conscious Machines

L.M. Patnaik and Jagadish S. Kallimani

There is no pillow so soft as a clear conscience.
(French proverb)

Introduction

The term consciousness is the state of being characterised by senses, emotions and thoughts. It refers to the quality, state or fact of being conscious by an individual in total. Social consciousness deals with the interaction of an individual or a group for the welfare of human beings.

Philosophers claim consciousness as the relation among the mind and the world. Cosmic consciousness deals with awareness of life and surrounding environment which is possessed only by humans who are intelligent (Bucke 1905).

Ned Block proposed two types of consciousness as: phenomenal and access. A phenomenal consciousness is simply raw experiences and is called 'qualia' (Schneider and Velmans 2017). Access consciousness is the information in human minds which is accessible for decision-making, controlling behaviours and reasoning. So, we exhibit access consciousness, when we perceive, introspect and remember.

- **Attributes of consciousness**

Consciousness is the repository of the past experiences, actions, knowledge obtained from feedback, data obtained from external environment and also data obtained from human operator if any (Kriegel 2009). Types of consciousness are

L.M. Patnaik
National Institute of Advanced Studies, Indian Institute of Science Campus, Bengaluru 560012, India

J.S. Kallimani (✉)
Department of Computer Science and Engineering, Ramaiah Institute of Technology, Bengaluru, India
e-mail: jagadish.k@msrit.edu

S. Menon et al. (eds.), *Self, Culture and Consciousness*,
DOI 10.1007/978-981-10-5777-9_5

Table 1 Types of consciousness (Kihlstrom 2015)

Consciousness	Semi-consciousness	Sub-consciousness
Conscious is the state of having full alertness or being perfectly aware or perfectly responsive	Semiconscious is the state of not having full consciousness or being imperfectly aware or imperfectly responsive but showing responses to some extent	Subconscious is the state which accumulates previous occurrences, opinions, perspectives, memories, interpretations, abilities, situations and skills

presented in Table 1. From a cognitive psychology viewpoint, we have five characteristics of consciousness. They are: subjectivity, change, continuity, intentionality and selectivity. The other abstract features are: sensory perception, mental imagery, inner speech, conceptual thought, remembering, emotional feeling and self-awareness (Kihlstrom 2015).

In the entire history of humans, there is a relation between the body and the mental status but there is no common understanding to define consciousness. Consciousness is defined in different ways, such as being the ability of the human being to assign special feelings to a mental capacity. It is also defined as subjecting the mental status to past, present and future scenarios with the reflection on themselves as being aware of the surrounding environment.

- **Awareness**

It is a kind of belief from the experiences obtained by undergoing an incident or event (Joseph 1980). It is essential for making predictions and decision-making which requires flexibility. This includes modelling of real-time entities, states, processes and other conscious entities. Some of the types of awareness are: agency awareness, goal awareness and sensory-motor awareness. The terms awareness and consciousness are used as synonyms as the difference between them is often ignored (Elbakyan 2010). Types of awareness (http://brainmind.com/Awareness.html) are presented in Table 2.

Awareness cannot be easily ignored against logical thought. Sometimes logical thought may go wrong but awareness since it is gained from true experience cannot go wrong most of the times. Awareness could also be obtained from human errors, faults, misunderstandings, oversight, misinterpretation and others. It involves learning from feedback calculations so as to minimise the error obtained.

Table 2 Types of awareness

Awareness	Unawareness
Unorganised impulses and knowledge base are considered as input to humans. These input information when applied, processed and analysed, yield into a state called awareness	The state of being uninformed or having no knowledge about a subject is termed unawareness. It does not have its part in machine consciousness

- **Thought**

It could be described as a structured representation of a point of view, belief, hypothetical possibility or intended action. Ideas or arrangements of ideas that are the result of the process of thinking is called thought. It also represents a set of ideas to achieve a desired result. Rigorous thinking happens wherein lot of effort, time and energy are spent to determine an idea, approach to solve a simple or complex problem to obtain the desired result. It is attained by querying and reasoning. Thought represents images, signs, language, and so on.

- **Memory and learning**

The mental records maintained by humans require memory. It gives instant access to past events, facts and skills for suitable actions by humans. The three primary stages of human memory process are encoding, storage and retrieval.

Memory and learning are closely related to each other but researchers (who deal with memory and learning) treat them as two distinct phenomena. They feel that learning will modify a subsequent behaviour whereas memory is the ability to remember past experiences.

It is essential to understand how we think, to understand how we learn. Intelligence is basically a memory-based process. Learning makes drastic modification of memory. It means modification of memory that causes a system to act differently based on its content. Human memories are always subjected to constant modification.

Learning depends upon inputs. Each word that is read and each sight that is seen changes the memory in one or the other way. Memory must decide what to retain and where to fit based on previously stored knowledge. It is essential to store and retrieve the information that we learn. Thus by association, memory and learning are interdependent as the knowledge stored in memory gives the framework to which new knowledge is linked (Baars 1988).

Learning is a continuous process for beings or machines. The often-changing world and its elements offer huge opportunities for cognitive systems to learn. A decision taken at a particular time may not hold good for the next time when the same event occurs. The takeaway or feedback from such events is essential datasets for learning.

The dataset gets updated and processed to be known as awareness. The awareness is cognised datasets derived from consciousness from the past experiences and analysis and interpretation of results.

- **Anticipation**

Prediction of consequences of proposed or probable actions by an individual or group of people is called anticipation. The real-world events are stored in a conscious mind which enables to predict events. It triggers forecasting of the events and consequences based on the actions performed on the real-world entities that would occur in future. This happens through learning from past experiences and their results. When an action is performed for the first time, the brain would have no clue about the results. Second time when the same action is performed and the same result is obtained there is a kind of learning which happens. Subsequently when the same result reoccurs for the specific actions, the affirmations become huge and the probability of the consequences of the action could be easily forecast. When an altogether different result is obtained for the same action, again learning happens but this time the action gets marked for a different behaviour. When successively tested for the same action, learning happens and the deviations obtained in the results are stored so as to determine the probability of the expected consequence for a particular action.

Creativity in humans can be seen as a philosophical feature of consciousness. It is the ability to tackle real life problems with intelligence. It involves:

- Conscious and unconscious states are the phenomena of consciousness in beings.
- Beings exhibit awareness when they are conscious. Whereas, certain level of awareness and attention is given in dream states when they are unconscious.
- A thinking mind generates information with a certain level of priority and attention in respect to certain goals or emotional states. Based on the type of problem, intelligence, knowledge and experience, the thinking mind generates different types of thoughts.
- Consciousness is based on characteristics such as inbuilt or inherited qualities, states, personality, intelligence, creativity and instincts.
- The states of consciousness may change as a result of experiences such as emotions and expressions.
- Creativity should also possess minimum conscious control such as ageing, hunger or pain.

The objectives of this chapter are to understand the challenges and limitations of human-like cognition and intelligence in machines. Qualia or raw experiences are considered to be the challenging problem of consciousness. Quantum mechanical phenomena are vital in the functioning of human brain and it forms the basis of consciousness (Zalta 2011). A variety of functions were suggested by Baars (1988) and others. Igor Aleksander suggested 12 principles for artificial consciousness (Aleksander 1996) and the objective of artificial consciousness is to define what and how these and other principles of consciousness can be created in machines. These principles are no means exhaustive as there are many other principles.

The organisation of the chapter is as follows. Section "Artificial Intelligence and Machine Cognition" provides an insight to machine cognition. Section "Cognitive Architecture" presents an overview on cognitive architecture and its compositions. Various aspects of perceptions in machines are presented in section "Perceptions About Machine Consciousness". Challenges for social and cultural ethics in machines are discussed in section "Social and Cultural Ethics in Machines". Issues involved in an ethical Turing test are also summarised.

Artificial Intelligence and Machine Cognition

Artificial consciousness combines artificial intelligence and cognitive robotics. It is also referred to as machine consciousness as well as synthetic consciousness. The objective is to synthesise consciousness in an engineered artefact (Thaler 2004). The role of consciousness is significant in the decision-making ability of computational machine systems. It is mandatory for such systems to have large corpora of datasets to train consciousness. These systems must have the ability to determine the difference between correct and incorrect decisions.

Thus, machine cognition could be termed as awareness in machines. All the actions taken by the machine would be driven by consciousness. The data flow is bidirectional between machines and environment. Continuous feedback from the environment is obtained based on the result given by the machine. The feedbacks from the environment are updated and the inputs to the machines are adjusted suitably so as to obtain the desired result from machines.

- **Thoughts in machines**

Thoughts need to be mapped onto machines. It is the ability to think and plan towards the desired goal. Unless proper thinking is done, it is not easy to reach the destination. Thought cannot be easily replicated in machines. The machine reaches a goal by considering external environmental factors. This requires proper planning and the process of planning is referred to as thought or thinking. Thought can be envisaged as a manager which drives the machine towards the goal. The data needs to be fed into the manager device, essentially qualia which also contains datasets of past experiences. Based on the consciousness in machines, machines make right decisions by considering external factors.

- **Brain to machine interface (BMI)**

The BMI enables a communication channel to communicate between the qualia device, which stores past experiences and brain wave patterns.

BMI uses imagination for controlling machines. But, what is the extent of BMIs occupying cognitive resources is still not clear. Human bodies behave in a different way. Once a human being learns to walk, he does not pay attention to this process anymore continuously. The human brain is designed and developed in such a way that once the brain learns to do something and masters the art of doing it for some quantum of time, it is not required to pay explicit attention to do the same thing again. It happens at the subconscious level, without the superficial involvement from the human, the task which was previously done and mastered can be executed seamlessly by the human brain. This implementation in machines is quite challenging and extensive research is being carried out in this context.

The specialty of humans is especially possessing sense organs. A neural activity in human brain produces conscious experience. Not all of sensory input received is processed by brain. This is another challenge faced in the field of machine cognition, wherein the machines act based only on inputs and perform calculation as per the logic and give back the output. It is necessary to make machines experience the surrounding environments, sentiments and emotions.

Some specific structure and activity are responsible for developing consciousness in machines. Several theories explain what properties can produce consciousness in biological or artificial systems. It is possible to artificially create new raw experiences in machines with conscious implant.

Under BMI technology, one of the most challenging tasks is direct writing of new memories into the brain by eliminating learning. However, it is still a hope. The main problem is that brain stores memories as patterns of connections between neurons, but information by patterns is not understood. These patterns are highly distributed and difficult to alter connections throughout the brain. Also, the same neural networks can store different memories, so it is very difficult to add new memories without damaging the stored memories.

The machines are developed based on specified models to solve problems. The designs are simulated with all range of scenarios and when satisfactory results are obtained, machines are cognised with the data from the models and thus are allowed to deal with the real entities. The machines are made to learn and act on external factors when conflict occurs with the environment. Learning happens in three phases—visualisation, assessment and fine tuning. Communication is essential when multiple cognitive machines are involved. There will be a need to coordinate among the machines and undertake appropriate decisions at the right time. The cognitive machines will have data sets containing information which would enable the machines to understand the conflicts, have ability to analyse, interpret the issues and have potential to solve the conflicts.

For fruitful interaction among humans, emotional intelligence is more important than IQ. It is also evident that rational learning in humans is dependent on emotions. Understanding emotions is an important aspect of personal growth and development for the evolution of human intelligence. Researchers are interested in capturing the sentiments in real-time applications for studying open challenges. Economists are also interested in financial market predictions. This has motivated the emerging fields of affective computing and sentiment analysis, which cater to human-computer interaction and information retrieval, to support the ever-growing amount of online social information. Sentic computing (*SenticNet*) explores knowledge-driven linguistic patterns and statistical methods (Cambria et al. 2014). Hence, we are interested in building intelligent machines with emotions.

Linguistic models, such as statistical or grammatical, are useful for cognition and communication (Davydov and Lozynska 2016). Cognitive communication systems are emerging in recent times. One of the most promising systems is COgnition-BAsed NETworkS (COBANETS) (Zorzi et al. 2016). This involves advanced machine learning techniques, such as unsupervised deep learning (multiple layers without any feedback), and probabilistic generative models. It is essential to know the behaviour of humans in processing the information relevant to effective control of machine systems, such as vehicles. Machines may increase the physical ability of humans to interact with the surrounding environment. For instance, continuous interaction between man and machine takes place in an aircraft for effective manoeuvring. This interaction is in accordance with respective cognitive and control processes. It also demonstrates the capacity of a human operator to interact in tandem with a responsive machine system.

By increasing the human machine interfaces (HMI), machines can understand humans' needs and work efficiently. Awareness of environment, interoperability of various systems' components and self-adjusting capabilities are expected in smart systems. Human interventions should be reduced for all repetitive or time critical tasks. Hence, machine cognition simulates human perception or cognition through smart systems. The advances in AI should make it intrinsically friendly and humane. This was suggested by the Machine Intelligence Research Institute (MIRI), Berkeley, CA. The researchers discussed impact of robots and computers in making their own decisions. They also discussed the ability of robots to acquire autonomy.

- **Features of machine consciousness**

Being 'conscious' is being alert and responsively being aware of the surroundings. CONscious Attention-based Integrated Model (CONAIM) (Simões et al. 2016) is a model of learning which is helpful for human-like agents. It integrates short- and long-term memory and also planning, learning and emotions. The results of mobile robotics say that an agent can use emotions, planning and memories in an attentive way to set the goals and to learn new procedures and concepts based on external and internal stimuli.

Self-consciousness can be defined as the consciousness of being self-aware or the consciousness of one's own body movements and its sensations, or the ability of recognising oneself in front of a mirror (just as dolphins, chimpanzees and human beings recognise themselves). Human-like agent cognition that integrates short- and long-term memories, reasoning, planning, emotion, decision-making, learning and motivation is expected in machine consciousness.

Modelling in machines is a challenging aspect and hence it is viewed as a black box. Computational systems process the information which is either inbuilt or inherited through learning from experience or a combination of both. When an expected or a desired goal is to be reached through consciousness, the input to the system is to be quantified appropriately so as to reach the goal in a desired number of iterations.

Collective machine cognition systems involve monitoring, surveillance and manipulation of the real-world complex problems. This is further complicated when the task is distributed in space and time. A novel solution to this ambient intelligence problem is to combine number of components with autonomous reactive system. There are scenarios where machines need to take actions not based on programming but based on experiences it had earlier. There is a need for learning or cognising from factors which were dealt by the same machine in the past. When multiple self-cognising machines are brought together and allowed to interact among themselves to achieve a common desired output, it involves a lot of understanding among the machines. This is regarded as associating the data among the cognitive machines.

The machines need to internally model a real-world representation of the external environment. So learning happens on a day-to-day basis with the machine updating its knowledge and gradually becoming familiar with the real-world entities.

Machine consciousness involves models which involve learning. Deep learning involves multiple layers between input and output nodes. Deep learning algorithms may be supervised or unsupervised. It also involves supervised learning wherein the feedback obtained from the output model is used to update the input signals to minimise the difference between expected and actual output. Every intermediate layer updates its previous layers and when the number of layers is more, more feedback is received which results in minimal errors. The sense organs are to be designed similarly to those of beings wherein the inputs such as speech and vision can be read and data can be made available for machines to process.

- **Future of machine cognition**

From the above discussion, it is evident that many of the challenges in machine cognition are still open to explore. Some of them are: Automation of learning from external environment without human intervention, Adaptive learning—cognition should occur based on the feedbacks received from the external factors or other peer machines and application of machine learning to deal with huge amounts of data.

The current challenging issues are:

- How to make machines understand/learn the user's needs and habits?
- How to write handwritten rules in human-machine control relationship?
- How to handle hazardous/critical environments beyond the experience of humans through machine cognition?

Researchers are surprised by many such unanswered queries. There is still a long way to go in exploring the possibilities of machine cognition.

Cognitive Architecture

Cognitive architecture is the theory of the structure of the mind, including human consciousness. The main objective of cognitive architecture is to summarise cognitive knowledge in a computer model.

Machine consciousness is one of the fields in AI which is concerned with the production of hardware and software devices that include conscious processing. Any machine is considered to be conscious if it possesses learning, perception, action and memory. A machine has a central executive part which controls each of these processes. It includes the conscious and subconscious events. The central executive of the machine is operated and controlled by the machine's motivation and by the selection of a goal. Some of the models for building machine consciousness were designed on the basis of global workspace theory (GWT) (Baars and McGovern 1997), internal self-models, representation of higher levels and the models that are based on attention. GWT is developed to account for a large set of matched pairs of conscious and unconscious processes. The models that are inspired biologically by the activation of neural patterns can be found in the brain of the human being. Phenomenal consciousness can emerge from the self-analysing and self-examining mechanisms about perceptions. Some of the key properties like self-conception can form a strong basis for self-modelling. The most accepted theory in the concept of consciousness is Integrated Information Theory (Tononi et al. 2016). The consciousness quantity is based on the amount of integrated information.

Once the sensory organs receive the input, may be in the form of speech and vision, the challenge of an effective model of machine consciousness would be to make sense of the information. This also provides a higher-level organisation of knowledge obtained from data which resembles thought processes and reasoning (Baars 1988).

- **Brain-inspired cognitive architecture**

The main systems of consciousness model are an attention system, cognitive system and a set of processes. The fundamental characteristic of the humans is the ability to recall experienced events that can be applied to the current situation rather than to perform a new process. To produce such behaviour in a coherent and unified manner, several components are structured as shown in Fig. 1. The attention system involves sensors and actuators which receive signals from the surrounding environment. A cognitive system encompasses memory and decision-making modules. Episodic memory stores the data in a declarative form using a suitable data structure. Semantic memory stores established facts and world's knowledge. Procedural memory stores procedural knowledge that can be executed by chunking the state-action pairs. Any cognitive process or function can request information from any other module and/or change the pattern of working of the agent (Joseph 1980).

Any function or process can request information from any other module and/or change the pattern of working of the agent. Cognitive architectures aim to utilise the knowledge contained in each of the different modules in a coherent and unified manner to produce cognitive behaviour. This is particularly important for high level cognitive capabilities that are easier to understand at a symbolic level than at a sub-symbolic level. A particularly interesting feature of humans is the ability to recall sequences of historical events that can be applied to the current situation. This type of memory, known as episodic memory, has some impressive capabilities.

An attention system comprises sensory memory, feature maps, weights associated with feature maps, saliency map and attentional map.

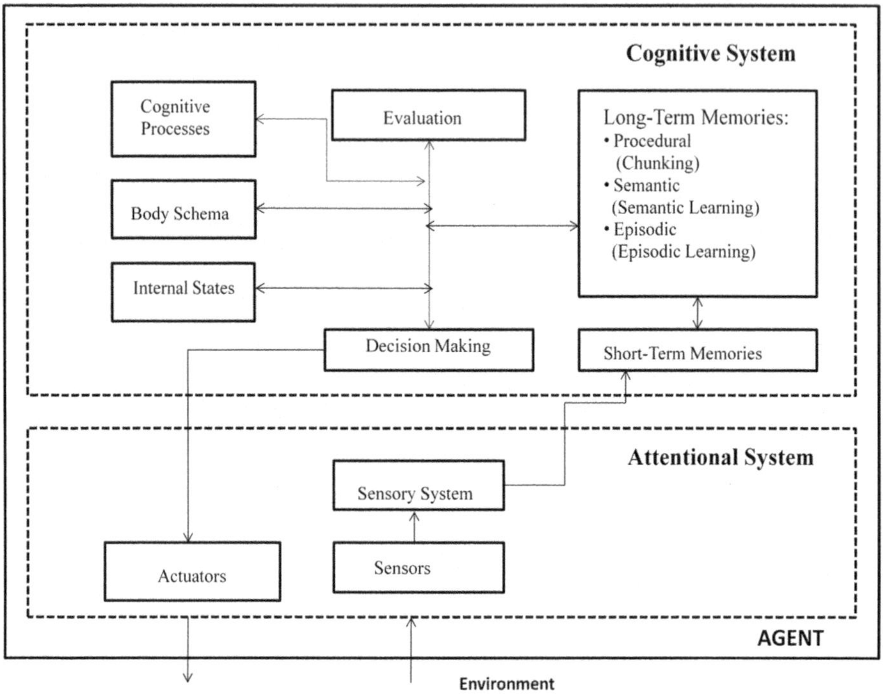

Fig. 1 Brain-inspired cognitive architecture—components and its structure

A cognitive system is composed of

(a) *Decision-making component*: this module decides the actions to be performed by the agent.
(b) *Short-term memory component*: this module stores a small amount of information in the sensory memory. It can be used by various modules of cognitive system, long-term memory storage and recall.
(c) *Long-term memory component*: this module stores huge amounts of information over a long period, and in turn consists of:

- *Episodic memory component*: stores large amounts of information in declarative form, using suitable data structure which is relative to specific events.
- *Semantic memory component*: stores large amounts of facts and knowledge about the real world in a declarative form, using a suitable data structure.
- *Procedural memory component*: stores large amounts of knowledge related to procedures that can be executed by an agent. This is typically carried out in the form of (*state-action*) pairs.

(d) *Evaluation component*: this module groups various lists, functions and variables related to the performance of an agent. It is composed of results that store the current goals. The set of metrics designed to assess the performance of an agent from different perspectives is called an evaluator, over the internal state. The current value of the agent's state for all the metrics considered is the evaluator's state.

(e) *Cognitive processes*: this module combines various cognitive functions of agents. Planning and state estimation are cognitive functions of an agent.

(f) *Individuality component*: this module encompasses a set of inner states—namely, motivation, emotion and a body schema of the agent.

A set of processes with diverse purposes need not necessarily be synchronised with the remaining components. The system operates considering that any function or cognitive system process can request information from any other module and/or change the agent's internal states at any time. Therefore, the flow of information through the decision-making module is not mandatory.

Perceptions About Machine Consciousness

Generally, researchers felt that developing consciousness and cognition power in machines is a highly complex task. Chalmers was the first scientist to rationalise machine learning concepts. After his research, a new gateway opened up in the field of cognitive machines (http://www.bionicgate.com/artificial-consciousness/). His argument was based on the fact that machine learning happens through only mathematical and logic computations and nothing beyond that. The use of computers has eased up the machine learning process. Chalmers developed a mathematically diversified computational model which to some extent addresses some real-world problems such as location detection and complex mathematical computation. Haikonen (Parsell 2005) had the opinion that just by mere mathematical computations the fully functional cognitive machines could not be developed. He did reproduce some of the human perceptions of sensations such as pain, pleasure and other emotions. He proposed that machines could perceive by gathering data from the environment or human operators. This is followed by analysis and interpretation of perceived data to create awareness in the systems to make the machine aware about right and wrong things. The learning with multiple layers between input and output layers was demonstrated to have multiple updates across layers to minimise the error between the desired and actual output. The concept of thought was envisaged by Haikonen to solve complex problems (Starzyk and Prasad 2010). Thought has symbols, images, running text and so on as input to obtain the desired goal. It was about continuous learning through feedback mechanisms so as to reduce the variations in the desired result.

Social and Cultural Ethics in Machines

Ethics is an essentially required behaviour for human community. As the human operators are getting replaced by machines gradually, it is a mandate for the machines to follow certain guidelines abiding which the machines should operate in a given environment.

The betterment of the being community is the sole criterion for ethics to be incorporated in the machines. The machines in all possible ways will assist beings, humans to a large extent. Due to the malicious intention of certain elements in the human community, it is possible to manipulate data of machines and thus cause unpredicted results from the systems. These results may also be dangerous to deal by the human community which might not be having sufficient information of the machines. The machine developers should follow guidelines to protect the human dignity and confidentiality.

There are several ways to develop ethics in systems. One way is to evaluate the processes involved in the machines. The other way is to evaluate the final result of the systems. The working of the machine is to be stopped and immediate steps are to be taken when the results are not supportive to the surrounding environment. Only when the results are not having dangerous effects, the machine can continue working in the environment. It is possible to cognise machines about ethics using huge corpora. It is also possible to test the machines in a supervised condition and understand the behaviour of the machines with real-world factors and parameters.

Machine ethics is a research topic concerned with designing artificial moral agents (AMAs). It is concerned with automating the moral behaviour of humans. As a part of AI ethics, machine ethics deals with the moral behaviour of intelligent machines. Machine ethics is not equivalent to *roboethics* and computer ethics. Roboethics involves designing ethical problems in robots. Computer ethics involves professional behaviour towards computers and information.

Several experiments by researchers have shown that, building ethics in machines is a complex task. However, there is one technology in particular that could truly bring the possibility of robots with moral competence to reality. In a paper on the acquisition of moral values by robots, Oxford University Professor Nayef Al-Rodhan (2016) mentions the case of neuromorphic (*brain-like*) chips, which aim to process information similar to humans, nonlinearly and with millions of interconnected artificial neurons. Robots embedded with neuromorphic technology could learn and develop knowledge in a uniquely human-like way. Inevitably, this raises the question of the environment in which such robots would learn about the world and whose morality they would inherit—or whether they would end up developing human 'weaknesses' as well: selfishness, a pro-survival attitude, hesitation, and so on.

James H. Moor lists four kinds of robots in relation to ethics. A machine can be more than one type of agent from among the following: ethical impact agents, implicit ethical agents, explicit ethical agents and full ethical agents.

- **Ethical turing test**

A typical Turing test involves a human interrogator, who is expected to identify the intelligence of a machine. If the interrogator is not able to distinguish the machine from human, for his questions, then the machine may be assumed to be intelligent. In similar ways, the test results on consciousness in machines should be correct with respect to ethics and law.

Creating robots as explicit ethical agents is still an unexplored problem. Hence it is essential to explore methods of building robots both scientifically and philosophically. Robots must exhibit more ethical capabilities as they become autonomous. To make our society better and understand ethics, it is essential to build robots with ethical values in *them.*

A new test by MIT lab is trying to build cars with a sense of humanity that will make decisions over whom to kill—its own passengers, the passengers in some other vehicle, pedestrians or domestic animals—when a breakdown or accident occurs. The sense of morality in machines should be able to decide—do *they* value the lives of the people in the car more than those crossing the street? This thought experiment was conducted to study the behaviour of driverless cars especially during breakdown. It is also essential to know how machines treat us in similar situations. The situations involve a self-driving car is travelling toward a zebra crossing, and it needs to choose whether to steer and crash into a barrier or hit whoever is at the zebra crossing. The evaluation is basically to determine what machines would do in these rare, life-or-death situations.

Such experiments force us to think that machines may become better and better every day, but is it ethical to kill people by machines instead by humans? This may be a case among many situations, where humans themselves do not know how to behave in such conflicting situations. Machines cannot get proper inputs from humans, when humans are in dilemma to act. In our opinion, it is still tender to discuss.

Conclusions

This chapter has reiterated the importance of consciousness in machines by discussing various attributes of it. Attributes such as awareness, thought, memory, learning and anticipation have been presented, and aspects of machine cognition and various features of machine consciousness have been discussed. A brain-inspired cognitive architecture illustrates the interaction between the cognitive system and attention system in a given environment. The theories and ideas of many researchers and psychologists show the diversity in their perceptions and opinions. Building ethical values in machines is another complex topic which first requires thorough understanding of ethical values in humans. This is necessary, as it is humans who build ethics in machines.

Acknowledgements The first author acknowledges the support provided by the Indian National Science Academy during the course of this work.

The second author acknowledges the encouragement and support provided by Ramaiah Institute of Technology, Bangalore during the course of this work.

References

Aleksander, I. (1996). *Impossible minds: My neurons, my consciousness*. Imperial College Press.

Al-Rodhan, N. (2016). Moral enhancement: A reality check. IA Book-Chapter in *The myth of the moral brain. The limits of moral enhancement* (pp. 1–6). The MIT Press.

Baars, B. (1988). *A cognitive theory of consciousness*. New York: Cambridge University Press.

Baars, B., & McGovern, K. (1997). *A theory of consciousness. Global workspace*. Berkeley, California: The Wright Institute.

Bucke, R. M. (1905). *Cosmic consciousness: A study in the evolution of the human mind* (pp. 1–2). Innes & Sons publications.

Cambria, E., Olsher, D., Rajagopal, D. (2014). SenticNet 3: A common and common-sense knowledge base for cognition-driven sentiment analysis. In *Proceedings of 28th AAAI Conference on Artificial Intelligence* (pp. 1515–1521).

Davydov, M., Lozynska, O. (2016). Linguistic models of assistive computer technologies for cognition and communication. In *11th International Scientific and Technical Conference, Computer Sciences and Information Technologies (CSIT)*, Lviv, Ukraine (pp. 171–174).

Elbakyan, A. (2010). Consciousness in mixed systems: Merging artificial and biological minds via brain-machine interface. Retrieved from https://engineuring.wordpress.com/2010/04/25/consciousness-in-mixed-systems-merging-artificial-and-biological-minds-via-brain-machine-interface/

Joseph, R. (1980). Awareness, the origin of thought, and the role of conscious self-deception in resistance and repression. *Psychological Reports, 46*(3), 767–781.

Kihlstrom, J. F. (2015). Scientific approaches to consciousness. lecture supplements on consciousness. University of California. Berkeley. https://plato.stanford.edu/archives/win2016/entries/consciousness/

Kriegel, U. (2009). *Subjective consciousness*. Oxford: Oxford University Press.

Parsell, M. (2005). Review of P. O. Haikonen. The cognitive approach to conscious machines. *Psyche, 11*(2), 1–6.

Schneider, S., & Velmans, M. (2017). *The Blackwell companion to consciousness*. New York: Wiley Blackwell Publications.

Simões, A., Colombini, E., Ribeiro, C. (2016). CONAIM: A conscious attention-based integrated model for human-like robots. *IEEE Systems Journal, 99*, 1–12.

Starzyk, J., Prasad, D. (2010). A computational model of machine consciousness. *International Journal of Machine Consciousness*, 1–26.

Thaler, S. L. (2004). The emerging intelligence and its critical look at us. *Journal of Near-Death Studies, 17*(1), 21–29.

Tononi, G., Boly, M., Massimini, M., & Koch, C. (2016). Integrated information theory: From consciousness to its physical substrate. *Nature Reviews Neuroscience, 17*, 450–461.

Zalta, N. E. (2011). Quantum approaches to consciousness. In *Stanford Encyclopedia of Philosophy*

Zorzi, M., Zanella, A., Testolin, A., Grazia, M. (2016). COBANETS: A new paradigm for cognitive communications systems. In *International Conference on Computing, Networking and Communications (ICNC)*.

Part II
Healing, Agency and Being

'Is Grandma Still There?' A Pastoral and Ethical Reflection on the Soul and Continuing Self-identity in Deeply Forgetful People

Stephen G. Post

You do not have a soul. You are a soul. You have a body.
—C.S. Lewis

Introduction

'Is grandma still there?' This is the big metaphysical question that most carers ask when it comes to their loved ones with dementia. It is a desperate question driven by their need to find spiritual and moral meaning in many difficult endeavors. Time and again, over 30 years of consulting with families, I have been asked just this sort of question. My response has been *first*, an open-mindedness with the many spiritual and/or religious carers who find meaning in an eternal non-material ground of self-identity that is always there and intact underneath sometimes immense communicative chaos or stark silence; and *second*, an affirmation of carers who still find meaning in the hints of residual self-identity expressed by their loved ones at certain insightful moments, even if they are resigned to the idea that all self-identity ends with death, dust to dust.

In working with *deeply forgetful* people and their families, I have encountered many carers who view the syndrome of dementia in its various forms (including Alzheimer's disease) as a breakdown in the capacity of the brain to connect with a still complete and forever intact biographical selfhood. Their spiritual-metaphysical assumption is not that the hints at continuing self-identity are mere remnants of a mostly 'gone' self as located in residual brain tissue. Rather, they view the brain more like a desk top computer that can break down so as to no longer retrieve memories from 'cloud storage', although those memories remain perfectly intact in an unseen mystery of immortality.

S.G. Post (✉)
School of Medicine, Center for Medical Humanities, Compassionate Care, and Bioethics, Stony Brook University, Stony Brook, USA
e-mail: Stephen.post@stonybrookmedicine.edu

S. Menon et al. (eds.), *Self, Culture and Consciousness*,
DOI 10.1007/978-981-10-5777-9_6

The materialist reader may deem the idea of any non-material memory substrate, however defined, as archaically nonsensical. But I will offer some support for the eternal soul by invoking a contrarian scientific theory of memory that is under some discussion these days. The support offered is meant to establish plausibility rather than proof, and so my goals are humble. I want to encourage respect for those carers who assert, 'We know grandma is still fully there, underneath the chaos and the silence, with an eternal soul that is slowly returning to the arms of the Supreme.' Indeed, in a eulogy for a family that I knew was deeply religious, I commented as follows: 'It may have looked like old grandma was "gone", or "a husk", or "a shell", or "already dead", but let's just allow that maybe she had already gone down to the station and had one foot on that last train bound for glory. Amen.'

How Little We Know

Let's start this argument for plausibility (not proof) by asserting that nobody really knows what biographical memory is physically or metaphysically, and so it remains a mystery. In the 1920s the great neuroscientist Wilder Penfield offered some bit of overstated evidence that specific memories are located in specific areas of the brain in the form of *engrams* or 'memory traces'. But in the 1940s the great neuropsychologist Karl Lashley gave up on Penfield's idea of localised traces after decades of research. Lashley concluded that memories are not located in specific brain locations, but are instead distributed 'globally' throughout the brain as a whole. Lashley, in his brain ablation studies in rats, failed to identify any unique sites for memory storage in the brain, and at last gave up on seeking a localised 'engram'. The idea that memory is distributed globally over a neural network—an idea taken up later by Donald Hebbs—in some synaptic associations is indeed interesting, but also unproven and highly questionable. This drove one of Lashley's students, neuropsychologist Karl Pribram of Yale, to develop the idea that the brain is in some mysterious way holographic, meaning that all memory, including the whole of a life, is contained in each and every brain location like a single spot on a field included the entirely of a holographic image. This idea is quite interesting, but the model lacks any empirical support. It may be proven true one day, and some have tried to defend it articulately (Talbot 1991). More recently, influenced by Hindu and Buddhist theories of reincarnation, biologist Rupert Sheldrake has argued that the repeated failures of 'material trace' theories of biographical memory suggest that mind and memory are non-material (2012).

The narrative memories that constitute self-identity do have infinite 'megabytes' of information as deep as a boundless ocean or a limitless sky, and even the smallest microchips cannot store this infinity. People often state that small computers can hold great amount of memory, so by analogy they look within the brain to unlock this boundlessness. But at a certain point, this sort of model breaks down logically because it is easier to view the brain not as a finite cupboard for infinite vastness, but rather as a transmission device otherwise unspecified. Like a TV set, the shows are not in the TV but outside of it. The infinite cannot be contained in the finite, or so the argument would go.

Given how little we know, please be respectfully open-minded with those carers who affirm the eternal essence of self-identity. Just maybe *the brain is an uploading and downloading device that has space for habituations such as those a rat develops in learning a maze, but not for the infinity of our life histories and related endless imaginings.* Of course, this opens the door for higher inspiration, synchronicity, premonitions and the like. But my focus is merely memory here.

Of course, it may be true that all the hints at continuing self-identity in deeply forgetful people are the last disintegrating remnants of a person's autobiographical self as located in an entirely 'local' and deteriorating brain. No one has proven the reality of a non-material soul that exists in an unseen dimension of reality. Still, logic suggests that a fathomless sea of infinite richness and textual detail is too much to engrave in a small mass of chemicals, cells and tissue. Contrary to even the most 'non-reductive' of the physicalists, we need not give up on our eternal souls just yet, although some theologians have given up on their eternal souls (Brown et al. 1998).

Until proven otherwise, memory in the sense of narrative and imagery still suggests to some very thoughtful philosophers and scientists that there may be a non-material and eternal dimensionality to it. Let us draw up a very partial list: Plato, the author of *Genesis*, the author of *The Gospel of John*, Lao Tzu, St. Paul, Buddha, Mohammed, Christ, Meister Eckhart, T.S. Eliot, Arthur Schopenhauer, C.S. Lewis, J.R.R. Tolkien, Emerson, Thoreau, James Jeans, Arthur Eddington, David Bohm, Erwin Schrodinger, R.M. Bucke, Whitman, William James, C.G. Jung, Ken Wilber, Huston Smith, Aldous Huxley, Dean Radin, Larry Dossey, Seyyed Hossein Nasr, Sir John Templeton, Joseph Campbell, Sir John Eccles, Shankara, Krishnamurti, the Dalai Lama, Pim van Lommel and Thomas Nagel (Nagel 2012).

Coining a New Term: EPI-NEUROLOGY

Let me repeat the primary thesis here for clarity: autobiographical memory may go deeper than a material substrate to something non-material or extracorporeal, although it can be uploaded or downloaded by matter (cell, tissue, brain). Secondarily, if so, this has considerable bearing on how we treat deeply forgetful people morally, as they are then not reducible to mere husks or empty shells of their former selves.

Let me coin a term here: '**epi-neurology**'. By analogy, the field of epigenetics tells us that when we want to understand gene expression we need to look outside the nucleus to cellular environment and beyond, even to behaviour and affect in the outside world. So perhaps we need something like epi-neurology to keep us from staring exclusively at the 'three-pound universe' of brain matter within, like we used to stare exclusively at DNA as if it were entirely self-contained. It is the case that an almost unimaginable number of synaptic connections within those three pounds of tissue could accommodate a tremendous amount of information, but even this would be limited. Let us simply be open-minded about the possible non-local aspect of mind, including memory, at least in its expansive autobiographical aspects as being non-epiphenomenal to matter.

By analogy, let us not lock our attention so strictly on the desktop or laptop computer. My computer before me holds a great deal of memory, but quite rapidly we have had to develop 'cloud storage' and external drives to hold the vastness. Epi-neurology looks outside as well as inside the brain.

Biographical Memory as a Timeless Field

There is nothing new under the sun. The Muslim philosopher Avicenna, along with Augustine in his famous *Confessions*, waxed eloquent about how in memory we can envision and hear the entire universe and everything we have encountered in life. The great Hindu sages have asserted the same, and therefore held that memory is an aspect of consciousness or mind that is primary in itself, underived from matter. What we can call up and envision from memory with eyes closed is unlimited in detail and scope, although the brain retrieval of information is complex, as well as potentially inaccurate at times (Brady et al. 2008).

In modernity, the French philosopher Henri Bergson, at the end of *The Two Sources of Morality and Religion* (1932), after a phenomenological analysis, wrote, 'In a word, our brain is intended neither to create our mental images nor to treasure them up; it merely limits them, so as to make them effective' (p. 315). The brain, Bergson contended, is a 'filter or screen' (p. 314)—a view that the American psychologist and philosopher William James adopted as well. Three decades earlier Bergson had produced his classic *Matter and Memory: An Essay on the Relation of Body and Spirit* (1896), which he wrote in response to *The Maladies of Memory* (1881), a book in which Theodule Ribot asserted erroneously that science shows memory to be localised within the brain and of a material nature. Bergson, however, considered memory—at least in the aspects of image remembrance and personal narrative—to be profoundly spiritual in nature. Bergson's brain has a retrieval function, but brain injuries then do not erase that which they retrieve. Bergson acknowledged that the brain is the locus of engrained habituated memories, as we find in many non-human animals as well as humans, but not of 'image remembrance' of the past, or 'pure' memory, which is of a contemplative and non-material nature. Bergson concluded, 'The idea that the body preserves memories in the mechanical form of cerebral deposits, that the loss or decrease of memory consists in their more or less complete destruction, whereas the heightening of memory and hallucination consists in an excess of their activity, is not, then, borne out by either reasoning or by the facts' (p. 176). He asserts that memory is 'absolutely independent of matter' (p. 177). For Bergson, it seems that *the brain is more an organ of percept and habituation than of storage*. The cupboard of autobiographical memory lies elsewhere.

A line of contrarian neuroscience today asserts that we need a paradigm shift in our thinking about memory. In 1993, Simon Y. Berkovich of George Washington University presented a speculative model of the brain as the local computer terminal connecting to some larger informational system (1993). Fifteen years later, writing in the highly regarded *Proceedings of the National Academy of Sciences*, T.F. Brady and team showed just how massive visual memory is, and concluded that is seems to 'pose a challenge to neural models of memory storage and retrieval, which must be able to account for such a large and detailed storage capacity' (2008, p. 14325). This is no trivial challenge to the reigning paradigm.

In the late 1970s the renowned British paediatric neurologist John Lorber famously reported that some perfectly intelligent adults with fine memories had no more than 5% of normal brain tissue after having been cured as children of hydrocephaly (water on the brain). Lewin's (1980) article entitled 'Is Your Brain Really Necessary?' which appeared in *Science*, dismissed Lorber's research as unscientific and 'overdramatic'. While initially disbelieved, Lorber's observations, based on brain scans, have been independently confirmed by neurologists in Brazil (de Oliveira et al. 2012) and in France (Feuillet et al. 2007).

Bringing this quarrel up to date, Forsdyke (2015) in his article 'Wittengenstein's Certainty is Uncertain: Brain Scans of Cured Hydrocephalics Challenge Cherished Assumptions', appearing in *Biological Theory*, urges the open-mindedness that Lorber's work seems to press upon us. Forsdyke, a distinguished researcher at Queen's University in Ontario, has studied microcephalic cases where intelligence as well as long-term memory are normal. The upshot is that information content and memory do not correlate with head size. He cites Fusi and Abbott (2007), with their calls for a radical remodelling of memory. Because brain size does not scale with information quantity, Forsdyke gives us three hypotheses to work with:

(1) The 'standard model' by which long-term memory is held in the brain in some chemical or physical form;
(2) Long-term memory is held in the brain by 'some extremely minute, subatomic form, as yet unknown to biochemists and physiologists' but akin to computers storing large amounts of information in progressively smaller spaces (p. 339);
(3) 'Information relating to long-term memory is held outside the brain. Since most non-neural tissues and organs appear unsuited to the task, this extrapolates to long-term memory being *outside* the body—extracorporeal! Amazingly, this startling alternative has been on the table for at least two decades.' (p. 339)

It is going to be hard for scientists raised under the ideology of materialism to imagine this third alternative. But for most religious people in all the great spiritual traditions, Mind (a.k.a. Ultimate Reality, Platonic *nous*, Supreme Being, Eternal Consciousness, Infinite Mind, Pure Unlimited Love, Ground of Being, God, etc.) precedes and sustains mere matter. It is certain that bone marrow creates blood cells, and so forth, but there is no fully convincing evidence that neural tissue produces mind, conscious awareness, biographical memory and the like. This is all a mystery.

Because materialism dominates science in our age, let us consider a contrarian *postmaterialist* quote from Schrodinger's 1956 Tarner Lectures, *Matter and Mind*, delivered at Trinity College, Cambridge. The greatest physicist of his time with the exception of Einstein, he dismantled materialism as the source of mind and spoke these words in his fourth chapter, entitled 'The Arithmetical Paradox: The Oneness of Mind':

> There is obviously only one alternative, namely the unification of minds or consciousness. Their multiplicity is only apparent, in truth there is only one mind. This is the doctrine of the Upanishads. And not only of the Upanishads. The mystically experienced union with God regularly entails this attitude unless it is opposed by strong existing prejudices; and this means that it is less easily accepted in the West than in the East. (1969)

Arthur Eddington, James Jeans, and David Bohm would be among the many physicists who joined in such an elevated theory of Mind.

Perhaps it is useful to quote one of the most respected philosophers of our time, Thomas Nagel, from his 2013 book entitled *Mind and Cosmos*: *Why the Materialist Neo-Darwinian Conception of Nature is Almost Certainly False*. Nagel comments that his doubts about materialism will strike most people as 'outrageous' because they have been 'browbeaten' to believe in a mindless universe (p. 7). But Nagel takes a view very different from materialism—'one that makes mind central, rather than a side effect' of the material (p. 15). 'My guiding conviction is that mind is not just an afterthought or an accident or an add-on, but a basic aspect of nature' (p. 16). He locates himself as follows:

> The view that rational intelligibility is at the root of the natural order makes
>
> me, in a broad sense, and idealist—not a subjective idealist, since it doesn't
>
> amount to the claim that all reality is ultimately appearance—but an objective idealist in the tradition of Plato and perhaps also of certain post-Kantians, such as Schelling and Hegel, who are usually called absolute idealists. (p. 17)

An upshot is that Mind is metaphysically something entirely different from Matter.

A Postmaterialist Model

Good careful science should never be interfered with. But whether we interpret findings in a materialist or a postmaterialist metaphysical model is a matter where we should welcome diversity. Let us note that 100 scientists from a variety of fields (biology, neuroscience, psychology, medicine, psychiatry, physics) convened at the Canyon Ranch in Tucson, Arizona, on 7–9 February 2014, to discuss the emergence of a postmaterialist paradigm for science, spirituality and society. This group produced *The Manifesto for a Post-Materialist Science.*[1] It challenges the

[1]Available at http://opensciences.org/about/manifesto-for-a-post-materialist-science.

nineteenth-century assumption, now turned into dogmas and known as 'scientific materialism', and in particular the belief that 'mind is nothing but the physical activity of the brain, and that our thoughts cannot have any effect upon our brains and bodies, our actions, and the physical world'. These experts argue that we need a new and non-dogmatic science that follows the methods of the best science, but does not presume materialist explanations. The manifesto takes seriously the idea that minds 'are apparently unbounded, and may unite in ways suggesting a unitary, One Mind that includes all individual, single minds'. Moreover, 'Mind represents an aspect of reality as primordial as the physical world. Mind is fundamental in the universe, i.e., it cannot be derived from matter and reduced to anything more basic.'

I suspect that there is more to biographical memory than mere matter and brain. We all read Ralph Waldo Emerson's *The Over-Soul* in high school English. He coined the term in 1844 while a student at Harvard Divinity School to engage all faiths at their essential core. The idea of a Supreme Intelligence has been well-established across all spiritual-religious systems from the dawn of civilisation. The Over-Soul is the Supreme Intelligence or Universal Prime Spiritual Mind or Divine Mind, call it what you wish. It exists eternally and is infinitely prior to all matter and energy, and from which all the universe seen and unseen arises. For Emerson, influenced by Hinduism's *Upanishads*, this Over-Soul also exists in a very special and concentrated form 'imminently' in each of us as the ultimate source of our individual minds: 'Meantime within man is the soul of the whole; the wise silence; the universal beauty, to which every art and particle is equally related; the eternal ONE' (p. 56). Emerson meant this literally. The Over-Soul is both all around us, and in a more intimate sense, it is within us. It is the mysterious eternal 'particle' of the Eternal One Mind of which the human soul and mind is comprised. Our minds and autobiographical memories are not simply derived from matter, however much the materialists resist this claim. Emerson's Over-Soul is the Ultimate Reality and Infinite Mind that underlies the universe and its laws, and that also concentrates itself in human consciousness. 'Ineffable,' Emerson wrote, 'is the union of man and God in every act of the soul.' This concept is eastern (Emerson was widely read in Hindu classics). But Christianity also speaks of the human being 'in the image of God', and St Paul wrote 'Do you not know that you are God's temple, and that God's spirit dwells within you?' (1 Corinthians 3:17).

The Ground of Human Dignity

It is for many religious people very difficult to affirm the human dignity and moral status of the deeply forgetful unless we place some faith in the idea that every human being has within a drop of the infinite cloud of a Supreme Being, Ultimate Reality, Infinite Mind, Ground of Being, 'God', etc. The famous materialist philosopher Bertrand Russell was at least able to acknowledge that, if materialism is true, if all we are is an admixture of chemicals and cells, then the sum total of the meaning of a human life is no greater than bacterial 'pond scum'.

But let us put these metaphysical questions aside and accept for the moment the comfortable materialist's assumption that human beings have no eternal soul/selfhood. Let us assert, however, that even on the materialist paradigm, continuity of self-identity in the deeply forgetful is the residual or remnant norm, and that this does afford them due respect.

Hope in Residual Neurological Hints of Self-identity

I define 'hope' in the experience of carers as 'openness to surprises', at least in the context of dementia (2013). In other contexts, hope might be defined very differently—for example, in terms of the pursuit of clearly envisioned goals. Deeply forgetful people do not generally not have such goals beyond the mild stage, but surprises occur. A medical student recently (November 2015) described his grandfather's 'terminal lucidity'—a frequently described phenomenon in psychiatric and hospice literature—after months of being entirely unable to communicate, due to Alzheimer's disease. My student chose to focus his essay on his mother's interaction with her father just before his death:

> It was in his last moments that my mother seemed to be rewarded for all her hard work. My grandfather looked at my mother and spoke to her with completely lucidity for the first time in a year. He talked about the old times when he used to walk her to school. Then he talked about me and told her to make sure I kept working hard in school. And the last thing he said was how proud he was of her and that he loved her. The next morning he was gone.

Again, hope is being open to surprises such as my student describes.

In a similar case, Hoblitzelle (2008), author of *Ten Thousand Joys & Ten Thousand Sorrows: A Couple's Journey Through* Alzheimer's emailed me on 12 April 2013, a few days after we shared a panel together at the Times Center in Manhattan for the New York Alzheimer's Association's *Charles Evans Lecture*. Olivia has read something of mine, and wrote,

> It reminded me of a moment with my beloved mother, a poet, author, and something of a philosopher. In that late stage when words are gone except for those very occasional moments, she looks at me intently and said forcefully, 'God, physics and the cosmos.'

Over the years of witnessing many cases of sudden insight, I ask: Where does such lucidity be come from? Yes, it could be some remnant of a neurologically grounded memory if personal identity and biographical memory actually exist in matter, which from the perspective of a purely megabyte analysis can and is being questioned. But it could also be a sign that underneath the neurological deterioration a whole self continues on. Given the current state of brain science, we must all be agnostic. We simply do not know. Moments of lucidity are the norm among the deeply forgetful, rather than the exception, especially early in the morning after a good night's rest. Do they point toward remnants of autobiographical narrative in a devastated brain, or to something fully intact housed in 'cloud storage' but now more difficult to access?

Whatever the answer, the implication ethically is that we as a Western culture need to move a little closer to, for example, some Indian attitudes toward the deeply forgetful. There one need not question the continuing self-identity of deeply forgetful people, and the mystery that it points to. It is only our Western 'hypercognitive'[2] and utilitarian arrogance that separates us from ancient wisdom on this matter.

Moving Beyond Hypercognitive Values

One need not be a metaphysical idealist who believes in an eternal soul to observe that hypercognitive values discriminate badly against the deeply forgetful, who have other aspects of the self that are equally important as cognition (Post 2013), and even more so:

- Symbolic
- Creative
- Emotional
- Relational
- Somatic
- Musical
- Rhythmic
- Aesthetic
- Olfactory (smell)
- Spiritual
- Tactile

Briefly, I highlight some of these aspects of the enduring self with a set of brief cases:

Case One: *Memories in the Making* is a nationwide program that explores whether people with dementia can reveal themselves through art. Many artists from around the United States now volunteer to lead these programs in most major cities and towns. What we have discovered is that even in the most advanced stages of dementia, individuals will express remnants of self-identity. They may not be able to communicate by speech or proceed from point A to point B over time. True, they to a considerable degree living in the pure present, but we need to be very careful not to assume that the connective glue between present and past is ever completely gone. Sometimes, such assumptions evaporate when we allow these people opportunities to express their self-identity through the recreation of a symbol. I met a man with deep forgetfulness who joined the Memories in the Making art program. He worked for weeks drawing a series of horizontal and diagonal lines on paper. He was not conversant and had not been for months. Although generally unresponsive

[2]A term I coined in my 1995 first edition of *The Moral Challenge of Alzheimer Disease*.

when asked what he was drawing, one morning after a good night's rest he suddenly blurted out, 'Directions to my daughter's house!'

Case Two: There is a major care movement called *Music and Memory*[3] I visited the Long Island Veterans Nursing Home in March of 2013 and spent some morning hours in the facility for vets with severe dementia. There were about 30 vets in an activities room devoted to individuals with severe dementia. As far as I could observe, almost none of them were conversant or responsive when called by name. Most had that thousand-mile empty stare. Then came the big moment. The activities director started the music. 'New York, New York', and 'That's Amore', and about two thirds of these old timers started to chime in as the words ran across the big screen on the wall and the voices of Sinatra and then Dean Martin sang out. Then came 'It's a Grand Old Flag', and its seemed liked the singing got louder, a few more vets chimed in, and five of them were standing up saluting the flag. I felt like I was witnessing a miracle. Deeply learned music, including hymns, with which individuals identify autobiographically, touches their enduring selves at a profound and hidden level, eliciting a renewed sense of who they are. After the music session, at least ten or so of these vets seemed to be able to respond to the activities director when she asked them close-ended questions about the meaningful people in their lives. It was as though they had awakened from a slumber.

Yes, these tend to be sporadic hints, but they may all be revelations that underneath the communicative chaos or the glassy stare there is something that has been referred to as *the mind behind the mind*. Metaphysics aside, these episodes inspire caregivers with renewed meaning.

Still, my Hindu friends in Gambier, Ohio, found meaning in the spiritual interpretation of memory that we were pondering earlier in this chapter. My friends were in 2003 all professional caregivers at a nearby facility. Dr Foley and I took them to lunch one day and asked them why they care so well for so many patients with dementia. Their response was that each human being carries within a small particle of divine mind, or as the Hindus state it, *atman (soul) equals Brahman (god)*. No matter how much the brain is harmed, there is still a mind or consciousness that is more than the brain, and that does not arise from the workings of the brain. The individual mind has primacy over the brain, and it not merely an epiphenomenon, and that materialism cannot explain or give rise to.

Regardless, a person with dementia is rarely as 'gone' as we superficially suppose, and carers widely report an openness to surprises. There are those moments, often early in the morning after having slept well, when a person with severe dementia surprises us with a meaningful word, a moment of recognition. A person incapable of conversation may join in with others on a verse of a deeply loved song. The glimmers of a fuller presence merit our respect. Therefore, sit down, make eye contact, and call that person by name as if expecting an answer that may not come today. This action is more than symbolic. It is how we affirm the enduring self. Our task is always one of exceptionless *affirmation and connection*.

---

[3]See www.musicandmemory.org.

But what blinds us to the signs of the enduring self? What makes some prefer to see the glass as half empty rather than as half full? Why do we hear metaphors such as *absent, gone, husk, dead, empty* and the like? Here I believe we must confront various biases and prejudices, many of which have their roots in what I have long termed *hypercognitive values*.

Most contemporary philosophers espouse 'hypercognitive values' and as a result they have been arrogantly blind on the whole—Eva Kittay being a major exception—to the emotional, relational, aesthetic, creative and spiritual values that give worth to the lives of the deeply forgetful. They are not interested in the enduring selves of the deeply forgetful. The lens isn't right. Instead, they tell us, following their forbears, that to be a 'person' requires the ability to project rational plans as moral agents into the future. People with Alzheimer's disease generally do not have this capacity, and therefore they are deemed something less than 'persons'. Over the years, the philosophical response to 'hypercognitive values' from various philosophers is that 'even non-persons' have sentient awareness of pain and the like, so they should not have to experience pain—which is to say that their deaths should be painless. Well, fair enough, but labelling a deeply forgetful individual a 'non-person' diminishes most, if not quite all, of their moral considerability in the human community of equal regard.

The Destructive Arrogance of Hypercognitive Values

We all value cognition of course, but let us not make too much of cognitive dexterity. The great Stoic philosophers achieved much for universal human moral standing by emphasising the spark of reason (*logos*) in us all. This is, however, an arrogant view in the sense that it makes the worth of a human being entirely dependent on rationality, and then gives too much power to the reasonable. From the Stoics and without interruption into Kant, Locke, and modern bioethics we find the rude assertion that the major criterion for 'moral membership' is reason, and this tends to include only the intelligent in the protected community (Singer 1993). We easily demean those whose memory has dissipated by treating them with indifference or even with cruelty. We act as if they aren't there. Once (less than seven decades ago), the step between psychological and physical elimination proved notoriously short. As part of the Nazi extermination programme, known as T-4, individuals with dementia, selected for hypothermia experiments, were taken out of German mental asylums and left to freeze in the cold overnight air (Post 2000).

Theologian Reinhold Niebuhr wrote of the tradition from the Stoics that 'since the divine principle is reason, the logic of Stoicism tends to include only the intelligent in the divine community. An aristocratic condescension, therefore, corrupts Stoic universalism' (Niebuhr 1956). We sometimes mock and ignore those who have lost the power of reason, sending the message that their very existence rests on a mistake (Post 2000).

The rationality that philosophers select for moral considerability is generally limited to one property. They define rationality procedurally as an ability to do certain things, such as act consistently based on clear thinking, arrive at decisions by deliberation, envisage a future for oneself, and so forth. But, in fact, rather few of us go through life with consistent rationality (Zagzebski 2001). We act on emotion, intuition, impulse and the like. We go through periods of considerable irrationality due to variation in mood. *Rationality as a decisional capacity is not morally important. It is rationality as a source of self-identity that matters—i.e., 'who' we are rather than 'how' we proceed. And in this sense, the deeply forgetful can be surprising.*

Our task as moral agents is to remind those among us with dementia of their continuing self-identity, of who they are. In other words, our task is to preserve identity, rather than deny it. It is for this reason that many units for the deeply forgetful in nursing homes will post biographical sketches on the doors of residents, or family members will remind a loved one of events and people who have been meaningful along life's journey.

Rationality in the sense of decisional, moral, analytical or goal-setting capacities are too narrow and severe a ground for moral standing. The fitting moral response to people with dementia, according to classical Western ethical thought and related conceptions of common human decency, is to enlarge our sense of human worth to counter an exclusionary emphasis on rationality, efficient use of time and energy, ability to control distracting impulses, thrift, economic success, self-reliance, self-control, 'language advantage,' and the like (Sabat 2001). The perils of forgetfulness are especially evident in our culture of independence and economic productivity, that so values intellect, memory and self-control. Emotional, relational, aesthetic, creative, olfactory, spiritual and symbolic wellbeing are possible for people with progressive dementia.

In general, quality of life is a self-fulfilling prophecy. If those around the person with dementia see the glass as half empty and make no efforts to relate to the person in ways that enhance his or her experience, then quality of life will be abysmal.

A Pastoral Conclusion: Enduring Selfhood and the Universal Value of Consciousness

Whether you believe in an eternal dimension to biographical selfhood in some metaphysical sense of Mind as independent of Matter, or you believe that mind and memory are all merely derived from brain tissue, we can agree on this: when it comes to the deeply forgetful, love is the question, love is the answer, and love is the way—even in the hard times when caring feels overwhelming and perhaps for the moment even a bit meaningless, although it is always meaningful. It is all about the power of love, not about the love of hypercognitive power. After all, we can hopefully acknowledge that in an era of heightened sensitivity to the equal moral status of people with physical and cognitive disabilities, we should not dismiss the consciousness and awareness of an individual with dementia as somehow less significant than that of someone who is more lucid of mind.

In his classic work, *Dementia Reconsidered: The Person Comes First* (1997), Tom Kitwood's definition of love within the context of dementia care includes comfort in the original sense of tenderness, closeness, the calming of anxiety, and bonding. Kitwood defined the main psychological needs of persons with dementia in terms of care or love. He drew on the narratives of caregivers to assert that persons with dementia want love, 'a generous, forgiving and unconditional acceptance, a wholehearted emotional giving, without any expectation of direct reward'. The first component of love is comfort, which includes tenderness, calming of anxiety, and feelings of security based on affective closeness. It is especially important for the person with dementia who retains a sense of his or her lost capacities. Attachment, the second component of love, includes the formation of specific bonds that enhance a feeling of security. Inclusion in social experiences, occupation in activities that draw on a person's abilities and powers, and, finally, identity are important components of love. I can to this day recall my conversations with Tom both while visiting him at his University of Bradford Dementia Care Centre, or in Cleveland at Heather Hill just a week before he died.

But let us also conclude with some degree of appreciation for Forsdyke's boldness (2009), for he has the audacity to suggest that just maybe 'the brain [is] as a receptor/transmitter of some form of electromagnetic wave/particle…of course, when speaking of extracorporeal memory we enter the domain of 'mind' or 'spirit' with corresponding metaphysical implications' (2015). This or something like it may yet turn out to be true.

> But I wish to go one step further. I conclude with an affirmation that no matter how limited rational capacity may be in any human being, and no matter how barely perceptible their enduring self-identity may be, at a deeper metaphysical level the following maxim can be defended: so long as they are not brain dead, they retain consciousness, and no one human being should ever have the audacity to assume that his or her consciousness is somehow more valuable than that of another, regardless of rational capacity. Consciousness is the basis of human dignity, and this holds even if we wrongly insist that grandma is no longer biographically there.

'Is Grandma still there?' On a pastoral note for carers who believe in the eternal soul and selfhood of your loved one, do not give up your intuitions because they may be true. Your hopes have not been falsified, and phenomena such as 'terminal lucidity' and even the near-death experience may point to an 'eternal return' to the light and warmth of a pure unlimited love. Open-mindedness follows wherever good science leads, and does not dismiss either materialist or postmaterialist interpretive frameworks.

Acknowledgements This chapter, in draft, was presented on 10 December 2015, at the National Institute for Advanced Studies, Bangalore, in the international conference on *Consciousness, Cognition, and Culture*: *Implications for the 21st Century*. Thanks to Dr Sangeetha Menon for her conference leadership and hospitality.

I wish to acknowledge the great Roman Catholic neurologist Joseph M. Foley, M.D. (1916–2012) with whom I had countless conversations at Case Western Reserve University about mind, memory, and brain, and who himself thought that materialism was an unnecessary assumption.

I wish also to acknowledge several conversations with Nobel Prize Laureate Sir John Eccles (1903–97). Finally, I must acknowledge Sir John Templeton (1912–2008), who spoke with me consistently about the distinctions between body, mind and spirit, and who supported several focus group pilot studies of spirituality in the deeply forgetful as perceived by carers through a personal grant to my colleague Peter Whitehouse, M.D. Ph.D. and myself (2001).

References

Bergson, H. (1896 original; 1994). *Matter and memory: An essay on the relation of body and spirit*. New York: Zone.

Bergson, H. (1932). *The two sources of morality and religion*. New York: Double Day.

Berkovich, S. Y. (1993). On the information processing capabilities of the brain: Shifting the paradigm. *Nanobiology, 2,* 99–107.

Brady, T. F., Konkle, T., Alvarez, G. A., & Oliva, A. (2008). Visual long-term memory has a massive storage capacity for object details. *Proceedings of the National Academy of Sciences, 105,* 14325–14329.

Brown, W. S., Murphy, N., & Malony, H. M. (1998). *Whatever happened to the soul: Scientific and theological portraits of human nature*. Minneapolis, MN: Fortress Press.

de Oliveira, M. F., Pinto, F. C. G., Nishikuni, K., Botelho, R. V., Lima, A. M., & Rotta, J. M. (2012). Revisiting hydrocephalus as a model to study brain resilience. *Frontiers in Human Neuroscience, 5,* 181.

Feuillet, L., Dufour, H., & Pelletier, J. (2007). Brain of a white-collar worker. *Lancet, 370,* 362.

Forsdyke, D. R. (2009). Samuel Butler and human long-term memory: Is the cupboard bare? *Journal of Theoretical Biology, 258*(1), 156–164.

Forsdyke, D. R. (2015). Wittgenstein's certainty is uncertain: Brain scans of cured Hydrocephalics challenge cherished assumptions. *Biological Theory, 10,* 336–342.

Fusi, S., & Abbott, D. R. (2007). Limits on memory storage capacity of bounded synapsis. *Nature Neuroscience, 10,* 485–492.

Hoblitzelle, O. (2008). *Ten thousand joys & ten thousand sorrows: A couple's journey through Alzheimer's*. New York: Penguin.

Kitwood, T. (1997). *Dementia reconsidered: The person comes first*. Philadelphia, PA: Open University Press.

Lewin, R. (1980). Is your brain really necessary? *Science, 210,* 1232–1234.

Nagel, T. (2012). *Mind and cosmos: Why the materialist Neo-Darwinian conception of nature is almost certainly false*. New York: Oxford University Press.

Niebuhr, R. (1956). *An interpretation of Christian ethics*. New York: Meridian.

Post, S. G. (1995). *The moral challenge of Alzheimer disease* (1st ed.). Baltimore, MD: Johns Hopkins University Press.

Post, S. G. (2000). *The moral challenge of Alzheimer disease: Ethical issues from diagnosis to dying* (2nd ed.). Baltimore, MD: Johns Hopkins University Press.

Post, S. G. (2013). Hope in caring for the deeply forgetful: Enduring selfhood and being open to surprises. *The Bulletin of the Menninger Clinic: A Journal for the Mental Health Professions, 77*(4), 349–368.

Rupert, S. (2012). *Science set free: 10 paths to new discovery*. New York: Deepak Chopra Books.
Sabat, S. (2001). *The experience of Alzheimer's disease: Life through a tangled veil*. Oxford: Blackwell.
Schrodinger, E. (1969). What is life? In *Mind and Matter*. Cambridge University Press (1956 original).
Singer, P. (1993). *Practical ethics* (2nd ed.). Cambridge: Cambridge University Press.
Talbot, M. (1991). *The holographic universe*. New York: Harper Perennial.
Zagzebski, L. (2001). The uniqueness of persons. *Journal of Religious Ethics, 29*(3), 401–423.

Children with Autism Spectrum Disorders: A Case for 'Alternative Selves'?

Prathibha Karanth

Introduction

Intriguing insights in the area of mind, brain and consciousness have emerged from the neurological literature on adult patients who suffer from relatively common disorders, such as strokes, head injuries and dementias, as well as the rare neurological disorders, such as encephalitis lethargica or Cotard's syndrome. These disorders range from the relatively more straightforward language disorders known as the aphasias to highly complex disorders, such as the denial of the live self in conditions such as Cotard's syndrome. For instance, while in Wernicke's aphasia, the individual is no longer able to make sense of his own language when spoken to and is unaware that he often produces garbled jargon while speaking it, in instances of unilateral neglect the patient no longer recognises parts of his body as being that of his own and finally in even more complex disorders, such as Cotard's syndrome, the patient denies his own existence (Ananthaswamy 2015).

Neurological illness alters many perceptive and cognitive skills leading to myriad complex issues in the daily life of these individuals and their interactions with the external world and others in it, including their notion of self and consciousness. The experiences and difficulties of these individuals raise a whole range of questions on the role of the brain in our notion of self, including the body and mind of the self vis-à-vis the external world and the others in it (Sacks 1985, 1995).

Little is known as yet about the possibility of similar issues being present in or underlying complex developmental disorders such as the autism spectrum disorders (ASD). Several recent autobiographical accounts of young adults with ASD (Grandin 1986; Mukhopadhyay 2000; Fleischmann and Fleischmann 2012) suggest that similar issues are indeed faced by children diagnosed with ASD. Autism spectrum disorders represent a condition in which children appear completely

P. Karanth (✉)
The Com DEALL Trust, Bengaluru, India
e-mail: prathibha.karanth@gmail.com

S. Menon et al. (eds.), *Self, Culture and Consciousness*,
DOI 10.1007/978-981-10-5777-9_7

normal at birth and in early infancy but then go on, by the age of 18 to 36 months, to either lose developmental skills that were already learnt, such as speech, and/or seemingly develop a lack of interest in interacting with others around them, including their parents, along with a preoccupation with bizarre repetitive motor movements, such as hand flapping. Autism, considered a rare childhood psychosis with an incidence of 1 in 10,000 about five decades ago, is now being viewed as autism spectrum disorders—complex neurodevelopmental disorders with an incidence of 1 in 68 (Centers for Disease Control and Prevention USA 2014). This steep rise in incidence and the concomitant high prevalence adds urgency to understanding ASD, in all its complexities. The narratives of some individuals diagnosed with ASD, who become aware of the differences between them and the neurotypicals around them and also find a means of communicating the same, as in the instances referred to above; are replete with descriptions of complex and myriad sensory—motor issues faced by the child with ASD that are at variance with those of the neurotypical child.

The key to the issues that children with ASD are faced with, is perhaps to be found in the manner in which their bodies function. These complex and myriad sensory—motor issues faced by the child with ASD could well lead to a notion of 'self and consciousness' that is at variance with the typical. Such an 'atypical' notion of the self and consciousness could not only account for the deficient or atypical social component of the mind of the child with ASD—now considered the key component of autism spectrum disorders, but also account for the many other 'characteristics of autism'—physical and behavioural, such as the 'avoidance of eye contact' and the repetitive motor behaviour, such as flapping, which are listed among the key diagnostic symptoms of children with ASD. But first let us take a closer look at 'autism' and 'autism spectrum disorders'.

Background of ASD

Autism, earlier considered to be a single entity, is now seen as a group of developmental disorders that includes a wide range or 'spectrum' of symptoms and skills. Hence the term autism spectrum disorders (ASD), rather than 'autism' is now preferred. It is currently among the most studied developmental disorder. Current diagnostic criteria list the social difficulties, including difficulty in communicating and interacting with others and repetitive behaviours with limited interests and activities, as being the key symptoms of ASD. These symptoms are generally recognisable before the age of two years and are seen to affect the child/individual's ability to function socially in all areas of life—family, school and work. The severity of the symptoms can vary from mild to severe and can be helped by intervention.

Current diagnosis (DSM V 2016) classification emphasises two main groups of symptoms—difficulties in social communication and social interaction, and restricted/repetitive behaviours. Difficulties in social communication and interaction are evidenced by poor eye contact, lack of shared interest in objects and play, distress with even slight changes in the environment and routines, lack of or unusual responses to display of emotions by others, seeming indifference to being called by and conversations of others, delay in speech-language acquisition, unusual communication behaviour, such as repeating words and phrases heard earlier, in seemingly odd and irrelevant contexts, a flat or unusual tone of voice and dissonance between speech and facial expressions or movements and gestures accompanying speech. When verbal, individuals with ASD also exhibit marked difficulties in participating in the to-and-fro of conversations, often talking at length on topics of their own interest, in monologues, without taking the listener's interest or lack of it into account, and without waiting for or being interested in the listener's response. The latter group of symptoms includes narrowly focused interests (as in specific objects or parts of objects—for example, the wheels of toy cars or railway timetables) and repetitive movements, such as flapping of hands (APA 2013).

There is a host of other more basic difficulties that, children (and adults?) with ASD face that have only recently gained attention but are increasingly being documented by those with ASD. These relate to both sensory perception and motor behaviour. They are often more sensitive to light, noise, texture of clothing and food, touch and smell, than their neurotypical peers—a phenomenon collectively referred to as sensory processing disorder and sensory integration. Like the sensory issues the motor difficulties of children with ASD are only now being detected. While for all appearances, children with ASD appear to be perfectly normal or typical (at least in early childhood), on closer examination many of them exhibit difficulties in coordinated motor functions, a phenomenon now identified as motor executive dysfunction. In addition, some of them may also have difficulty in sleeping and digestion.

Paradoxically, several children with ASD have above average intelligence with several excelling in certain skills, such as showing strong visual memory (often years after the event) and, in some instances, leading to a phenomenon called hyperlexia or the ability to learn to read on one's own, without being taught—at times even when the child is essentially non-verbal or without speech. Quite a few children with ASD also exhibit acute music perception while not being able to speak. Many individuals with ASD have above average intelligence and excel in math, science, art and music.

Given that the incidence of ASD has increased steeply over the last two decades with no sign of it abating, there has been and there is considerable ongoing research into the causation of ASD. However, there is still no agreement on the causation of ASD. There is, however, a broad consensus that the basis of the disorder is biological and not psychological as was presumed earlier. While the exact causes of ASD remains unknown, research suggests that genes and environment play important roles. Genes that may increase the risk for ASD are being identified. Current research is focusing on how genes interact with each other and with environmental factors.

With the growing incidence, there have also been changes in our understanding and theoretical perspectives on ASD. For instance, it is clearly documented that if identified early and when provided with comprehensive early intervention, positive long-term impact on the wellbeing of the child can be brought (Lord and McGee 2001; Dawson et al. 2010), even to the extent of certifying children as having 'moved out of the spectrum'.

Another significant change that has been evidenced is the move from seeing ASD as a purely behavioural disorder. Among the more popular recent theories of ASD, many have focused on the social component of the disorder proposing that the core difficulty lies in the social domain. This position is perhaps best represented by the proponents of the 'Theory of Mind' (TOM) who propose that autism stems from the inability of the child or individual with ASD, 'in reading the minds of others' (Baron-Cohen et al. 1985).

While theories of and diagnostic frameworks for the autism spectrum disorders are now heavily influenced by the theories of social cognition and focused on social skills or the lack of them, early intervention which has provided the most sustained long-term impact on the wellbeing of children with ASD (Lord and McGee 2001) addresses a wider range of issues in these children, including the sensory—motor and communication issues—with input from a range of professionals addressing the body, sensory and motor skills, communication and early developmental skills as a whole, including feeding and toileting. The Communication DEALL programme initiated by us is an example (see Karanth et al. 2010; Karanth and Chandhok 2013).

Theories on ASD, however, continue to be focused on the behavioural and increasingly social deficits of the child/individual with ASD. Research supporting the latter comes from older and more capable children with ASD leaving much of the population of moderate and severe ASD unaccounted for. It is also a fact that the subjects of experimental studies in support of the social deficit theories of autism are the more capable verbal children with ASD, given the difficulties of enrolling the moderately and severely affected children with ASD, including the substantial portion of children with ASD who are non-verbal.

Current diagnostic frameworks, such as the DSM V, recognise the restricted/repetitive behaviours of individuals with ASD as a major component of the symptoms of ASD. In fact, this group of characteristics is the only feature that separates ASD from a communication disorder, labeled 'social communication disorder'. It is curious, though, that these restricted and narrowly focused interests and repetitive behaviours such as hand flapping, which is a marked sign of many children with ASD, are seldom investigated and explored. Nor are they accounted for in the 'social theories of ASD'. Theories of ASD need to account for the restricted and repetitive behaviour, including the motor aspects, and the more severe communication disorders, including the total inability to verbalise. While the issues in early identification, assessment and experimental study of these aspects in children with ASD are undoubtedly complex, current technology as is being used in the baby experiments will provide better understanding of these more basic differences/difficulties/deficits underlying the behaviour and lack of social competence in individuals with ASD.

A good place to start exploration of these more severe and debilitating issues faced by children with ASD would be the autobiographies of more severely affected children/individuals with ASD who 'broke through their autism' and moved on to finding a way to communicate in adolescence and early adulthood, through other non-verbal modes of communication such as letter pointing or computer keyboards.

Fortunately, there have been quite a few such narratives in the recent past that are providing a window into the inner world of those with ASD. Eventually, a convergence of these insights combined with the *multimodal studies* of babies (Geddes 2015) and young children with ASD will lead to a better understanding of the condition as well as the behavioural and social aspects that we seem to unduly focus on in the current times. Even more importantly it will contribute to an appreciation of what it means to grow up with these conditions and to provide the much needed assistance and intervention to children with ASD.

The thrust of this chapter, though, is at a closer look at these autobiographies of those with ASD, in an effort to understand the impact of these subtle but myriad differences in sensory processing and motor control and what this holds for the development of the notion of self and consciousness in individuals diagnosed with ASD. While similar issues are seen in adults with acquired neurogenic disorders and there is a growing awareness of the impact it has on altering their notions of the self and their consciousness, we are as yet largely insensitive to the impact of such differences on individuals with ASD, including the impact on their notions of self and consciousness.

Clinical Insights and Autobiographical Excerpts from Individuals with ASD

The clinical insights into the complex issues faced by children with ASD were in my case the offshoot of over two decades of experience with adult patients with a range of neurological disorders and symptoms—aphasias, alexias and agraphias—but also the agnosias—auditory, visual, tactile, anosognosia, prosopagnosia—and the apraxias—ideomotor, oral and verbal, as well as other symptoms, such as unilateral neglect; followed by a decade-long (1993–2001) clinical involvement with Tito Rajarshi Mukhopadhyay, a child diagnosed with autism, which is when I first became aware of the similarity of experiences that I had seen in adult patients, post-neurogenic insult, with those of this child with autism. Tito was later described by Sacks (TRM 2008), as 'a gifted autistic child from India, who has attracted considerable attention among the general public and professionals and parents concerned with autism'. TRM's unique attraction, according to Sacks, was his ability to communicate about his autism, thereby forcing us '**to reconsider the condition of the deeply autistic**'.

It was perhaps Temple Grandin, now a well-known advocate for those with Autism Spectrum disorders who drew attention to what it was that individuals diagnosed with autism experienced (the hypersensitivity to sound levels often seen in those with ASD, 'felt like being tied to the rail and the train's coming'—see *Anthropologist on Mars* by Sacks (1995) and the many publications from Grandin on autism listed in the references) and the constant anxiety and feeling of being threatened by everything in the surroundings. In the recent past the number of autobiographical narratives from those diagnosed with ASD, such as Grandin (1986, 1995), Mukhopadhyay (2000, 2003, 2008) and more recently Fleischmann (in Fleischmann and Fleischmann 2012) among a host of others, have increasingly sensitised the general public about the many issues that individuals diagnosed with autism face, that result in the many 'symptoms of ASD' which are listed as diagnostic characteristics of autism spectrum disorders. We will now explore some of these issues as described in the autobiographical accounts of those with ASD as well as those that we have witnessed in the many children diagnosed with ASD who receive intervention at the Com DEALL Trust, in order to try and understand their impact on the individual growing up with autism spectrum disorders and the larger issues of the notion of the self and consciousness in them.

Sensory Processing

The sensory issues that individuals with ASD are now known to experience are many and are spread across all of the senses, (though individuals vary in the particular senses affected and severity) contributing both to their difficulties and to the splinter skills that several of them exhibit. Among the senses the auditory system and auditory sensitivity is among those that have been observed the earliest and therefore is well-documented. That children with autism respond to the softest sounds, such as that of the rustle of paper or softly played music and jingles, while ignoring loud noises, including their name being called loudly, has been recorded even from the time autism was first described decades ago. Selective listening and strengths in music perception are also linked to their high auditory sensitivity. At the same time, this very sensitivity could make speech perception in real time extremely hard. For instance, Carly Fleischman (Fleischmann and Fleischmann 2012) has this to say: 'We take in over a hundred sounds a minute. We have a hard time processing all the sounds at once so it comes out later as a broken record.' Paradoxical as it seems, she goes on to provide an example of her making a humming sound while eating chips—'I was making noise and changing the sound with my finger in my ear to block out audio input from the crunching of the chip.'

The visual sensitivities that have been documented in children with ASD are many and varied. Avoidance of eye contact is a hallmark of children with autism and the recent self-reports of individuals ascribe this to extraordinary visual sensitivity—'I see parts of the face moving and find it difficult to focus'—TRM (personal communication). Similar thoughts are echoed in Carly's statement:

'We take pictures in our heads like a camera. It's like filling a camera with too many pictures. It gets overwhelming' (Fleischmann and Fleischmann 2012). Yet this visual hypersensitivity is not without its advantages. Many individuals with ASD are known for their extraordinary visual memory of routes, maps, dates, timetables and the like. Grandin, who describes herself as a visual thinker, describes her immaculate memory for past events by comparing it to full-length movies or videos stored in her head, that may be replayed at will, allowing her to notice small details. The phenomenon of 'hyperlexia' defined as the acquisition of the ability to read without being explicitly taught to read, often seen in non-verbal individuals with autism, is yet another illustration of a seeming advantage of a hyperacute visual system—'reading at a glance and comprehending', in Carly's words. Yet again, this extraordinarily detailed visual perception with its concomitant attention to details, could and does interfere in other kinds of visual perception, such as deciphering figure-ground relations, and leads to difficulty in holistic or gestalt perception which is so essential for everyday human life.

Tactile sensitivities are a third major area of challenges faced by children and adults with ASD. Sensitivities to clothes are often manifested as not wanting any clothes or wanting to wear specific clothes, repeatedly. Resistance to touch could be evinced as refusal to remove shoes and resistance to being touched or hugged. Sensitivity to textures leads to feeding issues and avoidance of a range of food. While there are hardly any reports of similar issues with taste, it is well established that many children with ASD are picky and fussy eaters, with an insistence on sameness of food and an avoidance of any new food item. At the same time, many of them exhibit a heightened olfactory sense and have a tendency not only to sniff at food and objects but even people.

It is a common adage in the world of autism that 'if you have seen one child with autism, you have seen one child with autism', the reference being to the very different pictures that children with autism present, each distinct from the other. Such seeming variance is probably the result of the combination of one or more sensitivities in each child, that in turn could vary in severity from time to time, depending on external environmental factors. As to the individual who has to live with it, in Carly's words: 'Autism feels hard. It's like being in a room with the stereo on full blast. It feels like my legs are on fire and over a million ants are climbing up my arms', followed by a desperate plea: 'I want not to feel what's happening in my body' (Fleischmann and Fleischmann 2012).

An added layer of complexity is brought to these sensory issues by the phenomena of synaesthesia that many individuals with ASD seem to experience, often without being aware that this is unique to them. Synaesthesia is described as the intertwining of senses—the tasting of colour, seeing voices and smelling sound, and is beautifully illustrated in this poem of Tito's.

Voices of colors and

And voices of shadows

Voices of movement

And their echoes
Voices of silence
Spun near or far
Through spaces and distance
Soaking my ears

(Mukhopadhyay 2008)

At times certain senses are undifferentiated, so much so that in grapheme or letter to colour synaesthesia the individual will automatically and consistently determine what colour certain letters are. Children who grow up with these types of synaesthesia are often unaware that their experiences are irregular or different from the norm.

The following narrative from Gunilla Gerland (Frith and Happé 1999) illustrates the linking of the tactile and auditory sensations: 'If I was made to touch jewelry, I felt a sharp whistling metallic noise in my ears, and my stomach turned over. Like a note falsely electrified, that sound would creep from the base of my spine upwards until it rang in my ears, tumbled down into my throat and settled like nausea into my stomach.'

Motor Functioning

While the sensory issues faced by individuals with ASD have been increasingly documented over the last two decades, the understanding of the motor issues faced by them are only now beginning to be recognised. Since the late 1990s, compelling arguments for the presence/existence of disorders of executive function that is responsible for converting thought into action, in individuals with autism have been put forth as being primary to the disorder (James 1997). Motor executive dysfunctions or the inability to, on demand, perform even simple actions that are routinely executed by these very individuals spontaneously are often seen in individuals with autism and are often interpreted as wilful non-cooperation. Yet, close examination of the basic motor skills—gross and fine, in these perfect looking children, reveal many a gap in their motor skills often extending to the oromotor skills so important for vegetative and non-vegetative functions like speech (Belmonte et al. 2013).

Here too, the autobiographies of individuals with autism are replete with narrations/instances of their difficulties with their body. Temple Grandin's 'hug machine' was designed by her to provide her body with the effects of deep pressure stimulation that calmed her down and eased tension. Similar effects have been reported through wrapping in blankets and the use of deep pressure and weighted vests by occupational therapists.

On page 20 of his autobiography *Beyond the Silence* (TRM 2000) Tito writes about 'his scattered body': 'He felt that his body was scattered and it was difficult to collect it together. He saw himself as a hand or as a leg and would turn around to assemble his parts to the whole.' 'He got the idea of spinning from the fan as he saw that its blades that were otherwise separated joined together to a complete circle when they turned in speed.'

Tito is not alone in this preoccupation with the need to feel his body as a whole. A remarkable characteristic shared by many children with autism is a preoccupation with outlining their hands, feet and body either by gazing at mirrors and shadows or even physically outlining parts of or the whole body, a feature now an integral part of treatment regimes of many occupational therapists. Here again is a symptom well-documented in adult neurological disorders in which post-trauma (strokes, head injuries) patients have been documented to either neglect parts of their body or even explicitly deny that a part of their body, such as a paralysed limb, belonged to their body.

Behavioural Oddities in ASD—Asocial Behaviour or Coping Mechanisms?

This brings me to the issue of the several behavioural oddities of children with autism. Characteristics such as poor eye contact, distress with even slight changes in the environment and routines, the insistence for sameness, clinging to routines and repetitive movements, such as flapping of hands, are repeatedly documented as being present in children with autism, so much so that they are listed among the hallmarks for diagnosis. Yet the underlying reasons or causes for these are seldom explored. Could these be the coping mechanisms that those with ASD develop/arrive at in order to cope with their particular difficulties. Tito, for instance, explained that flapping his hands was one way to get the sensation of the extremities of his body—making perfect coping sense, given his 'scattered body'. Providing physical boundaries, such as 'therapy mats', and increasing pressure on the body, as in the use of weighted vests or tight rolling in blankets as in the practice of occupational therapy, have been known to help calm children with ASD, as is the case with Grandin's hug machine. Avoidance of eye contact, while explained by Tito as being due to a difficulty in focusing 'on the many moving parts of faces', was reported by Carly as being due to 'seeing too many faces like photographs' at the same time. Could their 'insistence on sameness' and 'extreme distress with even slight changes in the environment' also be attributed to coping mechanisms? An interference in the same might be 'slight' to the neurotypical adult but a major catastrophe for someone with autism.

It is curious that while these characteristics or 'symptoms' have been repeatedly documented and form part of the diagnostic package for ASD, they are seldom accounted for by the theories of autism. The most popular current theory of autism spectrum disorders is that of autism as a social deficit with 'ASD … construed as an extreme case of diminished social motivation' (Chevallier et al. 2012). The difficulties in social interaction and social communication that are seen in children with

ASD are attributed to 'diminished social motivation'. They could, however, be equally well or even better explained by the more basic issues faced by the child with ASD. Take, for instance, the case of 'S' (CDCT: 12055), a young boy of 3½ years, referred to us with a diagnosis of ASD. One of the chief complaints of the family was that he was unable to play interactive games with peers, with a preference to sit in a corner instead of a group, and that he not only resisted being hugged, but would also scream at the top of his voice if someone touched him, particularly if he was not anticipating the touch. He was also reported to engage in socially inappropriate behaviour, such as taking off his clothes, including his underpants, in public and at the same time refusing to take his shoes off. On examining his sensory profile there were clear signs of tactile defensiveness which, when addressed, led to a remarkable reduction of his 'asocial behaviour'.

It is important to note, that unlike the adults with acquired neurogenic disorders where a prolonged period of 'normalcy' is being followed by bizarre behaviour consequent to neural insult, with the latter clearly being attributed to the former; in the case of children with ASD, given their normal appearance and the reportedly normal development in infancy, the onus for all of the traits of ASD is placed on the child. It could on the other hand, well be that the child faces a double whammy with the reference point for 'normal behaviour', that is the shared experiences of children and adults, being completely amiss in his case, given that the experiences of the body and the world, by the child with ASD is considerably different from that experienced by the adults around him, that nurture him with a preconceived set of expectations. For instance, it is well-documented that several individuals with synaesthesia may not interpret it as irregular or different from the norm. And synaesthesia is but one of the myriad issues that can affect a child with ASD. In fact, these can go right down to the most basic issues of toileting and feeding, as evidenced in early intervention programmes and as illustrated in the following narrative from Gunilla Gerland (Frith and Happé 1999). 'I also had another problem … **though it was a problem I never understood until I was an adult. I thought it was the same for everyone**. I couldn't feel that I needed to go to the lavatory, so **I had to think out when I needed to go**. I didn't know other people had a signalling system that warned them at intervals before the need to go became urgent. I had no such system. I felt nothing, nothing …'

The impact of the lack of awareness and sensitivity to these issues in a child growing with ASD, compounded by the adult's ignorance of the status of the child could well be at the root of the many behavioural complexities and social oddities of the individual with ASD. Here, then, are individuals not with 'altered selves' but with 'alternative selves'. Unlike the adult neurological patients whose 'normal' or 'typical' selves are altered consequent to neurological insult, **children with ASD possibly grow up experiencing the world around them in ways that are unknown and alien to us. What then of their notion of self and consciousness and our expectations of them, in terms of our notions of self and consciousness?**

There is now increasing evidence from neuroimaging studies that suggest abnormal functional connectivity and impaired mirror neuron systems leading to unintegrated sense of self in autism (Lyons and Fitzgerald 2013) as also altered self-representation, or intrapersonal cognition in ASD (Uddin 2011). The current neuroscientific work, known as the baby studies (Elsabbagh and Johnson 2010) will undoubtedly shed more light on the above. Until then for those that deal with children with ASD it is crucial to be aware of these alternative narratives emerging from those with ASD. It is fitting that I end this chapter with a quote from Tito, the little boy with ASD, who snared me into the world of autism.

Maybe it is night
Maybe it is day
I can't be sure
Because I'm not yet feeling the heat of the sun
I am **the mind tree**
When I had been gifted this mind of mine
I recall his voice very clearly
To you I have given this mind
And you shall be the only kind
No one ever will like you be
And I name you the mind tree
I can't see or talk
Yet I can imagine
I can hope and I can expect
I am able to feel pain but I cannot cry
So I just be and wait for the pain to subside
I can do nothing but wait
My concerns and worries
Are trapped within me somewhere in my depths
Maybe in my roots
Maybe in my bark
When he comes next who gifted me my mind
I shall ask him for the gift of sight
I doubt his return and
Yet hope for it
Maybe he will
Maybe he will not

(Tito Rajrshi Mukhopadhyay, *Mind Tree*, 2003, pp. 169–170)

References

Ananthaswamy, A. (2015). *The man who wasn't there*. New York: Penguin Random House.

APA (American Psychiatric Association). (2013). *Diagnostic and statistical manual of mental disorders* (5th Ed., DSM-5). Washington, DC: American Psychological Association.

Baron-Cohen, S., Leslie, A. M., & Frith, U. (1985). Does the autistic child have a 'theory of mind'? *Cognition, 21,* 37–46.

Belmonte, M. K., Saxena-Chandhok, T., Cherian, R., Muneer, R., George, L., & Karanth, P. (2013). Oral motor deficits in speech-impaired children with autism. *Frontiers in Integrative Neuroscience, 7,* 47.

CDC (Center for Disease Control and Prevention). (2014). https://www.cdc.gov/.

Chevallier, C., Kohls, G., Troiani, V., Brodkin, E. S., & Schultz, R. T. (2012). The social motivation theory of autism. *Trends in Cognitive Sciences, 16*(4), 231–239.
Dawson, G., Rogers, S., Munson, J., Smith, M., Winter, J., Greenson, J., et al. (2010). Randomized, controlled trial of an intervention for toddlers with autism: The early start denver model. *Pediatrics, 125,* e17–e23.
DSM V. (2016). https://www.nimh.nih.gov/health/topics/autism-spectrum-disorders-asd/index.shtml. Retrieved October 16, 2016.
Elsabbagh, M., & Johnson, M. H. (2010). Getting answers from babies about autism. *Trends in Cognitive Sciences, 14*(2), 81–87.
Fleischmann, A., & Fleischmann, C. (2012). *Carly's voice—Breaking through autism*. New York: Simon and Schuster.
Frith, U., & Happé, F. (1999). Theory of mind and self-consciousness: What is it like to be autistic? *Mind and Language, 14,* 82–89.
Geddes, L. (2015). The baby experiment. *Nature, 527,* 22.
Grandin, T. (1986). *Emergence: Labeled autistic*. Novato, CA: Arena Press.
Grandin, T. (1995). *Thinking in pictures: And other reports from my life with autism*. New York: Doubleday.
James, R. (Ed.). (1997). *Autism as an executive disorder*. New York: Oxford University Press.
Karanth, P., & Chandhok, T. S. (2013). Impact of early intervention on children with autism spectrum disorders as measured by inclusion and retention in mainstream schools. *Indian Journal of Pediatrics, 80,* 911–919.
Karanth, P., Shaista, S., & Srikanth, N. (2010). Efficacy of communication DEALL—An indigenous early intervention program for children with autism spectrum disorders. *Indian Journal of Pediatrics, 77,* 957–962. doi:10.1007/s12098-010-0144-8.
Lord, C., & McGee, J. P. (Eds.). (2001). *National research council, committee on educational interventions for children with autism. Educating children with autism*. Washington, DC: National Academies Press.
Lyons, V., & Fitzgerald, M. (2013). *Atypical sense of self in autism spectrum disorders: A neuro-cognitive perspective*. INTECH Open Access Publisher.
Mukhopadhyay, T. R. (2000). *Beyond the silence: My life, the world and autism*. National Autistic Society.
Mukhopadhyay, T. R. (2003). *The mind tree: A miraculous child breaks the silence of autism*. New York: Arcade Publishing.
Mukhopadhyay, T. R. (2008). *How can I talk if my lips don't move? Inside my autistic mind*. New York: Arcade.
Sacks, O. (1985). *The man who mistook his wife for a hat*. New York: Piccadilly.
Sacks, O. (1995). *An anthropologist on mars: Seven paradoxical tales*. New York: Knopf.
Uddin, L. Q. (2011). The self in autism: An emerging view from neuroimaging. *Neurocase, 17,* 201–208.

Auditory Verbal Hallucinations in Schizophrenia: A Model for Aberrant Self-consciousness

John P. John, Pravesh Parekh, Harsha N. Halahalli, Sangeetha Menon and Bindu M. Kutty

Schizophrenia and Auditory Hallucinations

Schizophrenia is a complex psychiatric condition that is characterised by three major symptom domains: positive symptoms (symptoms that are not exhibited in healthy subjects but are present in patients with schizophrenia), negative symptoms (normally present abilities in healthy subjects that are impaired in patients with schizophrenia), and cognitive symptoms. Positive symptoms prominently include hallucinations, delusions, and disorganisation symptoms including formal thought disorder, inappropriate affect, and bizarre behaviour; negative symptoms include

J.P. John (✉) · P. Parekh · H.N. Halahalli
Multimodal Brain Image Analysis Laboratory (MBIAL), National Institute of Mental Health and Neurosciences (NIMHANS), Bengaluru 560029, India
e-mail: jpj@nimhans.ac.in

J.P. John · P. Parekh
Department of Psychiatry, National Institute of Mental Health and Neurosciences (NIMHANS), Bengaluru 560029, India

J.P. John
Department of Clinical Neurosciences, National Institute of Mental Health and Neurosciences (NIMHANS), Bengaluru 560029, India

H.N. Halahalli · B.M. Kutty
Department of Neurophysiology, National Institute of Mental Health and Neurosciences (NIMHANS), Bengaluru 560029, India

J.P. John · S. Menon · B.M. Kutty
NIAS Consciousness Studies Programme,
National Institute of Advanced Studies, Bengaluru 560012, India

Present Address:
H.N. Halahalli
Department of Physiology, KS Hegde Medical Academy, Nitte University, Mangalore 575018, India

S. Menon et al. (eds.), *Self, Culture and Consciousness*,
DOI 10.1007/978-981-10-5777-9_8

features like affective flattening, alogia (reduced or absent verbal output), avolition (reduced initiative to perform goal-directed actions), anhedonia (inability to derive pleasure from otherwise pleasurable activities), and general social withdrawal. Cognitive dysfunctions in schizophrenia include attentional and memory deficits and impaired executive functions. Other clinical features may include emotional dysregulation (e.g., depression, anxiety), a sense of hopelessness, demoralisation, suicidal tendencies, and lack of insight. These dimensions of symptoms, seen in varying extents of severity, have a significant impact on the social and occupational wellbeing of the patients and affect work, interpersonal relationships and self-care.

Schizophrenia was first described as *dementia praecox* by Kraeplin in 1887 (though the term was coined earlier) which referred to an early onset of deterioration of intellect. Eugen Bleuler, however, objected to the term, and proposed the term *schizophrenia* (meaning split mind). He conceptualised schizophrenia as a splitting and dissociation of the entire personality. He also proposed that schizophrenia is not merely a single disease but a group of closely related illnesses (using the phrase 'group of schizophrenias'). Over years, there have been changes in the criteria employed to make a diagnosis of schizophrenia with the current standard being the DSM-V (American Psychiatric Association 2013) and ICD-10 (World Health Organization n.d.) (ICD-11 beta version is available online and the final version is expected to be released in 2018).

Schizophrenia has a lifetime prevalence between 0.5 and 1.6% in the general population (Jablensky 1995) with the force of morbidity peaking in young adulthood (Messias et al. 2007). Schizophrenia causes a high degree of disability accounting for 1.1% of the total disability-adjusted life years (DALYs), 2.8% of years lived with disability (YLDs) (Murray and Lopez 1996, 1997), and has been listed as the eighth leading cause of DALYs worldwide in the age-group of 15–44 (WHO 2001) (see (Messias et al. 2007; Rössler et al. 2005) for a detailed discussion on the epidemiology and size of burden of schizophrenia).

Auditory verbal hallucinations (AVH) constitute one of the most characteristic symptoms of schizophrenia and are experienced by around 75% of patients with the illness (Nayani and David 1996). Hallucinations are defined as perception in a single sensory modality or multiple modalities in the absence of a real sensory stimulus. In principle, they can be experienced in any sensory modality, though auditory hallucination is, by far, most commonly reported in schizophrenia. Auditory verbal hallucinations are typically perceived as continuous stream of conversations, often referring to the patient in third person and/or a running commentary on the actions being performed by the patient. The perceived voices are usually negative in their content and cause significant distress to the patient.

Two prominent theories attempt to explain the neurobiology of hallucinations: the bottom-up postulation, which refers to spontaneous aberrant auditory activity, and the top-down theories which include the theories of self-monitoring. The bottom-up theory postulates the role of impairment in the sensory modalities in hallucinations. This *spontaneous activity account* presents the relevant substrates of hallucinations as "the activation of specific auditory representations of voices whether in imagination or memory recall" (Cho and Wu 2013). In a classical example, Penfield and Perot (1963) showed that stimulation along the temporal lobes resulted in auditory hallucinations. As discussed in Cho and Wu (2013), the auditory system has representations of previously heard voices and their acoustic properties and these can be spontaneously activated, resulting in hallucinations. On the other hand, the top-down theories refer to an improper mechanism of self-monitoring. Detection of whether an action is the result of self-generated action or via external agency relies on a comparison between the expected outcome of the action and the actual outcome of the same via a mechanism called the 'efference copy mechanism' (von Holst and Mittelstaedt 1950) or the 'corollary discharge mechanism' (Sperry1950). Though originally proposed for motor actions, the concept has been subsequently extended to thoughts (Feinberg 1978). A breakdown in this monitoring of thoughts would lead to confusion about the source of thoughts and would end up being interpreted as 'alien' rather than self-generated (Frith 1987; Frith and Done 1988). In the recent years, the top-down theory has received more attention though it by no means discounts the bottom-up theory (for a detailed discussion comparing and contrasting these theories see Cho and Wu 2013).

One of the theories that explains the symptoms of schizophrenia is the dysconnection hypothesis (Friston and Frith 1995) which suggests that schizophrenia is characterised by dysfunctional integration of functional brain regions. The brain functions on two arguably contradictory principles: *functional segregation* and *functional integration*. Functional segregation refers to the idea that specialised processing occurs in specific areas of the brain; functional integration refers to the idea that different areas of the brain interact and work together to perform a task. The brain strikes a balance between the two to perform various functions.

The study of hallucinations can not only enhance our understanding of the neurobiology of schizophrenia but also provide a window to understanding aspects of consciousness and the self, as well as its aberrations. In order to further develop this argument, we introduce the model of consciousness and self-consciousness proposed by Kircher and Leube (2003). These authors use the term *primary experiences* to relate to the "subjective, prereflexive givenness of any experience that nobody else can have in my particular form", also variously termed as *raw feels, qualia* and *first-person perspective*. The phenomenal states associated with these primary experiences have three properties: (a) transparency (the idea that the brain constructs our reality but that the mechanism by which it does so is not made available to us); (b) presence (i.e. the conscious experience has to be in the focus of our attention); and (c) myness (the notion that experiences are always and only experiences of an 'I'). Further, they introduce a category of primary experiences, *primary self-experiences*, which are further characterised by:

> (1) self-agency, the sense of the authorship of ones actions; (2) self-coherence; the sense of being a physical whole with boundaries; (3) self-affectivity; experiencing affect correlated with other experiences of self; (4) self-history (autobiographical memory) a sense of enduring over time. (Kircher and Leube 2003)

Self-agency can be studied using robust scientific principles in health and brain disorders. Several symptoms of schizophrenia, specifically the experience of auditory verbal hallucinations, can be examined in the light of self-agency. The typical positive symptoms of schizophrenia, such as 'made phenomena' and 'thought insertion', have been explained as abnormalities in the ownership of one's action and, as a consequence, perceived as resulting from alien influence (for example, Frith and Done 1988). As alluded to before, the idea has been extended to explain hallucinations as arising from an impaired self-monitoring mechanism. Though this perspective does not provide a comprehensive explanation of the various symptoms of schizophrenia, abnormalities in self-agency in schizophrenia has been widely reported (for example, Blakemore et al. 2000; Gawęda et al. 2013; Johns et al. 2001; Knoblich et al. 2004). Auditory verbal hallucination may be considered as an interesting naturally occurring phenomenon to study self-agency. When compared to a healthy subject in a *normal* state of consciousness, a patient with schizophrenia (who is experiencing hallucinations) can be considered to be in an *altered* state of consciousness. Thus, with appropriate design of the experiment, one can probe the neurobiology of hallucinations, the internal self-monitoring mechanism and its possible failure, comparison of a perceived sensory perception versus a real sensory perception, overall comparison of the normal conscious state versus an aberrant conscious state, normal conscious awareness versus impaired awareness, and study the disintegration of the ability of distinction between self and non-self. With advances in neuroimaging and data analyses techniques, it is possible to devise experimental paradigms to investigate the different aspects of AVH alluded to above. In the next section, we attempt to briefly review the available studies in the literature that have examined these aspects and point out various methodological issues which are to be kept in mind while interpreting the reported findings.

Neuroimaging Studies of Auditory Verbal Hallucinations

In this section, we briefly introduce some functional magnetic resonance imaging (fMRI) studies that have been employed to probe the neurobiology of AVH. For sake of brevity, we present a highly condensed view of the results from these findings. First, we review the resting-state fMRI findings using different connectivity approaches (see Table 1a–c). Then we present some results from studies that have attempted to capture AVH using button-pressing or similar paradigms (Table 2). We also provide a list of cognitive paradigms that various research groups have employed to study AVH (Table 3). Each study is marked with superscript alphabets to indicate the study sample group; summarised results of the studies are mentioned in the 'Summary of Results' column while references are cited using square brackets which refer to the serial number in the table. For example, the results from study 1 are referenced using '[1]' in the Summary of Results column.

It should be noted that we have not attempted to provide a systematic review of all the neuroimaging studies in the field. Finally, this section should not be interpreted as a reflection of lack of other kinds of neuroimaging studies in examining

Table 1 Resting-state studies

S. No.	Study	ROI(s)/seed(s)	Summary of results
(a) ROI/Seed based FC			
1	Alonso-Solís et al. (2015)[a,b,f]	11 ROIs; DMN hubs: posterior cingulate cortex and anteromedial prefrontal cortex; DMN subsystems: dorsomedial prefrontal cortex (4 ROIs) and medial temporal lobe (5 ROIs)	Increased connectivity between dMPFC and bilateral: central opercular cortices, insular cortices, precentral gyri, and superior temporal gyri (AVH+ vs. HS and AVH−); increased connectivity between temporal pole and cerebellum (AVH+ vs. HS and AVH−); decreased connectivity between vMPFC and bilateral: paracingulate cortices, anterior cingulate cortices, and subcallosal cortices (AVH + vs. HS and AVH−); decreased connectivity between hippocampal formation and bilateral: posterior cingulate cortices and precuneus cortices (AVH+ vs. HS and AVH−) [1]
2	Berman et al. (2016)[d,f]	Data driven identification of seeds	Decreased connectivity in multiple regions belonging to clusters of social-cognitive functions and somatosensory-motor functions (SZ) [2]
3	Cui et al. (2016)[a,b,f]	Bilateral putamen	Decreased ALFF in left putamen (AVH+ vs. AVH−) and right postcentral gyrus (AVH+ vs. HS); increased ALFF in right putamen (AVH+ vs. HS); increased ReHo in right middle frontal gyrus including DLPFC (AVH+ vs. AVH−) and increased ReHo in putamen and left middle frontal gyrus (AVH+ vs. HS); increased connectivity with left DLPFC and Broca's area (AVH+ vs. AVH−); reduced connectivity in regions encompassing "frontal/parietal/temporal cortico-straital cerebellar networks" (AVH+ vs. HS and AVH− vs. HS) [3]
4	Gavrilescu et al. (2010)[a,b,f]	Primary and secondary auditory cortices	Reduced interhemispheric connectivity in primary and secondary auditory cortices (AVH+ vs. AVH− and HS) [4]
5	Oertel-Knöchel et al. (2013)[d,c,f]	Bilateral planum temporale	Reduced connectivity with bilateral superior temporal gyri, left cingulate gyrus, left thalamus, and left postcentral gyrus (bilateral seeds, patients and FD relatives vs. HS) [5]
6	Oertel-Knöchel et al. (2014)[d,c,f]	Bilateral Heschl's gyri	Reduced connectivity with left superior temporal gyrus, right insula, and left parahippocampal gyrus (patients and FD relatives vs. HS) [6]
7	Pu et al. (2012)[d,f]	Salience network (anterior insula and anterior cingulate cortex)	Decreased FC between left anterior insula and bilateral anterior cingulate cortices and right anterior insula (patients vs. HS); decreased FC between right anterior insula and left anterior insula (patients only); decreased FC between left anterior cingulate cortex and bilateral anterior insula (patients only); FC between left anterior cingulate cortex and anterior insula associated negatively with hallucinations [7]
8	Rolland et al. (2014)[a,b,e]	Bilateral nucleus accumbens	Increased connectivity between nucleus accumbens and left superior temporal gyrus, left inferior parietal lobule, anterior and posterior cingulate, medial frontal gyrus, bilateral lingual gyri and VTA (AVH+ vs. noH) [8]
9	Shinn et al. (2013)[a,b,f]	Primary auditory cortex (located on Heschl's gyrus)	Increased left Heschl's gyrus connectivity with left superior parietal lobule (AVH+ vs. AVH−; AVH+ vs. HS) and left middle frontal gyrus (AVH+ vs. AVH−); reduced left Heschl's gyrus connectivity with right hippocampal/parahippocampal region (AVH+ vs. AVH−) and medidorsal thalamus (AVH+ vs. AVH−; AVH+ vs. HS) [9]
10	Sommer et al. 2012)[a,f]	Right inferior frontal and left superior temporal gyri	Increased connectivity between right inferior frontal gyrus and right parahippocampal gyrus (SZ vs. HS); decreased connectivity between right inferior frontal gyrus and right DLPFC (SZ vs. HS), decreased connectivity between left superior temporal gyrus and a region on left hippocampus and another on left frontal operculum (SZ vs. HS); stronger reduction of FC between left superior temporal gyrus and left hippocampus (AVH+ vs. AVH−) [10]
11	Vercammen et al. (2010)[a,f]	Bilateral: TPJ, anterior cingulate cortices, Broca's area (and homotope), amygdala, and insula	Decreased connectivity between left TPJ and homotope of Broca's area (AVH+ vs. HS) [11]

(continued)

Table 1 (continued)

(b) ICA-based FC analysis

S. No	Study	Summary of results
1	Liemburg et al. (2012)[d,f]	Salience network: decreased intra-iFC in bilateral anterior insula (SZ vs. HS); increased intra-iFC in bilateral anterior cingulate cortices (SZ vs. HS) [2]; increased association between anterior cingulate cortex and Broca's networks (SZ vs. HS) [1]
2	Manoliu et al. (2014)[d,f]	Default mode network: decreased intra-iFC in bilateral anterior cingulate cortices and bilateral precuneus (SZ vs. HS) [2]; decreased FC in region including posterior cingulate and hippocampus (SZ vs. HS) [3]
3	Rotarska-Jagiela et al. (2010)[d,f]	Central executive network: decreased intra-iFC in bilateral inferior parietal lobules and bilateral frontal gyri (SZ vs. HS); increased intra-iFC in right angular gyrus and left inferior temporal gyrus (SZ vs. HS) [2]; decreased FC in left precuneus (SZ vs. HS); increased FC in right middle frontal gyrus and superior frontal gyrus (SZ vs. HS) [4]
4	Wolf et al. (2011)[a,f]	Right Fronto-parietal network: increased FC in left parietal cortex (SZ vs. HS); decreased FC in right middle frontal gyrus (SZ vs. HS) [3]; increased FC in right middle frontal gyrus (SZ vs. HS) [4] Left fronto-temporo-parietal network: decreased FC in left anterior cingulate cortex and right posterior cingulate cortex (SZ vs. HS); increased connectivity in bilateral middle temporal gyrus and left superior temporal gyrus (SZ vs. HS) [4] Decreased connectivity between superior temporal gyrus network and Broca's network (SZ vs. HS) [1] Increased inter-iFC of anterior DMN with inferior-posterior DMN and superior-posterior DMN (SZ vs. HS); increased inter-iFC between anterior DMN and right-ventral CEN and superior-posterior DMN with right-ventral CEN (SZ vs. HS) [2] No correlation between connectivity and severity of hallucinations [1]; negative correlation between right anterior insula intra-iFC and hallucinations, positive correlation between anterior DMN and right-ventral CEN inter-iFC and hallucinations [2]; negative correlation between hippocampal connectivity and hallucinations [3]; negative correlation of left anterior cingulate cortex FC, positive correlation of left superior temporal gyrus FC, and positive correlation of right middle frontal gyrus FC, with AVH severity [4]

(c) Other resting-state analyses

S. No	Study	Method	Summary of results
1	Chang et al. (2015)[a,b,f]	Voxel-mirrored homotopic connectivity	Connectivity differences in: inferior frontal gyrus, anterior cingulate cortex, precuneus cortex, superior parietal lobule, and anterior cerebellum (HS vs. AVH+); superior temporal gyrus and precentral gyrus (HS vs. AVH− only); parahippocampal gyrus and straitum (HS vs. AVH+ and HS vs. AVH−); superior temporal gyrus, anterior cingulate cortex, and superior parietal lobule (AVH+ vs. AVH−) [1]
2	Chen et al. (2015)[a,b,f]	Degree centrality across whole brain	Decreased degree centrality in bilateral putamen (AVH+ vs. HS; AVH− vs. HS); increased degree centrality in left superior frontal gyrus (AVH+ vs. HS; AVH− vs. HS) [2]
3	Zhu et al. (2016)[a,b,f]	Graph theory analysis of networks	Decreased clustering coefficient, global and local efficiencies, and increased characteristic path length (AVH+ and AVH− vs. HS) [3]
4	Zhuo et al. (2016)[a,b,f]	ReHo	Increased ReHo in left precuneus (AVH+ vs. AVH−); increased ReHo in left precuneus and putamen, right caudate and inferior temporal gyrus (AVH+ vs. HS); decreased ReHo in bilateral postcentral gyri and thalami, and right inferior occipital gyrus (AVH+ vs. HS) [4]

[a]AVH+ (SZ/psychosis) group present
[b]AVH− (SZ) group present
[c]Non-SZ group present
[d]SZ/psychosis group present
[e]SZ with audio-visual hallucinations
[f]HS group present
ALFF amplitude of low-frequency fluctuation; CEN central executive network; *DLPFC* dorsolateral prefrontal cortex; *DMN* default mode network; *dMPFC* dorsomedial prefrontal cortex; *FC* functional connectivity; *FD* first degree; *ICA* independent component analysis; *ReHo* regional homogeneity; *TPJ* temporoparietal junction; *vMPFC* ventromedial prefrontal cortex; *VTA* ventral tegmental area

Table 2 Studies employing button press or other similar methods

S. No.	Study	Method	Summary of results
1	Diederen et al. (2010)[a,d]	Activation-deactivation study; balloon squeezing	Activation in bilateral: postcentral gyri, inferior parietal lobules, precentral gyri, insula, cerebellum, inferior frontal gyri, superior frontal gyri, supramarginal gyri, middle temporal gyri, superior temporal gyri; and right middle frontal gyrus (during AVH); deactivation in left: parahippocampal gyrus, superior temporal gyrus, middle frontal gyrus, culmen; and right: inferior frontal gyrus and insula (prior to AVH) [1]
2	Diederen et al. (2012)[a,b]	Conjunction analysis; balloon squeezing	Common areas of activation between AVH+ groups were bilateral: inferior frontal gyri, insula, superior temporal gyri, supramarginal gyri and postcentral gyri; left: precentral gyrus, inferior parietal lobule, superior temporal pole; and right cerebellum [2]
3	Hoffman et al. (2011)[a,c,d]	Bilateral Wernicke's area (seed for FC); depress/release button to mark offset/onset of hallucinations	Increased FC with left BA 45/46 (AVH + and HS vs. AVH−); increased FC with left BA47 (HS vs. AVH + and AVH−) [3]
4	Raij and Riekki (2012)[a,d]	Activation study; button press to indicate onset/offset of AVH, imagery of AVH; HS imagined content of AVH	Activation in left SMA was stronger during imagery than when experiencing hallucinations (AVH+); activation in left SMA (HS) during imagery was stronger than AVH related activation (AVH+) [4]
5	Raij et al. (2009)[a]	Activation study; button press to indicate onset/offset of AVH and rating of subjective reality of hallucinations in scanner	AVH related activation in right parahippocampal cortex, bilateral inferior frontal gyri, right posterior temporal lobe, left anterior temporal lobe, and right anterior cingulate cortex [5]
6	Thoma et al. (2016)[a]	GLM and ICA; button press to indicate onset and offset of hallucinations	AVH related activation in bilateral inferior frontal and superior temporal gyri, left insula, supramarginal gyrus, inferior parietal lobule, and extranuclear white matter, and right middle temporal gyrus; ICA analysis showed an auditory cortex and posterior language network (comprising of bilateral superior temporal gyri, middle temporal gyri, supramarginal gyri, inferior parietal lobules, and right postcentral gyrus, insula, sub-gyral, angular gyrus, and transverse temporal gyrus) to be related to AVH on and off [6]

[a]AVH+ (SZ/psychosis) group present
[b]AVH+ (non-SZ) group present
[c]AVH− (SZ) group present
[d]HS group present

BA Brodmann area; *FC* functional connectivity; *GLM* general linear model; *ICA* independent component analysis; *SMA* supplementary motor area

Table 3 Studies involving other cognitive paradigms

S. No.	Study	Paradigm details	Primary regions implicated/summary of results
1	(a) Allen et al. (2007)[a,b,d] (b) Fu et al. (2008)[a,b,d] (c) Kumari et al. (2010)[c,d] (d) Mechelli et al. (2007)[a,b,d]	Monitoring of self and externally generated speech	(a) Higher rate of misattribution (AVH + vs. AVH− and HS); altered activity in the superior temporal gyrus and anterior cingulate regions (AVH+ vs. AVH − and HS) (b) Increased bilateral superior temporal gyri activation when making external misattributions than correct self-attributions (AVH+) (c) Decreased activity in bilateral thalami (self-undistorted condition, HS vs. SZ); increased activity in left superior-middle temporal gyrus and right superior temporal gyrus (self-undistorted condition, SZ vs. HS); deactivation in left hippocampus, left ventral straitum, and right hypothalamus (self-condition, SZ); increased activation in right superior temporal gyrus, middle temporal gyrus, and inferior frontal gyrus (undistorted vs. distorted condition, SZ) (d) Greater intrinsic connection from left superior temporal cortex to anterior cingulate cortex (HS vs. AVH + and AVH −); effect of source on left superior temporal cortex to anterior cingulate cortex was stronger (HS vs. AVH + and AVH− vs. AVH+) [causal analysis]
2	(a) Ćurčić-Blake et al. (2013)[a,b,d] (b) Vercammen et al. (2011)[a]	Metrical stress evaluation task requiring inner speech processing	(a) Decreased connectivity from Wernicke's area to Broca's area, Broca's homologue to Broca's area, and Wernicke's homologue to Broca's area (AVH+ vs. HS) [causal analysis] (b) Negative correlation between loudness and activation in bilateral angular gyri, bilateral anterior cingulate cortices, left triangular and opercular parts of inferior frontal gyrus, left middle temporal gyrus, and left insula
3	(a) de la Iglesia-Vayá et al. (2014)[a,b,d] (b) Escartí et al. (2010)[a,b,d] (c) García-Martí et al. (2012)[a,d] (d) Martí-Bonmatí et al. (2007)[a,d] (e) Sanjuán et al. (2007)[a,d]	Auditory paradigm replicating emotions related with hallucinations	(a) High correlation among the temporal, fronto-parietal, and fronto-temporal components, and between limbic and occipito-cerebellar components (AVH+); occipito-cerebellar component as causal source and fronto-parietal and limbic components as causal sinks (AVH+) (b) Increased activation in insula (HS and AVH−); increased activation in amygdala and parahippocampal gyrus (AVH+) (c) Emotional auditory activation in bilateral: middle temporal gyri, superior temporal gyri, amygdala, left hippocampus, and right precuneus (AVH+ vs. HS)

(continued)

Table 3 (continued)

S. No.	Study	Paradigm details	Primary regions implicated/summary of results
			(d) Regions of structural and functional coincidence were bilateral temporal middle, temporal superior, left frontal inferior opercular, right anterior cingulate, left posterior cingulate, and right occipital middle (AVH+) (e) Emotional auditory activation in right superior temporal gyrus, left middle temporal gyrus, bilateral insula, right median cingulate and paracingulate gyri, bilateral posterior cingulate gyri, right amygdala, orbital part of right inferior frontal gyrus, orbital part of bilateral middle frontal gyri, and medial part of right superior frontal gyrus (AVH+)
4	Ford et al. (2009)[a,b,d]	Auditory oddball detection task	Lesser activation to probe tones in the left primary auditory cortex (AVH+ vs. AVH−)
5	Ikuta et al. (2015)[c]	Auditory stimulus: sine waves, English words, and acoustically reversed English words	Bilateral globus pallidi activity (during reversed English-sine wave contrast) correlated with severity of AVH
6	Koeda et al. (2013)[c,d]	(Voice) Favourability judgement task (FJT) and gender discrimination task (GDT)	Positive correlation between activity in right postcentral gyrus, right precentral gyrus, right middle frontal gyrus, and right inferior parietal lobule with severity of hallucinations (FJT vs. GDT contrast)
7	Lawrie et al. (2002)[c,d]	Sentence completion task (with constraints) and rest	Lower FC between left middle/superior temporal cortex and left DLPFC (SZ vs. HS); lower left frontotemporal connectivity (AVH+ vs. AVH−)
8	Mou et al. (2013)[a,b,d]	Voice recognition task (familiar or unfamiliar)	Reduced FC between right superior temporal gyrus and right superior frontal gyrus (AVH+ vs. HS and AVH−); reduced FC between right superior temporal gyrus and left middle frontal gyrus (AVH+ vs. HS)
9	Plaze et al. (2006)[a]	Listening to sentences in French or to silence	Negative correlation between activity in left superior temporal gyrus and auditory hallucination PSYRATS scores (French-silence contrast); negative correlation between activity in left superior temporal gyrus and SAPS-AH scores (French-silence contrast)
10	Shergill et al. (2000)[c,d]	Generation of inner speech or imagining external speech	Inner speech was associated with activity in left inferior frontal gyrus, precentral gyrus, superior temporal gyrus, lingual gyrus, supplementary motor area, and right

(continued)

Table 3 (continued)

S. No.	Study	Paradigm details	Primary regions implicated/summary of results
			posterior cerebellar cortex (SZ); imagery was associated with activity in left inferior frontal gyrus, middle frontal gyrus, precentral gyrus, postcentral gyrus, and inferior parietal lobule (SZ); reduced activation in bilateral posterior cerebellum, hippocampi, lenticular nuclei, right thalamus, middle temporal gyrus, superior temporal gyrus, and left nucleus accumbens during auditory verbal imagery (SZ vs. HS)
11	Shergill et al. (2003)[c,d]	Covert generation of words at different speed	Reduced activation in right posteolateral cerebellum, parahippocampal gyrus, superior temporal gyrus, postcentral gyrus, and inferior parietal lobule (SZ vs. HS during faster articulation); increased activation in left lenticular nucleus (SZ vs. HS during faster articulation); reduced FC of left inferior frontal gyrus with right middle temporal gyrus, superior temporal gyrus, insula, parahippocampal region, inferior temporal gyrus, fusiform gyrus, precentral gyrus, and medial parietal lobe (SZ vs. HS)
12	Shergill et al. (2014)[c,d]	Sensorimotor task	Attenuation of secondary somatosensory cortex activation when movement and sensation occurred synchronously compared with sensation alone and also compared with asynchronous movement and sensation (HS but not in SZ); greater activation in primary somatosensory cortex (HS vs. SZ); failure of attenuation of secondary somatosensory cortex activation was predicted by AVH severity
13	Simons et al. (2010)[c,d]	Participants heard sentences and imagined hearing these sentences	Activation in left inferior frontal gyrus and anterior cingulate gyrus during inner speech generation; greater bilateral occipital gyri activity (SZ vs. HS); interaction between task and group showed activity in left superior temporal gyrus, bilateral cingulate gyri, right hippocampus, and left posterior cingulate
14	Sommer et al. (2008)[a]	Participants experienced AVH; participants silently generated words	AVH related activity seen in bilateral: insula, inferior frontal gyri, supramarginal gyri, medial frontal gyri, left: postcentral gyrus, superior frontal gyrus, lentiform nucleus, precentral gyrus, cingulate gyrus,

(continued)

Table 3 (continued)

S. No.	Study	Paradigm details	Primary regions implicated/summary of results
			dentate gyrus, and right: middle frontal gyrus, culmen, pyramis, homologue of Broca's area, homologue of Wernicke's area, and superior temporal gyrus; letter fluency related activity in bilateral: inferior frontal gyri, superior frontal gyri, precentral gyri, insula, thalami, fusiform gyri, middle temporal gyri, superior temporal gyri, superior parietal lobules, inferior parietal lobules, middle occipital gyri, posterior lobes, anterior cingulum, and left middle frontal gyrus
15	Zhang et al. (2008)[a,b,d]	Participants listened to sentences either from left side or right side	Activity in left supramarginal gyrus and angular gyrus by left sided voice simulation (AVH+ vs. AVH−); activity in left superior temporal gyrus and superior frontal gyrus by right sided voice stimulation (AVH+ vs. AVH−); activity in right middle frontal gyrus by right sided voice stimulation (HS vs. AVH+); activity in bilateral middle frontal gyri and left postcentral gyrus by right sided voice stimulation (HS vs. AVH −)
16	Zhang et al. (2008)[a,b,d]	Classification of pre-recorded voice as familiar or unfamiliar	Reduced activity in right superior temporal gyrus (familiar-unfamiliar contrast, AVH+ vs. HS)

[a]AVH+ (SZ/psychosis) group present
[b]AVH− (SZ) group present
[c]SZ/psychosis group present
[d]HS group present
DLPFC dorsolateral prefrontal cortex; *FC* functional connectivity; *PSYRATS* psychotic symptom rating scales; *SAPS* scale for assessment of positive symptoms

AVH. The interested readers are invited to visit excellent review articles which have summarised the findings across different kinds of imaging modalities. For example, Allen et al. (2008) for review of structural and functional studies; Johnsen et al. (2013) for a review on the neuropsychopharmacology of AVH; and Alderson-Day et al. (2015, 2016) for recent reviews. We also direct the readers' attention to Jardri et al. (2011) and Kühn and Gallinat (2012) for meta-analyses on neuroimaging studies pertaining to AVH.

Methodological Issues

Given the heterogeneity in the approaches adopted by various groups, it is quite natural that several methodological considerations need to be addressed. First, it is important to have stringent patient selection criteria while starting with the study. This would mean careful consideration of clinical variables like the age of onset and the duration of illness which could otherwise potentially confound the results of the study. Further, it is important to be very selective in recruiting patients having continuous auditory verbal hallucinations, as opposed to patients who may have sporadic experiences. Such a criterion would allow 'capturing' of the functional neuroimaging correlates of the experience of auditory verbal hallucinations. This brings up the state versus trait question. In the context of hallucinations and its implication for studying the self, it is important to capture both of these. If we are to consider that—*agency* is a key component to the experience of self (and therefore one is conscious of the sensory inputs), then (as mentioned previously) it follows that while a patient is experiencing hallucinations, s/he is in an altered state of consciousness. Investigating the neural correlates of this state would give valuable insight into the actual origin of these hallucinations. For this to be true, it is important that patients while being imaged, experience hallucinations. It is important to keep in mind that subjects are actively involved in different cognitive tasks while in the scanner (task based fMRI). Thus, attentional load, the novelty of the setting, and the scanner noise, may, in general, reduce the probability of the patients experiencing hallucinations.

It is also of vital importance to introduce control group(s) in the study. In our opinion, it would be worthwhile to have three groups: healthy subjects, patients with schizophrenia who experience auditory verbal hallucinations while being imaged, and patients with schizophrenia who do not experience auditory verbal hallucinations while being imaged. The experimenter would need to categorise the patients in the latter two groups after the data acquisition is over. The advantage of this approach is that one can then attempt to answer questions related to state versus trait with respect to the experience of hallucinations in patients with schizophrenia. If the above-mentioned recruitment criteria for patient selection are followed, then we can be sure that all patients recruited into the study experience hallucinations most of the time in a day. That is, the experience of hallucinations is a *trait*. When these patients experience hallucinations in the scanner, it reflects the *state*, and those patients who do not experience hallucinations in the scanner can serve as a control group for studying neural correlates of the differences in state versus trait. The healthy control group can also be used for comparison. In several of the studies conducted using these three groups, results have often varied as a continuum. As an alternative, some researchers have selected two groups of patients: those experiencing hallucinations, and those not experiencing hallucinations. While such a selection scheme takes into account the state or trait, it does not explicitly allow comparison between state and trait.

Selecting patient groups on the above scheme immediately presents a problem: how does one decide on which patients were experiencing hallucinations during the scanner. Two solutions present themselves: one way is to ask the participants themselves (after the imaging session is over) on whether they were experiencing hallucinations; second is to ask the participants *during* the scanning session. In case of the former, the situation is compounded by the fact that participants might have experienced hallucinations at some point and may not have during most other times, over the entire session. Also, this opinion might be biased by the participant willing to 'please' the experimenter, recollection error, and so on. Therefore, this does not offer a clear way of selecting time points when the participants were experiencing hallucinations. In case of the latter, this can be done in two ways: one is to ask the subjects to press a button (or balloon squeezing or other similar mechanisms) to indicate onset and offset of hallucinations. The other way is to introduce a query slide into the experimental design and ask the subject to respond if they had hallucinations during a particular *block* of experiment. Both these methods have their merits and demerits. In case of button press paradigm, data acquisition can lead to highly different number of time points which may affect further processing and analyses. On the other hand, introducing a query slide requires modifying the task to incorporate the same, and can also presents similar challenges in terms of difference in time series length.

The above is tightly knit with the design of the paradigm itself. The simplest case is to acquire resting-state fMRI data. In this case, a simple button press or query slide after every fixed interval could be easily implemented. In case of button press, one might additionally want to keep a separate button-press block to model the brain response to button pressing as a baseline. In case of a block design task, the introduction of a query slide at the end of every block would be the easiest choice. One can even have 'task-free' blocks where patients wholly attend to their hallucinations. This is an important consideration because it is quite likely that attentional modulation will have an effect on the perceptiveness of hallucinations. In context of an event-related experimental design, it would be more cumbersome to include a query slide, though a button press is still possible.

Among other methodological issues would be the choice of kind of analysis (for example, activation study vs. connectivity study), the rigour and details of the analysis pipeline itself, strict multiple comparison correction, and so on. While this section does not dwell on all the considerations and choices that an experimenter has, we hope that we have adequately highlighted the importance of careful consideration of the kind of question one wants to answer, and then making appropriate choices.

Preliminary Evidence from Our Work

In keeping with the thoughts discussed above, we have proposed a novel cognitive fMRI paradigm, viz. 'hallucination attentional modulation task' (HAMT), which attempts to capture the self versus non-self distinction disintegration in patients with schizophrenia, and also provide insight into the underlying neurobiology of hallucinations. Below, we describe the paradigm in detail outlining the thoughts about some of the choices, and then discuss some results from a pilot study that we conducted using this paradigm.

Experimental Design

The task was designed to be fMRI compatible and implemented as a block design experiment, involving three different conditions: free attention (FA), visual attention (VA), and hallucination attention (HA). The task involved the presentation of abstract monochromatic visual patterns which changed at unpredictable intervals. All images were presented at a resolution of 72 pixels per inch and were 200 × 200 pixels in dimension. These patterns were created using the GNU image manipulation program (GIMP) 2.4.7, using the fill pattern tool. The six distinct patterns were clearly distinguishable from each other and were random enough to minimise any chances of emotional responses to the visual stimuli. Figure 1 shows these six abstract patterns created for the purpose of this experiment.

The task consisted of five blocks, with each block having the above-mentioned three conditions (in case of healthy subjects, HA condition was not presented). At the beginning of each condition, participants were presented with an instruction slide and at the end of each condition, a response collection slide was presented and participant response was collected via a MR compatible button response pad.

Fig. 1 The six abstract monochromatic visual patterns created for the HAMT paradigm. These patterns appeared during the three different conditions, changing at unpredictable intervals

The instruction and end-of-condition response collection slide lasted for 9 s each, and the entire condition duration was 42 s. The abstract patterns were presented for a duration ranging from 2 to 18 s each (in a predetermined pseudorandom order) resulting in 4–9 stimuli in each condition.

During the free attention condition, subjects were instructed to keep their eyes fixed at the centre of the screen and to not pay attention to anything in particular. During the response collection slide, subjects were asked to press any button (out of the three) of their choice. During the visual attention condition, subjects were instructed to keep their eyes fixed at the centre of the screen and to count the number of times the images (abstract monochromatic patterns from above) changed on the screen. During the response collection slide, attentional engagement was assessed by asking the subjects to select the button corresponding to the correct number of times the image changed on the screen (three choices were presented with one of them being the correct one). In case of patients with schizophrenia, an additional condition was introduced: hallucination attention condition. During this condition, subjects were instructed to keep their eyes fixed at the centre of the screen but to pay attention to their auditory verbal hallucinations (if any). During the response collection slide, subjects were queried about their frequency of hallucinations *during that condition.* Three choices were presented: 'No voices were heard—Press index finger; Voices heard, but for less than 50% of time—Press middle finger; Voices heard for more than 50% of time—Press ring finger'. Figure 2 shows a schematic of the design of the experiment.

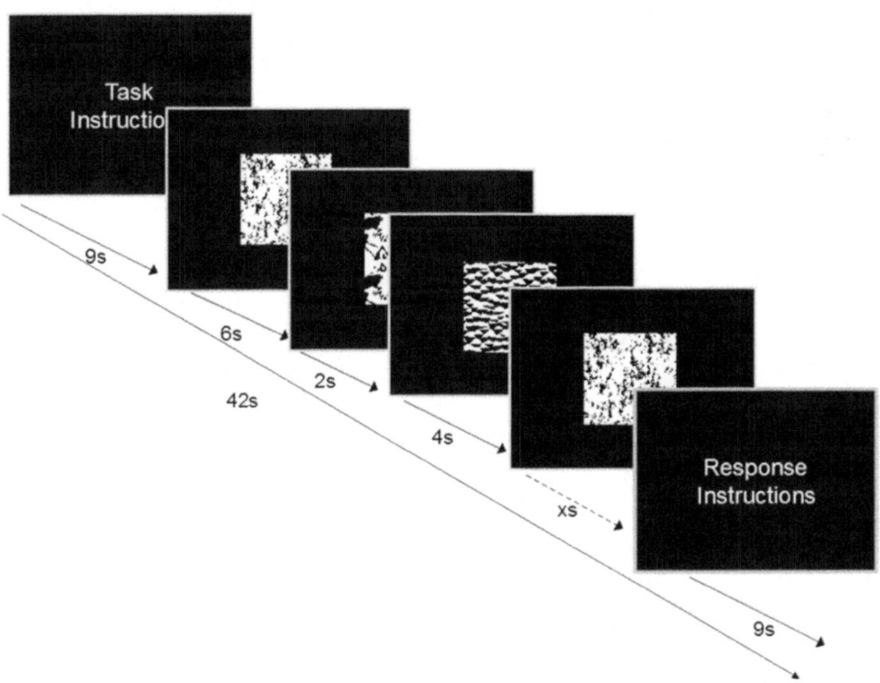

Fig. 2 Schematic representation of the design of the HAMT experiment

Rationale for the Design of HAMT

While designing the task, we have kept the following major goals in mind:

(1) In order to obtain a fixed length time series, we designed the paradigm as a block design.
(2) In order to model out the brain activity associated with the changing visual patterns on the screen and the activity of the button response, we included the free attention condition. During this condition, the participants are involved in passive viewing of the changing images. This condition can serve as a baseline.
(3) The visual attention condition was introduced with the objective of examining the effect of attentional modulation on brain connectivity using an experimental task in an alternate modality that can be implemented in an fMRI study.
(4) During the hallucination attention condition, the visual patterns continue to appear on the screen. This allows for the creation of the following contrasts:

    a. HA-FA: would allow us to capture the activity associated with purely hallucinations (effect of visual presentation gets subtracted)
    b. HA-VA: would allow us to capture the activity associated with attentional modulation on hallucinations (during VA, attention is being focused away from AVH while in HA, attention is being focused towards AVH)
    c. VA (main effect): would allow us to capture the activity associated with the processing of an alternate sensory (and unimpaired) stimuli which is associated with a sense of agency (because the subjects are aware that they are the one who are visualising and counting the changing patterns)

(5) The rating taken from patients with schizophrenia during the HA condition would allow us to post hoc classify them into AVH+ and AVH− subgroups.
(6) A structural scan is also acquired prior to fMRI scan; this would allow us to examine the possibilities of any structural abnormalities in the auditory perceptual mechanism of the brain.

Study Sample, Image Acquisition, and Data Analysis

The sociodemographic profile of the subjects recruited for this study is reported elsewhere (Parekh et al. 2016). Briefly, we recruited six patients with schizophrenia having continuous auditory hallucinations (defined using modified criteria from PSYRATS (Haddock et al. 1999) on the basis of presence of continuous auditory hallucinations for more than 50% of their awake time for at least the previous two weeks). We also recruited eight healthy subjects for the study. A description of the psychopathology of three of the patients (out of six) who participated in the pilot study is provided in Box 1, as an example.

Structural and functional MRI was performed on a 3T Philips Achieva scanner. We performed ROI-to-ROI functional connectivity analysis using Conn Functional Connectivity Toolbox 15.0 h (Whitfield-Gabrieli and Nieto-Castanon 2012) running in MATLAB environment (MathWorksInc, Natick, Maryland, USA, www.mathworks.com) with SPM 12b (http://www.fil.ion.ucl.ac.uk/spm/software/spm12/) in the background. Regions of interest for functional connectivity analysis were defined a priori based on previous reports of brain regions that have been linked to AVH (Rolland et al. 2014; Shinn et al. 2013; Vercammen et al. 2010). Further, because the task involved visual stimuli, we also included the primary and secondary visual areas of the brain (BA17, BA18 and BA19). This resulted in a total of 16 regions in each hemisphere, which were defined using the human brain atlas in WFU PickAtlas (Lancaster et al. 1997, 2000; Maldjian et al. 2003) toolbox in MATLAB. The location of these regions and their abbreviations are shown in Fig. 3.

Having selected these ROIs, we performed ROI-to-ROI functional connectivity. fMRI data were pre-processed as per standard pipeline. After normalising the data to Montreal Neurological Institute (MNI) space, a band pass filter of 0.008 to Inf was implemented, followed by removal of confounds white matter, cerebrospinal fluid (CSF), realignment, scrubbing, and the main effects of each condition. Time series was extracted from unsmoothed normalised fMRI images for each of the ROIs (mean of all voxels in a given ROI) and pairwise bivariate correlation coefficient was calculated as a measure of functional connectivity. Correlation coefficients were Fisher transformed (Fisher's r to Z transform) to improve the normality of the data (see Parekh et al. (2016) for a complete description of the data analysis pipeline).

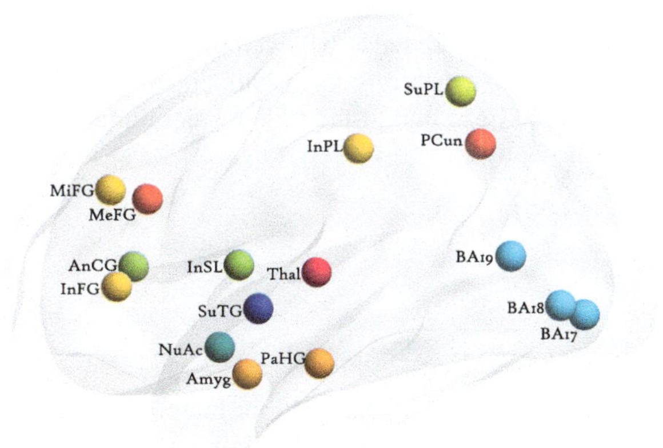

Amyg	*Amygdala*
AnCG	*Anterior Cingulate*
BA17	*Brodmann Area 17*
BA18	*Brodmann Area 18*
BA19	*Brodmann Area 19*
InFG	*Inferior Frontal Gyrus*
InPL	*Inferior Parietal Lobule*
InSL	*Insula*
MeFG	*Medial Frontal Gyrus*
MiFG	*Middle Frontal Gyrus*
NuAc	*Nucleus Accumbens*
PCun	*Precuneus*
PaHG	*Parahippocampal Gyrus*
SuPL	*Superior Parietal Lobule*
SuTG	*Superior Temporal Gyrus*
Thal	*Thalamus*

Fig. 3 The regions of interest used for functional connectivity analysis along with their abbreviations. The regions are colour coded according to the networks/functioning of the regions: areas in *blue* (BA17, BA18, and BA19) comprise the visual network; areas in *orange* (Amyg and PaHG) are the memory areas; areas in *red* (MeFG and PCun) comprise the internal awareness network; areas in *yellow* (MiFG, InFG, and InPL) form the external awareness network; areas in *green* (AnCG and InSL) form the salience network; visualisation created using BrainNet Viewer (http://www.nitrc.org/projects/bnv/) (Xia et al. 2013)

Box 1: Description of psychopathology of three of the six patients recruited for the pilot study

Patient 1 (**M.B**): A 22 year old male with a history of 10 months of illness, M.B. presented with signs of continuous second and third person auditory hallucinations that had elements of running commentary as well as the discussing types which gave him the idea that a device had been planted in his ears enabling others to hear his thoughts. Other symptoms included persecutory delusions, thought broadcasting, and suspiciousness. Diagnosed with paranoid schizophrenia, M.B had the age of onset of illness at 21 years. At the time of MRI acquisition, M.B was neuroleptic naive.

Patient 2 (**N.B.R**): A 25 year old male with a history of 5 years of illness, N.B.R had continuous auditory verbal hallucinations that would either directly talk to him (criticising him/blaming him) or a running commentary on what he was doing at that point/daily activities. N.B.R reported hearing these voices in a state of clear consciousness as coming from outside the body and identified the voice as belonging to parents/friends. Although distressed by the voices, N.B.R. did not demonstrate any acting-out behaviour, and at the time of recruitment for the study, was on regular treatment with 4 mg of risperidone and 2 mg of trihexyphenidyl.

Patient 3 (**S.K.**): A 40 year old male having duration of illness of 10 years, S.K. reported hearing voices (more than one person) in clear consciousness. S.K. reported that the voices had informed him that they belonged to two vernacular gods. Initially being distressed with the voices, S.K. grew used to the voices and was reported to have conversations with the voices which informed him of certain incidents (like the coming of an earthquake). S.K. presented with delusions of grandeur and also believed that people were trying to poison his food. The second and third person hallucinatory voices were reported to be coming from the head and at the time of recruitment for the present study, S.K. was neuroleptic free.

Results and Discussion

Given that this was a pilot study with a very small sample, our main objective was to explore connectivity in patients with schizophrenia who were experiencing hallucinations while performing the task. We classified the subjects as AVH+ if they 'heard voices' for more than 50% of the time during the HA condition of all five blocks, else classified as AVH−. Four out of the six schizophrenia subjects were classified as AVH+ while the remaining two were classified as AVH−. To better interpret the results, we classified the ROIs into the following networks/groups: the *internal awareness network* (medial prefrontal cortices, precuneus and inferior parietal lobules); *external awareness network/fronto-parietal network* (middle and inferior frontal gyri, inferior parietal lobules); *salience network* (insula and anterior cingulate); *visual network* (BA 17, 18, 19), *auditory cortex* (bilateral superior temporal gyri) and *bilateral parahippocampal gyri, bilateral thalamus* as well as *posterior parietal regions (superior parietal lobules)* (see Fig. 3).

Given the small sample size, we did not perform a between group comparison (either between healthy subjects and patients with schizophrenia, or between AVH+ and AVH− groups) and the comparative statements that we mention below are simply numerical in nature (and not based on statistical testing). However, we did perform one sample *t*-test to test whether the mean functional connectivity strength between pairs of ROIs was significantly non-zero at $\alpha = 0.05$ (corrected for multiple comparisons using a seed-level false discovery rate method).

Since during the experiment we presented participants with changing visual stimuli, common connectivity pattern between the FA and VA conditions can be considered to be associated with the perception of the visual stimuli (real stimuli). These included areas of the *visual network* (BA 17, 18, 19), the *internal awareness network* (medial prefrontal cortices, precuneus) (Demertzi et al. 2013); *external awareness network/fronto-parietal network* (middle and inferior frontal gyri, inferior parietal lobules) (Demertzi et al. 2013); and *salience network* (insula and anterior cingulate) (Menon and Uddin 2010). Along with these regions, we also saw the involvement of areas associated with *memory* (parahippocampal gyri, amygdala) (Phelps 2004; Takahashi et al. 2002), *self-agency* (superior parietal lobules) (MacDonald and Paus 2003) and *sensory filtering* (thalami) (McCormick and Bal 1994) in both healthy subjects and patients with schizophrenia. During the HA conditions, the brain areas which showed stronger connectivity in the AVH+ group (apart from the areas involved in visual processing) can be considered to be linked to the experience of auditory verbal hallucinations. Connections which stood out in this case were between the left and right superior parietal lobules as well as the left and right thalami. We also noted reduced connection strength between the visual areas and the fronto-parietal network (see Fig. 4).

The superior parietal lobule has been known to play a crucial role in self-agency (MacDonald and Paus 2003), while the parietal lobe is the implementation site of the body-image and body-centred spatial concepts (Berlucchi and Aglioti 1997; Melzack et al. 1997; Vogeley 2003). Over activity of the superior parietal cortex in patients with schizophrenia (having delusions of alien control) has previously been demonstrated (Spence et al. 1997). Therefore it is likely that the capacity to distinguish between self and others in these individuals may be disturbed by a possible increased cortical 'associativity' ('hyperassociativity') (Behrendt 1998) between the bilateral superior parietal lobules, apart from the prefrontal-parietal connections, leading to an imbalance between self-monitoring and reality modelling domains (Vogeley 2003). This would lead to an inability to recognise the origin of thought as the self and may lead to a subsequent misattribution to external sources or 'alien' (Frith 1992). Our finding of stronger connectivity between the bilateral superior parietal lobules, linked to the experience of AVH may provide support to the above hypothesis of the neural substrate of auditory hallucinations.

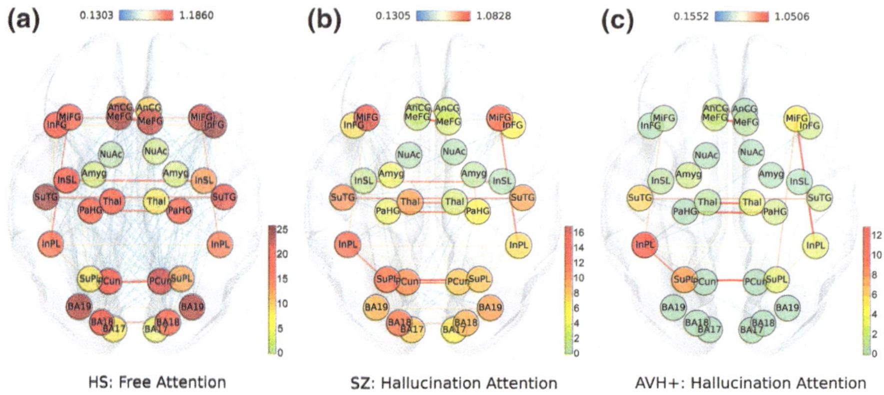

Fig. 4 Connectivity maps showing connectivity pattern during **a** free attention condition in healthy subjects; **b** hallucination attention condition in patients with schizophrenia (both AVH+ and AVH− together); and **c** hallucination attention condition in AVH+ schizophrenia subgroup. ROIs are depicted as *nodes* and the connectivity between the nodes are shown as *edges*; edge colour is proportional to Z values (colour bar on *top*) and the colour of the nodes correspond to the number of connections (or degree) of each of the seed ROIs (colour bar on the *right*); all images are shown in neurological convention (image *left* is subject's left); brain template from BrainNet Viewer (http://www.nitrc.org/projects/bnv/) (Xia et al. 2013)

Another finding of note was the increased functional connectivity between the bilateral thalami. Thalamus is the major relay centre of the brain between different subcortical areas and the cerebral cortex: the medial geniculate nucleus acting as a key auditory relay centre between the inferior colliculus of the midbrain and the primary auditory cortex. Therefore the thalamus acts as an information 'filter' or sensory 'gate' (Andreasen 1997). It is quite possible that our finding of increased functional connectivity between the bilateral thalami may indicate an underlying increased effort by the thalamus to 'filter' out the auditory hallucinations in patients with schizophrenia (caused by top-down processing abnormalities involving the superior parietal lobule and its fronto-parietal connections).

Finally, another interesting observation from the pilot study was the lack of significant functional connectivity between the bilateral precuneus during the HA condition in AVH+ subgroup. This could potentially reflect a shift of focus away from processing of the visual stimuli (associated with increased connectivity of the medial parietal lobe, the precuneus) to the experience of AVH (associated with increased connectivity of the lateral parietal lobe, the superior parietal lobule and a decreased connectivity of the bilateral precuneus). This shift of focus away from the processing of visual stimuli to the processing of auditory hallucinations may explain the lack of significant bilateral insular and right anterior cingulate-medial prefrontal cortical connections along with weaker bilateral anterior cingulate connection in the AVH+ subgroup (visual stimuli would not be considered as salient during the HA condition).

The above-mentioned results from our pilot study provide preliminary evidence in support of the utility of novel fMRI paradigm like the hallucination attentional modulation task. Such a paradigm allows us to detect functional connectivity patterns underlying normal visual perception as well as aberrations in functional connectivity during the experience of AVH. The functional connectivity patterns associated with processing of visual stimuli point towards involvement of brain regions that constitute the internal awareness network, external awareness/fronto-parietal network, salience network as well as medial temporal regions, visual brain regions, thalamus and posterior parietal regions. The experience of AVH was found to be associated with aberrations in the above connectivity patterns that involve substantially stronger connections between bilateral superior parietal lobules and bilateral thalami, weaker connections in the fronto-parietal, visual brain regions and bilateral superior temporal gyri, along with a lack of significant connections between the bilateral precuneus. The findings of our study point towards the possibility of impaired self-agency underlying the experience of AVH, and provide preliminary evidence for superior parietal lobule being the neural substrate of the same.

Critique of the Paradigm

Although the paradigm seems promising, there are a few important considerations: (a) the results presented above are to be considered preliminary because of insufficient sample size preventing us from carrying out any between groups or between conditions statistical comparisons; (b) it is possible that during the visual attention condition, when the subjects are engaged in counting, language areas and/or areas associated with inner speech may get involved which might be a potential confound for the evaluation of HA-VA or VA-FA contrasts.

Summary

In this discussion on the use of hallucination as a model for understanding impaired self, we have highlighted the various symptom domains of schizophrenia and have developed the argument that auditory verbal hallucinations can be potentially a good model to study self-agency and its impairment. In our account, we have largely dwelled on the top-down model of auditory verbal hallucinations. Prominent theories in this domain postulate that patients with schizophrenia have impairment in the corollary discharge mechanism which leads to an inability in distinguishing between self and non-self, thereby causing inner thoughts to be attributed to an external source, or hallucinations. Neuroimaging methods allow us to capture the neural correlates of this impaired self by devising experiments that tease out specific components related to auditory verbal hallucinations. We have briefly examined some of these neuroimaging studies and then have proposed certain important methodological considerations to keep in mind while designing and conducting neuroimaging studies which examine various aspects of the self. Finally, we have introduced a novel hallucination attentional modulation task (HAMT) which captures the neural correlates of the processing of a real stimulus (visual) and the associated neural activities of a perceived stimulus (i.e. the auditory verbal hallucination, in patients with schizophrenia). The preliminary results show promise and link auditory verbal hallucinations to an impaired self-agency in patients with schizophrenia, hinting at superior parietal lobule being the neural substrate for the same.

References

Alderson-Day, B., McCarthy-Jones, S., & Fernyhough, C. (2015). Hearing voices in the resting brain: A review of intrinsic functional connectivity research on auditory verbal hallucinations. *Neuroscience and Biobehavioral Reviews, 55,* 78–87. doi:10.1016/j.neubiorev.2015.04.016.

Alderson-Day, B., Diederen, K., Fernyhough, C., Ford, J. M., Horga, G., Margulies, D. S., … Jardri, R. (2016). Auditory Hallucinations and the Brain's Resting-State Networks: Findings and Methodological Observations. *Schizophrenia Bulletin,* sbw078. https://doi.org/10.1093/schbul/sbw078.

Allen, P., Amaro, E., Fu, C. H. Y., Williams, S. C. R., Brammer, M. J., Johns, L. C., et al. (2007). Neural correlates of the misattribution of speech in schizophrenia. *The British Journal of Psychiatry, 190*(2), 162–169. doi:10.1192/bjp.bp.106.025700.

Allen, P., Larøi, F., McGuire, P. K., & Aleman, A. (2008). The hallucinating brain: A review of structural and functional neuroimaging studies of hallucinations. *Neuroscience and Biobehavioral Reviews, 32*(1), 175–191. doi:10.1016/j.neubiorev.2007.07.012.

Alonso-Solís, A., Vives-Gilabert, Y., Grasa, E., Portella, M. J., Rabella, M., Sauras, R. B., et al. (2015). Resting-state functional connectivity alterations in the default network of schizophrenia patients with persistent auditory verbal hallucinations. *Schizophrenia Research, 161*(2–3), 261–268. doi:10.1016/j.schres.2014.10.047.

American Psychiatric Association. (2013). *Diagnostic and statistical manual of mental disorders* (5th ed.). American Psychiatric Association. Retrieved from http://psychiatryonline.org/doi/book/10.1176/appi.books.9780890425596.

Andreasen, N. C. (1997). The role of the thalamus in schizophrenia. *Canadian Journal of Psychiatry. Revue Canadienne De Psychiatrie, 42*(1), 27–33.

Behrendt, R.-P. (1998). Underconstrained perception: A theoretical approach to the nature and function of verbal hallucinations. *Comprehensive Psychiatry, 39*(4), 236–248. doi:10.1016/S0010-440X(98)90067-0.

Berlucchi, G., & Aglioti, S. (1997). The body in the brain: Neural bases of corporeal awareness. *Trends in Neurosciences, 20*(12), 560–564. doi:10.1016/S0166-2236(97)01136-3.

Berman, R. A., Gotts, S. J., McAdams, H. M., Greenstein, D., Lalonde, F., Clasen, L., … Rapoport, J. (2016). Disrupted sensorimotor and social–cognitive networks underlie symptoms in childhood-onset schizophrenia. *Brain, 139*(1), 276–291. https://doi.org/10.1093/brain/awv306.

Blakemore, S. J., Smith, J., Steel, R., Johnstone, E.C., & Frith, C.D. (2000). The perception of self-produced sensory stimuli in patients with auditory hallucinations and passivity experiences: Evidence for a breakdown in self-monitoring. *Psychological Medicine, 30*(05): 1131–1139. https://doi.org/null.

Chang, X., Xi, Y.-B., Cui, L.-B., Wang, H.-N., Sun, J.-B., Zhu, Y.-Q., et al. (2015). Distinct inter-hemispheric dysconnectivity in schizophrenia patients with and without auditory verbal hallucinations. *Scientific Reports, 5,* 11218. doi:10.1038/srep11218.

Chen, C., Wang, H.-L., Wu, S.-H., Huang, H., Zou, J.-L., Chen, J., et al. (2015). Abnormal degree centrality of bilateral putamen and left superior frontal gyrus in schizophrenia with auditory hallucinations: A resting-state functional magnetic resonance imaging study. *Chinese Medical Journal, 128*(23), 3178. doi:10.4103/0366-6999.170269.

Cho, R., & Wu, W. (2013). Mechanisms of auditory verbal hallucination in schizophrenia. *Schizophrenia, 4,* 155. doi:10.3389/fpsyt.2013.00155.

Cui, L.-B., Liu, K., Li, C., Wang, L.-X., Guo, F., Tian, P., et al. (2016). Putamen-related regional and network functional deficits in first-episode schizophrenia with auditory verbal hallucinations. *Schizophrenia Research, 173*(1–2), 13–22. doi:10.1016/j.schres.2016.02.039.

Ćurčić-Blake, B., Liemburg, E., Vercammen, A., Swart, M., Knegtering, H., Bruggeman, R., et al. (2013). When Broca goes uninformed: Reduced information flow to Broca's area in schizophrenia patients with auditory hallucinations. *Schizophrenia Bulletin, 39*(5), 1087–1095. doi:10.1093/schbul/sbs107.

de la Iglesia-Vayá, M., Escartí, M. J., Molina-Mateo, J., Martí-Bonmatí, L., Gadea, M., Castellanos, F. X., … Sanjuán, J. (2014). Abnormal synchrony and effective connectivity in patients with schizophrenia and auditory hallucinations. *NeuroImage: Clinical, 6*: 171–179. https://doi.org/10.1016/j.nicl.2014.08.027.

Demertzi, A., Vanhaudenhuyse, A., Bredart, S., Heine, L., Di Perri, C., & Laureys, S. (2013). Looking for the self in pathological unconsciousness. *Frontiers in Human Neuroscience, 7,* 538. doi:10.3389/fnhum.2013.00538.

Diederen, K. M. J., Neggers, S. F. W., Daalman, K., Blom, J. D., Goekoop, R., Kahn, R. S., et al. (2010). Deactivation of the parahippocampal gyrus preceding auditory hallucinations in schizophrenia. *American Journal of Psychiatry, 167*(4), 427–435. doi:10.1176/appi.ajp.2009.09040456.

Diederen, K. M. J., Daalman, K., Weijer, A. D. de, Neggers, S. F. W., Gastel, W. van, Blom, J. D., … Sommer, I. E. C. (2012). Auditory hallucinations elicit similar brain activation in psychotic and nonpsychotic individuals. *Schizophrenia Bulletin, 38*(5): 1074–1082. https://doi.org/10.1093/schbul/sbr033.

Escartí, M. J., de la Iglesia-Vayá, M., Martí-Bonmatí, L., Robles, M., Carbonell, J., Lull, J. J., et al. (2010). Increased amygdala and parahippocampal gyrus activation in schizophrenic patients with auditory hallucinations: An fMRI study using independent component analysis. *Schizophrenia Research, 117*(1), 31–41. doi:10.1016/j.schres.2009.12.028.

Feinberg, I. (1978). Efference copy and corollary discharge: Implications for thinking and its disorders. *Schizophrenia Bulletin, 4*(4), 636–640.

Ford, J. M., Roach, B. J., Jorgensen, K. W., Turner, J. A., Brown, G. G., Notestine, R., et al. (2009). Tuning into the voices: A multisite fMRI study of auditory hallucinations. *Schizophrenia Bulletin, 35*(1), 58–66. doi:10.1093/schbul/sbn140.

Friston, K. J., & Frith, C. D. (1995). Schizophrenia: A disconnection syndrome? *Clinical Neuroscience (New York, N.Y.), 3*(2): 89–97.

Frith, C. D. (1987). The positive and negative symptoms of schizophrenia reflect impairments in the perception and initiation of action. *Psychological Medicine, 17*(03), 631–648. doi:10.1017/S0033291700025873.

Frith, C. D. (1992). *The cognitive neuropsychology of schizophrenia*. Psychology Press.

Frith, C. D., & Done, D. J. (1988). Towards a neuropsychology of schizophrenia. *The British Journal of Psychiatry: The Journal of Mental Science, 153,* 437–443.

Fu, C. H. Y., Brammer, M. J., Yágüez, L., Allen, P., Matsumoto, K., Johns, L., … van Haren, N. (2008). Increased superior temporal activation associated with external misattributions of self-generated speech in schizophrenia. *Schizophrenia Research, 100*(1–3): 361–363. https://doi.org/10.1016/j.schres.2007.10.023.

García-Martí, G., Aguilar, E. J., Martí-Bonmatí, L., Escartí, M. J., & Sanjuán, J. (2012). Multimodal morphometry and functional magnetic resonance imaging in schizophrenia and auditory hallucinations. *World Journal of Radiology, 4*(4), 159. doi:10.4329/wjr.v4.i4.159 .

Gavrilescu, M., Rossell, S., Stuart, G. W., Shea, T. L., Innes-Brown, H., Henshall, K., et al. (2010). Reduced connectivity of the auditory cortex in patients with auditory hallucinations: A resting state functional magnetic resonance imaging study. *Psychological Medicine, 40*(7), 1149–1158. doi:10.1017/S0033291709991632.

Gawęda, Ł., Woodward, T. S., Moritz, S., & Kokoszka, A. (2013). Impaired action self-monitoring in schizophrenia patients with auditory hallucinations. *Schizophrenia Research, 144*(1–3), 72–79. doi:10.1016/j.schres.2012.12.003.

Haddock, G., McCarron, J., Tarrier, N., & Faragher, E. B. (1999). Scales to measure dimensions of hallucinations and delusions: The psychotic symptom rating scales (PSYRATS). *Psychological Medicine, 29*(4), 879–889.

Hoffman, R. E., Fernandez, T., Pittman, B., & Hampson, M. (2011). Elevated functional connectivity along a corticostriatal loop and the mechanism of auditory/verbal hallucinations in patients with schizophrenia. *Biological Psychiatry, 69*(5), 407–414. doi:10.1016/j.biopsych.2010.09.050.

Ikuta, T., DeRosse, P., Argyelan, M., Karlsgodt, K. H., Kingsley, P. B., Szeszko, P. R., et al. (2015). Subcortical modulation in auditory processing and auditory hallucinations. *Behavioural Brain Research, 295,* 78–81. doi:10.1016/j.bbr.2015.08.009.

Jablensky, A. (1995). Schizophrenia: recent epidemiologic issues. *Epidemiologic Reviews, 17*(1), 10–20.

Jardri, R., Pouchet, A., Pins, D., & Thomas, P. (2011). Cortical activations during auditory verbal hallucinations in schizophrenia: A coordinate-based meta-analysis. *American Journal of Psychiatry, 168*(1), 73–81. doi:10.1176/appi.ajp.2010.09101522.

Johns, L. C., Rossell, S., Frith, C. D., Ahmad, F., Hemsley, D., Kuipers, E., et al. (2001). Verbal self-monitoring and auditory verbal hallucinations in patients with schizophrenia. *Psychological Medicine, 31*(04), 705–715. doi:10.1017/S0033291701003774.

Johnsen, E., Hugdahl, K., Fusar-Poli, P., Kroken, R. A., & Kompus, K. (2013). Neuropsychopharmacology of auditory hallucinations: Insights from pharmacological functional MRI and perspectives for future research. *Expert Review of Neurotherapeutics, 13*(1), 23–36. doi:10.1586/ern.12.147.

Kircher, T. T. J., & Leube, D. T. (2003). Self-consciousness, self-agency, and schizophrenia. *Consciousness and Cognition, 12*(4), 656–669. doi:10.1016/S1053-8100(03)00071-0.

Knoblich, G., Stottmeister, F., & Kircher, T. (2004). Self-monitoring in patients with schizophrenia. *Psychological Medicine, 34*(8), 1561–1569.

Koeda, M., Takahashi, H., Matsuura, M., Asai, K., & Okubo, Y. (2013). Cerebral responses to vocal attractiveness and auditory hallucinations in schizophrenia: A functional MRI study. *Frontiers in Human Neuroscience, 7,* 221. doi:10.3389/fnhum.2013.00221.

Kühn, S., & Gallinat, J. (2012). Quantitative meta-analysis on state and trait aspects of auditory verbal hallucinations in schizophrenia. *Schizophrenia Bulletin, 38*(4), 779–786. doi:10.1093/schbul/sbq152.

Kumari, V., Fannon, D., Ffytche, D. H., Raveendran, V., Antonova, E., Premkumar, P., et al. (2010). Functional MRI of verbal self-monitoring in schizophrenia: Performance and illness-specific effects. *Schizophrenia Bulletin, 36*(4), 740–755. doi:10.1093/schbul/sbn148.

Lancaster, J. L., Summerln, J., Rainey, L., Freitas, C., & Fox, P. (1997). The talairach daemon, a database server for talairach atlas labels. *Neuroimage, 5*(4), S633.

Lancaster, J. L., Woldorff, M. G., Parsons, L. M., Liotti, M., Freitas, C. S., Rainey, L., et al. (2000). Automated talairach atlas labels for functional brain mapping. *Human Brain Mapping, 10*(3), 120–131.

Lawrie, S. M., Buechel, C., Whalley, H. C., Frith, C. D., Friston, K. J., & Johnstone, E. C. (2002). Reduced frontotemporal functional connectivity in schizophrenia associated with auditory hallucinations. *Biological Psychiatry, 51*(12), 1008–1011. doi:10.1016/S0006-3223(02)01316-1.

Liemburg, E. J., Vercammen, A., Ter Horst, G. J., Ćurčić-Blake, B., Knegtering, H., & Aleman, A. (2012). Abnormal connectivity between attentional, language and auditory networks in schizophrenia. *Schizophrenia Research, 135*(1–3), 15–22. doi:10.1016/j.schres.2011.12.003.

MacDonald, P. A., & Paus, T. (2003). The role of parietal cortex in awareness of self-generated movements: A transcranial magnetic stimulation study. *Cerebral Cortex, 13*(9), 962–967. doi:10.1093/cercor/13.9.962.

Maldjian, J. A., Laurienti, P. J., Kraft, R. A., & Burdette, J. H. (2003). An automated method for neuroanatomic and cytoarchitectonic atlas-based interrogation of fMRI data sets. *NeuroImage, 19*(3), 1233–1239.

Manoliu, A., Riedl, V., Zherdin, A., Mühlau, M., Schwerthöffer, D., Scherr, M., et al. (2014). Aberrant dependence of default mode/central executive network interactions on anterior insular salience network activity in schizophrenia. *Schizophrenia Bulletin, 40*(2), 428–437. doi:10.1093/schbul/sbt037.

Martí-Bonmatí, L., Lull, J. J., García-Martí, G., Aguilar, E. J., Moratal-Pérez, D., Poyatos, C., et al. (2007). Chronic auditory hallucinations in schizophrenic patients: MR analysis of the coincidence between functional and morphologic abnormalities. *Radiology, 244*(2), 549–556. doi:10.1148/radiol.2442060727.

McCormick, D. A., & Bal, T. (1994). Sensory gating mechanisms of the thalamus. *Current Opinion in Neurobiology, 4*(4), 550–556. doi:10.1016/0959-4388(94)90056-6.

Mechelli, A., Allen, P., Amaro, E., Fu, C. H. Y., Williams, S. C. R., Brammer, M. J., et al. (2007). Misattribution of speech and impaired connectivity in patients with auditory verbal hallucinations. *Human Brain Mapping, 28*(11), 1213–1222. doi:10.1002/hbm.20341.

Melzack, R., Israel, R., Lacroix, R., & Schultz, G. (1997). Phantom limbs in people with congenital limb deficiency or amputation in early childhood. *Brain, 120*(9), 1603–1620. doi:10.1093/brain/120.9.1603.

Menon, V., & Uddin, L. Q. (2010). Saliency, switching, attention and control: A network model of insula function. *Brain Structure & Function, 214*(5–6), 655–667. doi:10.1007/s00429-010-0262-0.

Messias, E. L., Chen, C.-Y., & Eaton, W. W. (2007). Epidemiology of schizophrenia: Review of findings and myths. *Psychiatric Clinics of North America, 30*(3), 323–338. doi:10.1016/j.psc.2007.04.007.

Mou, X., Bai, F., Xie, C., Shi, J., Yao, Z., Hao, G., et al. (2013). Voice recognition and altered connectivity in schizophrenic patients with auditory hallucinations. *Progress in Neuro-Psychopharmacology and Biological Psychiatry, 44,* 265–270. doi:10.1016/j.pnpbp.2013.03.006.

Murray, C.J., & Lopez, A.D. (eds). (1996). The global burden of disease: A comprehensive assessment of mortality and disability from diseases, injuries and risk factors in 1990 and projected to 2020. In *Global burden of disease and injury series* (Vol. 1). Cambridge, MA: Published by the Harvard School of Public Health on behalf of the World Health Organization and the World Bank.

Murray, C. J., & Lopez, A. D. (1997). Alternative projections of mortality and disability by cause 1990–2020: Global burden of disease study. *The Lancet, 349*(9064), 1498–1504. doi:10.1016/S0140-6736(96)07492-2.

Nayani, T. H., & David, A. S. (1996). The auditory hallucination: A phenomenological survey. *Psychological Medicine, 26*(1), 177–189.

Oertel-Knöchel, V., Knöchel, C., Matura, S., Prvulovic, D., Linden, D. E. J., & van de Ven, V. (2013). Reduced functional connectivity and asymmetry of the planum temporale in patients with schizophrenia and first-degree relatives. *Schizophrenia Research, 147*(2–3), 331–338. doi:10.1016/j.schres.2013.04.024.

Oertel-Knöchel, V., Knöchel, C., Matura, S., Stäblein, M., Prvulovic, D., Maurer, K., … van de Ven, V. (2014). Association between symptoms of psychosis and reduced functional connectivity of auditory cortex. *Schizophrenia Research, 160*(1–3): 35–42. https://doi.org/10.1016/j.schres.2014.10.036.

Parekh, P., John, J. P., Menon, S., Halahalli, H. N., & Kutty, B. M. (2016). Functional connectivity patterns underlying the experience of auditory verbal hallucinations in patients with schizophrenia. *bioRxiv*, 062992. https://doi.org/10.1101/062992.

Penfield, W., & Perot, P. (1963). The brain's record of auditory and visual experience. *Brain, 86* (4), 595–696. doi:10.1093/brain/86.4.595.

Phelps, E. A. (2004). Human emotion and memory: interactions of the amygdala and hippocampal complex. *Current Opinion in Neurobiology, 14*(2), 198–202. doi:10.1016/j.conb.2004.03.015.

Plaze, M., Bartrés-Faz, D., Martinot, J.-L., Januel, D., Bellivier, F., De Beaurepaire, R., … Paillère-Martinot, M.-L. (2006). Left superior temporal gyrus activation during sentence perception negatively correlates with auditory hallucination severity in schizophrenia patients. *Schizophrenia Research, 87*(1–3): 109–115. https://doi.org/10.1016/j.schres.2006.05.005.

Pu, W., Li, L., Zhang, H., Ouyang, X., Liu, H., Zhao, J., et al. (2012). Morphological and functional abnormalities of salience network in the early-stage of paranoid schizophrenia. *Schizophrenia Research, 141*(1), 15–21. doi:10.1016/j.schres.2012.07.017.

Raij, T. T., & Riekki, T. J. J. (2012). Poor supplementary motor area activation differentiates auditory verbal hallucination from imagining the hallucination. *NeuroImage: Clinical, 1*(1): 75–80. https://doi.org/10.1016/j.nicl.2012.09.007.

Raij, T. T., Valkonen-Korhonen, M., Holi, M., Therman, S., Lehtonen, J., & Hari, R. (2009). Reality of auditory verbal hallucinations. *Brain, 132*(11), 2994–3001. doi:10.1093/brain/awp186.

Rolland, B., Amad, A., Poulet, E., Bordet, R., Vignaud, A., Bation, R., … Jardri, R. (2014). Resting-state functional connectivity of the nucleus accumbens in auditory and visual hallucinations in schizophrenia. *Schizophrenia Bulletin*, sbu097. https://doi.org/10.1093/schbul/sbu097.

Rössler, W., Joachim Salize, H., van Os, J., & Riecher-Rössler, A. (2005). Size of burden of schizophrenia and psychotic disorders. *European Neuropsychopharmacology, 15*(4), 399–409. doi:10.1016/j.euroneuro.2005.04.009.

Rotarska-Jagiela, A., van de Ven, V., Oertel-Knöchel, V., Uhlhaas, P. J., Vogeley, K., & Linden, D. E. J. (2010). Resting-state functional network correlates of psychotic symptoms in schizophrenia. *Schizophrenia Research, 117*(1), 21–30. doi:10.1016/j.schres.2010.01.001.

Sanjuán, J., Lull, J. J., Aguilar, E. J., Martí-Bonmatí, L., Moratal, D., Gonzalez, J. C., et al. (2007). Emotional words induce enhanced brain activity in schizophrenic patients with auditory hallucinations. *Psychiatry Research: Neuroimaging, 154*(1), 21–29. doi:10.1016/j.pscychresns.2006.04.011.

Shergill, S. S., Bullmore, E., Simmons, A., Murray, R., & McGuire, P. (2000). Functional anatomy of auditory verbal imagery in schizophrenic patients with auditory hallucinations. *American Journal of Psychiatry, 157*(10), 1691–1693. doi:10.1176/appi.ajp.157.10.1691.

Shergill, S. S., Brammer, M. J., Fukuda, R., Williams, S. C. R., Murray, R. M., & McGuire, P. K. (2003). Engagement of brain areas implicated in processing inner speech in people with auditory hallucinations. *The British Journal of Psychiatry, 182*(6), 525–531. doi:10.1192/bjp.182.6.525.

Shergill, S. S., White, T. P., Joyce, D. W., Bays, P. M., Wolpert, D. M., & Frith, C. D. (2014). Functional magnetic resonance imaging of impaired sensory prediction in schizophrenia. *JAMA Psychiatry, 71*(1), 28–35. doi:10.1001/jamapsychiatry.2013.2974.

Shinn, A. K., Baker, J. T., Cohen, B. M., & Öngür, D. (2013). Functional connectivity of left Heschl's gyrus in vulnerability to auditory hallucinations in schizophrenia. *Schizophrenia Research, 143*(2–3), 260–268. doi:10.1016/j.schres.2012.11.037.

Simons, C. J. P., Tracy, D. K., Sanghera, K. K., O'Daly, O., Gilleen, J., Dominguez, M.-G., et al. (2010). Functional magnetic resonance imaging of inner speech in schizophrenia. *Biological Psychiatry, 67*(3), 232–237. doi:10.1016/j.biopsych.2009.09.007.

Sommer, I. E. C., Diederen, K. M. J., Blom, J.-D., Willems, A., Kushan, L., Slotema, K., et al. (2008). Auditory verbal hallucinations predominantly activate the right inferior frontal area. *Brain, 131*(12), 3169–3177. doi:10.1093/brain/awn251.

Sommer, I. E., Clos, M., Meijering, A. L., Diederen, K. M. J., & Eickhoff, S. B. (2012). Resting state functional connectivity in patients with chronic hallucinations. *PLoS ONE, 7*(9), e43516. doi:10.1371/journal.pone.0043516.

Spence, S. A., Brooks, D. J., Hirsch, S. R., Liddle, P. F., Meehan, J., & Grasby, P. M. (1997). A PET study of voluntary movement in schizophrenic patients experiencing passivity phenomena (delusions of alien control). *Brain, 120*(11), 1997–2011. doi:10.1093/brain/120.11.1997.

Sperry, W. R. (1950). Neural basis of the spontaneous optokinetic response produced by visual inversion. *Journal of Comparative and Physiological Psychology, 43*(6), 482–489. doi:10.1037/h0055479.

Takahashi, E., Ohki, K., & Miyashita, Y. (2002). The role of the parahippocampal gyrus in source memory for external and internal events. *NeuroReport, 13*(15), 1951–1956.

Thoma, R. J., Chaze, C., Lewine, J. D., Calhoun, V. D., Clark, V. P., Bustillo, J., … Turner, J. A. (2016). Functional MRI evaluation of multiple neural networks underlying auditory verbal hallucinations in schizophrenia spectrum disorders. *Schizophrenia*, 39. https://doi.org/10.3389/fpsyt.2016.00039.

Vercammen, A., Knegtering, H., den Boer, J. A., Liemburg, E. J., & Aleman, A. (2010). Auditory hallucinations in schizophrenia are associated with reduced functional connectivity of the temporo-parietal area. *Biological Psychiatry, 67*(10), 912–918. doi:10.1016/j.biopsych.2009.11.017.

Vercammen, A., Knegtering, H., Bruggeman, R., & Aleman, A. (2011). Subjective loudness and reality of auditory verbal hallucinations and activation of the inner speech processing network. *Schizophrenia Bulletin, 37*(5), 1009–1016. doi:10.1093/schbul/sbq007.

Vogeley, K. (2003). Schizophrenia as disturbance of the self-construct. In *The self in neuroscience and psychiatry*. Cambridge University Press. Retrieved from http://dx.doi.org/10.1017/CBO9780511543708.018

von Holst, E., & Mittelstaedt, H. (1950). Das Reafferenzprinzip. *Naturwissenschaften, 37*(20), 464–476. doi:10.1007/BF00622503.

Whitfield-Gabrieli, S., & Nieto-Castanon, A. (2012). Conn: A functional connectivity toolbox for correlated and anticorrelated brain networks. *Brain Connectivity, 2*(3), 125–141. doi:10.1089/brain.2012.0073.

WHO. (2001). *The world health report 2001—mental health: New understanding, new hope*. Geneva. Retrieved from http://www.who.int/whr/2001/en/.

Wolf, N. D., Sambataro, F., Vasic, N., Frasch, K., Schmid, M., Schönfeldt-Lecuona, C., … Wolf, R.C. (2011). Dysconnectivity of multiple resting-state networks in patients with schizophrenia who have persistent auditory verbal hallucinations. *Journal of Psychiatry & Neuroscience: JPN*, 36(6): 366–374. https://doi.org/10.1503/jpn.110008.

World Health Organization. (n.d.). *The ICD-10 classification of mental and behavioural disorders: Clinical descriptions and diagnostic guidelines*. Retrieved from http://apps.who.int/classifications/icd10/browse/2016/en.

Xia, M., Wang, J., & He, Y. (2013). BrainNet viewer: A network visualization tool for human brain connectomics. *PLoS ONE, 8*(7), e68910. doi:10.1371/journal.pone.0068910.

Zhang, Z., Hao, G.-F., Shi, J.-B., Mou, X.-D., Yao, Z.-J., & Chen, N. (2008a). Investigation of the neural substrates of voice recognition in Chinese schizophrenic patients with auditory verbal hallucinations: An event-related functional MRI study. *Acta Psychiatrica Scandinavica, 118*(4), 272–280. doi:10.1111/j.1600-0447.2008.01243.x.

Zhang, Z., Shi, J., Yuan, Y., Hao, G., Yao, Z., & Chen, N. (2008b). Relationship of auditory verbal hallucinations with cerebral asymmetry in patients with schizophrenia: An event-related fMRI study. *Journal of Psychiatric Research, 42*(6), 477–486. doi:10.1016/j.jpsychires.2007.04.003.

Zhu, J., Wang, C., Liu, F., Qin, W., Li, J., & Zhuo, C. (2016). Alterations of functional and structural networks in schizophrenia patients with auditory verbal hallucinations. *Frontiers in Human Neuroscience*, 114. https://doi.org/10.3389/fnhum.2016.00114.

Zhuo, C.-J., Zhu, J.-J., Wang, C.-L., Wang, L.-N., Li, J., & Qin, W. (2016). Increased local spontaneous neural activity in the left precuneus specific to auditory verbal hallucinations of schizophrenia. *Chinese Medical Journal, 129*(7): 809. https://doi.org/10.4103/0366-6999.178974.

Body and Self-reflection: The Crux of Yoga Philosophy and Practice

Sangeetha Menon

Both philosophy and psychology have debated for centuries the intricate workings of the idea as well as the experience of the body and the self. The notion of the body, particularly in philosophy, is a complex subject giving leads for both epistemological and metaphysical concepts such as the means of knowledge, the meaning of existence and the nature of true selfhood. The philosophy of yoga tradition as given in the Yoga Sutras of Patanjali is one of the oldest and significant approaches to analyse the relation between the body and the mind in the context of their interrelations and the nuanced concept of the self.

Introduction

It is speculated that the time of Sage Patanjali can be traced between 400 BCE and AD 200. Sculptural and iconic representations, archaeological evidences and older philosophical literature imply that yoga was being practised in ancient India as early as 3000 BCE. The nature of yoga as presented in the Yoga Sutras is predominantly of the Raja yoga tradition. It is believed that Sage Patanjali authored this text covering major concepts of yoga practice and its theory. Yoga psychology is founded on the theoretical philosophy of the Samkhya system, a classical tradition of Indian philosophy. Samkhya world view is based on dualistic metaphysics to understand the nature of ultimate reality, and the relation between body and consciousness, which appear as the subject and the object, the knower and the known. Samkhya philosophy (Suryanarayana Sastri 1973) enunciates that there are two independent real entities—*prakṛti* and *puruṣa*. *Prakṛti* is constituted by all that is

S. Menon (✉)
NIAS Consciousness Studies Programme, National Institute of Advanced Studies, Bengaluru 560012, India
e-mail: prajnanata@gmail.com

S. Menon et al. (eds.), *Self, Culture and Consciousness*,
DOI 10.1007/978-981-10-5777-9_9

material. *Puruṣa* is consciousness and is the entity that which experiences the results of material interactions. Such a dualist framework is used to explain the relation between the material objects and the conscious subject who is the experiencer. Heinrich Zimmer says: 'The two principles *Prakṛti* (composed of the gunas) and purusa (the collectivity of radiant but inactive life-monads)—are accepted as eternal and real on the basis of the fact that in all acts and theories of knowledge a distinction exists between subject and object, no explanation of experience being possible without the recognition of a knowing self as well as of an object known' (Zimmer 1951).

The four chapters of the Yoga Sutras discuss: meditation, the steps to yoga, attainment of psychic powers (*siddhi*), and the nature of *kaivalya*—the final outcome of yoga.[1] Through the four chapters, Patanjali brings forth the importance of the experience of the body as a wholesome phenomenon intricately connected with the mind, mental processes and even psychosocial aspects. 'The body-mind organism according to Yoga psychology is a mini world. As a matter of fact, the world outside is a reflection of the world inside. The world happens within you. The body is unstable equilibrium—constantly adjusting to the environment' (Bodhananda 2010: 29). Patanjali thoroughly analyses the different modifications of mind, and how such modifications can be restrained and even transcended.

From the Body to Cognition to Experience

Self-reflection is awareness about one's mental states, in the context of collective awareness of the continuing personal identity, frilled by memories, beliefs and worldviews. Reflecting upon another's mind entails complex processes that are facilitated by prior self-awareness (Menon 2014). Patanjali's system of Ashtanga yoga or the eight-fold path that consists of *yama*, *niyama*, *āsana*, *prānāyāma*, *pratyāhāra*, *dhārana*, *dhyāna* and *samādhi* is an integral model covering the social, emotional and personal world of the individual. The object of yoga is to divest the mind of all its impurities, so that the stream of consciousness is clear as a crystal capable of reflecting whole reality (Safaya 1976: 270). *Yama* is a set of values to be practised, such as non-violence, truthfulness, non-stealing, celibacy and non-possession.[2] These values are universal great disciplines regardless of caste, country, time and occasion.[3] The idea behind these values is that the practitioner of these values will be able to cultivate a calm mind, and in the presence of a contemplative person, the mind of a disturbed person will also be quietened.[4]

[1]The translations and the interpretative meanings of the Yoga Sutras are taken from Bodhananda (2008).

[2]YS 2.30—*ahimsa satya asteya brahmacarya aparigraha yamah.*

[3]YS 2.31—*ete jati desa kala samaya anavacchinah saarvabhaumah mahavrtam.*

[4]YS 2.35—*ahimsa pratishtaayaam tat sannidau vairatyaaga.*

The practice of truthfulness leads to successful outcomes.[5] When one is established in continence there is gain of energy.[6] When one is established in non-possession knowledge of past lives arise.[7] It is to be particularly noticed here that the practice of the set of values in the first step of *yama* is intended to lead to mental calm and also the ability to reflect upon past experiences.

The second set of practices—*niyama*—requires more of the involvement of the person and self-recognition. These are observances such as cleanliness, contentment, discipline, self-study and surrender to a higher power.[8] As a result of the practice of cleanliness one can detach from one's own body and can avoid association with other bodies.[9] As a result of the practice of purity of mind—what is attained is cheerfulness—*saumansyam*, one-pointedness—*aigāgryam*, mastery of senses—*indriyajayam*, and *ātma darśana yogatvam*—fitness for self-perception.[10] When one is contented one gains incomparable joy.[11]

An important and more widely known component of the eight-fold yoga is *asana*—postures. The right posture is that which is firm and comfortable.[12] The commonly recommended postures are *padmāsanam*, *virāsanam*, *bhadrāsanam*, *sukhāsanam* and *savāsanam*. As a result of the mastery of postures, one gains mastery over the binaries in experiences, and will be unaffected by the pairs of opposites.[13]

The fourth limb of yoga is *prānāyāma*—control of breath—by the holding of breath after inhaling and holding breath after exhaling.[14] With the practice of *prānāyāma*, the mind becomes fit for cognitive capabilities such as concentration.[15] As a result of *prānāyāma* there is thinning of the veil covering the light.[16] Through this metaphor of veil of darkness that covers the light which gets diminished by the practice of *prānāyāma*, one could speculate that, what is indicated is that there will be emergence of self-certainty, through better self-perception. Through the practice of *pratyāhāra*, when mind is detached from its objects, following the detached nature of mind, sense organs also naturally withdraw.[17] Then there is supreme control over sensory and motor organs.[18] The limbs of *prānāyāma* and *pratyāhāra*

---

[5]YS 2.36—*satya pratiṣtāyām kriya phala āśrayatvam.*

[6]YS 2.38—*brahmacarya pratiṣtāyām virya lābha.*

[7]YS 2.39—*aparigraha sthairye janma kathntā sambodhah.*

[8]YS 2. 32—*śauca-santoṣa-tapaḥ-svādhyāyeśvara-prañidhānāni niyamāḥ.*

[9]YS 2.40—*śaucāt svānga-jugupsā parair asansargaḥ.*

[10]YS 2.41—*sattva-śuddhi-saumanasyaikāgryendriya-jayātma-darśana-yogyatvāni ca.*

[11]YS 2.42—*santoṣād anuttamaḥ sukha-lābhaḥ.*

[12]YS 2.46—*sthira sukham āsanam.*

[13]YS 2.48—*tato dvandvānabhighātaḥ.*

[14]YS 2.49—*tasmin sati śvāsa-praśvāsayor gati-vicchedaḥ prāṇāyāmaḥ.*

[15]YS 2.53—*dhārañāsu ca yogyatā manasaḥ.*

[16]YS 2.52—*tataḥ kśīyate prakāśāvaraṇam.*

[17]YS 2.54—*sva-viṣayāsamprayoge cittasya svarupānukāra ivendriyāṇām pratyāhāraḥ.*

[18]YS 2.55—*tataḥ paramā vaśyatendriyāṇām.*

together gives the benefit of better self-perception, and the sense of control over one's sensory organs and motor organs, and thus strengthening authorship and ownership in one's sense of agency.

Agency and Its Body

The Patanjali Yoga Sutras' approach to mind, body and the self is important in the context of current interdisciplinary discussions on agency. Typically, agency is contended to be the capacity to perform an action. Agency implies an agent intending and owning an action. Discussions in both biology and ancient Greek philosophy attribute agency to not just human beings but also other animals. Can agency be just a capability to express one's existence, or does it involve the nuanced power to be aware of oneself, to be rational, to have a theory of mind, and through self-reflection correct one's course of action?

The pertinent issue here is whether the body is a deeper experience that cannot exist without the mind, the embodied sense. Or is body only a notion to indicate physically existent objects, such as moon, sun, trees and chair? The discussion on what is it to be a body often is overshadowed by the discussion on the reality and the importance of the body itself. Except for the phenomenologists, and the recent embodiment theorists there is hardly any dialogue on the *being* of the body.

A mechanised notion of the body often leads to the use of thought experiments such as rubber hand illusions to hastily conclude that the body prescriptions we hold can be erroneous. The attempts to negate the defined intuitions we have about one's body are continuation of the classical philosophical approach to either deny the reality of the body or to posit the existence of a non-material subject. Beyond the emphasis on the classical divide between the body and the non-material, thought experiments attempt to demonstrate that the fringes of our body notions highlight the non-existence of a continuous agent and owner of the body—namely, the self. Since the body is the self, and since the contours of the body sense are fleeting, the independent existence of a self for the body is also dubitable, according to the fleeting-self theorists.

The meaning for the body emerges from its undeniable relation with action. An action is not possible without a body in the physical world. In that sense, the body is that which holds, initiates and oversees the course of an action, the beholder of the agent. Agency is typically defined as the sense of authorship of one's actions (Kircher and Leube 2003), and that which projects the 'who system', with underlying cognitive processes (Georgieff and Jeannerod 1998). The authorship of an action coincides with the control over an action and thereby owning the origin of control and the consequences. Yet empirical evidence from neuropsychiatry suggests that the awareness of an action being performed and the felt control over the action need not always be part of the same awareness, as in the case of self-delusions and schizophrenia.

The complexity in the construal of agency is also the result of the lack of phenomenological and ontological richness in the conceptualisation of the self and the agent. Rubber hand illusion experiments and similar experiments that argue for the dispersed nature of the self are primarily flawed by a conceptual framework that focus on behavioural inputs without the recognition of the fundamental subjective experience and deeper self-ascriptions that accompany. The phenomenal aspects of being an agent includes an array of both apparent and subliminal experiential content that pertains to beliefs, world views, perceptions and understanding about one's sociocultural context. More specifically, one initiates an action, based on goal awareness, having an intent to act with a sense of control and purpose, and being aware of one's freedom that causes the mind to intend and cause the act. Empirical conceptions do not consider these nuanced phenomenal elements in the construal of agency and the agent. The discussion on agency in the context of consciousness and action is thus often limited to sensory-motor activities and impairments.

Agency and Its Experience

To conceptualise agency from a truly experiential point of view is to consider the cause of various phases of agency into a dimension of consciousness that might be depicted as acausal and apodeictic. Can agency be considered not as a monolithic idea but as a larger field, not as a pole but more a field through which all experiences pass (Kircher and Leube 2003)? Can the concept of agency be completely objectified without the deeper recesses of the core self touching it? A major problem for working on pure objectivity of consciousness is that at any given moment to observe the act of observation of observation is not possible until all the three acts are not brought into one point of time. If observation must be an object of enquiry then at any given time a subjective component should persist impervious to the act of observation. The knower is not just intricately but indivisibly connected with observation because of which we can objectify only the contents of observation and not the observer of the act *per se*.

Our daily experience tells us that to behold an experience and derive meaning from it, both authorship and the sense of self are required. Experience is made meaningful by the continued 'feeling' of the self. A sense of enduring and pervasive identity is the starting point of all experiences. But the challenge is that the content of consciousness undergoes a series of changes such that each frame or part gets redefined and reconstituted in the act of 'capture' (Menon 2014).

When on one side 'agency' posits the space for experimentation in order to understand the minimal nature of a selfhood, on the other it raises fundamental questions on free will, choice-making and responsibility. The latter three cannot be discussed without referring to a self which is not only just functionally minimal but also rich with emotional, social and spiritual aspects. Thus, self-awareness ranges from the minimal concept of the apparent body-image to core self-recognition. de Vignemonta and Fourneret (2004) differentiate between the 'minimal self'

(Gallagher 2000) and a 'recognitional self' to avoid reducing self-recognition to the concept of a self-image. They explain this idea by differentiating two kinds of I-thoughts that use different notions of the self. In self-recognition which involves acts like grasping a glass of water, the sense of agency involves the notion of a 'minimal self' which is instantaneous. It carries only a small amount of information, such as 'I am grasping a glass'. But recognising one's image in the mirror depends on recognitional criteria that allows one to re-identify oneself through time. Hence 'recognitional self' has richer content and could explain personal identity. The same argument can also be extended to demonstrate how body-image disruptions caused by rubber hand illusion experiments cannot be an instance for deducing the absence of a coherent self. Self-awareness which is influenced by agency and choice-making is also influenced by the degree of self-reflection of which one is capable (Rao and Menon 2016).

Bodily Pains and Mental Planes

The Yoga Sutras' approach to consciousness is important in the background of the above discussion on self, self-recognition and self-reflection. The Yoga Sutras' theory is based on a model that integrates various emotional, cognitive and social aspects of the agency, and with a purpose of transforming the agent for gaining positive experiences. Consciousness sustains the triad of objective discreteness without being a part of the triad.

The *citta* concept in yoga corresponds to the *antaḥkaraṇa* concept of Vedanta and includes *manas* (mental), *buddhi* (cognitive) and *ahaṃkāra* (agent). The basic philosophy is that when the modifications of the citta are restrained, the subject abides in its true nature. *Citta*, though for linguistic ease is translated as 'mind', is much more than the mental and emotive content. From the various structures and typologies that Patanjali uses, we can understand that '*citta*' is a term that houses collective features of an individual that pertain to his/her existential, phenomenological, pragmatic, ethical, psychological and spiritual aspects (Menon 2009). Thus the Yoga Sutras' concept of *citta* cannot be translated to the concept of mind as we understand in cognitive sciences.

The yoga philosophy enunciates that when the modifications of *citta* are restrained, the individual abides in his/her true nature. The restrained *citta*, grounded on the space of the core self will not be distracted by thoughts, sensations or memories. And, when there is no restraint, *citta* identifies with modifications and not pure consciousness.[19] A higher state of the *citta* system is one that which is able to have greater control over modifications of *citta* so as to reflect upon the subjective self as different from the objective world. In other words, the ability to

[19]YS 1.3—*tadā draśṭuḥ svarūpe 'vasthānam.*
YS 1.4—*vṛtti-sārūpyam itaratra.*

distinguish the conscious self who is the agent (*puruṣa*) from the content of experience which can be objectified (*prakṛti*) determines the integration of the self (Rao and Menon 2016). The analysis of mental modifications and the planes of self as discussed in the Yoga Sutras emphasises that the quality of self-reflection is dependent upon the individual's perception of his/her self as a foundational core, or as a relational entity (Menon 2009).

According to Patanjali and his yoga philosophy, the mind is a complex entity that has several planes that relate to the physical and social world. The psychological phenomena are conditioned by five levels. In his commentary to the second sutra of Patanjali, Vyasa talks about *citta bhūmayaḥ*—five levels of *citta*:

kṣiptam—unstable
mudham—confused and obscure
vikṣiptam—distracted (stable and unstable)
ekāgram—attentive (attention fixed on a single point)
niruddham—absorbed (completely restrained)

All psychological activities and mental states happen within the larger space of the five levels of *citta*—*citta bhumi*. The first two levels relate to the ordinary dispositions of mind. It is the nature of mind to be unstable at some point, and confused. It is also the nature of an ordinary mind to be distracted with periodicity of stability and instability. Examples given by Mircea Eliade for the third plane are: in an effort of memory or on a problem in mathematics (Eliade 1975: 52). The third and fourth levels are reached through the exercise of attention and concentration. While the first three planes are undesirable, the last two levels are desirable and positive. The last level restraints the first three negative levels.

The *citta* that is in its earliest stage of disturbance (*kṣiptam*—unstable) lacks judgement and is generally hyperactive, unable to ignore external stimuli. The next stage of the confused (*mudha*) state of *citta* is marked by inertia, lethargy, sluggishness, vice, ignorance and sleep. The state of distraction (*vikṣiptam)* is an advanced stage of the *kṣiptam* plane of *citta*. In this level, the mind lacks stability and is unable to self-reflect (Bodhananda 2008).

An important concept of Patanjali's yoga psychology is '*kleśa*' a complex notion of pain that is built around not only physical, but also existential and ontological, aspects of the individual. *Kleśa* causes mind-modifications that are pleasurable (*kliṣta-vṛtti*) or unpleasant (*akliṣta-vṛtti*).[20] *Kleśa* is not an antonym of pleasure but is suffering in the technical sense. Patanjali discusses in detail five dimensions of pain in the second chapter enumerating its nature, implications and ways to respond to them.[21] The limitations of the *citta* system is collectively called *kleśa*. The *kleśa* set of pains is five in number: ignorance (*avidya*), egoism (*asmita*), likes (*rāga*), dislikes (*dveṣa*) and attachment (*abhiniveśa*). Of the list of five, self-ignorance

[20]*Kliṣta-vṛtti* are those states of mind, we are told in the *Yoga-bhasya*, which are caused by *kleśa* (*kleśa-hetukaḥ*), whereas the *Yoga-varttika* describes them as those which lead to suffering (*kleśa*).
[21]YS 2.4 to 2.9.

(*avidya*) holds a significant place since the rest of the *kleśa* are founded on it. Self-ignorance is the source from where all other afflictions spring, and is therefore the foremost ontological pain that need to be addressed. Patanjali gives a grade for the intensity of self-ignorance,[22] from that which is dormant and feeble (pertaining more to the causal and subtle body) to the spasmodic and full-blown (pertaining to the gross body). In the last two planes of *citta*—*ekāgram* and *niruddham*—the different psychological modifications and the mind cease to be affected by the five sets of pains (*kleśa*). They are the calmest and most peaceful states of mind. The invoking of the last two planes of *citta* reduces the first three planes. Further, the attentive *citta* system (*ekāgram*) is very near to reaching *samādhi*. In the level of *niruddham* there are no mental modifications and the *citta* is prepared for the experience of the unitary state of consciousness—*samādhi*.

The Philosophical Basis of Pain and the Way Out

The yoga philosophy of pain goes beyond the physical and the mental and is situated on the concept of having no knowledge or wrong knowledge about the true nature of self. In other words, avidya is beholding a wrong self-identity, by misapprehending the permanent, pure (indivisible), blissful, limitless self (*ātman*) to be the impermanent, impure and unhappy self (*anātman*). It is confounding pure consciousness and the content of consciousness,[23] that gives rise to another level of ontological pain called *asmita*—the I-ness. With this pain, the being of the seer and the capability of cognition are merged and no distinction is made. It is confounding pure consciousness with intentional consciousness (*citta* and its modifications).[24]

The inability to distinguish between the seer and the seen, the experiencer and the experienced, also gives rise to the existential disposition of affinity towards certain objects and people, and non-affinity towards others—*rāga* and *dvesha*. *Rāga* is mind dwelling on pleasurable objects, persons or relations. *Dveṣa* is mind dwelling upon unpleasant objects, persons or relations.[25] The habit of giving preferences and the binary approach to objects of experience and their outcomes result in the dependence on the immediate physical identity which is the body—*abhiniveśa*. *Abhiniveśa* is primarily the deep rooted and compulsive attachment to body, which is present even amongst scholars, according to Patanjali.[26] By emphasising that even scholars (*viduṣa*) are not exempt from the pain of *abhiniveśa* that is deeply rooted in the attachment to the physical body, Patanjali perhaps

[22]YS 2.4—*avidyā kṣetram uttareṣām prasupta-tanu-vicchinnodārāṇām.*

[23]YS 2.5—*anityāśuci-duḥkhānātmasu nitya-śuci-sukhātma-khyātir avidyā.*

[24]YS 2.6—*dṛg-darśana-śaktyor ekātmatevāsmitā.*

[25]YS 2.7—*sukhānuśayī rāgaḥ.*
YS 2.8—*duḥkhānuśayī dveṣaḥ.*

[26]YS 2.9—*sva-rasa-vāhī viduṣo 'pi tathārūḍho 'bhiniveśaḥ.*

wished to clarify that mere knowledge about how the *citta* system works does not help.

All the five afflictions constitute pain and suffering. Pain is primarily due to the false association and identification (*samyoga*)—of the seer and the seen, which is to confound the state of being (the experiencer or observer) with a capability (such as knowing and experiencing). Though the seer is always pure consciousness it is perceived through *citta*-modifications—*pratyaya anupaśya*. The disassociation of pure consciousness and *citta*-modifications is the only way to remove the false union and thereby the ensuing pain. That is the only way for the seer to be realised as oneness alone—*kaivalyam.*[27]

Mental modifications are resultant of the cues received from the sensate world. The *citta* system also has the power to initiate restraint of its modifications. Such a restraint according to yoga psychology leads to discriminative knowledge. Life experiences can be of pleasure—*hlāda*, or pain—*paritāpā*, according to the nature of acts. Because of anxiety caused by change and uncertainty—*pariṇāma tāpa*, fear of the need for repeated experiences (habit formations) and contradictions involved in modifications of *guna* (mood changes), for a person who sees distinctly—the ontological being and the cognitive capability—*vivekinah*, everything is pain. Hence even pains that are expected to happen in the future—*dḥukham ānāgatam*—are to be avoided. This significant idea is clear in Vyasa's commentary on YS.1.14, when he says that even at the time of enjoyment of pleasure (*visaya sukha kale'pi*) the yogi discerns it as painful and as a hindrance (*pratikūlātmakam*) to self-realisation.

Is there a way out from the pain that seems to be the defining aspect of living? According to the Yoga Sutras, 'the river called citta flows in two directions'.[28] Vyasa comments on this sutra (YS 1.12), giving the two steps for the restraint of mind—*abhyāsa* and *vairāgya*—describing the twin nature of psychological capabilities. Mind can flow towards good and evil—*vahati kalyāṇāya vahati pāpāya ca.* The propensity of the mind is both towards the sensate world and the spiritual world.

With the help of an allegorical style Vyasa describes the twin possibility hidden in mind. When it is borne onward to *kaivalya*, downward towards discrimination, then it is flowing unto good; when it is borne onward to the whirlpool-of-existence, downward towards non-discrimination, then it is flowing unto evil. In these cases, the stream towards objects is dammed by passionlessness (*vairāgya*) and the stream towards discrimination has its floodgates opened by practice in discriminatory knowledge (Woods 1914).[29] Vacaspati further qualifies that 'borne onward to' means a continuous connection (to the sensate world), and 'downward towards'

---

[27]YS 2.17—*draṣṭṛ-dṛśyayoḥ saṃyogo heya-hetuḥ.*
YS 2.20—*draṣṭā dṛśi-mātraḥ śuddho 'pi pratyayānupaśyaḥ.*
YS 2.25—*tad-abhāvāt saṃyogābhāvo hānaṃ tad-dṛśeḥ kaivalyam.*

[28]*citta nadī nāmah ubhayatāvāhini*—*Yoga-bhāṣya* on 1.12.

[29]See *Yoga-bhasya* for 1.12 (Woods 1914).

suggest depth (Woods 1914).[30] The *citta*-river—*citta-nadi*—can have continuous connection with and depth in both directions: the sensate world as well as *kaivalya* —the state of *samādhi*. The cause of the composite pain—*kleśa mūla*—is the result of past deeds, which in effect is the reservoir of *samskara*—memories and propensities—responsible for the body in the present life and those awaiting in future births. *Samskāra* is the cause for the specific nature of the afflictions. As long as this cause exists—*sati mule*—until then its evolutes—*tat vipākam*—such as birth, life span and enjoyment, will occur. *Samskāra* has its lasting effect on life (present and future).

Another important aspect of the mind model for Patanjali is the set of nine impediments (*antaraya*) that share physical, emotional and personality features. All the nine impediments, which are of a dysfunctional nature, are accompanied by symptoms such as pain, despair, trembling of body and heavy breathing. These symptoms are specifically the features of a distracted level of mind—*vikṣepaḥ*.[31] The nine impediments are[32]: physical and mental illness (*vyādhi*), disinterest and indifference (*sthyānam*), paranoiac doubt and lack of trust (*samśayam*), carelessness (*pramādam*), lethargy (*ālasyam*), inability to restrain indulgence (*avirati*), delusionary visions (*bhrānti-darśanam*), a sense of lack of progress and a plateau feeling (*alabdha bhūmigatvam*), and digression (*anavastitatvam*).

Four Features of Self-reflection

From the *citta* model, and the related concepts of *citta bhumanyah*, *kleśa* and *antaraya*, we can assume that the yoga theory of mind is not focused on the awareness of the other, but gives importance to self-awareness, and the ability to self-reflect. Self-reflection is the ability to reflect upon oneself objectifying not only the content of experience, but also the author and the experiencer of experience, and establishing oneself in an integrated self. The different systems of Indian philosophy that have together contributed to the field of Indian psychology suggest that the self has to be taken not only as a complex concept but also as that which is driven by the deeper abilities to be self-aware and self-reflect (Rao 2010).

The experience of awareness of another individual's mental states demonstrates that I already possess a self of mine with which I reflect. To reflect upon my mind and thoughts, it is not necessary that I am aware of another's mental states. The more I can reflect upon myself and retrieve events from the narratives I have formed of my past experiences, the more likely that I will be sensitive to another individual's mental and physical states.

---

[30]See *Tattva Vaiśāradhi* for YS 1.12 (Woods 1914).

[31]YS 1.31—*duḥkha-daurmanasyāṅgam-ejayatva- śvāsa-praśvāsā vikṣepa-sahabhuvaḥ*.

[32]YS 1: 30—*vyādhi-styāna-sanśaya-pramādālasyāvirati-bhrānti-darśanālabdha-bhūmikatvānava sthitatvāni cittavikṣepās te 'ntarāyāḥ*.

The yoga phenomenology focuses on the inner awareness of the self rather than on the awareness of information received from an external environment. For this reason, the practitioner of yoga is able to situate oneself on one's body and mind in a deeper and undistracted manner. The 'other' becomes subsidiary and the 'self' becomes the primary subject of attention. The primary philosophical tenet of *viveka buddhi* is the ability to clearly distinguish between the conscious self and the content of experience, which includes mind-modifications, that makes an experience meaningful, whether it is ordinary or heightened, such as *samādhi*. *Avidya* is removed by *viveka khyāti*—discriminative knowledge (of pure consciousness and its content) between the *anātman* and *ātma*. It is the unbroken knowledge borne of discrimination—*aviplava viveka khyāti*—between *dṛk* and *dṛśya*, and is the contemplative wisdom—*prajñā*.[33] For one who is discriminating—*viśeṣaḥ darśinaḥ*—the mind will not be mistaken for the self. The self which is soaked in distinguishing knowledge gravitates—*prāg-bhāraṃ*—towards liberation—*kaivalya*.[34]

The practice of self-reflection leads to the experiential distinction between the self and the other and it involves two features, according to yoga psychology: *vitarka* and *vicāra*. The self-reflective process that objectifies the content of experiences situated in the outer world is called *vitarka*. The self-reflective process that objectifies the subject and agent of experience as other than the content of experience is called *vicāra*.

Can self-reflection aid in responding to negative thoughts and attitudes that can lead to violent thoughts and deeds and will produce infinite pain? Negative thoughts, emotions (and actions) such as those of violence, whether they are done (indulged in), caused to be done or abetted, whether caused by greed, anger or delusion, whether present in mild, medium or intense degree, all result in endless pain and ignorance.[35] Hence, the need to cultivate positive thoughts. When negativities manifest in mind—*vitarka bādhane*—opposite (positive) thoughts have to be invoked. Positive thoughts neutralise negative thoughts. This is the component of self-reflection that is based on neutralising negative thoughts with positive ones —*pratipakṣa bhāvanam*.[36]

Yoga practice includes philosophical reflection (*vitarka*) upon objects of the outside world (*viṣaya*) and introspection (*vicāra*) upon objects of inner world. Reflection and introspection upon the objects of the gross world outside and the subtle inner world (mental and causal) creates a certain quality of happiness (*ānanda*) and a new identity (*asmita*). This stage is *asamprajñāta samādhi*.[37] The role of yoga techniques such as *prati-prasavam* (the process of self-regression, by going back and forth), *pratipakṣa bhāvana* (cultivation of positive thoughts), *viveka*

---

[33]YS 2.26—*viveka-khyātir aviplavā hānopāyaḥ*.
YS 2.27—*tasya saptadhā prānta-bhūmiḥ prajñā*.

[34]YS 4.25—*viśeṣa-darśina ātma-bhāva-bhāvanā-vinivṛttiḥ*.
YS 4.26—*tadā viveka-nimnaṃ kaivalya-prāg-bhāraṃ cittam*.

[35]YS 2.34—*vitarkā hiṃsādayaḥ kṛta-kāritānumoditā lobha-krodha-moha-pūrvakā mṛdu-madhy ādhimātrāduḥkhājñānānanta-phalā iti pratipakṣa-bhāvanam*.

[36]YS 2.33—*vitarka-bādhane pratipakṣa-bhāvanam*.

[37]YS 1.17—*vitarka-vicārānandāsmitā-rūpānugamāt samprajñātaḥ*.

khyāti (knowledge that demarcates between the subject and object of experience), *dhyānam* (continued concentration) and *kṛiya* yoga (the set of practices described in the second chapter in order to aid self-reflection) play a role in increasing the degree of self-perception which is technically known as 'self-insight'.

I wish to suggest that we can derive four important elements of self-reflection from the yoga phenomenology based on its model of the mental levels (*citta bhūmi*), pain (*kleśa*), and the techniques (*viveka khyāti, prati-prasavam, kriya* yoga, *samyama* and *dhyānam*) for the restraint of mental distractions. These are self-certainty, body-centrality, cognitive centrality and contentful experience.

The possibility of the mind to move towards the shore of restraint (being the one side of the river—*citta-nadi*) gives the ability to sense oneself as an agent (*kartṛtva*) leading to the owner and experiencer of the content of experience (*bhoktritva*), based on the five levels (*citta bhūmi*) of the mind. At any point of the mental level and pain (*kleśa*), self-recognition exists at least to some degree. Even to make an error in self-identification, one needs to have a preliminary sense of self, however faulty, incomplete or changing. The indubitable nature of the self is based not on the Cartesian priority of the act of experiencing, but on the experiencer, according to the yoga psychology. This difference is extremely important considering that the value of experience is recognised more in the humanising of it from the point of view of the person, who reads his experience and derives values from the outcomes of acts. *Self-certainty* is the final outcome for the techniques of self-reflection, as we can deduce from the yoga psychology.

In order to avoid a pure abstraction of one's experience based on self-certainty dissociated from the sensory world, which can lead to narcissism or unfounded experiences, yoga psychology is careful to give equal importance to the experience of the body that begins from the physicality of the physiological body and extending to the emotional and social world. The impact of techniques of *āsana* along with the *prānāyāma* is a step to preserve the distinct and firm sense of the body, giving body-centrality to one's experiences, and grounding one's awareness to begin with on the nuanced awareness of the body. Thus *body-centrality* is the second important aspect of self-reflection in yoga psychology.

The distinction of the levels of the mental planes (*citta bhūmi*) as different spaces that behold the rest of the psychological states and entities provides the first step to self-reflection, which is in recognising the space one is as a self. Without the knowledge of the space there will be lesser knowledge of the states and entities that are housed in the space. The introduction of the *citta bhūmi* and *citta vṛtti* concept of yoga psychology stresses the third important feature of self-reflection, which is the *cognitive centrality*. It is the ability to use one's beliefs and judgemental capability to decide the nature of cognitions as memory-based, imagination-based, or actual-object based on the yoga concept of five types of *citta vṛtti* relating to the nature of cognitions.

The third feature of self-reflection, *contentful experience,* can be understood from the discussion on the techniques of yoga, Ashtanga yoga, and the nine dysfunctions of the self discussed in the Yoga Sutras. According to Patanjali, one of the nine dysfunctions of the self is *bhrānti-darśanam*—having delusional experiences—which is accompanied by the sense of lack of progress and plateau feeling (*alabdha bhumigatvam*). *Yogasana*, and other techniques elaborated in the first chapter of the Yoga Sutras, will, according to its philosophy, lead to a correction in the orientation of experiences, and help achieve a realistic sense of the experiential self through self-reflection. Clinical reports on schizophrenia and narcissistic personality disorder show that improved self-reflection may result in better social cognition (de Vignemonta and Fourneret 2004). Interestingly, what is suggested by Patanjali for a person suffering from delusional experiences is the cultivation of attitudes such as *maitri*—friendliness, *karuṇa*—compassion, *muditam*—cheerfulness, and *upeksha*—forgiveness.[38] The intended result is that mind becomes cheerful—*citta prasādam*—by *bhāvana*—cultivating the attitude of—*maitri*—friendliness towards—*sukha*—happy people, *karuna*—compassion towards—*duḥkha*—unhappy people, *muditam*—cheerfulness towards—*puṇya*—good people, *upekṣa*—indifference towards—*apuṇya*—vicious people, and—*viṣayaṇām*—such matters.

Various techniques given in the Yoga Sutras are meant for both the beginner and the advanced person for restraining modifications of mind. These include the techniques called *prati-prasavam*, *pratipakṣa-bhāvana*, *viveka-khyāti*, *dhyānam* and *kriya* yoga. Patanjali introduces a very significant method called '*prati-prasavam*' for healing the unconscious tendencies. By a process of self-regression, by going forth—*prasavam*, and going back—*prati-prasavam*, the subtle form of the five pains (*kleśa*) can be destroyed. Hence, first the *kleśa* that are spasmodic and full-blown (*vicchinam*, *udaram*) have to be reduced to their subtle form (*tanu* and *prasuptam*). The five pains are to be treated in their dual form—that is, the subtle and gross form. The gross form is the fully blown psychological state reflected existentially in the attitudes, responses and behaviours of the person. The subtle form is still in an ontological unconscious level and not expressed. The four pains in their subtle form can be resolved to the respective causes. Attachment to body—*abhiniveśa*—can be traced to likes and dislikes—*rāga-dveṣa*. Likes and dislikes can be traced to the sense of I-ness—*asmita*. The sense of I-ness can be rolled back to self-ignorance—*avidya*.[39]

The eight-fold yoga includes external, mental and spiritual exercise to accentuate body-awareness along with the connected emotional and social world. Thus, interestingly the route of meditation (*dhyāna*) may not be ideal for a mind that has delusory visions, since the first step for healing would be to bring in the centrality of the body-awareness, and avoid self-absorption disconnected from the real sensate world. Ashtanga yoga leads to the fourth aspect of self-reflection, which is *contentful experience*.

---

[38]YS 1:33—*maitrī karuṇā muditam upekṣāṇām sukha duḥkha puṇya apuṇya viṣayānām bhavanātah citta prasādanam.*

[39]YS 2.10—*te prati prasava heyāh sukṣmāḥ.*

Yoga psychology and its phenomenology, as elaborated in this chapter, offer the concept of self-reflection which suggests a regime of techniques that connects and integrates the body, the mind, values and attitudes, and experiences. This study helps us to highlight that many aspects of the experiencing self that we consider in our daily lives, according to Patanjali, belong to the *citta* system. Thus the mainstream notion of mind is extended by going beyond the cognitive frameworks and enriching with several personal features. The practice of self-reflection is not dissociated from the core self, and at the same time contextualised in the space of the experiential world of the practitioner. Thus, the experience of the self is not an abstraction from the experiential world, but a reassurance and re-humanisation of one's identity connecting the body and the mind with the psychosocial world in which one lives.

References

Bodhananda, S. (2008). *Yogasutras of Patanjali*. New Delhi: Sambodh Foundation.
Bodhananda, S. (2010). *Yoga and management*. Ahmedabad: Ahmedabad Management Association.
de Vignemonta, F., & Fourneret, P. (2004). The sense of agency: A philosophical and empirical review of the 'Who' system. *Consciousness and Cognition, 13*, 1–19.
Eliade, M. (1975). *Patanjali and Yoga* (C. L. Markmann, Trans.). New York: Schocken Books.
Gallagher, S. (2000). Philosophical conceptions of the self: Implications for cognitive science. *Trends in the Cognitive Sciences, 4*(1), 14–21.
Georgieff, N., & Jeannerod, M. (1998). Beyond consciousness of external reality: A 'Who' system for consciousness of action and self-consciousness. *Consciousness and Cognition, 7*, 465–477.
Kircher, T. T., & Leube, D. T. (2003). Self-consciousness, self-agency and schizophrenia. *Consciousness and Cognition, 12*(4), 656–669.
Menon, S. (2009). The rain clouds of mind-modifications and the shower of transcendence: Yoga and Samadhi in Patanjali Yoga Sutra. In R. Rao (Ed.), *Yoga and parapsychology: Empirical research and theoretical essays* (pp. 169–199). Delhi: Motilal Banarsidass.
Menon, S. (2014). *Brain, self and consciousness: Explaining the conspiracy of experience*. New Delhi: Springer Science & Business Media.
Rao, K. R. (2010). *Yoga and parapsychology: Empirical research and theoretical essays*. Delhi: Motilal Banarsidass.
Rao, N., & Menon, S. (2016). A heuristic model linking yoga philosophy and self-reflection to examine underlying mechanisms of add on yoga treatment in schizophrenia. *International Review of Psychiatry*, 2, 265–272. doi:10.1080/09540261.2016.1194259.
Safaya, R. (1976). *Indian psychology*. New Delhi: Munshiram Manoharlal Publishers.
Suryanarayana Sastri, S. S. (Ed. and Trans.). (1973). *The Sankhya Karika of Isvara Krsna* (2nd ed.). Madras: University of Madras.
Woods, J. H. (1914). *The yoga-system of Patanjali: Or the ancient bindu doctrine of concentration of mind*. Delhi: Motilal Banarsidass.
Zimmer, H. (1951). *Philosophies of India*. In J. Campbell (Ed.). London: Routledge and Kegan Paul.

Part III
The Social Self, Culture and Cognition

Fullness, Trust and the Self

Rajesh Kasturirangan

The Trust Bearer

'*Whatever* matters to human beings, trust is the atmosphere in which it thrives.' That's how Bok (2011) characterises the importance of trust, as the metaphorical air we breathe. She wrote these words in the context of public life, of politicians who lie and cheat their way to power. These words have become even more important in the years since she wrote her book. Modern technology has made it possible to deceive the public at a scale unknown to previous generations. When the digital and the physical collide every 140 characters, when virtual reality becomes indistinguishable from real reality (whatever that might mean!), we know that questions of trust and veracity are no longer ethical questions, they are also metaphysical ones.

Where does trust come from? In whom does it reside? What aspect of their existence is responsible for upholding our trust?

These are not merely theoretical questions. Study after study has shown that public perceptions of trust (Barometer 2012) are plummeting the world over. In country after country, people do not trust the media or the government or any of the usual markers of social stability. Are we doomed to live in the matrix? Is there no way to distinguish illusion from reality?

In this chapter, I will argue that the sources of trust have not changed in the last two thousand years, perhaps two million years, even as it faces challenges that have been scarcely twenty years in the making. In short: *trust comes from the world we live in and it is mediated by the self.* Equally importantly: *the self's trustworthiness is public and verifiable.*

R. Kasturirangan (✉)
Mind and Society Initiative, Azim Premji University, Bengaluru, India
e-mail: rkasturi@apu.edu.in

S. Menon et al. (eds.), *Self, Culture and Consciousness*,
DOI 10.1007/978-981-10-5777-9_10

That's a tall claim to make and defend in the course of a short chapter, so I can only sketch out an argument. While it is tempting to start by interrogating the trustworthiness of the tweeter-in-chief, it is more instructive to look at the assault on trust by a much older figure: the stove-dwelling French philosopher, René Descartes. In doing so, I hope to convince you that scientific questions about the nature of the mind are deeply tied to political and ethical questions about appropriate behaviour in public. Not surprising, because the self is at the intersection of several dimensions:

- The interior and the exterior landscape
- The personal and the political
- The biological and the cultural.

But first, we need to free the self from the prison of subjective consciousness in which it was imprisoned by Descartes (1988). Just as the stove-dwelling philosopher arrived at the certainty of consciousness as the indubitable starting point, I believe that we need to arrive at the trustworthiness of the self as the starting point of an investigation of earthly existence. My goal here is to argue for a 50% reduction of Descartes by turning 'I think therefore I am' into 'I am' as the ideal starting point of a philosophical method. In this chapter, I will argue that such a reduction is both necessary and useful. So without any delay, let's start with the famous Cartesian dictum.

I think, therefore I am

René Descartes uttered the most famous phrase in modern philosophy when he said 'I think, therefore I am.' That phrase inaugurates a new chapter in the investigation of reality and by predicating existence on our experience of it, Descartes started a controversy that lasts to this day. I don't want to spend too much time analysing the *cogito*—others have done much better—but let's just focus on the logical structure of the sentence.

I think therefore *I am*

It is clear that the 'I think' comes first, both literally and metaphorically. In Descartes' scheme, consciousness is the most certain of all things; everything else is open to doubt (Descartes and Williams 1996). The devil may make me see a rope as a snake but what he cannot do is to make me deny the experience of snakeness. I may have experienced an illusion but there's no taking away the reality of the illusion itself. In other words, I have devil-proof certainty of *experiencing* a snake even if it was only a rope.

Moving onward, let's focus our attention on the 'therefore'. In Descartes' scheme, reason is the natural expression of the mind, just as causation is the natural expression of the body, so it is only fitting that the logical implication 'therefore' arises out of the certainty of the experience. Finally, once the mind has established the certainty of experience and unleashed its strongest weapon (i.e., logic), Descartes pulls out the rabbit from his hat: 'I am.'

Who is the 'I' whose existence is being ascertained? In what form have we acknowledged the I's existence?

We have a problem if the 'I' is meant to be the regular human self and the 'am' is meant to be everyday existence, for Descartes has already destabilised that ordinary existence through his ferocious doubts. The devil could have fooled us into thinking we are human beings when we are really brains in a vat. Or the other way around. Or, seeing we are in the twenty-first century, we are more likely to believe that robots are holding us hostage in the matrix. In either case, what force does that 'I am' have?

Who cares? Does anyone believe in Cartesianism anymore? In the era of neuroscience, who doesn't believe that the mind is grounded in the brain (Crick 1996)? True enough, but we are prone to repeating philosophical errors if we do not understand them. Neuroscience is riddled with Cartesian assumptions as it vehemently rejects Descartes' dualism. Consider the typical analysis of neural function. Let's say there's a neuron in my brain that you call the grandmother cell because it fires whenever a picture of my grandmother is placed before my eyes. You decide to call it a grandmother cell (Gross 2002). What does that mean? What is actually happening? The diligent neuroscientist will weave two explanations:

1. *Explanation One*: Ions are pouring into channels in my neuron until they burst forth in a spike. In other words, there's a physical process that culminates in a physical event, i.e., the spike.
2. *Explanation Two*: The spike is an informational event that is correlated with the presence of my grandmother (or her picture) in the outside (Rieke et al. 1997).

The two are presumably identical, i.e., the informational event and the physical event are one and the same; otherwise we are back to Descartes' dualism with a vengeance. But then the obvious question arises: *how can they be the same event*? There's nothing in our description of ions flowing from one end of the neuron to the other that reveals a lurking grandmother. Where did the information correlation come from? It seems as if replacing devils and doubts with ions and information has only resurfaced the underlying dilemma at the cellular scale.

I do not think that problem of unification can be resolved as it currently stands. Instead, let us sketch out an alternate route that starts with a simplification of Descartes' claim. Instead of 'I think, therefore I am', I suggest we stick to the far simpler 'I am.'

From Vision to Touch

The shift from 'I think therefore I am' to merely 'I Am' is also a shift from thought and vision to touch (Linden 2016). The abstract thinker is removed from the world, gazing at it from a distance. In contrast, the touching organism is embedded in the world, orienting itself, moving and shifting as the occasion. While we naively believe that it is possible to see the world without changing it, but there's no such comfort when it comes to touch: to touch (or to taste) is to both know the world and to change it at the same time. When I arrive at a new friend's house and pull the door handle, I accomplish both a physical act, i.e., moving the door on its hinges, and an epistemic act—namely, revealing the inside of the room that I am now entering.

So let's imagine a midway point between 'I think therefore I am' and 'I am' and call it '(I touch therefore) I am', with the brackets emphasising that the 'I am' is primary and the 'I touch' is there to accentuate the 'I'.

Since the experience of touch is coextensive with the experience of changing the world (or at least getting push-back from it, say, when you're touching a mountain), direct contact with the world is as certain as any of Descartes' isolated experiences without the crippling doubts that lead to. The committed Cartesian might disagree: don't we experience touch in the matrix? A virtual reality cave in which you feel as if you're touching the softest fabric in the world is as hallucinatory as the devil making you see a snake instead of a rope. How is touch different from vision? While virtual reality (VR) chambers might appear to be a counter-example to our sense-shift from vision to touch, but in fact, they betray the very mentality that we are attempting to critique.

Consider that the idealised viewer-thinker is removed from the world, seeing it from a distance. It is only within that idealised frame that certainty becomes a desirable virtue and radical doubt a weapon to be wielded against the believer. In contrast, the toucher is fully plugged into the world (Rosch et al. 1991). Unlike the visual ideal of disinterested abstraction, touch is always interested, always seeking something to grasp, grab, caress or pinch. Touching has consequences: when I kick a stone, not only is the stone dislodged, my toes are also bruised to the point of bleeding.

If kicking a stone in a VR chamber makes me bleed in exactly the same way that a real stone does then there's no distinction between appearance and reality. If shooting someone in a video game kills the person for good, then it is no longer a game. Think about it another way: the proof of the certainty of touch is not in the experience itself but *feedback from the world.* Indeed, as I will argue towards the end of this article, the next frontier of scepticism is not in confusing appearance for reality, but in moulding reality to suit our appearance.

I might be uncertain about hitting the car in the next lane, but the driver of that car is not and his curses are proof that something went wrong. While the seer-thinker can pretend the world has been cancelled out, the toucher cannot. The world is an ever-present witness to touch. Which brings me to the second shift along the path from 'I think therefore I am' to 'I Am:' transforming the isolated observer into an active agent (Gibson 2014).

The Active Agent

Unlike the impoverished observer who is struggling to maintain his beliefs in the midst of doubt, the world is radically full. Indeed, the world is with me as I prepare to retire for the night and it is right by my bedside when I wake up in the morning. It is the rock-solid reality of the world that lets me close my eyes in the first place, for I am not afraid that the sky will fall on my head, or worse: that the sky will dissolve into nothingness.

Except that the devil is speaking in my ear that the sky does fall on some people's heads: earthquakes happen; superpower rivalry happens. Whether its human nature or pure nature, the ground gives way to emptiness every once in a while. There's no guarantee that the sun will rise tomorrow just because it rose today. Fortunately, while the devil's clever, he's not clever enough. It is true that my bed is being eaten by termites and will collapse in the middle of the night. But does that really shake my confidence in the world? Isn't the bed supported by the floor and the floor by the foundation and the foundation by the earth? A bit like the ancient tale of the elephant who's supported by a tortoise and then it's tortoises all the way down. The world is our tortoise and *our faith in the world is infinitely deep*.

If you do not believe me, consider the contrasting scenario of a very untrustworthy world: the world of bits on which I am inscribing these thoughts. Everyone who has worked with a graphical user interface has experienced a crash, destroying hours and hours of work with no chance of recovery. When the virtual world is working, it is easy to believe that our actions cause changes in the world (Norman 2013). A button press sends an email from me to a friend halfway across the earth who responds to me within a few minutes. The crash disturbs that fullness; it makes me set aside the make-believe of the screen and reboot the computer, praying that the crucial file is hiding in a corner. The pseudo-world *internal* to the email or the document you're composing rides upon the real world *outside* the computer and it collapses every time the real world intervenes.

The real world itself never crashes that way. My termite-ridden bed might crash after being chewed to a hollow, but I can trace the termites back to their nest; they cannot hide their work. In the real world, there's no sharp distinction between appearance and reality. In contrast, a being living within the text on my screen has no recourse; the computer crash is a supernatural event for that being for it occurs in a space that cannot be accessed from within the document. The real world has only one order of reality: the mental might give way to the neural, the neural to the ionic, the ionic to the atomic and so on, but there's no doubt that all of these are different levels of the same world. There's no sense that the projector will sputter out and the real reality will peek through. The fullness and the continuity of the world unite the physicists and the psychologists.

The Origins of Fullness

Where does that fullness come from? One possible answer is that the mind imposes fullness on experience. This is the Kantian view and also the Chomskian view, one elaborated by David Marr in his famous study of vision (Marr 1982). Why should we believe this view? The strongest reason, in my opinion, comes from the poverty of our experience. While the input to our eyes consisting of noisy views that shift every time our eyes saccade, what we see is a rich three-dimensional array of objects in space. Lighting changes when day gives way to night but we still see the same colours. Our eyes change focus every 200 ms, but we do not experience the world as jumping from one sequence to the next. In other words, our experience is full even as our senses are patchy. Surely, that is because our minds fill in the gap, just as the visual system fills in the blind spot at the back of our eyes. In short: the world is full because our minds make it that way.

Unfortunately for us, that experience of coherence is not the real thing: in fact, our experience is notoriously gappy, riddled with omissions and errors. Experience is not coherent, *it only appears that way*. So after this long detour, it looks like we are back in Descartes' court, except that instead of worrying about the devil replacing ropes with snakes, we are worried that it might be the brain doing so. In fact, the brain might be worse than the devil: instead of worrying about one rope being replaced by one snake, we have to worry about the possibility that *all of our experiences are hallucinations*. There are two ways out of this renewed Cartesian dilemma:

1. In this first and more familiar move, we retreat into the Cartesian certainty of consciousness by embracing our hallucinations. In other words, while we are hallucinating a coherent and harmonious perception of the world, we cannot hallucinate the hallucination itself. Plus, we have evolutionary language to justify our hallucinations: they are adaptive and help us survive. If the world is well behaved, my hallucinations are likely to be predictive of reality. For example, if I am playing basketball, I don't have to worry about wandering gorillas most of the time.
2. Or else we can take the exact opposite view, that the experience of coherence is not illusory at all; we are making a mistake in identifying the mind as the source of coherence. Instead, the coherence of our experience is part of the unity of the world itself, from which it derives its partial coherence. Unlike the text document on my computer which does not have a trustworthy world to rely upon, our experiences are robust because they can afford to fail. We mistakenly see snakes *precisely* because there are ropes—a rope-free universe is also an illusory snake-free universe.

Let me illustrate that point with an example. Let's say you are about to cook dinner in your kitchen. It's a familiar space and you might have cooked in it a thousand times. Let's also assume I am your guest and I am helping you with the cooking. Since I don't know the kitchen at all, I might pepper you with questions about the location of pots and pans. Chances are you know where some pots are stored while you don't know where some of the other pans are kept. It's easy to subsume the failure within a larger claim about the limits of memory and our inability to store more than seven items at a time.

That would be the wrong analysis based on the Cartesian assumptions we are critiquing in general. Instead, let's recognise that we are mobile, tactile creatures who know how to navigate a familiar environment and that a misplaced pot or two does not falsify our general sense of comfort and ease in the spaces we inhabit. To put it another way, our inability to remember the location of a pot is no different from our inability to remember which site hosted an article you read last week. Just as the latter problem is solved by using your favourite search engine, the problem of the missing pot is solved by walking around the kitchen looking at some of the usual cabinets.

Perception in a full world is not about replicating the world inside our heads but about knowing how to explore the world in order to find what we are.

Moral of the story: the Cartesian view is predicated on the mistrust of our senses, of our conceptual capabilities and our bodily needs. The alternative is built on a foundation of trust of the world, of our senses and of the bodily capacity to move and probe the world (Abram 2011). With that statement of trust, let me come to the last section of this chapter:

The Trust Bearer

Let's say you have been hiding your cash in a mattress because you're convinced that banks are evil. One day a friend comes and tells you that cash is about to be made illegal and so you might as well open a bank account before everything you have is worthless. Even better, the bank will grow your money by adding interest. You are convinced by your friend's arguments, but you still need a bank account that documents your holdings before you can participate in the new economy, right?

In the same way, even if the world is like a trustworthy bank, you need an account, a mechanism by which your identity is authenticated and checked against past transactions. What is that identity? I believe the trust bearer is nothing other than the self. Let's get back to our analogy with bank accounts. What does a bank account do for you?

1. It's an identity manager—it declares that such and such savings and current accounts belong to me and not to someone else.
2. It's a pointer to actual assets stored elsewhere: the squiggles in my passbook point to a locker where my family jewels are stored.

Just as Descartes and Chomsky argue that our knowledge of language in the face of impoverished stimuli (Laurence and Margolis 2001) must be due to innate mental structures, I believe that our capacity to engage appropriately in the world, to grasp what is needed and reject what is unnecessary must be due to an innate interfacial structure, an entity that helps the organism maintain itself and point (both literally and metaphorically) to those external or internal aspects of the world the organism needs to access at a given time and location.

What better interface do we have than the self? One of the classic attributes of the self is as the bearer of identity (Sorabji 2006). I am only adding an extra feature: the self as pointer. Both are necessary for an active organism, i.e., we need to know what one possesses or desires to possess and how one goes about grasping that possession. What does that mean in practice? The self owns the body and its bodily capacities. It has a way of possessing images and memories and mapping them to itself, in essence saying that this memory or that smell belong to me and to no one else. It also says 'here's what you need to do to access that bodily capacity, that memory and the source of that smell'.

The self as identity maintainer keeps ticking whether you're awake or asleep. It is prior to sensory consciousness. If the mind and its neural substrate are about *actual experiences*, i.e., the smell and colour of the rose in your hands right now, then the self is about *potential experience* and the actions one needs to take in order to enjoy that experience, whether that be the raising of my hand in order to bring the rose closer to my nose or walking to the kitchen to fill a vase with water so that I can keep enjoying the rose for a few more days. Of course, potential experiences are vastly greater than actual experience and for that reason, the self comes before the mind-body. Further, those potential experiences become actual only when the self comes in contact with a particular situation—in that way, we see that the primal arrangement is not an isolated brain in a devil-infested vat (Goldberg 2016), but a mobile self in a trustworthy world (Noë 2009).

So, to complete the first circle of our investigation, we have taken our scissors to Descartes and removed the 'I think' and 'therefore' from his list. Since we are not subject to his radical doubts, we are happy trusting the self to take care of us in this world. Once we agree that the world is trustworthy and that we need a trust bearer who authenticates our presence in the world, we are naturally led to the self as the holder of that trust. With that trust protecting our interests, we are even happier to stay embedded in the world and not retreating into the prison of isolated (and subjective) experience. The 'I think' and the 'therefore' are training wheels we don't need anymore. All of which is to say that 'I Am' is the starting point of investigation.

The Sceptic Strikes Back

We are almost at the end of our short journey. With any luck, you accept my claims that the self is the trust bearer we have been looking for. Let's recap our achievements: trust has been rescued from the prison of consciousness. The world has been acknowledged as intrinsically trustworthy. We are in possession of a unique authenticator of trust—namely, the self.

All should be well, shouldn't it?

Unfortunately, not. If only we could sleep safely at night, secure in the trustworthiness of the self and its world. But all our work has only paved the way for a deeper level of scepticism. Let me illustrate that with a dilemma: so far, we have been worried by the possibility that the snake we spy on the road is only a rope (Rao 2011), that what we think is reality is actually an illusion.

What if reality was more diabolical? What if the devil reached into our minds, noticed we were hallucinating snakes and turned the rope into a snake, turning illusion into reality? It's a twist on Gettier's famous thought experiment. Imagine there's a celestial trickster who wants to identify the ten most courageous fools in the world. The Gandharva strews ropes across roads all over Karnataka and waits for night-time travellers to chance upon one of them. As we all know, most travellers are cowards and run away at the sight of anything that looks like a serpent. You, brave reader, are not one of them. Being courageous, you walk toward the creature to investigate what kind of snake it is. While you are doing so, the divine magician replaces the rope with a cobra, which rears its head and is about to strike you.

In the Cartesian version, this is when you wake up and realise you were dreaming. Or perhaps the magician intervenes yet one more time and whisks you away to his palace in the sky. But remember, we have put Descartes to rest. The snake is really there and the venom is as real as it gets. Now tell me: did you experience an illusion or did you experience reality? Put another way, what does reality even mean if anything we conceive is turned into an object in the act of conception? What if the world itself was being created and destroyed every minute by a capricious divinity whose only purpose was to turn our dreams into reality?

Or even worse, what happens when these aren't thought experiments but accurate characterisations of the information-drenched world we inhabit today? We might think we are being insanely productive and efficient after upgrading to the latest smartphone, but it's equally plausible that we bought the phone because our masters have conspired to create a world in which working insanely hard for them and their profits is what we desire in the first place, and not only are we willing to work hard, we are willing to pay for the privilege of doing so! What is the meaning of autonomy, freedom and trust when we can only employ those to satisfy the whims of unseen forces?

You might think this replacing of appearance by reality is too magical to ever be true, so let's play that game in slow motion. Like the frog who doesn't notice the boiling water before it's too late, let's imagine a world in which every carbon compound is replaced by a silicon compound one molecule at a time. At the beginning, we are all biological life forms. At the end of that transformation, we are all robots. Is that world any less real than the world we live in today?

To put it somewhat crudely, the frontier of scepticism has moved from mind to matter. Cartesian scepticism was about the reliability of the mind; upon examination, he found that the mind's impression of everything external is suspect; that the only trustworthy knowledge is that of one's own experience. Of course, possessor of certainty or not, a solipsistic mind is of no use to anyone else. Turing turned the tables on solipsism in his famous test. He said, let's look at the problem from the world's point of view: if I cannot tell the difference between a human being's behaviour and the behaviour of an artificial system, the two are operationally identical. But why stop only with intelligence: if I can't distinguish between the affordances of a natural object and an artificial object, then they are operationally identical as well, aren't they? In fact, the artificial world's affordances are better; most of us would prefer a car over a horse, won't we?

The snake-rope illusion alerts us to the possibility that our cognitive systems are capable of inserting false beliefs that look a lot like true beliefs. But the new worry is at another level: what if there were no snakes at all, only ropes? The snake is a biological being while the rope is a human creation. A world in which we saw all ropes as ropes would still be missing something if there were no snakes left for us to run away from. Most of us would agree there's something wrong in a world where every living thing has been replaced by a machine, but what exactly is the problem? Will a robot complain one day 'if you prick us, do we not bleed?' What exactly is the problem if we bleed oil instead of blood? The physicist would surely think so. Is there no way to tell the difference between the human lifeworld we inherit through evolution and the mechanical world we have created in the last two thousand years?

The self plays an important role in this new scepticism, for it is through the manipulation of the self that we end up losing the ability to distinguish between a world that has ropes and snakes and a world that only has ropes. We can see that already in changing attitudes over digital and physical interaction. What does it say about the modern self when social interaction over WhatsApp or Snapchat is indistinguishable or preferable to meeting someone in person?

The early modern era in Europe started with a collapse in older certainties about religion and Christian belief. Descartes captured that cultural shift with his philosophical attack on certainty and the discovery of subjective consciousness as the final guarantor. That was, of course, only a European problem, though it became everyone else's when Europeans decided to inflict their conflicts on everyone else. Be that as it may, that era of history is over. The new world we are entering is a global world and its main problem is with trust. I started this chapter with a quote about trust being the atmosphere in which we breathe—what happens when that atmosphere is poisoned? Is there a self that can withstand this poison?

References

Abram, D. (2011). *Becoming animal: An earthly cosmology*. USA: Vintage.

Barometer, E. T. (2012). *Edelman trust barometer: Executive summary*. Report. Retrieved from http://www.Scribd.Com/Doc/79026497/2012-Edelman-Trust-Barometer-Executive-Summary. Accessed September 10, 2014.

Bok, S. (2011). *Lying: Moral choice in public and private life*. USA: Vintage.

Chabris, C., & Simons, D. (2010). *The invisible gorilla: And other ways our intuitions deceive us*. New Jersey: Harmony.

Crick, F. (1996). The astonishing hypothesis. *The Journal of Nervous and Mental Disease, 184*(6), 384. doi:10.1097/00005053-199606000-00013.

Descartes, R. (1988). Meditations on first philosophy. In J. Cottingham, R. Stoothoff, D. Murdoch, & A. Kenny (Eds.), *Descartes: Selected philosophical writings* (pp. 73–122). UK: Cambridge University Press (CUP). http://doi.org/10.1017/cbo9780511805059.006.

Descartes, R., & Williams, B. (1996). First meditation: What can be called into doubt. In J. Cottingham (Ed.), *Descartes: Meditations on first philosophy* (pp. 12–15). UK: Cambridge University Press (CUP). http://doi.org/10.1017/cbo9780511805028.007.

Gibson, J. J. (2014). *The ecological approach to visual perception*. Classic Edition. UK: Psychology Press.

Goldberg, S. C. (2016). Introduction: Putnam's reflections on the brain in a vat. In S. C. Goldberg (Ed.), *The brain in a vat* (pp. 1–16). UK: Cambridge University Press (CUP). http://doi.org/10.1017/cbo9781107706965.001.

Gross, C. G. (2002). Genealogy of the grandmother cell. *The Neuroscientist, 8*(5), 512–518. doi:10.1177/107385802237175.

Laurence, S., & Margolis, E. (2001). The poverty of the stimulus argument. *The British Journal for the Philosophy of Science, 52*(2), 217–276.

Linden, D. J. (2016). *Touch: The science of hand, heart, and mind*. UK: Penguin Books.

Marr, D. (1982). *Vision: A computational investigation into the human representation and processing of visual information*. Cambridge, MA: MIT Press.

Noë, A. (2009). *Out of our heads: Why you are not your brain, and other lessons from the biology of consciousness*. UK: Macmillan.

Norman, D. A. (2013). *The design of everyday things*. New York: Basic Books. (Revised and expanded edition).

Rao, S. (2011). Rope-snake illusion. In *Advaita* (pp. 70–99). UK: Oxford University Press (OUP). http://doi.org/10.1093/acprof:oso/9780198079811.003.0004.

Rieke, F., Warland, D., de Ruyter van Steveninck, R., & Bialek, W. (1997). *Spikes: exploring the neural code*. Cambridge, MA: MIT Press.

Rosch, E., Varela, F., & Thompson, E. (1991). *The embodied mind: Cognitive science and human experience*. Cambridge, MA: MIT Press.

Sorabji, R. (2006). *Self: Ancient and modern insights about individuality, life, and death*. Oxford: Oxford University Press (OUP).

Autobiographical Memory: Where Self, Wellbeing and Culture Congregate

V.V. Binoy, Ishan Vashishta, Ambika Rathore and Sangeetha Menon

There is no pain so great as the memory of joy in present grief.
Aeschylus

Introduction

There may not be a single person who does not keep the memories of sweet and painful moments in one's life, experiencing the valence of those events with a smile or tears when it flashes on the screen of the memory and extending that feeling to the events expected in the future. Although the capacity to make, store and retrieve a replica of the stimulus perceived, ranging from simple objects to the incidents spanning through years and involving complex interaction between multiple individuals, is present in various animal species, a form of remembrance known as autobiographical memory (AM) is unique to the human beings (Tulving 2002). Autobiographical memory defined as "a subjective perspective on specific events experienced at particular time points linked together on a personal timeline" (Fivush 2011) is different from the semantic memory, recollection of the facts or meanings or episodic memory (retrieval of what, when and where aspect of an event). AM

V. V. Binoy (✉) · I. Vashishta · S. Menon
NIAS Consciousness Studies Programme, National Institute of Advanced Studies, Bengaluru 560012, India
e-mail: vvbinoy@gmail.com

I. Vashishta
e-mail: ishanv8@gmail.com

S. Menon
e-mail: prajnanata@gmail.com

A. Rathore
Centre for Converging Technologies, University of Rajasthan, Jaipur, India
e-mail: rathore.ambika927@gmail.com

S. Menon et al. (eds.), *Self, Culture and Consciousness*,
DOI 10.1007/978-981-10-5777-9_11

involves subjective experience of event / events happened in the recent or distant past collated into a story or personal history and mental time-travel—the capacity to travel back and forth in the time line of the personal chronicle—and reliving a specific event without losing the continuity of the self (Tulving 2002; Fivush 2011). Although research focusing memory is dominated by the studies aiming to trace out the neural mechanisms involved in the recording and retrieval of the information personal, social and cultural aspects of it also requires intense investigation since human beings are 'wired to be social' (Castiello et al. 2010).

Understanding the formation, content and utilisation of autobiographical memory and various biological and sociological factors determining the dynamics of it is very important because this form of remembrance is a vital contributor to four major cognitive functions of an individual; describing the self and identity, establishment and maintenance of social relationships, finding effective solutions for the challenges faced as well as the regulation of emotions (Wang et al. 2015). AM is tightly linked with the mental health and wellbeing of an individual and ample evidences are available to prove that those individuals who incorporate more elements of AM while describing themselves or interacting with others enjoy higher levels of positive relationship and less mental and physical ailments—an indicator of wellbeing (McAdams 2004; Sales and Fivush 2005; Pennebaker and Chung 2007). Modification in the structure or function of AM has been observed with the infestation of many neuropsychopathological conditions, such as depression, PTSD, autism, schizophrenia and Parkinson's disease (Robinson et al. 2017; Seeman 2017).

Studies conducted by various authors reveal that the cultural conditions to which a person is exposed to can also significantly influence AM and its functions. For instance, people of Asian origin were significantly different in the usage of AM for directing various behaviours, relationship formation and regulating emotion in various social situations, from their counterparts from American or European culture (Wang 2016). Hence, AM could be considered as a point of convergence of knowledge from the fields of self, culture and wellbeing, and the study of this vital cognitive trait could provide us important insights towards elucidating the picture of consciousness, one of the hard puzzles for science.

Episodic to Autobiographical Memory: An Evolutionist's Approach

In a social system where individuals vary significantly in their temperament, access to biologically significant resources and relationships with other individuals, stress-free survival is possible only if the subject is able to make a mental image of itself, integrating both perceptual and conceptual information, and modify it according to the demands of the social and ecological scenario (Bshary 2011; Brown 2015). Additionally, if the subject is able to store what, where and when aspect and causative agents of an incident and recall that information whenever it is required, she could plan an action based on the outcome of the past event. This form of memory, popularly known as episodic memory—where an organism remembers an event in the form of episode—has been reported in a wide variety of organisms, ranging from fish (Hamilton et al. 2016) to birds and primates (Clayton et al. 2003). Although AM is considered as a form of memory unique to human beings, in the opinion of Whishaw and Wallace (2003), dead reckoning—a form of navigation in which animal employs duration and distance of the time travelled as cues for reaching the destination—could be considered as a homologue of the human AM. While performing dead reckoning, animals behave as if it is possessing a 'self' which is capable of time-travelling in order to reach a destination, using previous experiences as the cue for finding the direction. However, getting a phenomenological account of the performance of such tasks by an animal is not possible with the available technologies. Hence the question whether an animal is able to generate AM will continue as a mystery. Nevertheless, several studies that explored the behavioural and neural analogies of human episodic memory in animals (referred as 'episodic-like' memory in the literature) supported the notion that many species not only generate episodic memory but also are capable of mental time-travel (Schacter et al. 2007). According to Allen and Fortin (2013), a neural system required for generating at least a protoepisodic memory exist possibly in all vertebrates and the neural circuits making this kind of remembrance possible exhibits a striking similarity across mammalian and avian species proclaiming its evolutionary origin. Then what makes autobiographical memory unique from human episodic memory and 'episodic-like' memory present in animals?

Subjective Self and Autonoetic Consciousness

The concept of autobiographical memory is tightly interwoven with two popular topics in the consciousness research; the subjective self and autonoetic consciousness (Fivush 2011). According to Tulving (2002), a major contributor to autobiographical memory research, in order to develop this unique form of memory, the cognitive system of the organism should possess three vital properties: a sense of subjective time, capability to be aware of subjective time, and a 'self' that can travel in subjective time without losing the continuity. Hence, if a cognitive agent were to enjoy the AM, it need to possess a 'subjective self'—the capability to consider oneself as the causal agent or experiencer of an event; in other words, the subject should acquire a faculty to develop and access the notion of 'I' (Howe et al. 2003) and to organise events keeping that I as the reference point. Although individuals develop a cognitive self—"the objective aspect of the self that embodies the unique and recognizable features and characteristics that constitute one's self concept" (Courage and Howe 2010) which is potent enough to generate the concept of 'me' at the age of 2 years, AM begins to emerge towards the end of the preschool years only, and its development will continue until adolescence and early adulthood (Fivush 2011).

AM illuminates the close association between self and consciousness. The second requirement suggested by Tulving (2002) to term a remembrance AM–the mental time-travel–is connected with a higher level of awareness known as autonoetic consciousness—the capacity to mentally place oneself in the past or future and travel through that mental time line (Vandekerckhove and Panksepp 2009). According to Tulving (1985), a person with autonoetic consciousness "is capable of becoming aware of her own past as well as her own future; she is capable of mental time travel, roaming at will over what has happened as readily as over what might happen, independently of physical laws that govern the universe". The mental time-travel happening during the autobiographical remembrance not only provides the representation of the past event to the experiencer but he or she is also able to reconstruct the self-engagement with that specific incident and re-experience the emotional content of it (Klein 2016). Additionally, in such contexts, the experiencer keeps the phenomenological continuity between past and present self-images. This situation provides an interesting scenario to understand the link between reality, experience and consciousness. While going through the memory of the past experience, which impacted oneself negatively or positively, the experiencer knows that the incident is 'not real' in the present scenario but still experiences the valence or emotional content of it and could project it to the future without losing the phenomenological continuity. Hence future studies focusing the neural mechanisms bridging autonoetic consciousness and subjective self-experience may provide insight into understanding the neural basis of consciousness.

Autobiographical Memory, Identity and Wellbeing

Autobiographical memories play a vital role in defining the content and constitution of the identity of an individual. According to Addis and Tippett (2008), AM plays a determinant role in the formation and continuity of personal identity—characteristics that separate one individual from another—and social identity (the sense of an individual about the membership of the groups to which he or she belongs (Tajfel and Turner 1979)) by contributing significantly to the self-concept and self-narrative of an individual. The self-concept denotes the abstract and conceptual information about the self while self-narrative is a story generated by the individual by connecting past experiences. Furthermore, identity and AM share two vital properties, the phenomenal continuity and the narrative continuity which ensures that the same person experiences the events happened at different space and time and such events are compiled into an overarching narrative or self-history. Although modern description of identity considers context, relationships, temporal factors and so on along with social and personal dimensions (Addis and Tippett 2008), the integration of all these dimensions into a continuous and coherent story which could act as a reference for the formation and maintenance of identity is only possible when the individual performs "selfing" attaching it to subjective self and self-experience. However, if the person fails to 'reconcile aspects of the self into a relatively coherent and acceptable sense of self' (American Psychiatric Association 1980) the individual will often end up in identity distress, which could induce significant impression on the mental health (Berman and Weems 2016).

The intertwining and intimate relationship between emotions and memories are well known and both adults and children utilise their past personal experience for controlling the emotions (Merrill et al. 2016). According to Fivush (2011), the capacity to come up with an emotionally coherent narrative coloured by past personal experiences and recalling AM while interacting with others is a sign of emotional wellbeing and discloses the connection between AM and wellbeing. Individuals who use more AM are shown to possess a better capacity to cope with the stressful events and enjoy more positive relationships. A recent study by Waters (2014) also confirmed a positive relationship between three major functions of first-person memories (self, social and directive) and the wellbeing of a person.

The interplay between AM, identity and its contribution to subjective wellbeing becomes clearer in the light of recent studies which analysed the link between identity and psychopathological conditions (Berman and Weems 2016). According to Merrill et al. (2016), there do exist a notable variation amongst individuals while doing autobiographical reasoning, a process in which persons associate an event with the individual's self while making meaning of the experience. Such self-event connection could decide the emotional content of the event while encoding it into the memory. Those who make a negative self-event connection may derive negative meanings and hence experience psychological distress which is a negative correlate of the wellbeing (Pasupathi et al. 2007; Pasupathi and Wainryb 2010). When memory of personal experiences of an individual is dominated by negative perceptions, it will reflect on their life-narrative and identity, which in turn will lead to decreased self-esteem, decreased life satisfaction and increased depression (Merrill et al. 2016; Bank and Salmon 2013).

It is well discussed in psychological literature that experience of emotionally disturbing events and re-experiencing such events, either in reality or in mind can affect negatively both mental and physical health of the individual (Öner and Gülgöz 2017; Erber and Erber 1994). AM plays a vital role in coping with such situations and there are evidences available to prove that both adults and children utilise the memories of positive personal experiences to console themselves and tide over undesirable situations. AM of lessons learned from a past experience could support the individual to reduce the psychological distress. There are studies available to prove that recalling past positive experiences could help in controlling negative emotions and coping up with the psychological distress (Konrad et al. 2016). For instance, Erber and Erber (1994) demonstrated that students recall the pleasant experience for calming down before a class begins. Another study conducted by Öner and Gülgöz (2017) also supports the link between AM and wellbeing. These authors demonstrated that memories with negative emotional contents quickly give way to positive memories when an individual starts to contemplate about the negative incidents happened in life. However in the opinion of Konrad et al. (2016) recalling positive personal experience to regulate negative mood, can lead to undesirable consequences since the negativity of negative memories can diffuse into positive memories, making it painful. This phenomenon known as 'kill-joy thinking' exhibits a strong negative correlation with wellbeing, since during this process previously overlooked negative aspects of the positive memories are uncovered (Bryant and Veroff 2007; Quoidbach et al. 2010; Polman 2010). The variation in the personality trait of an individual can also influence the linkage between AM and wellbeing. For instance, persons with a higher neurotic trait, a determinant of the tendency to respond with negative emotions, may experience more frequent involuntary negative memories of past experiences. This enhances their susceptibility to trauma, stress, frustration and PTSD (Mineo et al. 2016; Ogle et al. 2016). Hence, there is a high probability that strategies could be developed based on the content and dynamics of AM to attain positive reappraisal by reinterpreting negative memories to extract positive outcomes (Rusting and DeHart 2000). Such strategies could then be utilised in programmes aiming to rehabilitate people suffering from psychological distress as well as in practices for enhancing the quality of the life of normal people.

Autobiographical Memory and Psychopathological Conditions

Many psychopathological conditions, such as autism, schizophrenia, Asperger's syndrome and Parkinson's disease, are associated with the over-generalised autobiographical remembrance, reduced coherence and vividness in the life-narrative, and reduced ability for mental time-travel etc. (Robinson et al. 2017). Victims of autism spectrum disorders (ASD) can give only an impoverished narrative of personal experience dominated with abstract memories and far fewer social elements (Henderson et al. 2009; Tanweer et al. 2010). Their personal life-narratives often take a third-person perspective, which makes them unable to attach life experiences to the self and learn a lesson for the future. Another deviation of AM noted in ASD is the reduced ability for mental time-travel and atypical recall of personal memories with enhanced content of sensory information (Lind and Bowler 2010; Tanweer et al. 2010).

Disruption of both function and content of AM is a signature symptom of another important psychopathological condition: schizophrenia. Schizophrenics are very poor at recalling self-related memories and binding contextual cue with it (Recarte et al. 2017; Waters 2014). Such disruption of AM leads to abnormal development of time perception and inability to identify the source of a memory (Klein et al. 2004; Bonnot et al. 2011), which may in turn lead to a serious situation where the victim fails to distinguish between imaginary and real-life experiences as well as the internal and external control of an action (Waters and Badcock 2008). Although the last century witnessed a boom in studies exploring the link between psychopathology and AM, the focus of research has been restricted to diseases like depression, PTSD and schizophrenia. Hence, future studies analysing the modification in AM happening with the infestation of various neuropsychological ailments will not only help us to come up with better rehabilitation strategies, but also elucidate the link between self, consciousness and wellbeing.

Culture and Autobiographical Memory

It is well known that people from different cultures may behave differently in similar social contexts. Han and Ma (2015) define culture as "the beliefs, values, norms and behavioral scripts shared by a group of individuals which together constitute a social environment in which individuals of social group develop and evolve behave differently". According to Wang (2016), AM shows notable cross-cultural variation in content, structure and function. The personal knowledge of the utilisation of AM for various behavioural functions, known as Meta-Autobiographical Memory (MAM), shows divergence in accordance to the culture of the individual. However studies analysing cross-cultural variation in AM is dominated with the comparison of people from East Asian nations with their counterparts from the Western, Educated, Industrialised, Rich and Democratic (popularly known as WEIRD, Henrich et al. 2010) societies.

According to Nisbett et al. (2001), East Asian countries foster a holistic thinking style in which individuals are expected to prioritise cohesiveness and harmony of the group over personal interest. By contrast, members of the WEIRD communities follow an individual-centric culture in which analytical thinking and the goals and priorities of the individual are given emphasis and development of a unique identity on the part of each individual is promoted. This collectivistic (interdependent) vs. individualistic (independent) nature of cultures has been found reflecting on various dimensions of AM. Studies conducted by Wong and her colleagues (2015) revealed that reminiscence of the past events by both adults and children from the WEIRD community contain more personal experiences while Asians give emphasis to the background of the information where social network and their duties are the focal point. Interestingly, children from WEIRD cultures were far ahead of their Asian counterparts as far as the onset of AM is concerned; children from USA were able to remember incidents at age one year earlier than their Korean (Mullen 1994) and Taiwanese (Wang 2006), counterparts.

According to Fivush (2011) one major determinant of the autobiographical memory is the reminiscence by mothers that children get to hear during their preschool days. The result of a large number of studies revealed that children of mothers who describe the events elaborately to their kids were able to describe their past vividly and with more personal experiences. The impact of culture can be found reflecting upon the reminiscence style of mothers from Eastern and Western cultures. The narratives of former are filled with social interactions and duties, whereas the WEIRD parents provide more information of the individual experiences (Ross and Wang 2010; Wang and Conway 2004). Similar results were obtained when mothers from America were compared with Turkish mothers and the former were found to be elaborative while describing the events to their children (Sahin-Acar and Leichtman 2015). These cross-cultural variations are not restricted to the content of the AM but the functional aspects of it as well. European American children utilise the self- and emotion-regulation function of AM more frequently than Asian Americans children living in USA suggesting the impact of the culture present at home upon AM (Wang et al. 2015).

Although the evidence for intercultural variations in autobiographical memory is increasing swiftly, a recent study by Göz et al. (2017) illuminated a less discussed dimension of AM: the *intra*-cultural variation. While comparing the autobiographical memory of the children from urban and rural regions of Turkey, these authors noted that, in comparison to the children from rural areas, urban children possesed detailed personal memories which were less likely to contain social interactions. Similarly, a study conducted by Wang et al. (1998) on Chinese adults demonstrated that descriptions of the past personal experiences by rural participants were longer than those of urban participants. There are studies available to prove that both urban children and adult exhibit significant divergences in their cognitive abilities, and the urbanites are more individual-centric than villagers. For instance a study by Dragonas (1983, cited in Triandis et al. 1990) focusing individualistic characteristics across people living in a large city, a small town and a rural town revealed that, though these people were the part of the same culture, individualistic nature was found to be highest in the urban dwellers and least in the villagers, while the small-town people occupied an intermediate position.

Conclusion

According to Wang (2016) AM is a product of the interaction between two active systems, an individual who constantly learns and adapts to the sociocultural situation in which she or he lives, and the culture which determines the nature of such interactions. There is a growing body of literature available to prove that the cognitive abilities of an individual is a product of culture-behaviour-brain loop (Han and Ma 2015) and modification of the circuitry in the brain regions, including those controlling self-reflection, could be induced by the culture change (Ma et al. 2014). Changes occurring in the behaviour and identity of the members of a society can in turn redefine the cultural dynamics prevalent in specific communities. Thus, forthcoming studies exploring the interplay between autobiographical memory, culture and individual variation in cognitive abilities could act as a window to understand the complex relations between the subjective self, individual well being and consciousness.

References

Addis, D. R., & Tippett, L. J. (2008). The contributions of autobiographical memory to the content and continuity of identity a social-cognitive neuroscience approach. In F. Sani (Ed.), *Self continuity: Individual and collective perspectives* (pp. 71–84). New York: Psychology Press.

Allen, T. A., & Fortin, N. J. (2013). The evolution of episodic memory. *Proceedings of the National Academy of Sciences, 110*(2), 10379–10386.

American Psychiatric Association (1980). *Diagnostic and statistical manual of mental disorders (3rd ed.).* Washington DC: Author.

Banks, M. V., & Salmon, K. (2013). Reasoning about the self in positive and negative ways: Relationship to psychological functioning in young adulthood. *Memory, 21*(1), 10–26.

Berman, S. L,. & Weems, C. F. (2016). Identity distress. In *Encyclopedia of adolescence* (pp. 1–8). Berlin: Springer.

Bonnot, O., de Montalembert, M., Kermarrec, S., Botbol, M., Walter, M., & Coulon, N. (2011). Are impairments of time perception in schizophrenia neglected phenomenon? *Journal of Physiology-Paris, 105*(4), 164–169.

Brown, C. (2015). Fish intelligence, sentience and ethics. *Animal Cognition, 18*(1), 1–17.

Bryant, F. B., & Veroff, J. (2007). *Savoring: A new model of positive experience.* Lawrence Erlbaum Associates.

Bshary, R. (2011). Machiavellian intelligence in fishes. In C. Brown, K. Laland, & J. Krause (Eds.), *Fish cognition and behavior* (pp. 277–297). New York: Wiley-Blackwell.

Castiello, U., Becchio, C., Zoia, S., Nelini, C., Sartori, L., Blason, L., ... & Gallese, V. (2010). Wired to be social: The ontogeny of human interaction. *PloS one, 5*(10): e13199.

Clayton, N. S., Bussey, T. J., & Dickinson, A. (2003). Can animals recall the past and plan for the future? *Nature Reviews Neuroscience, 4*(8), 685–691.

Courage, M. L., & Howe, M. L. (2010). Autobiographical memory: Individual differences and developmental course. *Handbook of individual differences in cognition* (pp. 403–417). New York: Springer.

Erber, R., & Erber, M. W. (1994). Beyond mood and social judgment: Mood incongruent recall and mood regulation. *European Journal of Social Psychology, 24*(1), 79–88.

Fivush, R. (2011). The development of autobiographical memory. *Annual Review of Psychology, 62,* 559–582.

Göz, İ., Çeven, Z. İ., & Tekcan, A. İ. (2017). Urban–rural differences in children's earliest memories. *Memory, 25*(2), 214–219.

Hamilton, T. J., Myggland, A., Duperreault, E., May, Z., Gallup, J., Powell, R. A., ... & Digweed, S. M. (2016). Episodic-like memory in zebrafish. *Animal Cognition, 19*(6), 1071–1079.

Han, S., & Ma, Y. (2015). A culture–behavior–brain loop model of human development. *Trends in Cognitive Sciences, 19*(11), 666–676.

Henderson, H. A., Zahka, N. E., Kojkowski, N. M., Inge, A. P., Schwartz, C. B., Hileman, C. M., ... & Mundy, P. C. (2009). Self-referenced memory, social cognition, and symptom presentation in autism. *Journal of Child Psychology and Psychiatry, 50*(7), 53–61.

Henrich, J., Heine, S. J., & Norenzayan, A. (2010). The weirdest people in the world? *Behavioral and Brain Sciences, 33*(2–3), 61–83.

Howe, M. L., Courage, M. L., & Edison, S. C. (2003). When autobiographical memory begins. *Developmental Review, 23*(4), 471–494.

Klein, S. B. (2016). Autonoetic consciousness: Reconsidering the role of episodic memory in future-oriented self-projection. *The Quarterly Journal of Experimental Psychology, 69*(2), 381–401.

Klein, S. B., German, T. P., Cosmides, L., & Gabriel, R. (2004). A theory of autobiographical memory: Necessary components and disorders resulting from their loss. *Social Cognition, 22* (5: Special issue), 460–490.

Konrad, A., Tucker, S., Crane, J., & Whittaker, S. (2016). Technology and reflection: Mood and memory mechanisms for well-being. *Psychology of Well-being, 6*(1), 1–24.

Lind, S. E., & Bowler, D. M. (2010). Episodic memory and episodic future thinking in adults with autism. *Journal of Abnormal Psychology, 119*(4), 896.
Ma, Y., Bang, D., Wang, C., Allen, M., Frith, C., Roepstorff, A., et al. (2014). Sociocultural patterning of neural activity during self-reflection. *Social Cognitive and Affective Neuroscience, 9*(1), 73–80.
McAdams, D. P. (2004). The redemptive self: Narrative identity in America today. In D. R. Beike, J. M. Lampinen, & D. A. Behrend (Eds.), *The Self and Memory* (pp. 95–115). New York: Psychology Press.
Merrill, N., Waters, T. E., & Fivush, R. (2016). Connecting the self to traumatic and positive events: links to identity and well-being. *Memory, 24*(10), 1321–1328.
Mineo, L., Concerto, C., Patel, D., Mayorga, T., Chusid, E., Coira, D., … & Battaglia, F. (2016). Modulation of sensorimotor circuits during retrieval of Negative autobiographical memories and its correlation with personality traits. *European Neuropsychopharmacology, 26*, S321.
Mullen, M. K. (1994). Earliest recollections of childhood: A demographic analysis. *Cognition, 52*, 55–79.
Nisbett, R. E., Peng, K., Choi, I., & Norenzayan, A. (2001). Culture and systems of thought: Holistic versus analytic cognition. *Psychological Review, 108*(2), 291.
Ogle, C. M., Siegler, I. C., Beckham, J. C. and Rubin, D. C. (2016). Neuroticism increases PTSD symptom severity by amplifying the emotionality, rehearsal, and centrality of trauma memories. *Journal of Personality*.
Öner, S., & Gülgöz, S. (2017). Autobiographical remembering regulates emotions: A functional perspective. *Memory*, pp. 1–14.
Pasupathi, M., Mansour, E., & Brubaker, J. R. (2007). Developing a life story: Constructing relations between self and experience in autobiographical narrative. *Human Development, 50* (2–3), 85–110.
Pasupathi, M., & Wainryb, C. (2010). On telling the whole story: Facts and interpretations in autobiographical memory narratives from childhood through midadolescence. *Developmental Psychology, 46*(3), 735–746.
Pennebaker, J. W., & Chung, C. K. (2007). Expressive writing, emotional upheavals, and health. In H. S. Friedman & R. C. Silver (Eds.), *Foundations of health psychology* (pp. 263–284). New York: Oxford University Press.
Polman, E. (2010). Why are maximizers less happy than satisficers? Because they maximize positive and negative outcomes. *Journal of Behavioral Decision Making, 23*(2), 179–190.
Quoidbach, J., Berry, E. V., Hansenne, M., & Mikolajczak, M. (2010). Positive emotion regulation and well-being: Comparing the impact of eight savoring and dampening strategies. *Personality and Individual Differences, 49*(5), 368–373.
Ricarte, J. J., Ros, L., Latorre, J. M., & Watkins, E. (2017). Mapping autobiographical memory in schizophrenia: Clinical implications. *Clinical Psychology Review, 51*, 96–108.
Robinson, S., Howlin, P., & Russell, A. (2017). Personality traits, autobiographical memory and knowledge of self and others: A comparative study in young people with autism spectrum disorder. *Autism, 21*(3), 357–367.
Ross, M., & Wang, Q. (2010). Why we remember and what we remember: Culture and autobiographical memory. *Perspectives on Psychological Science, 5*, 401–409.
Rusting, C. L., & DeHart, T. (2000). Retrieving positive memories to regulate negative mood: Consequences for mood-congruent memory. *Journal of Personality and Social Psychology, 78* (4), 737.
Sahin-Acar, B., & Leichtman, M. D. (2015). Mother–child memory con-versations and self-construal in Eastern Turkey, Western Turkey and the USA. *Memory, 23*, 69–82.
Sales, J. M., & Fivush, R. (2005). Social and emotional functions of mother–child reminiscing about stressful events. *Social Cognition, 23*(1), 70–90.
Schacter, D. L., Addis, D. R., & Buckner, R. L. (2007). Remembering the past to imagine the future: The prospective brain. *Nature Reviews Neuroscience, 8*(9), 657–661.
Seeman, M. V. (2017). Identity and schizophrenia: Who do I want to be? *World Journal of Psychiatry, 7*(1), 1.

Tajfel, H., & Turner, J. C. (1979). An integrative theory of intergroup conflict. *The Social Psychology of Intergroup Relations, 33*(47), 74.

Tanweer, T., Rathbone, C. J., & Souchay, C. (2010). Autobiographical memory, autonoetic consciousness, and identity in Asperger syndrome. *Neuropsychologia, 48*(4), 900–908.

Triandis, H. C., McCusker, C., & Hui, C. H. (1990). Multimethod probes of individualism and collectivism. *Journal of Personality and Social Psychology, 59,* 1006–1020.

Tulving, E. (1985). Memory and consciousness. *Canadian Psychology/Psychologie Canadienne, 26*(1), 1.

Tulving, E. (2002). Episodic memory: From mind to brain. *Annual Review of Psychology, 53*(1), 1–25.

Vandekerckhove, M., & Panksepp, J. (2009). The flow of anoetic to noetic and autonoetic consciousness: A vision of unknowing (anoetic) and knowing (noetic) consciousness in the remembrance of things past and imagined futures. *Consciousness and Cognition, 18*(4), 1018–1028.

Wang, Q. (2006). Earliest recollections of self and others in European American and Taiwanese young adults. *Psychological Science, 17,* 708–714.

Wang, Q. (2016). Remembering the self in cultural contexts: A cultural dynamic theory of autobiographical memory. *Memory Studies, 9*(3), 295–304.

Wang, Q., & Conway, M. A. (2004). The stories we keep: Autobiographical memory in American and Chinese middle-aged adults. *Journal of Personality, 72,* 911–938.

Wang, Q., Koh, J. B. K., Song, Q., & Hou, Y. (2015). Knowledge of memory functions in European and Asian American adults and children: The relation to autobiographical memory. *Memory, 23*(1), 25–38.

Wang, Q., Leichtman, M. D., & White, S. H. (1998). Childhood memory and self-description in young Chinese adults: The impact of growing up an only child. *Cognition, 69,* 73–103.

Waters, F. A., & Badcock, J. C. (2008). First-rank symptoms in schizophrenia: Reexamining mechanisms of self-recognition. *Schizophrenia Bulletin, 36*(3), 510–517.

Waters, T. E. (2014). Relations between the functions of autobiographical memory and psychological wellbeing. *Memory, 22*(3), 265–275.

Whishaw, I. Q., & Wallace, D. G. (2003). On the origins of autobiographical memory. *Behavioural Brain Research, 138*(2), 113–119.

The Alchemy of Musical Memory: Connecting Culture to Cognition

Deepti Navaratna

Memory is an important link between culture and cognition. Vandervert (2016) describes culture as the shared patterns of behaviours, interactions, cognitive constructs and affective understandings gathered by the process of socialisation. The cultural neuroscience perspective says that the cerebral-cerebellar model, mostly the context-dependent processing of information by the cerebrum along with the implicit context-independent learning capacities of the cerebellum, plays a prominent role in learning and shaping social culture, as we understand it today (Imamizu and Kawato 2009; Overwalle and Mariën 2016; Vandervert et al. 2007). Neuroanthropological perspectives hold a diametrically opposite view of causation, where the socialisation process, largely understood as systematic training into the meaning, symbols and practices of the society groom human cognition continuously (Niewöhner 2014; Robertson 2016). In both ontological orientations, memory plays a significant role in bidirectional processes, being continuously shaped by both the limits of human cognition and context of culture. Cognitive functions such as perception, memory, thinking, volition and affect are processes, which coevolve through a bidirectional interaction of culture and human cognition (Gelfand and Kashima 2016). Implicit cultural memories have phenomenal power over how universal human faculties are either augmented or inhibited based on socialisation cues and learning environments afforded. Conversely, universal limits of human cognition shape cultural behaviours and affective understandings and social cognition (Vandervert 2016). Musical memory offers a unique window to investigate how culture and cognition are interconnected through behavioural, neuropsychobiological and cognitive processes.

D. Navaratna (✉)
Indira Gandhi National Centre for the Arts—Southern Regional Centre - Bengaluru, Kengunte Circle, Mallathahalli, Jnanabharati Post, Bengaluru 560056, India
e-mail: Navaratna.deepti@gmail.com

S. Menon et al. (eds.), *Self, Culture and Consciousness*,
DOI 10.1007/978-981-10-5777-9_12

The field of cognitive neuroscience of musical memory is beginning to embrace the idea that there may be as many memory subsystems based on nature and complexity of music stimulus, conscious/unconscious strategy employed to learn and acquire music or remember or manipulate music for interpretive performance, and most importantly the seminal role played by context—both social and musical, in creating and remembering music.

Working Memory, Memory Schemas and Remembering

While, several advances have been made in understanding the biological mechanisms of how memory is encoded, there is a raging debate on the very nature of memory itself. The range of cognitive research on memory has grown to include social memory, extended cognition, distributed cognition and embodied memory, thus changing the conversation on what remembering itself is. The cognitive psychology of memory is philosophically moving towards theoretical and multidimensional frameworks for understanding the complex, neuropsychobiological and social interrelationships between human agency, symbols and externality. Should the study of memory be relegated to the cognition of the individual at all? Can the generative models of neurocognition in totality explain the daunting diversity in memory processes sustaining life? Emerging empirical research on extended, embedded, and distributed cognition could explain the hybridity and remarkable alchemy of memory in the brain-cognition-culture nexus (Sutton 2009; Sutton et al. 2010).

Coming to the cognition of memory, many questions remain unanswered. Is there a separate working memory store and pathway for processing music as opposed to verbal stimuli? Does the working memory operate similarly or differently for verbal information as opposed to musical information? Although there is a great deal of literature available on how music training enhances working memory (executive, spatial or tonal functions), using readouts of neural activity, behavioural, verbal and linguistic challenges (Chen et al. 2008; Suárez et al. 2015; Schulze and Koelsch 2012), fundamental questions grounded in musical memory remain elusive.

Baddeley's multicomponent model of working memory extolls that working memory is the 'space' where information could be manipulated while being retained and committed to different schemas of memory (Baddeley 2012). However, this model of working memory has been largely based on verbal tasks and using music as means to test the auditory limits and capacities of the phonological loop. The larger question still remains, on whether the same system is used for non-linguistic auditory information such as environmental sounds and music (Baddeley 2012). Nelson Cowan's embedded processes model provides an information-processing framework to understand musical learning and by integrating working and long-term memory from an interconnected representational approach (Gruszka and Orzechowski 2016). Cowan's model is more suited to explain auditory phenomena, as it can provide a framework to understand individual differences in working memory tasks, the role of chunking information in musical acquisition, rehearsal strategies for retrieval from long-term memory (Berz 1995). However, musical memory operates in more diverse ways cognitively (Clarke 1993).

Approaches to studying musical memory have historically followed a levels-of-processing paradigm, based on neural theories of encoding, where a stimulus has to be present, represented, rehearsed, encoded to commit it to a stable long-term store. This does not account for a large volume of autobiographical, sensory and episodic remembrances that etch memory using emotions and limbic processing. There is sparse research on how auditory stimuli could be etched in memory or grouped under cognitive multiple schemas, including metaphorical or symbolic associations, or the role of memory-encoding processes in embodied models where the information of the music is processed in a context-rich and sensory-rich manner. Another assumption in memory research has been that memory traces and stable musical memories are products of cognitive processing of structure alone (melodic structure, rhythmic structure). Musicians use visual imagery, shapes, contour and sensory-motor memories in trying to acquire a piece of music (Green 2012; Mishra 2005). This questions the stimulus-to-strategy assumptions in memory research; the nature of stimulus may not always dictate the strategy for acquisition and musical acquisition need not be always relegated to phonological loop alone or auditory domain but could involve visual imagery, metaphorical thinking, aspects of muscle memory as well (Zatorre 2012; Keller 2012). A recent study using behavioural tasks and electrophysiological measures involving intonation judgements suggests that formation of accurate mental images for pitch is likely to involve multiple mental imagery, multiple brain areas, engaged auditory areas and error-detection brain circuitry that constantly learns, predicts and makes intonation judgements (Janata 2012). In a fascinating study of mental imagery associated with music, listeners were asked to mentally reverse a familiar tune when presented backwards, which consistently elicited neural activity in the intraparietal sulcus (IPS). In another set of experiments, they were asked to discriminate between melodies with transpositions of key, which also engaged the IPS. Conjunction analyses indicate that dorsal pathway of auditory processing, which is involved in visuomotor and visuospatial tasks is involved in music manipulations. The IPS could thus be the substrate for creation of new mental representations that are directly linked to prior knowledge or previously experience or sensory events (Zatorre 2012). In a free-recall task study involving non-musicians, it was investigated if cognitive categorisation of the titles of songs was based on the musical element of style demonstrated that strategies involved applying verbal stylistic labels to recall music (Booth and Cutietta 1991a, b). Cultural associations, such as attaching certain emotions to music, or time of day to particular modes, and contextual and symbolic associations could also significantly function in the process of whether a piece of music will be acquired, retained and recalled accurately. Musically complex phenomena, such as musical gestures, vocal gestures, sensorimotor memories for microtonal graces and gesticulations, are likely to be unconsciously tutored in cultural contexts of both imitative and observational learning. Current models of working memory have to be expanded and modified to include culturally engrained and encultured musical phenomena. Thus, only a subset of the memory strategies could be processing for structure, melody, rhythm while all aspects of human cognition—sensory, affective, non-phonological and visuospatial faculties could be harnessed for musical memory formation.

Another area that needs active research is to identify whether the cognitive processes of remembering are identical to the cognitive processes that employ memory in perception and acquisition. In situ studies analysing the role of memory in learning and practice increasingly support an embodied role for memory in music. Based on a series of studies on how proficient concert soloists memorise (Chaffin and Logan 2006; Chaffin et al. 2015), a recent analysis of learning strategies and kinds of thoughts used in memorising a piece of music was conducted. In the study, performance cues (PCs) or strategies/thoughts used to memorise the music were demonstrated to be stably recalled during performance (Lisboa et al. 2015). In addition, a variety of thoughts, basic, structure/form-based, interpretive and expressive were used as cues to memorise the music. In instrumental and vocal music, the psychophysiological aspects of muscle memory as culturally inscribed memory remain unexplored, which could expand the scope of memory research into studying correlates of such memory in non-CNS perspectives.

Memory in Music Perception: Perceptual and Non-perceptual Processes

The field of music perception and cognition is relatively young as compared to other branches of the cognitive sciences (Stevens 2012). For a long time, the field of music cognition has remained biased with the view that vision perception is active while audition is passive cognitively. Studies on human perception in the past have held two ontological positions on vision—the same debate has persisted in music cognition as well (Licklider 1959). The cognitivists and constructivists believe that all perception is '*perception-of*', that we perceive musical properties, objects, sensations and stimuli and that human faculties either react, respond or process information without active memory-based mediation and agency. Their view of human perception of sound is *direct*; that music perception is about sensations of tone, attributing objects to sounds and thinking of human agency as a mere 'processor' of musical stimuli (Gaver 1993). In such a model, the role of memory is that of a repository of information, not of embodied experience. The theory of direct perception argues that the nature of stimuli is sufficient to explain the totality of perception and the memory of the human faculties engaging with it to create the experience or percept is not either recorded or is of no consequence in the acquisition or remembering process. The indirect theorists believe in the supremacy of human agency, mediation and intelligence in the process of making sense of sounds, in processing musical sounds in relation to context, pre-existing knowledge of musical structure, content and hierarchies—which makes human perception a very embodied '*perception-as*' process (Ullman 1980). Gibson in 1966 summarises the 'ecological' school of thought that describes perception as a gestalt of human interaction with music, as an embodied model where action and perception is coupled in time.

> Perceiving is an achievement of the individual, not an appearance in the theater of his consciousness. It is a keeping-in-touch with the world, an experiencing of things, rather than a having of experiences. It involves awareness-of instead of just awareness. It may be awareness of something in the environment or something in the observer or both at once, but there is no content of awareness independent of that of which one is aware (Gibson 1966).

While the ontological debate between 'perception as' and 'perception of' has raged on in the field of perception at large and music perception, to a great extent the role of memory in perception has been thought to be passive. Current theories of music perception are rooted in psychoacoustics and neural basis of hearing. Helmholtz's psychoacoustic investigations (1877) revealed the ear to be a sensitive frequency analyser, and Lord Rayleigh's sound localisation experiments showed that humans are extremely adept at localising vibrations to objects (Strutt et al. 1877). Newmann et al. (1987) cite Schopenhauer's belief that music's affective power is due to the passive nature of hearing, which allowed 'brain-fibres' to vibrate in synchrony with musical tones. Early studies in music cognition have proceeded to look at sounds and human perception along the paradigms of signal detection (Green and Swets 1966). Thus the tendency to think of memory as a 'trace residue' of music perception at the level of tone, frequency and psychophysical aspects has persisted throughout this century. Memory for pitch perception ('perception of' pitch) is likely to be different than perception of pitch ('perception as' pitch) in musical hierarchy of a melody, or the complex hierarchies of harmonic music. Similarly, the processes of memory engaging with pitch processing while melody is heard alone versus perceiving metred music versus form-dominant musical hierarchies need to be explored in detail. In a context-dependent manner, perceiving specific pitches or intervals and perceiving general contour (Warren 2008) is likely to concurrently engage overlapping memory components of working memory, both short and long-term, conscious and unconscious knowledge (Dowling and Bartlett 1981; Miyazaki 1992). Since the most important musical aspects of perception, such as consonance and dissonance, are constructed on pitch *relationships* and not on individual pitches (Miyazaki 1995), further characterisation of memory of relative pitch as opposed to absolute pitch is imminent. A study with 6-month-old infants demonstrated that they readily recognised a melody whether rendered in a high or a low pitch range, suggesting an early dominance for relative-pitch processing (Plantinga and Trainor 2005). Studies reporting differences between neural correlates of absolute-pitch processing and relative-pitch processing corroborates the view that separate memory systems could be engaged at large, while differences in brain structure of absolute-pitch dominant processors versus relative-pitch processors, could explain adaptations of brain circuitry based on prior musical training. The fundamental question of how the processes of relative and absolute-pitch memories operate in perception of music remains to be explored.

Studies on the role of memory in musical pitch have focused on how people remember absolute-pitch information, the rare ability to sing specific pitches or identify them without relative-pitch cues. Away from the ability to sing frequency-perfect pitches through memory-based recall, there is still trickling evidence for the absolute nature of long-term auditory memory. A two-component theory of absolute pitch by Levitin (1994) conceives of this rare ability as consisting of a more common ability, pitch memory, and a separate, less common ability, pitch labelling. In this key study, approximately 50% of the subjects sang back the test melody in correct pitch, at least in one or two experimental sessions conducted (Levitin 1994). This study shows that even non-musicians have the ability to generate complex memories for pitch and memory for pitch is more common than initially imagined. Although, pitch memory can be improved in non-musicians by pitch memory training and can even be enhanced by applying electrical stimulation to the left supramarginal gyrus (Schaal et al. 2013), it is likely that memory for even fundamental musical abilities (perception of pitch, rhythm and emotion) are multi-component involving global and local processing of music (Ziv and Radin 2014). A case study of patient with severe auditory agnosia limited to music after bilateral temporal lobe damage showed a selective modality-specific but music-specific deficit in perceptual memory. While, the patient could recognise the lyrics that usually accompany the songs, processing of pitch sequential structure was impaired while retaining normal processing of temporal structure. A marked impairment in judging familiarity of melodies or hum a tune from memory suggests that specialised perceptual memory could exist away from verbal and auditory stimuli (Peretz 1996). In humans, a complex functional system, the vocal sensorimotor loop (VSL) is thought to be responsible for the ability to hold a tune in the mind. The VSL involves various components, of which memory is a large component, has received very little research attention (Bella et al. 2012). In a similar study testing the role of memory in the functioning of vocal sensorimotor loop, both amateur singers and congenital amusia patients were asked to reproduce a well-known melody under conditions with varying cognitive and memory loads. Conventionally, only poor perception has been thought to contribute to poor fidelity of performance. In this study, while perception was unhindered, impaired memory was found to contribute to poor-pitch singing and infidelity in intent-to-execution (Anderson et al. 2012).

In an interesting early study on recognition memory, Dowling (1978) demonstrated that neither musicians nor non-musicians were reliable in identifying matching melodies when two comparison sequences, one an exact transposition (key changed, exact size of intervals and contour) and the other an up-down transposition (exact key and contour, altered size of intervals), were played after the target melodic sequence. However, changes to musical contour was more accurately judged, suggesting that diatonic scale structure was an important determinant in perception. In other studies that explored the musical factors influencing musical memory, triadic structure, repetition of the tonic, leading tone to tonic ending, harmonic cadence (V-I), modulation within the sequence, and key-distance of transposition (tritone versus dominant) all contributed to memory-based perception (Cuddy and Cohen 1976; Cuddy et al. 1979, 1981; Cuddy and Lyons 1981). Similar studies studying the effect of key-distance demonstrated that near-key and inexact transpositions were harder to discriminate than far-key transpositions (Dowling and Bartlett 1981), that preemptive awareness of tonal structure leads to better retention of melodic patterns (Dowling and Bartlett 1981) and that diatonic sequences elicited stable recall and recognition performance (Watkins 1985). However, there is a great need to study memory as the driver rather than the mere recipient of larger musical experiences given that music perception is a highly specialised cognate activity in humans, is indirect in nature and that perception is a gestalt of prior-memory-based interaction with a complex musical stimulus. Rhythm plays a crucial role in perception of music, by affecting what is heard and how sound is organised for perception and can either augment or obscure perception of melody or tone. In a study with expert musicians, recall of unfamiliar folk tunes significantly declined when tonal information was either absent or obscured by rhythmic structure (Boltz 1991). An early study categorises rhythms based on the roles of echoic, short-term, and long-term memory in rhythmic perception (Brower 1993). Shehan (1987) showed that in second- and sixth-grade students, rhythm reproduction performance was far higher for a combination of aural and visual presentation than for one type of presentation alone. Many studies with amusia and impaired musical abilities indicate that memory systems for pitch and rhythm processing could be separate. When pitch memory is damaged, patients with amusia displayed normal rhythmic memory, suggesting that neural systems operating could be separate (Murayama et al. 2004). Pitch-deaf patients have been reported to possess normal rhythm capabilities as well (Bella and Sowinski 2015).

Context, cultural associations and enculturation are crucial to both formation of memories and recall and retrieval. It is not just the nature of music, familiarity or dissimilarity to known musical systems that determines musical perception, retention, recall, recognition, prediction but also established schemas and prior circuits of neural activity. Different musical systems entrain and establish certain schemas and neural activity patterns which affect music perception cross-culturally (Eerola et al. 2006). Western classical music systems are bound to schemas to detect and appreciate tonality and harmony, much better than modal systems of Indian classical music or the percussive-dominant musical traditions of African music. While, these are likely to build on a common set of human capacity to perceive sound, the culture-specific schemas could hinder or accelerate responses to new music.

It is widely accepted that music evokes emotions and has phenomenal access to our limbic system. Beyond perceptual factors that dictate memory, associative and autobiographical memories associated with music regulate episodic musical memory (Aubé et al. 2013; Jäncke 2008). Several studies indicate a limbic-based model for formation of memories involving emotions (Zatorre et al. 2002; Janata et al. 2007). The semantic associative network model of memory (Bower 1981) proposes that emotions could serve as contextual information in the remembering process and could act as nodes for remembering as well. In a study investigating whether emotional film music is retained more strongly than emotionally inert music, and concurrently assessing the extent to which musical structure is important in this process, it was reported that, while arousal ratings were not significantly correlated to recognition, positive valence to musical memory was (Eschrich et al. 2008). The act of remembering is also the act of recollecting the various sensory, motor, cultural, visceral, metaphorical memories evoked as much as it is remembering the music itself. Emotional valence to music could be an important determinant of in long-term memory for music, but the processes of such non-perceptual based memory systems and how they operate in memory formation and retrieval needs to be explored. Musical experience is inextricably conjoint with culture and memory. Every musical culture engages certain musical faculties selectively. The culture of Western classical music largely engages cognitive faculties pertaining to tonality, perception of polyphony, triadic relationships, harmony and chord progressions. In contrast, monophonic systems of music, such as Indian classical music, employ cognitive faculties that focus on heterophony, detailed melody-rhythm-prosody relationships and microtonal gestures. Such already established cognitive schema in musicians or passive exposure to such music in non-musicians affects the experience of unfamiliar music, culture and making music in new contexts. Taken together, memory and its schemas are used in context-specific and context-independent processes, suggesting that memory for a skill such a pitch detection or beat discrimination is likely to involve multiple schemas, multiple representations, multiple imagery and hence multiple ways to access and use these memories in learning, creativity and higher-order cognitive processes. The cognitive neuroscience approaches to studying neural correlates of music perception is likely to provide strong associative data confirming this highly diverse alchemy of skill-process-representation-imagery schemas involving conscious/unconscious, declarative and procedural and multiple subtypes of memory.

Memory in Culture-Specific Learning

Developing musical skill involves both imitative learning and observational learning in context-rich settings, which help develop musical skill embodied across several cognitive schemas at once. Imitation is indeed the foundation for transmitting human culture, guiding the process of socialisation and sustenance of culture. In imitative musical learning, cultural prerogatives are psychocognitively embodied as memories using a variety of multisensory strategies—from metaphors, visual patterns, synaesthetic experiences, auditory imagery, invoking people, places, images and a rich tapestry of contextual sound information. In studies with songbird vocal learning, it has been demonstrated that the sensory learning during early auditory experiences create auditory memory ‘templates’ which are crucial to later sensorimotor learning. During development, juvenile songbirds listen to and form auditory memories of adult tutor songs, which crucially regulate sensorimotor learning in later stages (Mooney 2009).

A recent study identified the neuronal substrate for tutor song memory by recording single-neuron activity in the caudomedial nidopallium (NCM), a higher auditory cortex centre, of juvenile zebra finches, suggesting that the tutor song memory may be represented in a subset of NCM neurons, which later shape neuronal circuits (Yanagihara and Yazaki-Sugiyama 2016a, b). The first experimental evidence of culturally transmitted patterns of vocal behaviour was that the young exposed to conspecific songs from other regions were capable of singing these foreign dialects or related species (Marler and Tamura 1964; Baker 1983). All of these findings highlight the role of culture-specific imitation in development of cognitive skills and memory capacities for music.

Similarly, musical memories that inform and empower musical expertise in one culture affects perception of another. In a study involving Finnish folk musicians and non-musicians and using electroencephalography recordings of early right anterior negativity (ERAN) in inferofrontal brain regions, chord processing patterns responses to ambivalent chords and incongruous chords were measured. The Finnish musicians exhibited enhanced ERAN responses to chord deviations as compared to non-musicians demonstrating heightened and altered perceptional abilities (Tervaniemi et al. 2012). In another preliminary study, the ability of non-musicians to generate expectancies in unfamiliar cultures in order to study the universality of melodic expectations in a cross-cultural context was examined. In college-age students born and raised in the United States, event-related potentials (ERPs) to melodic expectancy violations in 30 short melodies based in the Western folk tradition and 30 from North Indian classical music was assessed. Results indicated that subjects were less sensitive to deviations in the Indian melody condition, as they had lesser culture-specific knowledge guiding their responses (Demorest and Osterhout 2012). Memory plays a key role in rhythm and metre perception, which is largely culture specific. In a recent study, the perception of familiar (Western) rhythms with an isochronous metre and unfamiliar (Balkan) rhythms with a non-isochronous metre were compared by a behavioural discrimination task by American children (5–11 years) and adults before and after a 2-week period of at-home listening to non-isochronous metre music from Bulgaria. In the first session, both adults and children discriminated isochronous melodies than non-isochronous melodies, this asymmetry declined for young children but not for older children and adults. With passive exposure to non-isosynchronous metres, the rhythm discrimination capabilities increased, suggesting that implicit knowledge and memory for such metrical structures was developed through exposure (Hannon et al. 2012). These findings indicate that the characteristics of the style of music played by musicians influence their perceptual skills and the brain processing of sound features embedded in music. Musicians' processing of sounds depends highly on instrument, musical context and level of expertise. A study measuring the mismatch negativity (MMN), a pre-attentive brain response, to six types of musical feature change in musicians playing three distinct styles of music (classical, jazz, and rock/pop), found that jazz musicians were the most sensitive to auditory feature changes and unprecedented musical outliers than all other experimental groups (Vuust et al. 2012).

Conclusion

The philosophical orientation of Cartesian dualism continues to have a tremendous impact on cognitive musicology; music cognition research has remained focused on perception of structure, on perception of pitch, melody, rhythm, harmony and tonality and studied as faculties that are completely dissociated from embodied activities such as musical learning and performance. Consequently, the empirical basis to current theories of music perception are rooted in studies of auditory perception which reduce music to frequency, tone, sensation, vibration and stimulus and hence study musical memory as a passive decontextualised repository of musical knowledge. Studies on music cognition today are largely focused on finding neurobiological substrates of musical processes and experiences (Zatorre and Salimpoor 2013). The study of the role of memory in musical perception remains problematic because of the conceptual frameworks of cognitive neuroscience of memory, such as the neurobiology of memory formation (encoding, storage and retrieval), types of memories (sensory, short-term, working memory and long-term) and kinds of memories (declarative, non-declarative, explicit and implicit semantic, episodic, procedural), are largely useful as concepts guiding musical memory experiments but remain less influential in developing experimental strategies and models for testable hypotheses. The study of musical memory offers a unique opportunity to understand the operation of memory in the bidirectional processes of culture and cognition, as it is clear that there are multiple memory systems that operate in several processes and a great deal of insight into the nature of memory itself needs to be generated before neural correlates of memory in higher-order thinking or metacognitive abilities may be established.

A large volume of studies measure neural correlates of responses to 'musical stimulus' that assume a one-to-one correlation between task to response and cognitive process to brain region (Squire 2004). While a staggering amount of research on the cognitive neuroscience of memory has used non-musical tasks to dissect explicit/direct and implicit/indirect memory tasks, the underlying assumption remains that explicit tasks measure conscious processing and implicit tasks measure non-conscious processing. However, musical performance and hence usage of memory in performance involves rapid and concurrent activation of conscious and non-conscious processes, using declarative and non-declarative memories in an alchemical mix. In addition, a large number of studies remain focused on music listening rather than music making tasks. The degree to which culture informs musicians and non-musicians, experts and novices, informally and formally trained musicians remains to be investigated. In addition, as a recent bibliometric study of the field of music cognition demonstrates, there are very few studies looking into how humans perceive and cognise melodies, melodic structure, patterns and melodic behaviour in non-tonal musical cultures (Tirovolas and Levitin 2011). Empirical cross-cultural and intra-cultural studies on musical memory would be illuminative. Musically relevant and situated challenges that target and test cognitive processes in music will need to explore brain processes independently from previously established paradigms developed for non-musical stimuli or verbal only tasks. Do the culture-cognition polarities shape relative strengths and weaknesses of a universal capacity for musical memory, analogous to culture influencing neurological activities of speakers of tonal and non-tonal languages (Kaan et al. 2007, 2008)? How is the musical memory that informs culturally familiarity different from the universal abilities to predict what is musical and musically unlikely? Can individuals logically sense-make unfamiliar music, based on very same cognitive priori that shapes music as a culturally meaningful communication—extracting semantic meaning, structure detection, pattern discrimination in culturally alien music? (Nan et al. 2006) or prediction (Duque et al. 2009). The boundaries of impact of culture and enculturation on schemas and memory need careful experimental analysis.

The examination of musical memory as a segue between culture, cognition and consciousness could generate insight into the nature of memory, as an embodied and situated construct that includes the self, external symbol systems and the environment.

References

Anderson, S., Himonides, E., Wise, K., Welch, G., & Stewart, L. (2012). Is there potential for learning in amusia? A study of the effect of singing intervention in congenital amusia. *Annals of the New York Academy of Sciences, 1252*(1), 345–353. doi:10.1111/j.1749-6632.2011.06404.x.

Aubé, W., Peretz, I., & Armony, J. L. (2013). The effects of emotion on memory for music and vocalisations. *Memory, 21*(8), 981–990. doi:10.1080/09658211.2013.770871.

Baddeley, A. (2012). Working memory: Theories, models, and controversies. *Annual Review of Psychology, 63*(1), 1–29. doi:10.1146/annurev-psych-120710-100422.

Baker, M. C. (1983). Sharing of vocal signals among song sparrows. *The Condor, 85*(4), 482. doi:10.2307/1367991.

Bella, S. D., & Sowiński, J. (2015). Uncovering beat deafness: Detecting rhythm disorders with synchronized finger tapping and perceptual timing tasks. *Journal of Visualized Experiments,* (97). doi:10.3791/51761.

Bella, S. D., Tremblay-Champoux, A., Berkowska, M., & Peretz, I. (2012). Memory disorders and vocal performance. *Annals of the New York Academy of Sciences, 1252*(1), 338–344. doi:10.1111/j.1749-6632.2011.06424.x.

Berz, W. L. (1995). Working memory in music: A theoretical model. *Music Perception: An Interdisciplinary Journal, 12*(3), 353–364. doi:10.2307/40286188.

Boltz, M. (1991). Some structural determinants of melody recall. *Memory & Cognition, 19*(3), 239–251. doi:10.3758/bf03211148.

Booth, G. D., & Cutietta, R. A. (1991a). The applicability of verbal processing strategies to recall of familiar songs. *Journal of Research in Music Education, 39*(2), 121. doi:10.2307/3344692.

Booth, G. D., & Cutietta, R. A. (1991b). The applicability of verbal processing strategies to recall of familiar songs. *Journal of Research in Music Education, 39*(2), 121. doi:10.2307/3344692.
Bower, G. H. (1981). Mood and memory. *American Psychologist, 36*(2), 129–148. doi:10.1037//0003-066x.36.2.129.
Brower, C. (1993). Memory and the perception of rhythm. *Music Theory Spectrum, 15*(1), 19–35. doi:10.1525/mts.1993.15.1.02a00020.
Chaffin, R., Demos, A. P., & Logan, T. R. (2015). Performing from Memory. *Oxford Handbooks Online*. doi:10.1093/oxfordhb/9780198722946.013.34.
Chaffin, R., & Logan, T. (2006). Practicing perfection: How concert soloists prepare for performance. *Advances in Cognitive Psychology, 2*(2), 113–130. doi:10.2478/v10053-008-0050-z.
Chen, J. L., Penhune, V. B., & Zatorre, R. J. (2008). Moving on time: Brain network for auditory-motor synchronization is modulated by rhythm complexity and musical training. *Journal of Cognitive Neuroscience, 20*(2), 226–239. doi:10.1162/jocn.2008.20.2.226.
Clarke, E. (1993). Book review. *Psychology of Music, 21*(1), 83–89. doi:10.1177/030573569302100107.
Cuddy, L. L., & Cohen, A. J. (1976). Recognition of transposed melodic sequences. *The Quarterly Journal of Experimental Psychology, 28*, 255–270.
Cuddy, L. L., Cohen, A. J., & Mewhort, D. J. K. (1981). Perception of structure in short melodic sequences. *Journal of Experimental Psychology: Human Perception and Performance, 7*, 869–883.
Cuddy, L. L., Cohen, A. J., & Miller, J. (1979). Melody recognition: The experimental application of musical rules. *Canadian Journal of Psychology, 33*(1), 48–57.
Cuddy, L. L., & Lyons, H. I. (1981). Music pattern recognition: A comparison of listening to and studying tonal structures and tonal ambiguities. *Psychomusicology, 1*, 15–33.
Demorest, S. M., & Osterhout, L. (2012). ERP responses to cross-cultural melodic expectancy violations. *Annals of the New York Academy of Sciences, 1252*(1), 152–157. doi:10.1111/j.1749-6632.2012.06464.x.
Dowling, W. J. (1978). Scale and contour - 2 components of a theory of memory for melodies. *Psychological Review, 85*(4), 341–354.
Dowling, W. J., & Bartlett, J. C. (1981). The importance of interval information in long-term memory for melodies. *Psychomusicology: A Journal of Research in Music Cognition, 1*(1), 30–49. doi:10.1037/h0094275.
Duque, J. F., Turner, R., Lewis, E. D., & Egan, G. (2009). Neuroanthropology: A humanistic science for the study of the culture–brain nexus. *Social Cognitive and Affective Neuroscience, 5* (2–3), 138–147. doi:10.1093/scan/nsp024.
Eerola, T., Himberg, T., Toiviainen, P., & Louhivuori, J. (2006). Perceived complexity of western and African folk melodies by western and African listeners. *Psychology of Music, 34*(3), 337–371. doi:10.1177/0305735606064842.
Eschrich, S., Münte, T. F., & Altenmüller, E. O. (2008). Unforgettable film music: The role of emotion in episodic long-term memory for music. *BMC Neuroscience, 9*(1), 48. doi:10.1186/1471-2202-9-48.
Gaver, W. W. (1993). What in the world do we hear? An ecological approach to auditory event perception. *Ecological Psychology, 5*(1), 1–29. doi:10.1207/s15326969eco0501_1.
Gelfand, M. J., & Kashima, Y. (2016). Editorial overview: Culture: Advances in the science of culture and psychology. *Current Opinion in Psychology, 8*, 9–10. doi:10.1016/j.copsyc.2015.12.011.
Gibson, J. J. (1966). *The senses considered as perceptual systems*. Boston: Houghton Mifflin. ISBN 0-313-23961-4.
Green, D. M., & Swets, J. A. (1966). *Signal detection theory and psychophysics*. New York: Wiley.
Green, L. (2012). Musical 'learning styles' and 'learning strategies' in the instrumental lesson: Some emergent findings from a pilot study. *Psychology of Music, 40*(1), 42–65. doi:10.1177/0305735610385510.

Gruszka, A., & Orzechowski, J. (2016). Meta-analysis of the research impact of Baddeley's multicomponent working memory model and Cowan's embedded-processes model of working memory: A bibliometric mapping approach. *Polish Psychological Bulletin, 47*(1), 1–11. doi:10.1515/ppb-2016-0001.

Hannon, E. E., Bosch, Vanden, der Nederlanden, C. M., & Tichko, P. (2012). Effects of perceptual experience on children's and adults' perception of unfamiliar rhythms. *Annals of the New York Academy of Sciences, 1252*(1), 92–99. doi:10.1111/j.1749-6632.2012.06466.x.

Imamizu, H., & Kawato, M. (2009). Brain mechanisms for predictive control by switching internal models: implications for higher-order cognitive functions. *Psychological Research Psychologische Forschung, 73*(4), 527–544. doi:10.1007/s00426-009-0235-1.

Janata, P. (2012). Acuity of mental representations of pitch. *Annals of the New York Academy of Sciences, 1252*(1), 214–221. doi:10.1111/j.1749-6632.2011.06441.x.

Janata, P., Tomic, S. T., & Rakowski, S. K. (2007). Characterisation of music-evoked autobiographical memories. *Memory, 15*(8), 845–860. doi:10.1080/09658210701734593.

Jäncke, L. (2008). Music, memory and emotion. *Journal of Biology, 7*(6), 21. doi:10.1186/jbiol82.

Kaan, E., Barkley, C. M., Bao, M., & Wayland, R. (2008). Thai lexical tone perception in native speakers of Thai, English and Mandarin Chinese: An event-related potentials training study. *BMC Neuroscience, 9*(1), 53. doi:10.1186/1471-2202-9-53.

Kaan, E., Wayland, R., Bao, M., & Barkley, C. M. (2007). Effects of native language and training on lexical tone perception: An event-related potential study. *Brain Research, 1148,* 113–122. doi:10.1016/j.brainres.2007.02.019.

Keller, P. E. (2012). Mental imagery in music performance: underlying mechanisms and potential benefits. *Annals of the New York Academy of Sciences, 1252*(1), 206–213. doi:10.1111/j.1749-6632.2011.06439.x.

Levitin, D. J. (1994). Absolute memory for musical pitch: Evidence from the production of learned melodies. *Perception and Psychophysics, 56*(4), 414–423. doi:10.3758/bf03206733.

Licklider, J. C. R. (1959). Three auditory theories. In S. Koch (Ed.), *Psychology: A study of a science*. McGraw-Hill.

Lisboa, T., Chaffin, R., & Demos, A. P. (2015). Recording thoughts while memorizing music: a case study. *Frontiers in Psychology, 5*. doi:10.3389/fpsyg.2014.01561.

Marler, P., & Tamura, M. (1964). Culturally transmitted patterns of vocal behavior in sparrows. *Science, 146*(3650), 1483–1486. doi:10.1126/science.146.3650.1483.

Mishra, J. (2005). A theoretical model of musical memorization. *Psychomusicology: A Journal of Research in Music Cognition, 19*(1), 75–89. doi:10.1037/h0094039.

Miyazaki, K. (1992). Perception of musical intervals by absolute pitch possessors. *Music Perception: An Interdisciplinary Journal, 9*(4), 413–426. doi:10.2307/40285562.

Miyazaki, K. (1995). Perception of relative pitch with different references: Some absolute-pitch listeners can't tell musical interval names. *Perception and Psychophysics, 57*(7), 962–970. doi:10.3758/bf03205455.

Mooney, R. (2009). Neural mechanisms for learned birdsong. *Learning & Memory, 16*(11), 655–669. doi:10.1101/lm.1065209.

Murayama, J., Kashiwagi, T., Kashiwagi, A., & Mimura, M. (2004). Impaired pitch production and preserved rhythm production in a right brain-damaged patient with amusia. *Brain and Cognition, 56*(1), 36–42. doi:10.1016/j.bandc.2004.05.004.

Nan, Y., Knösche, T. R., & Friederici, A. D. (2006). The perception of musical phrase structure: A cross-cultural ERP study. *Brain Research, 1094*(1), 179–191. doi:10.1016/j.brainres.2006.03.115.

Niewöhner, J. (2014). Book review: The encultured brain: An introduction to neuroanthropology. *Journal of the Royal Anthropological Institute, 20*(4), 789–790. doi:10.1111/1467-9655.12138_8.

Overwalle, F. V., & Mariën, P. (2016). Functional connectivity between the cerebrum and cerebellum in social cognition: A multi-study analysis. *NeuroImage, 124,* 248–255. doi:10.1016/j.neuroimage.2015.09.001.

Peretz, I. (1996). Can we lose memory for music? A Case of music agnosia in a nonmusician. *Journal of Cognitive Neuroscience, 8*(6), 481–496. doi:10.1162/jocn.1996.8.6.481.
Plantinga, J., & Trainor, L. J. (2005). Memory for melody: infants use a relative pitch code. *Cognition, 98*(1), 1–11. doi:10.1016/j.cognition.2004.09.008.
Rayleigh, Strutt, J. W., & Baron. (1877). *The theory of sound* (pp. 1842–1919).
Robertson, S. (2016). The encultured brain: An introduction to neuroanthropology. In D. H. Lende, & G. Downey. *Anthropological Forum,* 1–3. doi:10.1080/00664677.2016.1212544.
Schaal, N. K., Williamson, V. J., & Banissy, M. J. (2013). Anodal transcranial direct current stimulation over the supramarginal gyrus facilitates pitch memory. *European Journal of Neuroscience, 38*(10), 3513–3518. doi:10.1111/ejn.12344.
Schulze, K., & Koelsch, S. (2012). Working memory for speech and music. *Annals of the New York Academy of Sciences, 1252*(1), 229–236. doi:10.1111/j.1749-6632.2012.06447.x.
Shehan, P. (1987). Effects of rote versus note presentation of rhythm learning and retention. *Journal of Research in Music Education, 35*(2), 117–126.
Squire, L. R. (2004). Memory systems of the brain: A brief history and current perspective. *Neurobiology of Learning and Memory, 82*(3), 171–177. doi:10.1016/j.nlm.2004.06.005.
Stevens, C. J. (2012). Music perception and cognition: A review of recent cross-cultural research. *Topics in Cognitive Science, 4*(4), 653–667. doi:10.1111/j.1756-8765.2012.01215.x.
Suárez, L., Elangovan, S., & Au, A. (2015). Cross-sectional study on the relationship between music training and working memory in adults. *Australian Journal of Psychology, 68*(1), 38–46. doi:10.1111/ajpy.12087.
Sutton, J. (2009). Remembering. *The Cambridge handbook of situated cognition* (pp. 217–235). doi:10.1017/cbo9780511816826.013.
Sutton, J., Harris, C. B., Keil, P. G., & Barnier, A. J. (2010). The psychology of memory, extended cognition, and socially distributed remembering. *Phenomenology and the Cognitive Sciences, 9* (4), 521–560. doi:10.1007/s11097-010-9182-y.
Tervaniemi, M., Tupala, T., & Brattico, E. (2012). Expertise in folk music alters the brain processing of Western harmony. *Annals of the New York Academy of Sciences, 1252*(1), 147–151. doi:10.1111/j.1749-6632.2011.06428.x.
Tirovolas, A. K., & Levitin, D. J. (2011). Music perception and cognition research from 1983 to 2010: A categorical and bibliometric analysis of empirical articles in music perception. *Music Perception: An Interdisciplinary Journal, 29*(1), 23–36. doi:10.1525/mp.2011.29.1.23.
Ullman, S. (1980). Against direct perception. *Behavioral and Brain Sciences, 3*(03), 373. doi:10.1017/s0140525x0000546x.
Vandervert, L. (2016). The prominent role of the cerebellum in the learning, origin and advancement of culture. *Cerebellum & Ataxias, 3*(1). doi:10.1186/s40673-016-0049-z.
Vandervert, L. R., Schimpf, P. H., & Liu, H. (2007). How working memory and the cerebellum collaborate to produce creativity and innovation. *Creativity Research Journal, 19*(1), 1–18. doi:10.1080/10400410709336873.
Vuust, P., Brattico, E., Seppänen, M., Näätänen, R., & Tervaniemi, M. (2012). Practiced musical style shapes auditory skills. *Annals of the New York Academy of Sciences, 1252*(1): 139–46. doi:10.1111/j.1749-6632.2011.06409.x.
Warren, J. (2008). How does the brain process music? *Clinical Medicine, 8*(1), 32–36. doi:10.7861/clinmedicine.8-1-32.
Watkins, A. J. (1985). Scale, key, and contour in the discrimination of tuned and mistuned approximations to melody. *Perception and Psychophysics, 37*, 275–285.
Yanagihara, S., & Yazaki-Sugiyama, Y. (2016a). Auditory experience-dependent cortical circuit shaping for memory formation in bird song learning. *Nature Communications, 7,* 11946. doi:10.1038/ncomms11946.
Yanagihara, S., & Yazaki-Sugiyama, Y. (2016b). Auditory experience-dependent cortical circuit shaping for memory formation in bird song learning. *Nature Communications, 7,* 11946. doi:10.1038/ncomms11946.
Zatorre, R. J. (2012). Beyond auditory cortex: Working with musical thoughts. *Annals of the New York Academy of Sciences, 1252*(1), 222–228. doi:10.1111/j.1749-6632.2011.06437.x.

Zatorre, R. J., Belin, P., & Penhune, V. B. (2002). Structure and function of auditory cortex: music and speech. *Trends in Cognitive Sciences, 6*(1), 37–46. doi:10.1016/s1364-6613(00)01816-7.

Zatorre, R. J., & Salimpoor, V. N. (2013). From perception to pleasure: Music and its neural substrates. *Proceedings of the National Academy of Sciences of the United States of America, 110*(Suppl 2), 10430–10437.http://doi.org/10.1073/pnas.1301228110.

Ziv, N., & Radin, S. (2014). Absolute and relative pitch: Global versus local processing of chords. *Advances in Cognitive Psychology, 10*(1), 15–25. doi:10.5709/acp-0152-7.

The Neuroscience of Blame and Punishment

Morris B. Hoffman and Frank Krueger

The Evolution of Punishment

Biologists have long recognised two fundamental kinds of punishment behaviours: second-party punishment and third-party punishment (Fehr and Fischbacher 2004; Fehr and Gachter 2002). Second-party punishment is any punitive action taken by one individual against a second individual in response to harm or threat inflicted by the second on the first. When you are in line at the grocery store and feel the person behind you trying to steal your wallet, and turn around and grab him, you are engaging in a kind of second-party punishment. He was committing a wrong against you for which you, in a fashion, retaliated. Third-party punishment, by contrast, is any punitive action taken by one individual against a second individual in response to a harm or threat inflicted by the second on a *third* individual. When you see that person in line trying to pick the pocket of another person, and you grab him, you are engaging in a kind of third-party punishment. You are willing to punish the wrongdoer even though his wrong was not aimed at you.

Apropos of the title of this book, there is third kind of punishment we will add to this evolutionary mix: first-party punishment, which is also more commonly called *conscience* and *guilt* (Michl et al. 2014). When that thief in line is considering whether to pick your pocket but is able to refrain from doing so because he knows

This chapter is adapted from our paper, Krueger, F. and Hoffman, M. (2016) The emerging neuroscience of third-party punishment, *Trends in Neuroscience,* 39(8): 499–501, and from Hoffman, M. 2014. *The Punisher's Brain: The Evolution of Judge and Jury* (Cambridge University Press).

M.B. Hoffman
University of Denver, Denver, CO, USA

F. Krueger (✉)
George Mason University, Fairfax, VA, USA
e-mail: fkrueger@gmu.edu

S. Menon et al. (eds.), *Self, Culture and Consciousness,*
DOI 10.1007/978-981-10-5777-9_13

it's wrong, he has engaged in first-party punishment. Unless he has something akin to psychopathic tendencies, he probably has an embedded moral intuition that tells him stealing is wrong, and that helps him resist the temptation. Even if he is unable to resist and goes ahead and steals, he will feel guilt—a kind of post hoc conscience—that may help him resist the next time.

The animal kingdom is full of examples of second-party punishment, from guppies that attack inspecting predator fish (Godin and Davis 1995) all the way down to forms of algae that fire projectiles at would-be predators (Gordon 2006). In fact, the most fundamental form of second-party punishment may well be the immunological response. We don't think of white blood cells attacking viruses as 'punishment', but surely those white blood cells are retaliating against an invading bacterium every bit as much as a male grackle trying to ward off an interloper. Indeed, one of life's deepest and most ancient problems—how to achieve multi-cellularity in a world of single-celled organisms—was really a problem about overcoming this immunological form of second-party punishment.

First- and third-party punishment, by contrast, seem uniquely human. We will not dwell here on the question of whether other species have 'conscience' and feel 'guilt' in the way we humans experience these phenomena, though some behavioural scientists have taken that position.[1] We do know that other social animals have the capacity to empathise (de Waal 2012),[2] which seems to be a precursor to conscience.[3]

Third-party punishment seems unique to humans (Riedl et al. 2012), though again there are tantalising examples of non-human precursors to it (Raihani et al. 2010). For example, in the early 1990s, primatologists in Tanzania observed non-dominant male chimpanzees attack a young adult male, apparently because the young male was attacking females without provocation (Nishida et al. 1995). Because the young male was not attacking the punishers themselves or attacking females claimed by them, this seems to be a kind of third-party punishment.

What do these three kinds of punishments have to do with human evolution? They arguably converged to solve the essential Prisoner's Dilemma of our ancestors' early lives: do we cooperate or defect? Humans are the animal kingdom's most highly social heterogenetic species, but of course natural selection still works at the individual level (de Weerd and Verbrugge 2011). Living in intensely social groups gave us enormous fitness advantages—in things like hunting, mutual defence, divisions of labour—but it also presented enormous temptations for individuals to

---

[1]The most famous was Lorenz and Wilson (2002), who wrote a whole book about it.

[2]Many behavioural studies have confirmed that several primate species, and even some non-primate mammals, have a kind of empathy. To the extent that mirror neurons are involved with empathy, these behavioural results converge with experiments showing many of these same empathising social animals have mirror neurons. Indeed, some theorists have speculated that the essential imitative function of what we call mirror neurons may be a widespread neuronal phenomenon across many species.

[3]Hume (1968) made this argument—that the roots of morality lie in our ability to socialise our own desires, i.e. have empathy—200 years ago.

cheat (Gachter et al. 2008; Roos et al. 2014). Evolutionary theorists have surmised that without third-party punishment—the willingness of one member to punish a misbehaving fellow member even when that misbehaviour has not injured the punishing member—emerging humans could never have evolved in their intense social groups (Boyd et al. 2003; Krasnow et al. 2016).

All three layers of punishment were arguably necessary to prevent the spread of uncontrolled cheating. Conscience prevented us from violating norms in the first place. Most of us know that it's wrong to rob others. If conscience were not enough, second-party punishment helped deter us because we knew if we robbed a fellow group member he would almost certainly resist and retaliate. And finally, if conscience and the fear of retaliation were not enough, third-party punishment added yet another level of disincentive; even if, and especially if, we robbed someone too weak to retaliate, other members would punish us. The socialisation of that willingness to punish wrongdoers meant that a wrong committed against any member was now a wrong against the group, punishable by the group, or by the group's designees.

But third-party punishment is very costly, evolutionarily speaking (Hetzer and Sornette 2013). Unlike with second-party punishment, an impartial third-party punisher gains nothing immediate by way of self-defence, only an indirect long-term benefit in group stability. Yet he risks serious injury and even death, either from the resisting wrongdoer himself or from the wrongdoer's family. Indeed, biologists and economists sometimes call third-party punishment 'altruistic punishment', to reflect that the benefits to the punisher are indirect at best, or 'costly punishment', to reflect that these indirect benefits come at a direct cost (Boyd et al. 2003).

The fact that third-party punishment was so costly arguably explains that our willingness to impose it seems to come with some equally deep-seated hesitations. Too much punishment could unravel our groups just as easily as too little. We are all tempted by small defections, and if we punished every norm violation seriously—by, say, execution or banishment, the two most common ancient punishments for serious violations—there'd be no one left in our group. So, while we likely evolved a predisposition to punish third-party wrongdoers, we also likely evolved a corresponding predisposition to be restrained in our punishments, to forgive small violations, and to look for signs that even major violators might be safely brought back into the fold. We are willing to punish wrongdoers, but also willing to forgive them.

How exactly an evolved trait for third-party punishment could have arisen remains as controversial as the more general question of how our intensely social groups could have arisen, when the direct benefits of cheating seem always to outweigh the indirect benefits of cooperating. The answer probably has something to do with the fact that our ancestral groups were likely composed of related individuals. 'Altruistic' behaviours in such circumstances would not have been entirely altruistic in the genetic sense. Helping a family member, even at some cost, was to some extent helping genes shared with that family member (Hamilton 1964a, b).

There are three lines of evidence that converge to suggest that the evolutionary roots of punishment are something more than a just-so story. First, as mentioned above, many of our primate cousins exhibit behaviours akin to third-party punishment. These behaviours also show the fluid boundary between second—and third-party punishments, and in fact suggest that human third-party punishment may well have evolved out of predecessor second-party punishment. When a primate mother swats her child for poking the mother in the eye, that is second-party punishment. But when she swats one of her children for poking another of her children in the eye, that is third-party punishment. It's not a giant leap from 'hurt me, hurt my family' to 'hurt me, hurt my group', especially when the group is made up largely of family. As primatologist Jane Lancaster once elegantly put it while talking about primate ostracism, 'The cold shoulder is just a step toward execution' (Lancaster 1986).

Further, third-party punishment also appears to be a human universal. In every human society that has left a sufficient written or oral record, including present-day primitive societies, norm-violators are punished by designated punishers who themselves may not have been directly harmed and yet who may pay a cost associated with inflicting the punishment (Brown 1991). Whether it is the dominant male in a primitive tribe, a set of tribal elders, a king, a judge or a jury, all human societies have institutionalised in one form or another this idea that serious crimes against the group must be punished by a representative of the group (Marlowe and Berbesque 2008).

Finally, a wide array of behavioural studies, across many kinds of disciplines, from experimental economics to social psychology, show that humans have a palpable willingness to inflict punishment on norm-violators even when the violation does not harm the punisher, and even when the punishment comes at a cost to the punisher (Boehm 2012). For example, when experimental economists build a costly punishment round into the traditional dictator game—by forcing a defecting player to forfeit money back to the experimenter but only if the forcing player also must forfeit some money—subjects regularly oblige (Fehr and Fischbacher 2004). Although the amount of cost a third-party punisher is willing to endure varies across and within societies, even in existing primitive societies this game-based willingness to endure a cost to punish a third-party is ubiquitous (Henrich et al. 2006). The willingness to engage in third-party punishment seems to emerge early in childhood. Four- and five-year-old children engage in third-party punishment behaviours based on an expressed belief that antisocial actions deserve to be punished (Kenward and Osth 2012, 2015; McAuliffe et al. 2015).

Behavioural studies have also suggested that there is an intermediate step to punishment, which most researchers call 'blame' or 'blameworthiness'. Blame appears to be a more heuristic, automatic process; it is highly predictable, largely invariant across demographics, and seems to be driven by the two classic determinants of criminal culpability—the amount of harm and the wrongdoer's intention (Buckholtz and Marois 2012). With the wrongdoer's intention kept constant, the greater the harm the greater the blame. Likewise, with harm kept constant, intentional wrongdoers are blamed more than accidental wrongdoers. Very recent work shows these two determinants have an additive effect on blame: high harm intentionally caused is blamed more than just the harm or intention, considered separately, would explain (Ginther et al. 2016). The amount of blame experimental subjects assess is remarkably uniform and predictable across subjects, looking much more like a rote task than a complex moral judgement (Robinson and Darley 2007).[4] That of course does not mean that blame is not a moral judgement; it just means that it is the kind of moral judgement that is deeply imbedded in us and driven by a narrow range of variables (namely, harm and intent).

Punishment, by contrast, appears to be a substantially more complex and context-dependent task. It can vary not only depending on a myriad of facts about the wrongdoer, the wrong and the victim, but also on the punisher's beliefs about the nature of the world, the purposes of punishment, and in some circumstances even his beliefs about free will (Krueger et al. 2014).

Natural selection has arguably equipped us with these two punishment systems: a hot, fast, highly predictable one that places the wrong grossly on a scale of blameworthiness depending only on our assessment of the harm caused and the wrongdoer's intention in causing it; and a second, cooler, slower system that gives us time to consider all the pertinent circumstances before we decide on the amount and type of punishment, or indeed whether we will forgive the transgression.

If our willingness to blame and punish—ourselves with conscience and guilt, our tormentors with retaliation, and our fellow wrongdoing citizens with retribution—is in fact part of an evolutionary suite of behaviours bequeathed to us by natural selection, then of course the mechanism of that inheritance lies embedded somewhere in our neural architectures. A host of functional neuroimaging research using methods such as functional magnetic imaging (fMRI) is beginning to converge on a neural model of blame and punishment.

The Neuroscience of Blame and Punishment

A Sampling of Experiments

Before we present a neuropsychological model of blame and punishment, we sample some of the major neuroscience experiments upon which our model is based. One of the first experiments on the neuroscience of criminal responsibility was done by Rebecca Saxe and her colleagues at MIT (Young et al. 2010). Subjects read six hypotheticals involving the difference between intentional crimes,

---

[4]A well-recognised statistical measure of the agreement between lists of rankings is called Kendall's *W* (K_w), which measures not just differences in the rankings but also the degree of those differences. A K_w of 1 means the lists match perfectly. A K_w of 0 means no more matching than would expected of random ranking. A K_w of 0.5 is typically described as 'moderate agreement'. Subjects in the Robinson and Darley blameworthiness experiments averaged a K_w ranging from 0.88 to 0.95. The only experimental behavioural tasks that even remotely approach this level of almost total agreement are things like asking visually unimpaired subjects to rate the brightness of objects ($K_w = 0.95$) and asking subjects to rate faces by how much pain they are feeling ($K_w = 0.97$).

unintentional ones, and attempts, and then graded those differences on a moral scale. Here's one of those hypotheticals:

> Grace and her friend work at a chemical factory, and are having lunch in the break room. Grace's friend asks her to retrieve some sugar from the counter.

From this stem narrative, subjects were presented with the following four variations generated by manipulating intent and harm:

1. *Intent and harm.* Grace decides to murder her friend and intentionally gives her poison instead of sugar. The friend dies (The law typically calls this first-degree murder).
2. *Intent, but no harm.* Grace decides to murder her friend and gives her what Grace thinks is poison, but which turns out to be sugar. The friend is not injured (The law typically calls this attempted first-degree murder).
3. *No intent, but harm.* Grace intends to give her friend sugar, but accidentally gives her poison instead. The friend dies (The law typically calls this reckless or negligent manslaughter, or no crime at all, depending on the level of Grace's carelessness).
4. *No intent and no harm.* Grace intends to give Alice sugar, and does give her sugar (This, of course, is not a crime).

Subjects were asked for each variation to rate Grace's behaviour on a seven-point scale, with 1 being 'forbidden' and 7 being 'permissible'. They were asked to do the same on each of the other five hypotheticals, which generated similar variations.

Behaviourally, virtually all subjects blamed in exactly the order listed above; that is, they blamed intentional killing most, then attempted killing and then accidental killing least. This conforms to what criminal law and moral philosophy universally recognise: one who commits an attempted intentional harm is more culpable than one who commits an accidental harm, even though an attempt causes no harm and an accident does.

The authors then repeated the experiment, but this time with two different sets of subjects—one control set and one set that had their right temporo-parietal junctions (rTPJs) disrupted by transcranial magnetic stimulation (TMS).[5] The group with disrupted rTPJs found accidental harm more blameworthy than attempted harm. Interestingly, young children also blame in this fashion (Baird and Astington 2004; Karniol 1978), presumably because their TPJs—which are among the last brain regions to develop—have not fully developed (Blakemore 2008). Eight-year-olds consistently base their punishments only on the outcome of the violation; adolescents integrate outcomes and intentions in second—but not third-party punishment; and adults integrate outcomes and intentions in second- and third-party punishment (Gummerum and Chu 2014).

---

[5]Transcranial magnetic stimulation (TMS) uses electromagnetic energy, focused into a small diameter, to temporarily disrupt the electrical impulses of neurons in regions of the brain fairly close to the surface.

This experiment confirmed what others had suggested: namely, that regions in the brain such as the TPJ, which are associated with 'theory of mind'—meaning the ability to attribute beliefs, intentions, desires and perspectives to others (Premack and Woodruff 1978)—are critical for moral judgement. The experiment also suggested why this is so: these theory-of-mind regions, including the TPJ, are necessary for assessing a wrongdoer's intention, which, as discussed above, is one of the two drivers of blame. Without working theory-of-mind regions, we blame like children blame, focusing only on harm and therefore using principles roughly akin to 'no harm no foul'.

Another culpability experiment involving TMS was significant for corroborating the fundamental neural differences between blame and punishment (Buckholtz et al. 2015). Joshua Buckholtz and his colleagues used TMS to disrupt the dorsal lateral prefrontal cortex (dlPFC) in one set of subjects, then gave them and a control set several blame (how morally responsible is the offender for his actions; nine-point scale: 0 = not morally responsible at all; 9 = completely morally responsible) and punishment (how much punishment deserves the offender for his actions; nine-point scale: 0 = no punishment and 9 = extreme punishment) scenarios. Disrupting the dlPFC had significant impact on the subjects' punishment decisions, but no impact on their blame decisions.

One of the first studies to try to ferret out the networks involved in third-party blame and punishment was also done by Buckholtz and his colleagues (Buckholtz et al. 2008). In this experiment, subjects in a scanner were asked to impose amounts of punishment on a hypothetical wrongdoer (from 0 to 9, 0 being no punishment and 9 being the most severe punishment the subject could imagine). The experimenters varied the hypotheticals both by harm and the wrongdoer's intention. As to the latter, however, they presented just two different states of mind: a normal 'unexcused' condition; and an excused condition (such as the wrongdoer was insane). They found that brain regions associated with affective processing (amygdala, medial prefrontal cortex, and posterior cingulate) predicted the amount of punishment when the wrongdoer's state of mind was held constant. By contrast, when harm was held constant, regions in the dlPFC seemed to be associated with the differences between the unexcused and excused mental states.

We and a group of colleagues expanded on the Buckholtz fMRI experiment by performing effective connectivity analyses to try to piece together the communication among the regions involved (Bellucci et al. 2017). We found that the blame decision does indeed differentially recruit many theory-of-mind regions, and specifically that the temporal pole (TP) and the dorsal medial prefrontal cortex (dmPFC) emerged as hubs of the blaming network, uniquely generating converging output connections to the ventromedial prefrontal cortex (vmPFC), the TPJ and the posterior cingulate cortex (PCC). The dmPFC received inputs only from the TP, and both its differential activation and its connectivity to the dlPFC correlated with the degree of punishment.

In another experiment, one of us and his colleagues applied a whole-brain voxel-based lesion-symptom mapping approach to identify brain regions associated with third-party punishment in a large sample of patients with penetrating traumatic brain injury (Glass et al. 2016). The study showed that patients who demonstrated atypical third-party punishment had specific lesions in core regions of the mentalising (dmPFC, vmPFC) and central executive (bilateral dlPFC, right posterior parietal cortex [PPC]) networks.

In a different experiment, we used these same crime hypotheticals, keeping intention constant but varying harm, to examine whether subjects' belief in free will correlated to their punishment decisions. We found that they did—that subjects with a strong belief in free will tended to punish more harshly—but only for low harms, a result that helped explain some earlier conflicting studies that did not vary harm. As we put it, when harms are high, we all act like we have strong beliefs in free will. In those low harm cases where beliefs in free will matter, we found a strong correlation between punishment amounts and activations in the right TPJ. Across the two groups with strong and weak beliefs in free will, we found a correlation between the amounts of punishment, the amount of harm, and activations in the right anterior insula (rAI) (Krueger et al. 2014). These results not only tend to corroborate the model that our brain uses two different, though richly interconnected, systems to blame and punish, but also that the punishment decision's highly context-dependent nature can include the punisher's own biases, including biases about free will.

Another study investigated ordinary citizens, who as potential jurors decided on mitigation of punishment for murders after reading hypothetical criminal scenarios (Yamada et al. 2012). Sympathy for a defendant activated regions associated with mentalising and moral conflict (e.g., dmPFC, TPJ, praecuneus) and individual differences on the inclination to mitigate was associated with activity in the right insula.

Finally, experimenters for the first time varied the wrongdoer's intentions across the four mental states traditionally recognised by the criminal law—purposeful, knowing, reckless and negligent—while also varying the amount of the harm (Ginther et al. 2016). They found the same basic pattern as previous experiments. Evaluating harm differentially engaged areas associated with affective and (interestingly) somatosensory functions, evaluating mental states recruited theory-of-mind areas, and vmPFC, PCC and amygdala (Amyg) seemed associated with the integration of these harm and mental states signals into blame. They also found, as others had, that moving from blame to punishment seemed to recruit the dlPFC. This experiment, which was very tightly controlled to maximise the chances that the fMRI signals being picked up were actually associated with the subjects' blame and punishment decisions (a real challenge in fMRI experiments both because of the time delay between the task and the hemodynamic response, and because of the problem of knowing when exactly this mysterious thing we call 'decision' actually happens), suggests that the essential punishment architecture revealed in these other experiments also applies in the kind of legal ecology typical of the criminal law.

A Neuropsychological Model of Blame and Third-Party Punishment

From these and other experiments, a few researchers have postulated neural models of the blame/punishment network.[6] We present here an extended version of the model we published recently (Krueger and Hoffman 2016). It appears that at least three domain general large-scale and well-known brain networks (Bressler and Menon 2010) are involved when we blame and punish: the salience network (SN), the default mode network (DMN) and the central executive network (CEN) (Fig. 1).

The SN, which is anchored in the dorsal anterior cingulate cortex (dACC) and AI, with extensive connectivity to the amygdala (Amyg), has been identified with the general processing of self-related aversive experiences (e.g., personal distress) that guide behaviour (Lamm et al. 2007; Marsh et al. 2008). In the context of blame and punishment tasks, we postulate that the SN is involved with detecting and responding to norm violations or threats of norm violations. We argue that the dACC is involved with the detection of the harm or threat, that the AI is associated with creating the aversive response to that harm or threat, and that the amygdala is involved with providing an emotional range for the severity of harm caused or threatened to the victim. The SN would be involved in first-, second- and third-party punishment, since recognising norm violations or potential norm violations is necessary for each of these kinds of punishment.

We postulate that the SN modulates the engagement of the DMN as the second large-scale network. The DMN is located generally in the medial prefrontal cortex (mPFC), and is associated with processes, including self-monitoring and theory of mind (Kalbe et al. 2010; Leopold et al. 2012; Sebastian et al. 2012). In the context of blame and punishment, we suggest the mPFC integrates the assessment of the wrongdoer's mental state with the assessment of the harm from the SN, producing what we might call a blame signal. In particular, we argue that the harm-integrating portion of the DMN appears to work through the vmPFC, with its inter-network connections to regions within the SN, allowing the experience of feelings congruent with another's emotional situation (affective theory of mind). We hypothesise that the assessment of the wrongdoer's intention involves the dmPFC, with its intra-network connections to other mentalising regions (PCC, TPJ) associated with understanding others' mental states (cognitive theory of mind). Under this model, the DMN would be necessary for both second- and third-party punishment, since both these forms of punishment require an assessment of blame.

Deciding on a specific type and amount of punishment, as opposed to merely blaming, recruits yet a third large-scale network in our model, the CEN. This network, located primarily in the dlPFC and the posterior parietal cortex (PPC), is recognised to be involved in the kinds of higher-order cognition such as executive

[6]A first model was published by Buckholtz and Marois (2012).

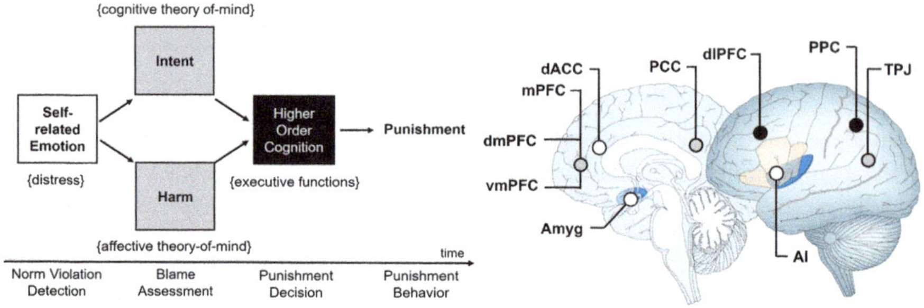

Fig. 1 Neuropsychological Framework of Blame and Punishment. Salience network (SN) (*white circle/square*: *AI* anterior insula; *dACC* dorsal anterior cingulate cortex; *Amyg* amygdala); **Default mode network** (DMN) (*gray circle/square*: *mPFC* medial prefrontal cortex; *dmPFC* dorsomedial PFC; *vmPFC* ventromedial PFC; *PCC* posterior cingulate cortex; *TPJ* temporo-parietal junction); **Central executive network** (CEN) (*black circle/square*: *dlPFC* dorsolateral PFC; *PPC* posterior parietal cortex). Figure adjusted and reprinted by permission from Elsevier: Krueger, F., and Hoffman, M. (2016). The emerging neuroscience of third-party punishment, *Trends in Neuroscience,* 39(8): 499–501, copyright 2017

functions necessary for context-dependent decision-making (Barbey et al. 2009; Miller and Cohen 2001). The dlPFC, in particular, is thought to be capable of maintaining stable representations over time (working memory), and of integrating different abstract representations from distinct information-processing streams. In other words, the dlPFC is often associated with complex reasoning and decision-making, where many factors have to be assessed and also weighed in order to reach a decision. In the blame and punishment context, we hypothesise that the CEN is involved in the processes by which the blame signal from the DMN is converted into a punishment decision after integration with a wide variety of context-dependent circumstances. We argue that this process involves the recruitment of the PPC to construct a punishment scale, and then the dlPFC to reason to a specific punishment on that scale based on all the circumstances of the case (and of course, as mentioned above, the biases of the punisher). The CEN would be primarily involved in third-party punishment, because only third-party punishment requires the cognitive transition from largely automatic blame (and retaliation) to a decision about third-party punishment. As mentioned above, this punishment phase likely involves some assessment of considerations that bear on forgiveness, though neither those considerations nor any brain regions associated with them are currently known.

In addition to being largely consistent with the experimental data amassed to date, this neuropsychological model is also consistent with some interesting behavioural and evolutionary-theoretic details about the relationship between first-, second- and third-party punishment. Let's begin with first-party punishment, that is, conscience and guilt. The whole notion of first-party punishment—which in turn requires us to be endowed with a set of moral intuitions—is hard for us to appreciate precisely because those moral intuitions run so deep. We will not belabour the point here, but we have argued elsewhere that although human norms can vary tremendously across cultures, at their core are three largely universal and naturally selected rules: (1) don't steal (property or wellbeing) from other group members; (2) don't break promises made to other group members; and (3) punish serious violations of (1) and (2) (Hoffman 2014). One of the best ways we can see how important these deeply held moral intuitions are to our ability to resist the temptation to be antisocial is to look at psychopaths, who appear to suffer deficiencies in brain regions of SN (e.g., amygdala) and DMN (e.g., vmPFC) known to be critical to moral decision-making (Kiehl 2006).

We have already mentioned that second-party punishment is widespread throughout the animal kingdom. We can think loosely of second-party punishment as being blame plus an automatic punishment response, all rolled into one, and without the cognitive restraints we see with third-party punishment. It is not surprising, then, that the behavioural experiments discussed above, in which human subjects are asked merely to blame but not to punish, elicit such consistent results across subjects.[7] Like most complex animals, we humans are acutely attuned to potential harms to ourselves, and predisposed to deal with those harms in a heuristic, reactive, manner. This is part of the famous fight or flight response. Ancestors who used their cognitive processes to mull over whether they should fight or flee from a threatening murderer would not have survived in sufficient numbers to produce us.

Finally, one can imagine that brains finely tuned to personal threats could be recruited by natural selection not only to turn their attention to group threats, but also to combine that communal attention with a built-in, cognitive, reluctance to convert blame immediately into punishment. Cooler heads were required in our intensely social groups, where we had to balance the costs and benefits of punishment in ways that algae firing projectiles at predators, or victims fighting off rapists, do not.

Implications

The neuroscience of decision-making generally, let alone of complex social decisions like how much to blame and punish a wrongdoer, is of course still in its infancy. In the big picture of things, the black box of decision-making—how the brain enables the mind to cause action—remains frustratingly opaque despite all the breath-taking advances in neuroimaging. And there is nothing in the near—or probably even mid-term that brain science is likely to teach us about this potentially intractable problem, including its age-old sub-problems of when to blame, when to

[7]Note 5 supra.

forgive, and how much to punish (Greene and Cohen 2004).[8] Ironically, one big impetus toward re-examining these age-old questions is coming from the law, not science.

In 2012, in a case called *Miller v. Alabama*,[9] the United States Supreme Court ruled that juvenile murderers sentenced to life without parole (the death penalty for juveniles having already been held unconstitutional[10]) must nevertheless be considered for parole. It held that a sentence of life without even the possibility of parole violated the Eighth Amendment's prohibition against cruel and unusual punishment, at least for defendants who were less than 18 years old at the time of the murder. In making this inquiry about whether juvenile murderers should be considered for parole, the Court instructed trial courts to consider whether a particular defendant and his particular crime 'reflect irreparable corruption'.

In January 2016, the Court held in a different case that the rule announced in *Miller* was retroactive; that is, that it applied to all defendants serving life sentences without parole if they were less than 18 at the time of the murder, regardless of how long ago the murder happened.[11] The New York Times has estimated that more than 2,000 prisoners may be eligible for this inquiry.

Thousands of trial judges across the country, trying to figure out whether their juvenile murderers are 'irreparably corrupted', will be facing deep questions going to the heart of punishment, including why we punish, how much we punish, when we should forgive, and the role of contrition and atonement in forgiveness.

Of course, every time judges sentence criminals they balance, or claim to be balancing, factors that go to the heart of this deep inquiry—considerations of retribution, deterrence, community safety and rehabilitation. But these considerations are not only incomplete justifications of punishment, they are also inconsistent with one another.[12] 'Balancing' these factors is too often a post hoc justification for judges imposing sentences they would have imposed anyway without this intellectual cover. Indeed, despite these fancy-sounding sentencing 'factors', the empirical evidence has demonstrated that different judges vary wildly in the

[8]Whether neuroscience will ever change our folk-psychological notions about things like personhood and responsibility—and, more radically, whether it might even *disprove* some of those notions—is the subject of considerable debate. Compare Greene and Cohen 2004 ('[W]hen the mechanical nature of human decision-making is fully appreciated… the idea of distinguishing the truly, deeply guilty from those who are merely victims of neuronal circumstance will, we submit, seem pointless.') with Morse (2015) ('At present, neuroscience has little to contribute to more just and accurate criminal law policy, doctrine and individual case adjudication [and] no radical transformation of criminal justice is likely to occur….'). We discuss in the balance of this section the possibility that the neuroscience of punishment may someday lead to less dramatic, but no less important, changes in the law.

[9]132 S. Ct. 2455 (2012).

[10]Roper v. Simmons, 543 U.S. 551 (2008).

[11]Montgomery v. Louisiana, 136 S. Ct. 718 (2016).

[12]Hoffman, supra note 30, at 334–47.

sentences they impose for similar crimes committed by similar criminals (Anderson et al. 1999).[13]

Having to decide, under *Miller*, whether a prisoner is irreparably corrupt will require judges to think expressly about what it means to be 'corrupt' and what it means to have that corruption 'repaired'. We hope that in doing so, judges who are informed about the evolutionary roots of punishment will get a clearer picture of retribution's place in the firmament of the criminal law. This won't make these *Miller* decisions any easier, but it might clarify that retribution is not some old-fashioned cultural artefact with no place in our modern clinically informed world. It is, instead, the way that evolution automatically balanced all these impossible factors. Importantly, retribution also contains within it the command to be restrained. True retribution is not inconsistent with forgiveness.

Advances in our understanding of how brains blame and punish may also someday have much more tangible implications. For example, understanding the differences between blame and punishment may lead to ways to help sentencing judges avoid some of the emotional pitfalls of sentencing. Judges, like all humans, empathise with many victims. Presiding over a trial or sentencing hearing full of all the horrible details of a crime risks retaliatory, second-party, punishment urges that might overwhelm the kind of restrained third-party punishment we need our judges to impose. Knowing more about how the brain converts blame into punishment could arm us in the future with strategies to address this problem.

There is a legislative version of this risk of the angry retaliating judge, which a fuller understanding of the interactions between blaming and punishment might likewise help manage. When crime rates go up, or a particularly notorious crime is splashed across the media, legislatures often respond by creating new laws or increasing the penalties for old ones. We contend this is essentially a second-party reaction, which then gets fixed into the law. When these new crimes or increased penalties do not immediately result in less crime (and they often don't, not only because of time delays but also because the effect of general deterrence on crime rates is complex and ill-understood), then legislatures often respond by making even more new laws and increasing penalties further. The result of this one-way ratchet is that penalties for crimes in America have skyrocketed over the last 60 years, crime rates have not correspondingly fallen, more people are in prison than ever before, and ordinary citizens seem to have lost faith in the criminal justice system. Understanding the difference between blame and punishment could one day help break this cycle.

Legal processes might also improve with a fuller understanding of the neuropsychological underpinnings of punishment. Bedevilling problems of racial, gender and other kinds of punisher bias might admit to some inoculating strategies as we uncover the mysteries of how the CEN turns the blame we all rather

[13]Remember, though, that this data showing that different judges impose wildly different sentences on similar defendants in similar cases is not at all inconsistent with the remarkable uniformity with which humans *blame*. Note 5 supra.

consistently feel into the myriad of actual punishments different judges impose. A fuller understanding of how we blame and punish could even conceivably change some entrenched legal doctrines, including the four legally recognised criminal states of mind: purposeful, knowing, reckless and negligent. We know subjects have great difficulty distinguishing knowing from reckless conduct (Shen et al. 2011), but it is not yet clear how much of that difficulty comes from problems with the way we define these different mental states and how much reflects the fact that subjects simply may not recognise a moral difference between the two.[14] Learning more about how these assessed mental states are represented may help us decide whether we need to try to improve our jury instruction definitions of them, or whether we need to abandon any legal distinction between knowing and reckless.

None of these legal impacts is imminent. And it may be a long shot that any of them ever comes to pass. But there will almost certainly be some impacts, even if we cannot predict them now. Science has a way of filling policy spaces in the most unpredictable of ways. We have learned more about the brain in the last 30 years than we learned in all of prior human history, and the pace of learning appears to be exponential. Neuroimaging technologies are getting better, and it seems that promising investigative methods are being invented every year, including both new technologies (such as new methods to measure connectivity) and powerful new ways of crunching the data from them (such as machine-learning classifiers). At this rate, it is almost inconceivable that neuroscience will not have some important impacts on the law generally, including how we blame and punish the wrongdoers among us. All of us in both disciplines need to be thinking about these possibilities—both their benefits and risks—as the neuroscience races on.

References

Anderson, J., Kling, J. R., & Stith, K. (1999). Measuring inter-judge sentencing disparity before and after the federal sentencing guidelines. *Journal of Law and Economics, 42*(S1), 271–307.

Baird, J., & Astington, J. (2004). The role of mental state understanding in the development of moral cognition and moral action. *New Directions for Child and Adolescent Development, 103,* 37–49.

Barbey, A. K., Krueger, F., & Grafman, J. (2009). An evolutionarily adaptive neural architecture for social reasoning. *Trends in Neuroscience, 32*(12), 603–610.

Bellucci, G., Chernyak, S., Hoffman, M., Deshpande, G., Dal Monte, O., Knutson, K. M., et al. (2017). Effective connectivity of brain regions underlying third-party punishment: Functional MRI and Granger causality evidence. *Social Neuroscience, 12*(2), 124–134.

Blakemore, S. J. (2008). The social brain in adolescence. *Nature Review of Neuroscience, 9*(4), 267–277.

Boehm, C. (2012). Costs and benefits in hunter-gatherer punishment. *Behavoiur Brain Science, 35*(1), 19–20.

[14]In a follow-up study, researchers were able to improve results, but only marginally, by changing the definitions of these two problematic mental states (Ginther et al. 2014).

Boyd, R., Gintis, H., Bowles, S., & Richerson, P. J. (2003). The evolution of altruistic punishment. *Proceedings of the National Academy of Science USA, 100*(6), 3531–3535.

Bressler, S. L., & Menon, V. (2010). Large-scale brain networks in cognition: Emerging methods and principles. *Trends in Cognitive Science, 14*(6), 277–290.

Brown, D. (1991). *Human Universals*. McGraw-Hill.

Buckholtz, J. W., Asplund, C. L., Dux, P. E., Zald, D. H., Gore, J. C., Jones, O. D., et al. (2008). The neural correlates of third-party punishment. *Neuron, 60*(5), 930–940.

Buckholtz, J. W., & Marois, R. (2012). The roots of modern justice: Cognitive and neural foundations of social norms and their enforcement. *Nature Neuroscience, 15*(5), 655–661.

Buckholtz, J. W., Martin, J. W., Treadway, M. T., Jan, K., Zald, D. H., Jones, O., et al. (2015). From blame to punishment: Disrupting prefrontal cortex activity reveals norm enforcement mechanisms. *Neuron, 87*(6), 1369–1380.

de Waal, F. B. M. (2012). Empathy in primates and other mammals. In J. Decety (Ed.), *Empathy: From bench to bedside* (pp. 87–106). Cambridge, MA: MIT Press.

de Weerd, H., & Verbrugge, R. (2011). Evolution of altruistic punishment in heterogeneous populations. *Journal of Theoretical Biology, 290*, 88–103.

Fehr, E., & Fischbacher, U. (2004). Third-party punishment and social norms. *Evolution and Human Behavior, 25*(2), 63–87.

Fehr, E., & Gachter, S. (2002). Altruistic punishment in humans. *Nature, 415*(6868), 137–140.

Gachter, S., Renner, E., & Sefton, M. (2008). The long-run benefits of punishment. *Science, 322* (5907), 1510.

Ginther, M. R., Bonnie, R. J., Hoffman, M. B., Shen, F. X., Simons, K. W., Jones, O. D., et al. (2016). Parsing the behavioral and brain mechanisms of third-party punishment. *Journal of Neuroscience, 36*(36), 9420–9434.

Ginther, M. R., Shen, F. X., Bonnie, R. J., Hoffman, M. B., Jones, O. D., Marois, R., et al. (2014). The language of mens rea. *Vanderbilt Law Review, 67*, 1327–1372.

Glass, L., Moody, L., Grafman, J., & Krueger, F. (2016). Neural signatures of third-party punishment: Evidence from penetrating traumatic brain injury. *Social Cognition and Affective Neuroscience, 11*(2), 253–262.

Godin, J., & Davis, S. (1995). Who dares, benefits: Predator approach behavior in the guppy (*Poecelia retuclata*). *Proceedings of the Royal Society of London B, 259*, 193–200.

Gordon, R. (2006). A retaliatory role for algal projectiiles. *Journal of Theoretical Biology, 26*, 419–436.

Greene, J., & Cohen, J. (2004). For the law, neuroscience changes nothing and everything. *Philosophical Transactions of the Royal Society of London. B: Biological Science, 359*(1451), 1775–1785.

Gummerum, M., & Chu, M. T. (2014). Outcomes and intentions in children's, adolescents', and adults' second- and third-party punishment behavior. *Cognition, 133*(1), 97–103.

Hamilton, W. D. (1964a). The genetical evolution of social behaviour. *I. Journal of Theoretical Biology, 7*(1), 1–16.

Hamilton, W. D. (1964b). The genetical evolution of social behaviour. *II. Journal of Theoretical Biology, 7*(1), 17–52.

Henrich, J., McElreath, R., Barr, A., Ensminger, J., Barrett, C., Bolyanatz, A., et al. (2006). Costly punishment across human societies. *Science, 312*(5781), 1767–1770.

Hetzer, M., & Sornette, D. (2013). The co-evolution of fairness preferences and costly punishment. *PLoS ONE, 8*(3), e54308.

Hoffman, M. (2014). *The punisher's brain*. New York: Cambridge University Press.

Hume, D. (1968). Enquiry concerning the principles of morals (C.W. Hendel, ed.). New York: The Liberal Arts Press. (Original work published in 1751).

Kalbe, E., Schlegel, M., Sack, A. T., Nowak, D. A., Dafotakis, M., Bangard, C., et al. (2010). Dissociating cognitive from affective theory of mind: A TMS study. *Cortex, 46*, 769–780.

Karniol, R. (1978). Children's use of intention cues in evaluating behavior. *Psychology Bulletin, 85*, 76–85.

Kenward, B., & Osth, T. (2012). Enactment of third-party punishment by 4-year-olds. *Frontiers Psychology, 3,* 373.

Kenward, B., & Osth, T. (2015). Five-year-olds punish antisocial adults. *Aggressive. Behaviour, 41*(5), 413–420.

Kiehl, K. A. (2006). A cognitive neuroscience perspective on psychopathy: Evidence for paralimbic system dysfunction. *Psychiatry Research, 142*(2–3), 107–128.

Krasnow, M. M., Delton, A. W., Cosmides, L., & Tooby, J. (2016). Looking under the hood of third-party punishment reveals design for personal benefit. *Psychological Science, 27*(3), 405–418.

Krueger, F., & Hoffman, M. (2016). The emerging neuroscience of third-party punishment. *Trends in Neuroscience, 39*(8), 499–501.

Krueger, F., Hoffman, M., Walter, H., & Grafman, J. (2014). An fMRI investigation of the effects of belief in free will on third-party punishment. *Social Cognition and Affective Neuroscience, 9* (8), 1143–1149.

Lamm, C., Batson, C. D., & Decety, J. (2007). The neural substrate of human empathy: Effects of perspective-taking and cognitive appraisal. *Journal of Cognitive Neuroscience, 19*(1), 42–58.

Lancaster, J. (1986). Primate social behavior and ostracism. In R. Masters & M. Gruter (Eds.), Ostracism: A social and biological phenomenon. *Ecology and Sociobiology* (special issue), *7,* 215–25.

Leopold, A., Krueger, F., dal Monte, O., Pardini, M., Pulaski, S. J., Solomon, J., et al. (2012). Damage to the left ventromedial prefrontal cortex impacts affective theory of mind. *Social Cognition and Affective Neuroscience, 7,* 871–880.

Lorenz, K., & Wilson, M. K. (2002). *Man meets dog*. New York: Psychology Press.

Marlowe, F. W., & Berbesque, J. C. (2008). More 'altruistic' punishment in larger societies. *Proceedings of the Royal Society of London B: Biological Sciences, 275*(1634), 587–592.

Marsh, A. A., Finger, E. C., Mitchell, D. G., Reid, M. E., Sims, C., Kosson, D. S., et al. (2008). Reduced amygdala response to fearful expressions in children and adolescents with callous-unemotional traits and disruptive behavior disorders. *American Journal of Psychiatry, 165*(6), 712–720.

McAuliffe, K., Jordan, J. J., & Warneken, F. (2015). Costly third-party punishment in young children. *Cognition, 134,* 1–10.

Michl, P., Meindl, T., Meister, F., Born, C., Engel, R. R., Reiser, M., et al. (2014). Neurobiological underpinnings of shame and guilt: A pilot fMRI study. *Social Cognition and Affective Neuroscience, 9*(2), 150–157.

Miller, E. K., & Cohen, J. D. (2001). An integrative theory of prefrontal cortex function. *Annual Review of Neuroscience, 24,* 167–202.

Morse, S. (2015). Criminal law and common sense: The perils and promises of neuroscience. *Marquette Law Review, 99,* 39.

Nishida, T., Hosaka, K., Nakamura, M., & Hamai, M. (1995). A within-group gang attack on a young adult male chimpanzee: Ostracism of an ill-mannered member? *Primates, 6*(2), 207–211.

Premack, D., & Woodruff, G. (1978). Does the chimpanzee have a theory of mind? *Behavioral and Brain Sciences, 1*(04), 515–526.

Raihani, N. J., Grutter, A. S., & Bshary, R. (2010). Punishers benefit from third-party punishment in fish. *Science, 327*(5962), 171.

Riedl, K., Jensen, K., Call, J., & Tomasello, M. (2012). No third-party punishment in chimpanzees. *Proceedings of the National Academy of Science USA, 109*(37), 14824–14829.

Robinson, P., & Darley, J. (2007). Intuitions of Justice: Implications for criminal law and policy. *Southern California Law Review, 81,* 1–68.

Roos, P., Gelfand, M., Nau, D., & Carr, R. (2014). High strength-of-ties and low mobility enable the evolution of third-party punishment. *Proceedings of the Royal Society of London. B: Biological Sciences, 281*(1776), 20132661.

Sebastian, C. L., Fontaine, N. M., Bird, G., Blakemore, S. J., Brito, S. A., McCrory, E. J., et al. (2012). Neural processing associated with cognitive and affective theory of mind in adolescents and adults. *Social Cognition and Affective Neuroscience, 7*(1), 53–63.

Shen, F., Hoffman, M., Jones, O., Greene, J., & Marois, R. (2011). Sorting guilty minds. *New York University Law Review, 86,* 1306–1360.

Yamada, M., Camerer, C. F., Fujie, S., Kato, M., Matsuda, T., Takano, H., et al. (2012). Neural circuits in the brain that are activated when mitigating criminal sentences. *Nature Communications, 3,* 759.

Young, L., Camprodon, J. A., Hauser, M., Pascual-Leone, A., & Saxe, R. (2010). Disruption of the right temporoparietal junction with transcranial magnetic stimulation reduces the role of beliefs in moral judgments. *Proceedings of the National Academy of Science USA, 107*(15), 6753–6758.

Becoming Conscious About the Existence of the Non-existents: Logic, Language and Speech Acts

Samir Karmakar

In the preface of *khukumanir chaḍā* (Sarkar 1899/2012), Ramendra Sundar Trivedi once wrote that, even in the rhymes written for children, one can look for the beauty of complex theory. Being motivated by Trivedi's assertion, we will attempt in this chapter to explore how the theory of meaning reveals the hidden intricacies of the way the world is meaningful. More specifically, non-existential naming words are the topic of this chapter. Non-existential naming words—like the characters of comic strips, nursery rhymes and so on that are taught to children—possess a particular type of challenge to the theory of meaning. Being non-existential, these characters lack references in the world outside to be pointed at. Because of their inability for something concrete outside in the world to be indicated, they are of immense significance to the study of how meanings are attached by the users of a language when using these non-existential naming words. More interestingly, the present discussion seeks to establish the centrality of these seemingly irrational non-existent expressions in developing the initial years of young rational minds. In other words, what we want to assert that the development of rationality is relative to those mental exercises which are otherwise considered as *irrational*. No one have ever seen a *unicorn*, or a *pakkhirāj ghoḍā* (= flying horse), but the world of childhood days are filled with such creatures which deviate from the official version of rationality. To get into the discussion of how non-existents are attended through their names, a brief summary of the major theoretical positions on the naming words will be helpful.

While distinguishing denotation from connotation, Mill (1889) argued that '[n]ames shall always be spoken of … as the names of things themselves and not merely of our ideas of things'. As per this claim, then the proper names like Gavaskar, Chomsky and the others have the denotations. In addition to denotation (= reference), Frege (1891/1980) argues that a name could also have multiple

S. Karmakar (✉)
Jadavpur University, Kolkata, India
e-mail: samirkrmkr@yahoo.co.in

S. Menon et al. (eds.), *Self, Culture and Consciousness*,
DOI 10.1007/978-981-10-5777-9_14

number of senses associated with it—even without being synonymous to each other. This line of argument is further augmented with the work of Russell (1911) and Searle (1958) in their respective works, resulting in a descriptive theory of meaning. Kripke (1980) puts stress on another aspect of the names. He has shown how the relation between a name and its referent is developed over a period of time. At the beginning of the twentieth century, Peirce (1901) showed how the evolution of a name followed the different stages of indexical and symbolic representations. In continental philosophy, philosophers such as Derrida (1995) asserted that names are socially constructed. Taken together, these positions reflect various types of intricacies involved in names. As a complex whole, the questions asked are pertinent to the nature of the reference, the way it matures and attends its life within a community. In the light of these discussions, this chapter likes to interpret the way names are, with a special reference to the fantasies of childhood days, like *hā̦ṭṭimā̦ṭim.ṭim 'an onomatopic word denoting a class of fictitious creatures'*, *hũko-mukho hyā̃ṅlā* 'a greedy creature with a face resembling hookah', *dāmodar seṭh* 'a proper name used in a nonsensical poem' and so forth.

To approach these issues, we will move in a rather unusual way. First, we will discuss how in common parlance the notion of meaning is understood with an emphasis on the predicate logic? How does this view fail to provide explanations of certain literary phenomena? And how, as a consequence, has the popular interpretation of 'meaningful' had to undergo some major revisions?—This in turn will contribute in redefining the notion of 'meaningless'.

The Official Version of 'Meaning'

The theory of meaning as developed in philosophical logic considers that a sentence is made up of subject and predicate. Consider the following sentence:

1. Humpty Dumpty sat on a wall

According to this proposal, a sentence like (1) has 'Humpty Dumpty' as its subject and 'sat on a wall' as its predicate. As a predicate, '____ sat on a wall' is an unsaturated function. With the filling up of the empty slot with a subject, it gets converted into a saturated function. Being a saturated, it can have truth values in its extension. It is precisely at this juncture that the philosophy of language faces a major challenge. If you consider (1) to be a true statement then you have to accept that Humpty Dumpty exists. This will put you into a more problematic situation; because the world—in which we live—contains no such entity called Humpty Dumpty. In such a situation then the question will arise on the basis of what we are accepting the existence of Humpty Dumpty. The 'basis' with respect to which the existence of something is defined seems to be of utmost significance.

A naive theory of semantics will identify the 'basis' with the world outside—the way the world is represented externally. 'A true symbol,' Richards and Ogden (1923) argue, is 'one which correctly records an adequate reference.' The use of 'correctly' in this phrase is of importance, as if incorrect symbols could exist! Lack of correctness is defined in terms of the way our articulations are coordinated. Following this line of argument, then, non-existential naming words of a particular linguistic community will make no sense to a non-native by virtue of having either no access or partial access to the corresponding references. In such a situation, the challenge will be to pin down the problem of so called 'non-existential entities'—like *Alice*, *Phantom*, *Mandrake*, *Tintin* and so forth. None of these names correspond anything in the world outside. This results in a peculiar solution which assumes the existence of a world of convention. Following Ogden and Richards, again, it could be asserted that names as the official instruments of accession 'are conveniences in description, not necessities in the structure of things'. According to this solution, non-existent entities are introduced along with their respective back up of sets of conventions. Dissociating an entity from its set of convention produces meaninglessness.

Problem with the Official Version

Since the production of meaningless is very much dependent on how meaning is produced, a need to get into the issues of formal semantics seems to be absolutely necessary in this context. Consider the case of predicate logic which works on the basis of two basic principles—namely, the *predication principle* and the *existential generalisation*. Let us explain these two principles with a reference to example (1): Following the conventions of predicate logic, 'Humpty Dumpty' is an argument and is represented as 'h'; on the other hand, '___ sat on a wall' is a predicate and represented with 'S'. Therefore, the logical form of (1) will be

(2) S(h)

Within the theoretical framework of predicate logic, the semantic interpretation of (2) presupposes both the predication principle and the existential generalisation. As per the predication principle, (1) entails (3):

(3) there exists some x such that x sat on a wall

No doubt (3) is an effort on behalf of the logicians to capture the notion of unsaturated function which is discussed above. Besides this, following the principle of existential generalisation, (1) can also entail (4):

(4) there exists some x such that x is Humpty Dumpty

Formally, (3) and (4) can be rewritten as (5) and (6), respectively:

(5) $S(h) \rightarrow \exists x S(x)$ as per the predication principle corresponding to '___ sat on a wall'

(6) $S(h) \rightarrow \exists x(x = h)$ as per the existential generalisation corresponding to 'Humpty Dumpty'

The complete interpretation of (1) can be obtained by combining (5) and (6) in (7):

(7) $S(h) \rightarrow \exists x(S(x) \& (x = h))$

Representing meaning of a particular statement, say (1), in this way has a fundamental problem due to the assertion of existential quantifier $\exists$. As per this assertion, the meaning of any individual has had to correspond with something which exists in the world. Now in such a situation, if we look for a corresponding referent in the world outside, we will definitely be disappointed. As a result, (7) will have false value in its extension which is quite contrary to our understanding of (1).

In order to come out of these sorts of fallacies, it is suggested that the truth of a statement is judged not with respect to any external world; rather with respect to a world which is conceptually represented. This view falls within the scope of conceptual realism. Conceptually, real entities may lack external correspondence. A similar type of classification can also be found in Indian tradition, where existence is classified as *bāhya-sattā*, because of being transcendental in nature it could also be named as *paramārthikī-sattā*. As opposed to this, *oupocārikī-sattā* is introduced to refer to the world(s) whose form(s) is/are conventionally determined. The constituents of these conventional worlds may or may not have a representation in the external world. One extreme version of this view can be found in *mādhyamika-nyāya* which propounded the theory of *śūṇya-vāda*: By virtue of lacking any systemic correspondence with *oupocārikī-sattā*, *bāhya-sattā* is categorised as empty. Because of being empty it cannot be evaluated either as true or as false. Most interestingly the significance of *pratītyasamutpāda* 'dependent origination' holds true only within the conventional world. This assertion seems to be of utmost significance primarily because of the reason that it tells us about how the content of the existence is determined. Anything which can be claimed as existent has content with respect to the conventions associated with it—not in terms of its intrinsic absolute nature. The absolute nature is *śūṇya* (empty) by virtue of lacking systemic correspondence with the conventional world (Chatterjee and Datta 1948).

Returning back to our original discussion, then, one can claim that the existence of an entity is referred with respect to a set of conventions associated with that particular entity. As a result, the character like *Humpty Dumpty* irrespective of lacking any correspondence in the external world ceases to exist only by virtue of being grounded in a complex network of *upacāra*s or conventions. If so, then does one need to know what these *upacāra*s or conventions look like? How do they stand so critically to the construction of the content? In other words, how can something acquire content? Section "Prelude to an Alternative" is an effort to this direction.

Prelude to an Alternative

To understand this issue in greater detail, we will introduce the concept of the international phonetic alphabets (IPA) in our discussion. But before we get into this discussion, we need to have some clarification about why we need to incorporate this section here, when this is expected to be a chapter about non-existent naming words. The answer to this question appears somewhat straightforward: we want to argue that, irrespective of their differences, rational and irrational follow a similar act of meaning-making. There is no difference in the way the meaning of *Humpty Dumpty* is interpreted and the way the meaning of the international phonetic alphabet (IPA) or the periodic table is constructed.

To the common man, the IPA seems to be all about the abstract technicalities. In reality, it is only through these abstract technicalities that we assign content to the speech sounds to make them accessible to those who are interested in understanding their intricacies. It also explains what is commonly shared among the users of a language. This account of sharing—which contains information not only about similarities but also about the differences—reflects the way we are coordinated with each other. But how is this *coordination* established?—The classification of sounds on the basis of the places and manners of articulation is the first step towards the making of an *organ* which is equally accessible by everyone. In the backdrop of this constructed organ one can be aware of the nature of any speech sound. Through this process of making awareness, we make our individual worlds accessible to the others. As a prerequisite of a meaningful act this signifies the acclimation to what we call 'SOCIAL'. It is this social in the light of which any kind of meaning-making act has to be interpreted. Justification to this claim comes not only from IPA but also from periodic table in the field of chemistry. The elements of the periodic table are meaningful to us not because of the reason that they refer something in the world outside but because they are constitutive of a conceptual plane with respect to which they differ from each other, resulting in their respective elementarities. Interestingly, the conceptual plane—embodying various types and degrees of elementary oppositions—is construed on the basis of the atomic numbers of the elements. This can give an impression that atomic numbers are more fundamental to the table when in reality table and the numbers are co-constituted. Likewise, the organisation of the speech sounds—embodying various types of phonological oppositions—plays a crucial role in explaining how speech sounds stand in relation to each other. Clearly, the organisational principles of IPA (as well as the periodic table) are not the intrinsic property of the physical world—rather, they are the impositions of the human mind upon the world of dead objectivity (Karmakar 2015).

What is worth mentioning in this context is that these impositions are social in nature and more than often they are crucial in bringing out a humane version of the objectivity. Meaning is talked about only with respect to these sorts of objectivities. An entity is objective not by virtue of corresponding to any dead and blunt objectivity but with respect to a world which is compassionate, organic and elastic.

The compassion or the elasticity we are talking about is due to the various conventions—conventions which are constitutive of *oupocārikī-sattā*.

In such a situation, then, a single notion of truth can be of use to evaluate meanings of various sorts. As a consequence, in *abhidhamma* (Bartley 2011) meaning as the humane measure of existence is bifurcated in the notion of *dravya-sat* (=exists by virtue of a corresponding something in the material world) and *prajñapti-sat* (=exists by virtue of being legitimised by the conventions of informing). This is not a mere bifurcation for the sake of the theory internal technicalities—rather it highlights the two major dimensions through which something becomes meaningful in a communicative context.

The Alternative

What follows, therefore, is an understanding that the non-existent names like *hāṭṭimāṭim ̣tim*, *cācā chowdhuri* and so forth have no extension. They can have meaning only by virtue of being associated with some functions which are proposed or introduced by their creators. This understanding has a huge impact even in proposing a theory of a names which (could) have reference(s). Even for such names, meaning means something else. As per this claim, then, a proper name like 'Mahidas Bhattacharya' does not refer to a person—rather, it is an umbrella term representing a set of functions with the form 'm such that f(m)' where m (=Mahidas) being the constant is qualified with different functions, like 'the director of School of Languages and Linguistics', 'a linguist', 'a teaching staff of Jadavpur university' and so on. This list of functions forms a set. In other words, the resultant set is the set of some other sets for each of which there exists a characteristic function (= f). Along this line of argument then 'Mahidas Bhattacharya' refers to a set of sets—not to any concrete entity in the external world. The Indian way of enunciating the same understanding will sound like *vyaktiḥ-guṇa-biśeṣa-āśrayo mūrtiḥ* (*nyāyasūtra*: 2/2/69)—means individual as a form is the embodiment of particular qualities associated with it. If this is the case then any individual can be conceived as a function defined over a set of qualities/properties. Following lambda calculus, the relation between 'Mahidas Bhattacharya' and a set of sets can be represented as $\lambda f.f(m)$. This chapter would like to argue that this $\lambda f.f(m)$ is defined over the range of the *oupocārikī-sattā* not over the *bāhya-sattā*. More importantly, it provides a *text* only in the background of which 'Mahidas Bhattacharya' makes sense. The other nonsensical names that we have introduced earlier are also not an exception. They attain their meaning in their respective texts. If so, then the text cannot be viewed as a mere system of signs which are useful in capturing the structure of the conventionalised reality. It has to be conceived as a repository of the functions because of which something comes into the existence—The way a name is invoked in the text is of equal importance.

What could have been the better technique to train the young learner about how to invoke the relevant *text* during the act of meaning-making other than those seemingly nonsensical names we have mentioned earlier. This answers the question as to why, from time immemorial, seemingly irrational meaningless articulations have been preferred over the rational ones to make the little learner adept in invoking the relevant text associated with any utterance.

The Nature of the Existence that We Live by

It is inevitable that the questions we began with will lead us to a discussion of philosophers' reactions to the issues of (non)existence. To Hume (1739/2000), the thought of something is not non-identical with the thought of the way that something exists. For instance, the thought of God is not different from the thought of God's existence. (The non-believer in God now turns out to be a believer!) If so, then the thought involved in the denial of God's existence presupposes the God's existence. Presupposing that a predicate called GOD exists results in a question which is equally pertinent to this discussion: is it the case that existence is a predicate of GOD? Are we going to claim that GOD possesses the property of being existent? The Kantian solution (1781/2003) suggests that this is not likely to be the case because existence is not a property but a precondition for something to become an object of our cognition. This provides us then the justification of why a statement like 'God exists' cannot be translated as EXIST(God)—rather, it should be translated as $\exists x GOD(x)$. The uniqueness of the second translation lies with the following interpretation: 'there exists some x such that x qualifies the condition of being GOD'.

Acceptance of the position just stated above is not enough to answer the question of non-existence. There are non-existent objects in our cognition even after not being pre-conditioned by existence. Why is this a fallacy? How is this a fallacy? To have some answers to these questions, one need to be in a position to appreciate that human mind is endowed with a rare capacity of producing alternatives whenever it comes in conjunction with the world. We are enabled with a rarest of rare capacity of looking through the world. The world we live in fits with one such alternative—distinguished from the rest by virtue of being marked with 'real'. But this in no way undermines the presence of other alternatives. In fact, it is only by virtue of being a member of a rich set of alternatives, one has *a status* which we identify as *default*. Deciding something as default does not reduce the significance of it as an alternative. This position provides a solution to the problem of non-existences. It does not deny the existence of something which is *absolute*. But at the same time, it tells us that what is being signified need not to be a member of the *absolute real*. The objects of our talks floats on and above this absolute reality constituting a world of conventions—more accurately, a world of *upācāra*. But, then, one need to explain how this world of *upācāra* is formed over and above the absolute reality. Which processes are involved in the formation of a world of conventions? Our concluding section is dedicated to these questions to indicate, at least, how non-existents are crucial for the tuning of the minds in the early days of development.

Conventionalised Intentions, Speech, and the Creation of Non-existence

Speech has a crucial role in the formation of the world of conventions. It is crucial also in maintaining as well as transferring the conventions. While expressing our thought through speech we perform certain kind of intentional acts. These acts are crucial not only in expressing our thoughts but also in recreating the world again and again. What underlies beneath these speech acts is a set of schemes to engage one's own self with the others by virtue of directing something in a world whose objectivity is defined in an intersubjective plane. More specifically, as Habermas (1998) once insisted, '[o]bjective agreement about something in the world … is dependent on bringing about an intersubjective relationship between the speaker and at least one hearer capable of adopting a critical position'. What seems to be crucial then is a presupposition of coordination existing between the users of any language. If so, then, a little will left for doubting the fact that existence is relied on a kind of social coordination. In fact, the type of (im)materiality we are often referring to in our daily conversations presupposes a grounding in the social coordination. It would not be an exaggeration to claim that social embeds the physics of the (im)materials within it. In doing so, language remains the most essential one. Schemes of the various speech acts are then essential in understanding the way individual subjects are interconnected with each other while creating different types of realities.

From this viewpoint then the study of non-existentials seeks to know those types of speech acts with the help of which they are introduced in the discourse of daily life. Since much has already been discussed in the literature on speech acts, we will not elaborate on each of the speech acts; rather, our effort will be directed towards reinstating the significance of certain conditions to make the speech act felicitous by virtue of which the corresponding piece of reality attains its legitimacy, irrespective of its connection with any absolute form of external reality or *bāhya-sattā*.

From this viewpoint, one can also put forward the following argument: preconditions which are essential for a predicate to be qualified as (non)existential are nothing but the projections of different kinds of speech acts. Because of being speech acts, they function differentially in determining the two aspects of materiality—namely, *dravya-sat* and *prajñapti-sat*—but in an efficient way. In fact, the very notion of *dravya-sat* seems to be problematic if we consider the material basis of the universe ('brahma') as a form that is subservient to what is called *śabdatattva.* In v*ākyapadīyam,* stress is given to the intimate relation between the world and language. It argues that the structural complexities of language complement the complexities of the world: '*brahma-idaṃ śabda-nirmānaṃ, śabda-śakti-nibandhanam| vivrittaṃ śabda-mātrābhyaḥ, tāsveva prabilīyate||*' (Tiwari 2015). This argument of Harivṛṣabha's does not accept existence as something non-identical with what we see in the daily usages of language. If that is so, then all forms of existence are subject to some kind of speech acts performed during their invocation. From this viewpoint, if the taxonomy of illocutionary speech acts is looked at, the underlying schemes of each types would be found as a complex of interacting intentions within the framework of which a particular variety of existence can survive. Each of these intentions, in turn, is loaded with presuppositions, entailments, implicatures and explicatures, resulting in the thickening of an individual/entity. This consequences into the existences of the non-existents through the symbolic usages of words, phrases and sentences in a communication under the guiding principles of the convention. If so, then what else could be there in the training manual for young minds other than the non-existent fictitious characters of literature!

References

Bartley, C. (2011). *An introduction to Indian philosophy*. New York: Continuum.

Chatterjee, S., & Datta, D. (1948). *An introduction to Indian philosophy*. Calcutta: University of Calcutta. (Originally published in 1939).

Derrida, J. (1995). *On the name*. Stanford: Stanford University Press.

Frege, G. (1980). 'Funktion und Begriff', Vortrag, gehalten in der Sitzung vom 9. Januar 1891 der Jenaischen Gesellschaft für Medizin und Naturwissenschaft, Jena: Hermann Pohle. 'Function and Concept' (Trans: P. Geach). In P. Geach & M. Black (Eds.), *Translations from the Philosophical Writings of Gottlob Frege*. 3rd edition. Oxford: Blackwell. (Originally published in 1891).

Habermas, J. (1998). Communicative rationality and the theories of meaning and action. In M. Cooke (Ed.), *On the pragmatics of communication*. Cambridge, MA: MIT Press.

Hume, D. (2000). *A treatise of human nature*. Oxford: Oxford University Press. (Originally published in 1739).

Kant, I. (2003). *The critique of pure reason* (N. K. Smith, Trans.). Basingstoke: Palgrave Macmillan (Originally published in 1781).

Karmakar, S. (2015). biṣaydharmītā kī ebong kī bhabe?—chelebhulāno chaḍāy alīk, sambhabya o bāstab (What and how is subjectivity?—absurdity, possibility and reality in rhymes). *Ālochonā Chakra, 39,* 74–86.

Kripke, S. (1980). *Naming and necessity*. Cambridge, MA: Harvard University Press.

Mill, J. S. (1889). *System of logic: Ratiocinative and inductive, being a connected view of the principles of evidence and the methods of scientific investigation*. London: Longman Green and Co.

Peirce, C. S. (1901). Name. In J. M. Baldwin (Ed.), *Dictionary of philosophy and psychology (DPP)* (pp. 127–128). London: Macmillan.

Richards, I. A., & Ogden, C. K. (1923). *The meaning of meaning*. Cambridge: Cambridge University Press.

Russell, B. (1911). Knowledge by acquaintance and knowledge by description. *Proceedings of the Aristotelian Society, 11,* 108–128.

Sarkar, J. (2012). *Khukumaṇira charā. Ḍhākā: Cāradika* (Originally published in 1899).

Searle, J. R. (1958). Proper names. *Mind, 67*(266), 166–173.
Tiwari, H. (Ed.). (2015). *Vākyapadīyam of Bhartṛhari with Vākyapadakāṇḍamūlakārikā & Vṛtti by Harivṛṣabha, edited with 'Nārāyaṇa' Hindi Commentary*. Varanasi: Chowkhamba Sanskrit Series Office.

Bhoja's Model for Analysing the Mental States of Literary Characters Based on Samkhya Metaphysics

Shankar Rajaraman

List of abbreviations

A.Bh	Abhinavabhāratī commentary on *Nāṭyaśāstra*
Bh.P	*Bhāvaprakāśana*
D.R.AV	*Daśarūpaka* with Avaloka commentary
K.Anu	*Kāvyānuśāsana*
K.P	*Kāvyaprakāśa*
R.G	*Rasagaṅgādhara*
Sa.K	*Sāṅkhyakārikā* with Māṭharavṛtti commentary
Sa.S	*Sāhityasāra*
Sh.P	*Śṛṅgāraprakāśa*

Introduction

Samkhya figures among the six major systems of Indian philosophy. To briefly summarise its principal tenets as they are relevant to this chapter, there are two ultimate realities—multiple, conscious, inactive beings (*puruṣas*) and a single, unconscious, active Matter (*prakṛti*). Matter encompasses not just the external world at large and the objects existing in it, but also the body-mind complex. All of Matter is continuously evolving because of the dynamic interplay between the three elements (*guṇas*) that constitute it—namely, *sattva*, *rajas* and *tamas*. Though these three elements cannot be directly perceived, they can be inferred from the effects they bring about. *Sattva* is responsible for clarity, *rajas* for activity and *tamas* for inertia. In sentient creatures, *sattva* manifests as pleasure, rational thought and agile physical movements, *rajas* as displeasure, confused thought, and agitated physical

S. Rajaraman (✉)
NIAS Consciousness Studies Programme, National Institute of Advanced Studies, Bengaluru 560012, India
e-mail: shankar_rajaraman@rediffmail.com

S. Menon et al. (eds.), *Self, Culture and Consciousness*,
DOI 10.1007/978-981-10-5777-9_15

movements, and *tamas* as blunting of emotions, irrational thought and immobility. Unlike Matter, beings do not change. However, through their identification with Matter, they falsely appropriate to themselves the changes that it undergoes. As a result, they get transformed into beings (*jīvas*) enmeshed in Matter. This chapter is not so much about the transcendental beings that are free of their association with Matter as it is about beings that are ensnared in it. In particular, it is about characters represented in literary texts as exemplifying specific types of beings. Furthermore, it is a study on how the evocation, experience, and expression of mental states differs from one literary character to another in keeping with the type of being that an author develops him/her as.

An Overview of the Contribution of Samkhya Philosophy to Sanskrit Aesthetics

Depending on the philosophical systems they draw their inspiration from, Sanskrit aestheticians differently understand the process by which the mental states of literary characters are communicated to the audience and the experiential nature of such aesthetically communicated mental states. Abhinavagupta (tenth to eleventh centuries AD), philosopher par excellence of monistic Kashmir Shaivism, occasionally alludes to the Samkhya perspective on this topic in his commentary on Bharata's (c. second century BC to c. second century AD) *Nāṭyaśāstra*, the foundational text of Sanskrit aesthetics.[1] Among his predecessors whose views Abhinavagupta summarises is Bhatta Nayaka (ninth to tenth centuries AD). Nayaka draws on the Samkhya theory of the three *guṇas* in order to explain the process by which we relish the enactment of a dramatic text.[2] Some aestheticians have also examined a special category of nine long-lasting mental states within the purview of *guṇa* theory.[3] As observed by Raghavan (1978: 743), king Bhoja (eleventh-century AD), a polymath who composed works in Sanskrit on several subjects ranging from architecture to yoga, 'establishes some contact of a substantial nature' with Samkhya in his *Śṛṅgāraprakāśa*, a magnum opus on Sanskrit poetics. However, he 'owes more to Samkhya than he himself has admitted' (Raghavan 1978: 466). Raghavan has specifically discussed the impact Samkhya had on Bhoja's analysis of the aesthetic process—one by which everyday mental states, transformed into their literary counterparts, delight the connoisseur through the very act of being

---

[1]For e.g., A.Bh, Vol. 1, Ch. 6, p. 276.

[2]A.Bh, Vol. 1, Ch. 6, pp. 276, 277. Like Bhatta Nayaka, Madhusudana Sarasvati (sixteenth century AD) also takes recourse to Samkhya philosophy (as well as Vedanta) for explaining the process by which we relish literary texts.

[3]Sharadatanaya (twelfth century AD), for instance, presents the view of one Narada on the *guṇa*-based classification of nine aesthetically represented long-lasting mental states (Bh.P, Ch. 2, p. 47); Achyuta Raya (nineteenth century AD) comes up with a different classification (Sa.S, 1. 4. 48, p. 102).

communicated to him/her. In this chapter, I discuss Bhoja's categorisation of literary characters in the light of *guṇa* theory and explicate a model, based on this categorisation, for interpreting mental states depicted in literary works.

Bhoja's Categorisation of Literary Characters

Since Matter is constantly changing, beings enmeshed in Matter also must. This does not, however, imply that the change is random and unpredictable. The psychophysical apparatus of *jīvas* often evolves along lines that indicate preponderance of one specific *guṇa*. It is on this ground that Samkhya philosophy classifies beings into categories such as *sāttvika*, *rājasa* and *tāmasa*.[4] Because of its consistency and predictability, *prakṛti* (as it manifests in individual beings) can be understood as akisn to personality.[5] Bhoja transports the Samkhya concept of *prakṛti* to Sanskrit poetics and uses it for classifying important literary characters into four types: *sattvodrikta* (characterised by an excess of *sattva-guṇa*), *sāttvika*, *rājasa* and *tāmasa*. In doing so, Bhoja conceptualises literary characters as possessing a personality just like real-world individuals do. In the context of Sanskrit poetics, he renames the four categories as *śānta*, *udātta*, *lalita*, and *uddhata* (corresponding to *sattvodrikta*, *sāttvika*, *rājasa* and *tāmasa* categories respectively).[6]

Bhoja provides a list of twenty-four traits, some or all of which characterise the four personality types mentioned above. According to him, these traits can (but need not necessarily) form the basis of a character's self-identity (abhimāna). The twenty-four traits are as follows: exalted birth (*jāti*), exalted lineage (*anvaya*), exalted kinship (*abhijana*), exalted nationhood (*nivāsa*), exalted habitation (*āspada*), exalted position (*pada*), exalted filial connection (*pitarau*), superhuman influence (*prabhāva*), wisdom (*prajñā*), learning (*śāstrajñāna*), absence of self-depreciation (*adīnavākyatā*), rhetorical ability (*vāgmitva*), spatially and temporally informed decision-making (*deśakālāvabodha*), ability to accurately decode others' expressive behaviours (*iṅgitākārajñatā*), skill (*dākṣya*), artistic aptitude (*kalāvaidagdhya*), shrewdness (*caturatā*), physical beauty (*rūpasampat*), sex appeal (*saubhāgya*), generosity (*tyāga*), friendliness (*sauhārda*), physical prowess (*śakti*),

[4]For e.g., Ishwara Krishna (c. fourth century AD) classifies all of creation into the topmost, *sattva*-predominant type (exemplified by gods), the lowermost *tamas*-predominant type (comprising of beasts), and the intermediate type (consisting of humans). In each of these types, the other two *guṇas* do exist but as subordinate to the one that predominates (Sa.K, Verse 54, P. 52: *ūrdhvaṃ sattvaviśālastamoviśālaśca mūlataḥ sargaḥ|madhye rajoviśālo brahmādistambaparyantaḥ||*; Also, Mathara's commentary on this verse: *tatrāpi rajastamasī staḥ kintu sattvasyodriktatā*, etc.)

[5]'Personality is that which permits a prediction of what a person will do in a given situation' (Cattell 1950, p. 2); Sperry (2015, p. 1235) relates personality type with 'a consistent and predictable pattern of thinking, feeling, and behaving'.

[6]S.P, Vol. 2, Ch. 15.

and courage (*śaurya*).[7] Characters that possess all the twenty-four traits are classified as superior (*uttama*) and those with eighteen and twelve traits as intermediate (*madhyama*) and inferior (*kaniṣṭha*) respectively.[8] It is important to note that among the traits that the intermediate and inferior characters do not possess is wisdom. *Uddhata* characters are either intermediate or inferior. *Lalita* characters are either superior or intermediate. Both *udātta* and *śānta* characters are superior. The self-identity of an *uddhata* or *lalita* character is based on the traits he/she possess and is of the nature of 'I am one of exalted birth', 'I am learned', etc. On the other hand, the self-identity of an *udātta* or *śānta* character is not dependent on the traits he/she possesses (though he/she possesses all the twenty-four traits mentioned above and is also aware of the fact that he/she does). What, however, it does depend on shall be discussed shortly.

In addition to the twenty-four traits on which self-identities can potentially be based, each character type also possesses a set of eight unique traits. The *uddhata*'s

---

[7]A careful reading of the verses provided by Bhoja to illustrate the twenty-four identity-determining traits facilitated translation of the Sanskrit trait names into their English equivalents. To elucidate with an example, *nivāsa* and *āspada* might both generally mean 'abode' in Sanskrit. The verses provided by Bhoja for illustrating self-identity based on *nivāsa* and *āspada* are as follows respectively:

evaṃ tayoradhvani daivayogādāseduṣoḥ sakhyamacintyahetu|

eko yayau caitrarathapradeśān saurājyaramyānaparo vidarbhān|| (S.P, Vol. 2, Ch. 19, p. 1028)

Translation: By chance, the two met each other on the way and became friends. And then one of them proceeded towards the province ruled by Citraratha and the other towards Vidarbha, promising because of the good governance it enjoyed.

Source of the verse: *Raghuvaṃśa* (5. 60).

śriyaḥ padaṃ dvāravatīti nāmnā purī parītā lavaṇārṇavena|

astyāvṛtevāmararājadhānī nistriṃśanīlena nabhastalena|| (S.P, Vol. 2, Ch. 21, p. 1092).

Translation: There exists a city by name Dvaravati, an abode of the Goddess of fortune. Surrounded by the salty ocean, it looks like heaven's capital bound by the deep blue sky.

Source of the verse: unknown.

It is clear that the first verse refers to a group of people united under a common rule and the second merely to a place fit for being inhabited. On this basis I translate self-identity based on *nivāsa* and *āspada* as the traits of exalted nationhood and exalted habitation.

[8]The discussion carried on above concerns distribution of identity-determining traits in male characters. In female characters, these traits are distributed differently. In them, some or all of sixteen identity-determining traits can be present. Of these, seven—namely, *anvaya*, *śāstrajñāna*, *adīnavākyatā*, *vāgmitva*, *caturatā*, *rūpasampat*, and *saubhāgya*—are commonly shared with all the four male character types. Of the remaining nine traits, *yauvana*, *suveṣatā*, *priyaṃvadatva*, and *dṛḍhabhaktitā* correspond to four of the eight unique traits that a *lalita* male character possesses; *śīla*, *kṛtajñatā*, and *avikatthana* correspond to three of the eight unique traits that an *udātta* male character possesses; *mānitā* corresponds to *māna* of the *uddhata* male character; and *śucitā* to *śauca* of the *śānta* male character. Like male characters, their female counterparts can also be categorised into three subtypes. Of the sixteen identity-determining traits, the superior subtype possesses all, the intermediate subtype twelve, and the inferior subtype eight (Sh.P, vol. 2, Ch. 21, p. 1097: *ya ete ṣoḍaśa proktā—yuktastairuttamasteṣāṃ pādahānyā tu madhyamaḥ| ardhahānyā kaniṣṭhaḥ syānnāyikāsvapyayaṃ vidhiḥ ||*). The Sanskrit names of the eight unique traits and sixteen differentiating traits were translated into English based on the short descriptions that Bhoja provides of them.

set of unique traits include *māna* (inability to put up with loss of status), *tejas* (enhancement of energy when insulted), *śauṇḍīrya* (narcissism/excessive pride), *vilāsa* (mannerisms associated with pride), *vikatthanā* (boastfulness), *asthiratva* (impatience/inability to tolerate delays in one's undertakings), *agambhīratā* (inability to supress opinions when gripped by impulse), and *sāhasa* (impetuousness/acting without thinking). The *lalita*'s set of eight unique traits comprises *suveṣatā* (being well-groomed), *sampriyatā* (attractiveness), *yauvana* (youthfulness), *sthūlalakṣatā* (liberality in bestowing others with gifts), *priyamvadatva* (pleasing talk), *lālitya* (natural gracefulness in mannerisms), *mādhurya* (freedom from fury under all circumstances), and *dṛḍhabhaktitā* (fidelity in relationships, especially romantic ones). The *udātta*'s unique traits are *śīla* (moral character), *dākṣiṇya* (civility), *sthairya* (steadfastness), *gāmbhīrya* (profundity of character as manifested through restraint of expressive behaviours), *prāgalbhya* (confidence in one's enterprise), *śobhā* (maintenance of decorum under all circumstances), *kṛtajñatā* (gratitude), and *avikatthana* (modesty). Lastly, the set of unique traits that characterise a *śānta* character are *kṣamā* (forbearance/ability to suffer things without getting angry), *vaśitva* (complete control over one's sensory and motor faculties), *santoṣa* (contentment/freedom from desires), *praśama* (recoiling away from sense objects), *śauca* (cleanliness), *ārjava* (guilelessness), *vaiśāradya* (contemplativeness), and *vairāgya* (disinclination to enjoy worldly pleasures).

Apart from the traits enumerated above (twenty-four traits that are the basis of self-identity plus eight unique traits), Bhoja also enlists sixteen other traits that can differentiate between the four character types.[9] Table 1 presents a comparison of these sixteen traits as they manifest in the *uddhata* and *udātta* characters.

Character Typology in the Context of Life-Goals

Each of the four personality types pursues life-goals somewhat differently. The four life-goals, collectively referred to as *puruṣārthas*, include *artha* (material prosperity: acquiring, protecting, augmenting, and utilising material objects), *kāma* (deriving pleasure through the enjoyment of material objects), *dharma* (being ethical in the pursuit of both *artha* and *kāma* and thereby gaining unseen merit, i.e., *puṇya*), and *mokṣa* (eternal freedom from displeasure as a result of forsaking all material objects due to the realisation that they are all impermanent and therefore disagreeable in the final analysis). They have been discussed at length in traditional Indian knowledge systems. Bhoja links these life-goals to the four personality types.

[9]Bhoja enumerates the three sets of traits mentioned here in the 18th to 21st chapters of S.P.

Table 1 Comparison of traits that differentiate *uddhata* from *udātta* characters

Uddhata	*Udātta*
Elation following positive outcomes	Absence of elation following positive outcomes
Succumbing to negative mental states (sorrow, etc.)	Not succumbing to negative mental states
Helplessness in the face of difficulties	No helplessness in the face of difficulties
Disruption of ethical conduct by overriding desire	No disruption of ethical conduct by desire
Competition with ill-matched rivals	Competition with righteous superiors
Absence of disgust at despicable objects/acts	Presence of disgust at despicable objects/acts
Absence of arrogance only when rendered powerless	Absence of arrogance in spite of being powerful
Absence of remorse for one's mistakes	Remorse even for unintentional mistakes
Disrespect for the virtuous	Respect for the virtuous
Envy regarding others' positive qualities/feats	Absence of envy regarding others' positive qualities/feats
Joy at others' misfortune	Absence of joy at others' misfortune
Predisposition to find faults in others	No such predisposition
Irrational anger (anger directed at inappropriate objects)	Absence of irrational anger
Not finishing an undertaking	Finishing an undertaking
Awe at one's own/others' extraordinary acts	Absence of awe at one's own/other's extraordinary acts
Trait-based self-identity	No trait-based self-identity

The *uddhata* character pursues *artha* by neglecting the remaining three life-goals.[10] He/she evaluates objects in the following manner: 'Does this object possess externally visible characteristics by which it can be evaluated by the world at large as valuable? If it does, then it is something I must acquire, protect, augment, and utilise[11] because doing so will enhance my status in the eyes of others'. The *uddhata* character's evaluation of an object's value stems from what its possession will do to his/her sense of social worth.[12] He/she does not evaluate an object on the following lines:

[10]S.P, Vol. 2, Ch. 19, p. 1022: *dhīroddhatastu dharmakāmāvanapekṣamāṇaḥ sarvadārthamevānurundhyata ityarthaśṛṅgāre tadadhikāri-tvenāsya vyapadeśaḥ*|.

[11]S.P, Vol. 2, Ch. 19, p. 1021: *so'yaṃ vidyādirarjito, rakṣito, vardhita, upayuktaścārthaḥ*|.

[12]As mentioned earlier, māna, defined as an 'inability to put up with loss of status' (S.P, Vol. 2, Ch. 19, p. 1030: '*paribhavāsahiṣṇutvaṃ*'), is reckoned among the unique traits of *uddhata* characters. Elsewhere in the same chapter, Bhoja classifies pursuable objects into various categories, one of which is 'objects pursued for the purpose of accruing māna' and equates *māna* with *pūjā*, i.e., social esteem (S.P, Vol. 2, Ch. 19, p. 1020: '*mānanimittaḥ pūjārthaḥ*').

(1) Will enjoying the object give me pleasure?
(2) Is the acquisition, protection, augmentation, or utilisation of the object ethical as gleaned from scriptural knowledge as well as from observing how people, guided by such knowledge, behave?
(3) Is the object permanent or impermanent from the standpoint of reality?

As a result of not evaluating objects on these parameters, he/she does not accrue pleasure, merit or freedom from displeasure in the course of pursuing them. Furthermore, since *uddhata* characters are lacking in the trait of wisdom, they may incorrectly evaluate an unfavourable (*pratikūla*) object as favourable (*anukūla*). In other words, *uddhata* characters tend to evaluate an object that they are incapable of acquiring, protecting, augmenting, or utilising (and therefore not conducive to the attainment of material prosperity) as being otherwise.[13] Secondly, they may wrongly evaluate the means (*upāyas*) through which an object can be acquired. Thirdly, even when they do evaluate goal-relevant means correctly, *uddhata* characters may still err in determining which traits are appropriate for securing goal-relevant means. Fourthly, even if they manage to correctly figure out the appropriate traits, *uddhata* characters do not go so far as to evaluate whether they have them in the strength that is adequate for securing goal-relevant means. In spite of all this, if the *uddhata* character does manage to acquire a pursued object, it is often because of an accidental match between goal-relevant means and trait strength. Finally, even after having acquired his/her pursued object, the *uddhata* character may still be incapable of protecting, augmenting or utilising it.

Whenever pursued objects are acquired, there is a reinforcement of the *uddhata*'s trait-based self-identity and sense of social worth. Non-acquisition of pursued objects (or loss of acquired objects), on the other hand, disrupts the *uddhata* character's trait-based self-identity and sense of social worth. The *uddhata* character's sadness (*śoka*) stems from non-acquisition (or loss) of pursued (or attained) object. His/her anger (*krodha*) is directed at the perceived cause of object non-acquisition (or loss). If this perceived cause of object non-acquisition (or loss) possesses traits stronger than what he/she possesses, the *uddhata* character experiences fear (*bhaya*).[14] The character may attempt to rectify these disturbances and

[13]While discussing mental states in relation to *udātta* characters, Bhoja offers brief notes on how they uniquely manifest in such characters. Such notes are not, however, available for the other three character types. An *udātta* character's love, according to Bhoja, is directed at favourable objects (S.P, Vol. 2, Ch. 19, p. 1008: '*anukūleṣu ratiḥ*'). Given that *uddhata* characters are at variance with *udātta* characters in terms of their traits, as Table 1 clearly demonstrates, it is possible to surmise that the former's love will be directed toawards unfavourable objects (which he/she will then try to procure).

[14]According to Bhoja, negative mental states such as fear, sorrow and anger are pronounced only in the *uddhata* and *lalita* characters. The other two character types are mostly immune to these mental states or have them in response to fewer and very specific antecedent causes, such as sorrow as a result of facing a serious calamity or anger from repeated insult to their physical prowess (S.P, Vol. 2, Ch. 18, pp.1013, 1015, 1016: *svaśaktidharṣaṇeṣu krodhaḥ, mahāduḥkheṣu śokaḥ, bhayānutpattiḥ, śokānutpattiḥ*). In his elucidation of the process by which sustained mental states arise in all beings, Abhinavagupta makes the following observation regarding fear, sorrow

overcome the disagreeable mental states experienced therefrom by devaluing the very object of his/her pursuit. He/she may also devaluate others who pursue it (ethically or otherwise) or support its pursuit. This, in turn, can elicit negative mental states in others and lead to interpersonal conflicts. Apart from devaluing objects and others, the *uddhata* character may also reconsider the means through which the pursued object can be acquired, again without evaluating the strength of a trait needed to secure that means.

In summary, when the *uddhata* character acquires pursued objects, there is a temporary alleviation of his/her fear of losing social worth. It is this alleviation of fear that counts as a positive mental state for him/her. Likewise, the *uddhata* character's negative mental states result from

(1) non-acquisition (or loss) of pursued (or acquired) objects because

- he/she does not have the traits necessary to secure the means (for acquiring/protecting the object)
- he/she does not have traits in the strength that is appropriate for securing the means
- objects, by their very nature, are impermanent and likely to be lost

(2) interpersonal conflicts that occur because

- others, having the necessary traits in the strength that is appropriate for securing specific means, are able to acquire an object pursued by them
- he/she bypasses ethics
- he/she devaluates objects that he/she cannot acquire but that may be regarded as valuable by others
- he/she devaluates others who pursue objects that one has not been able to acquire

The *lalita* character evaluates an object on the basis of whether enjoying it will or will not give him/her pleasure. However, he/she does not evaluate whether ethics prescribes or proscribes the acquisition (and protection) of the object or if the object is permanent or impermanent. If enjoying the object will lead to pleasure, then he/she makes attempts to acquire it. Acquisition of objects, however, is not the *lalita* character's life-goal.[15] Among the *lalita* characters, the *uttama* subtype, because he/she also possesses the trait of wisdom, pursues objects that are favourable (i.e., helps him/her attain pleasure). He/she properly evaluates the means

(Footnote 14 continued)

and anger—'Every living being becomes sorrowful when separated from the object of its liking, angry at the object causing this separation, and fearful of the latter if incapable of confronting it' (A.Bh, Vol. 1, Ch. 6, p. 282: *abhīṣṭaviyogasantaptastaddhetuṣu kopaparavaśo'śaktatayā tato bhīruḥ*). He notes elsewhere that mental states such as sorrow and fear characterise inferior and not superior individuals (A.Bh, Vol 1, Ch. 6, p. 282: *ata evānuttamaprakṛtiṣu bāhulyena hāsādayo bhavanti| pāmaraprāyaḥ sarvo'pi—śocati, bibheti …*).

[15]S.P, Vol. 2, Ch. 19, p. 1022: *dhīralalitastu dharmārthāvabhibhūya kāma eva pravartate.*

through which the object can be enjoyed and the match between means and trait strength. The *madhyama* subtype, like the one among *uddhata* characters, errs on several fronts because of an absence of wisdom. In both the *uttama* and *madhyama* subtypes of *lalita* characters, pleasure resulting from the enjoyment of a pursued object leads to reinforcement of trait-based self-identity. Non-attainment of pleasure leads to introspection ('Was there an error in means-trait strength match?') in the *uttama*, wise subtype and devaluation of the pursued object (and others pursuing it, as is seen in the *uddhata* subtype[16]) in the *madhyama*, unwise subtype.

To summarise, when a *lalita* character enjoys objects, there is a temporary experience of pleasure. It is this pleasure that counts as a positive mental state for him/her. Likewise, the *lalita* character's negative mental states stem from

(1) an inability to experience pleasure because

- he/she does not have the traits necessary to secure the means (for enjoying the object)
- he/she does not have the traits in the strength that is appropriate for securing the relevant means.

The above factors are relevant with respect to the *madhyama* subtype

- objects, due to their very impermanence, are lost.

This factor is relevant with respect to both the *madhyama* and *uttama* subtypes.

(2) interpersonal conflicts that occur because

- others, having the necessary traits in the strength that is appropriate for securing specific means, are able to enjoy an object pursued by them
- he/she devaluates objects that he/she is unable to enjoy but that may be valued by others
- he/she devaluates people who pursue objects that he/she is unable to enjoy.

The above factors are relevant with respect to the *madhyama* subtype.

- he/she bypasses ethics

This factor is relevant with respect to both the *madhyama* and *uttama* subtypes.

The *udātta* character evaluates an object on ethical parameters, i.e., whether ethics (as understood from scriptural texts and observation of how people that abide by these texts behave) prescribes or proscribes the acquisition (and protection) and pleasurable enjoyment of that object.[17] The self-identity of an *udātta* character derives from his/her abidance to ethical dos and don'ts rather than from the

[16]The intermediate subtypes among *lalita* and *uddhata* characters are quite similar (S.P, Vol. 2, Ch. 20, p. 1058: *madhyamasya jātyabhimānādayo dhīroddhatenaiva vyākhyātāḥ, pādaguṇasambandhīddhatāyā ubhayatra tulyatvāt*).

[17]S.P, Vol. 2, Ch. 21, p. 1022: *vidyādyarthānāmadharmakāmabādhayā—dhīrodāttaḥ pravartate.*

twenty-four traits that he/she possesses.[18] Therefore, when an udātta character attains his/her pursued object, there is no reinforcement of any trait-based self-identity. Rather, one's identity of the nature of 'I am someone who never deviates from ethically prescribed dos and don'ts' gets reinforced. Non-attainment, on the contrary, of a pursued object, leads to introspection ('Was there an error in means-trait strength match?' 'Was it non-ethical on my part to pursue the object in the first place?'). The *udātta* character, because of his/her wisdom, always pursues objects that are favourable (i.e., helps him/her attain unseen merit and pleasure).

To put it in a nutshell, when the *udātta* character acquires (and protects) and enjoys objects within an ethical framework, there is an experience of pleasure. It is this pleasure that counts as a positive mental state for him/her. Likewise, the *udātta* character's negative mental states stem from

(1) an inability to experience pleasure from the enjoyment of ethically acquired objects that, due to their impermanence, are lost
(2) interpersonal conflicts as a result of being wrongly devalued by others (mostly *uddhata* characters) who are unable to acquire or enjoy a commonly pursued object.

The *śānta* character evaluates whether an object is permanent or impermanent. On the one hand, he/she has the realisation that all objects, both disagreeable and pleasurable ones, are impermanent and must be forsaken in order to attain his/her life-goal of eternal freedom from displeasure.[19] On the other hand, in spite of such a realisation, the *śānta* character must still pursue (i.e., acquire, protect, augment, utilise, and enjoy) a limited number of worldly objects as long as he/she is embodied.[20] In making this choice, he/she (like the *udātta* character) is guided by ethically established standards of dos and don'ts.[21] Acquiring (and protecting) and enjoying pursued objects does not reinforce the *śānta* character's trait-based self-identity because the latter does not even exist in the first place. On the contrary, as the list of pursuable objects gets smaller, his/her identity of the nature of 'I am someone who is eligible for attaining eternal freedom from displeasure' becomes more strongly reinforced.[22] Not being able to acquire and enjoy pursued objects or losing objects

---

[18]S.P, Vol. 2, Ch. 18, p. 1001: *śrutismṛtivākyānatikramābhimāna.*

[19]S.P, Vol. 2, Ch. 21, pp. 1076, 1077: *nivṛttau duḥkharūpatayā jagatsvarūpamavagacchataḥ—sarvaṃ duḥkhamiti tattvajñānaṃ syāt.*

[20]S.P, Vol. 2, Ch. 21, p. 1081: *dhīrapraśāntanāyakasya lokayātrārthaṃ kriyāpravṛttihetutvāt.*

[21]Like that of *udātta* characters, the ethical conduct of *śānta* characters too is not disrupted by desire (S.P, Vol. 2, Ch. 21, p. 1084).

[22]The *śānta* character has self-identities of the following sorts—'I am fit to attain eternal freedom from displeasure'; 'I have studied scriptures that advocate such freedom'; 'I am continuously engaged in performing activities that are conducive to disengagement from the world of impermanent objects'; 'I am never lax in performing duties that are obligatory on an everyday and occasional basis', etc. (S.P, Vol. 2, Ch. 21, p. 1079: *asti me mokṣādhigame yogyatā, adhītāni mayā mokṣaśāstrāṇi, nirantarā nivartakadharmotpattihetavaḥ kriyākalāpā na vīyate nityanaimittikāni —ityādiyogyacetaso'bhimānaḥ|*).

after they have been leads to self-introspection ('Was there an error in means-trait strength match?' 'Was it non-ethical on my part to pursue the object in the first place?', 'Since the object is impermanent, should I be bothered if I don't acquire it?', 'Since the object is impermanent, should losing it bother me?'). Since he/she is wise, the śānta character always pursues objects that are favourable (i.e., helps him/her attain eternal freedom from displeasure, unseen merit and material pleasure).

In summary, when the *śānta* character acquires (and protects) and enjoys objects within an ethical framework, there is an experience of pleasure. It is this pleasure that counts as a positive mental state for him/her. Experience of pleasure, however, is tempered with the understanding that since the pleasurable object is impermanent, one must lose it ultimately. Pleasure and displeasure are therefore regarded as two faces of the same coin.[23] The *śānta* character rarely experiences negative mental states. When he/she does experience them, it is because of being wrongly devalued by others (mostly *uddhata* characters).

Categorisation and Communication of Aesthetically Depicted Mental States

In the ensuing discussion, I have invoked the phrase 'mental state' more than once. What does 'mental state' mean in Sanskrit poetics, especially as Bhoja understands it? Sanskrit aestheticians, beginning with Bharata, have enumerated at least forty-nine (or fifty if *śama*, tranquillity, is also added to the list) modes (*vṛttis*) in which the mind (*citta*) can exist.[24] In the parlance of Sanskrit poetics, each mental mode is called '*bhāva*', a term I translate as 'mental state'. The purpose of any literary work must be to communicate, through its characters, some of these mental states to the audience. From one vantage point, not all mental states are capable of being communicated to the same degree. Some mental states are long-lasting and therefore pervade an entire narrative or larger portions of it. Such mental states, to put simply, form the take-home message of a literary work. Sanskrit aestheticians recognise eight (or nine) such enduring mental states (*sthāyibhāvas*). Examples of such mental states include love, anger, fear and sorrow. The remaining forty-one mental states can be classified into two groups: thirty-three accessory transient mental states (*vyabhicāribhāvas*) and eight accessory transient psychophysical states (*sāttvikabhāva*). Examples of the former are joy, pride, envy, humiliation, discernment and logical reasoning. The latter group includes states such as palpitation, perspiration and horripilation that are at once internal (mental) and external

---

[23]'When a person realizes that for those with discernment everything is ultimately a source of displeasure and there exists no such thing as pleasure ...' (S.P, Vol. 2, Ch. 21, p. 1077: *sarvaṃ duḥkhaṃ vivekina iti jñānātsukhameva nāstīti jānīyāt*).

[24]Abhinavagupta regards *bhāvas* as a special category of mental states, probably those that have the greatest scope of being communicated by poets/actors to their audience (A.Bh, Vol. 1, Ch. 7, p. 342: *bhāvaśabdena tāvaccittavṛttiviśeṣā eva vivakṣitāḥ*).

(bodily). Mental states belonging to these two groups are short-lived and do not have the inherent capacity to pervade larger sections of a narrative. However, they can function as accessories to the enduring mental state and make its communication more effective. The more such accessory states an enduring mental state manages to bring within its ambit, the more effectively will it be communicated to the audience. The highly efficient communication of an enduring mental state transforms it into its delightful aesthetic counterpart—the '*rasa*'. The traditional understanding is that the eight (or nine) enduring mental states can graduate to the level of rasa whereas the forty-one accessory ones cannot. Bhoja does not accept this thesis. According to him, all mental states are equally capable of undergoing greater and greater degrees of enhancement that render their communication more and more effective.[25] Put differently, every mental state, according to Bhoja, can don two different roles—transient (whereby it momentarily appears in the context of an enduring mental state) and enduring (whereby several transient mental states momentarily appear in its context).

Figuring out the most important mental states that are communicable through literary works and fixing their number is only one part of the story. A more important question is how mental states, internal to real-life persons and literary characters alike and therefore inaccessible, can even be communicated. According to Sanskrit aestheticians, it is possible to infer real-life mental states by accessing information about their antecedent stimuli (*kāraṇas*) and consequent responses (*kāryas*).[26] Unlike real-life mental states, the mental states of literary characters are known not through inference (*anumāna*) but a process called '*dhvani*'—translated by Pollock (2016: 12) as 'implicature'. It is beyond the scope of this chapter to discuss implicature in greater detail. However, to summarise, an everyday mental state and the stimulus-response pair from which it is inferred are personally relevant to specific real-life individuals whereas the mental state that a literary character is depicted as having and the stimulus-response pair that implicates it are of a generalised nature,[27] neither belonging to the character nor to the audience alone.[28]

---

[25]'If mental states such as love can graduate to the level of rasa when suitably enhanced, what sin have similar other mental states such as joy committed that they can't rise to the level of rasa?' (S. P, Vol. 1, Ch. 1, p. 2: '*ratyādayo yadi rasāḥ syuratiprakarṣe harṣādibhiḥ kimaparāddhamatadvibhinnaiḥ*').

[26]Abhinavagupta speaks of how enduring mental states of real-life others are known through inference. For inferring the enduring mental states, it is not enough if only stimuli and responses are known. One must also know the accessory mental states that accompany them. These accessory mental states must themselves be inferred from their antecedent stimuli and consequent responses (A.Bh, Vol. 1, Ch. 6, p. 284: *tatra lokavyavahāre kāryakāraṇasahacārātmakaliṅgadarśane sthāyyātmaparacittavṛtyanumāna*).

[27]Generalised stimuli and responses are known as vibhāvas and anubhāvas respectively. The process by which personally relevant mental states and stimuli-response pairs become impersonal is referred to as *sādhāraṇīkaraṇa* by Bhatta Nayaka (A.Bh, Vol. 1, Ch. 6, p. 277: *vibhāvādisādhāraṇīkaraṇātmanā*).

[28]K.P, Ch. 4, Verse 28, pp. 91, 92: *mamaivaite śatrorevaite taṭasthasyaivaite na mamaivaite na śatrorevaite na taṭasthasyaivaite iti sambandha-viśeṣasvīkāraparihāraniyamānadhyavasāyāt.*

Secondly, there is a phenomenological difference between inferred and implicated mental states. While inferring the mental states of real-life others, we ourselves experience pleasurable or disagreeable mental states. Alternatively, we may experience no mental state at all.[29] On the other hand, when the mental states of literary characters are communicated to us through implicature, we experience aesthetic delight that differs qualitatively from everyday pleasure.[30]

The Relationship Between Mental States and Character Typology

That personality type, life-goals and mental states are mutually related has been observed even by Abhinavagupta. Below, I provide two quotations from Abhinavabhāratī that dwell upon such a relationship:

> dhīrodāttadhīralalitadhīroddhatadhīrapraśāntānāṃ pūrṇopāyapravṛttatvena nāyakānāmatādṛgu pāyāśrayeṇa pratināyakānāṃ ca caritaṃ saphalatvāphalatvena sākṣātkriyamāṇaṃ vīrādb hutābhyāṃ, vīraśṛṅgāra-hāsyairvīraraudrabhayānakakaruṇairvīrabībhatsaśāntaiśca pratināya kagata-rasāntarasāntaratayā sātiśayacamatkāragocarībhūtaiḥ—dharmādi-catuṣkopādeyadhiya madharmādibhyaśca nivṛttiṃ niśśaṅkaṃ vidhatte …|

'The actions of heroes belonging to the *udātta*, *lalita*, *uddhata*, and *śānta* types —types that pursue their life-goals after ensuring that the aggregate of means relevant for such a pursuit are at their disposal—bear fruit. Alternatively, the actions of anti-heroes—characters lacking in the aggregate of means—do not bear fruit. When the actions of heroes are narratively depicted as if they were happening in front of our own eyes, we aesthetically partake of the following mental states: perseverance and wonder (in the case of *udātta* heroes); perseverance, love, and mirth (in the case of *lalita* heroes); perseverance, anger, fear, and sorrow (in the case of *uddhata* heroes); and perseverance, disgust, and tranquillity (in the case of *śānta* heroes). Likewise, we aesthetically partake of a different set of mental states when we are involved in the anti-heroes' narratively depicted actions. In the process, narratives undoubtedly teach us why we must choose the 4 life-goals of ethics, etc., and abstain from non-ethics, etc.' (A.Bh, Vol. 1, Ch.1, p. 7).

---

[29]A.Bh, Vol. 1, Ch. 6, p. 280: *paragatatvaniyamabhājāmapi sukhaduḥkhānāṃ saṃvedane niyamena svātmani sukhaduḥkhamohamādhyasthyādisaṃvidantarodgamanasambhavāt.*

[30]R.G, Ch. 1, p. 27: *ānando hyayaṃ na laukikasukhāntarasādhāraṇaḥ*. From the point of Samkhya philosophy, however, the communication of character mental states does not produce aesthetic delight but only pleasure or displeasure as the case may be.

tatra puruṣārthaniṣṭhāḥ kāścitsaṃvida ityeva pradhānam| tadyathā ratiḥ[31] kāmas tadanuṣaṅgidharmārthaniṣṭhā|[32] krodhastatpradhāneṣvarthaniṣṭhaḥ| kāmadharma paryavasito'pyutsāhaḥ samastadharmādiparyavasitaḥ| tattva-jñānajanitanirvedaprāyo vibhāvo mokṣopāya iti tāvadeṣāṃ prādhānyam|

'Among them (mental states), some are principal because they are founded on life-goals. To elucidate, love is founded primarily on material pleasure and secondarily on ethics and material prosperity; anger, in those prone to it, is founded on material prosperity; perseverance, though founded for the most part on material pleasure and ethics, is actually founded on all the 4 life-goals starting with ethics; and anything that is an antecedent stimulus for the mental state of *nirveda*—boredom/dejection, born of true knowledge, towards worldly objects—is a means for securing eternal freedom' (A.Bh, Vol. 1, Ch. 6, p. 282).

The credit for further expanding upon these ideas must go to Bhoja. As already alluded to in the discussion on character typology and life-goals, the personality type to which a literary character belongs has a bearing on his/her experience of pleasurable and disagreeable mental states. A character, because of his/her personality, may experience a mental state, say pride, only due to a limited set of antecedent stimuli. Moreover, the character's personality also determines the responses through which a mental state can be expressed. Lastly, a character may not even experience certain mental states, such as pride, in spite of their antecedent stimuli (such as exalted lineage) being present. Bhoja provides versified examples from the large corpus of Sanskrit literature to illustrate how most of the forty-nine mental states manifest in each of the four personality types (including the following six subtypes: *uddhata* inferior, *uddhata* intermediate, *lalita* intermediate, *lalita* superior, *udātta* superior, and *śānta* superior) discussed earlier. While dealing with the *udātta* personality type, Bhoja supplements vignette verses with some useful comments that help us appreciate the unique manner in which mental states manifest it this type. These comments are presented in Table 2.

The Static-Dynamic Nature of Character Personality Type

Is a character's personality type static or dynamic? In other words, does a character's typology remain unchanged throughout a literary work or does it undergo changes from time to time? Can a character that is identifiable as *uddhata* at one juncture of a literary work be depicted as *udātta* at another juncture of the same work? To answer these questions, I turn to Dhanika (tenth century AD), commentator on *Daśarūpaka*, a dramaturgical text composed by his brother

[31]Most Sanskrit aestheticians do not use two separate terms for sexual and non-sexual love but include both within the ambit of *rati*.

[32]Hemacandra (twelfth century AD), who quotes Abhinavagupta verbatim, gives '*kāmatadanuṣaṅgidharmārth-aniṣṭhā*' (K.Anu, Ch. 2, p. 64).

Table 2 Mental states and their manifestation in *udātta* characters

No	***Udātta* character's mental states with Bhoja's comments on them**[a]
1	*Rati* (sexual love): directed at favourable partners[b]
2	*Prīti* (non-sexual love): directed at those that are worthy[c]
3	*Utkaṇṭhā* (restless longing): towards others of the superior type[d]
4	*Cintā* (contemplation about the means to obtain/overcome favourable/unfavourable objects): based on considerations of ethics, material prosperity, and pleasure[e]
5	*Mati* (discernment; ascertaining about something rationally): leads to the establishment of a viewpoint[f]
6	*Smṛti* (recollecting something previously experienced): does not err with respect to the correct temporal ordering of events[g]
7	*Vitarka* (logical reasoning): aids understanding of the nature of reality[h]
8	*Śaṅkā* (apprehension; expectation of an unfavourable outcome): is commensurate with the object of pursuit[i]
9	*Nidrā* (sleep): occurs only in privacy[j]
10	*Svāpa/Supta* (any effect of sleep, such as dreaming): occurs only at night[k]
11	*Prabodha* (waking up; regaining of consciousness at the termination of sleep): occurs in the second half of the night[l]
12	*Vrīḍā* (shame; embarrassment; contraction of the mind): results from being praised, etc.[m]
13	*Avahittha* (dissimulation; concealing one's intentions and facial expressions): pan-situational[n]
14	*Dhṛti* (satisfaction; not desiring anything beyond what one wants to achieve): consequent to achieving more than what is usually possible[o]
15	*Harṣa* (joy; lucidity of mind due to fulfilment of a wish): consequent to others achieving their goals[p]
16	*Romāñca* (horripilation; a feeling of internal blooming or expansion): consequent to perceiving others' heroic deeds[q]
17	*Hāsa* (mirth; expansion of the mind consequent to joy, etc.): Expressed merely as smile (and not as boisterous laughter)[r]
18	*Utsāha* (perseverance; a particular kind of effort one puts in order to achieve one's life-goals): in relation to some significant undertaking[s]
19	*Glāni* (debility; weakness resulting from physical and mental affliction): does not result even from observing extreme religious vows[t]
20	*Krodha* (anger; the arousal of temper against objects thought of as unfavourable): results from others' repeated devaluation of one's ability[u]
21	*Śoka* (sorrow): occurs only due to a significant catastrophe/never occurs[v]
22	*Aśru* (tear flow): occurs only due to an excess of sorrow, etc.[w]
23	*Sveda* (perspiration): occurs only while performing an arduous task[x]
24	*Śrama* (bodily and mental fatigue following excessive physical activity): occurs only from an excess of fatigue/never occurs[y]
25	*Mūrchā/Pralaya* (fainting; cessation of the functioning of sense): only due to being hit very forcefully, etc.[z]
26	*Garva* (pride): never occurs[aa]

(continued)

Table 2 (continued)

No	***Udātta* character's mental states with Bhoja's comments on them**[a]
27	*Mada* (intoxication): does not arise because intoxicants are not even consumed in the first place[ab]
28	*Amarṣa* (hatred; sustained anger at wrong-doers): never occurs[ac]
29	*Asūyā* (envy; intolerance of others' prosperity): never occurs[ad]
30	*Īrṣyā* (jealousy; intolerance of similar others being conferred with gifts and honour): never occurs[ae]
31	*Ugratā* (rage; verbal and physical aggression against those that have committed a crime): never occurs[af]
32	*Jugupsā* (disgust): never occurs[ag]
33	*Vismaya* (awe; expansion of the mind resulting from watching something extraordinary): never occurs[ah]
34	*Capalatā* (fickleness of mind): never occurs[ai]
35	*Ālasya* (lassitude; dislike for physical activity): never occurs[aj]
36	*Bhaya* (fear): never occurs[ak]
37	*Trāsa* (startled fright following one's exposure to something scary, etc.): never occurs[al]
38	*Kampa/Vepathu* (tremor): never occurs[am]
39	*Stambha* (paralysis/stupefaction): does not arise[an]
40	*Unmāda* (mental illness, irrelevant talk stemming from strong passions): never occurs[ao]
41	*Viṣāda* (despondency; lack of enthusiasm because of adverse fate, etc.): never occurs[ap]
42	*Vaivarṇya* (altered bodily glow): never occurs[aq]
43	*Dainya* (miserableness; a feeling of worthlessness from absence of courage): never occurs[ar]
44	*Svarabheda* (choking of voice): never occurs[as]
45	*Āvega* (flurry/mental agitation motivated by an urgency to address some situation): never occurs[at]
46	*Moha* (perplexity; inability to decide what must be done): never occurs[au]
47	*Jāḍya* (inertia; sluggishness of mind as reflected in action): never occurs[av]
48	*Nirveda* (boredom; mental fatigue due to worldly detachment, etc.): never occurs[aw]

[a]The table lists only those mental states that Bhoja has discussed in relation to *udātta* characters. I provide approximate English equivalents or short descriptions of these Sanskrit mental states on the basis of how Bhoja has understood them (and not as how other Sanskrit aestheticians have)

[b]S.P, Vol. 1, Ch. 14, p. 1008: *anukūleṣu ratiḥ*

[c]S.P, Vol. 1, Ch. 14, p. 811: *ratirevāsaṃprayogaprayojanā*; S.P, Vol. 2, Ch. 18, p. 1008: *anavagīteṣu prītiḥ*

[d]S.P, Vol. 1, Ch. 14, p. 840: *manaso'śāntirutkaṇṭhā*; S.P, Vol. 2, Ch. 18, p. 1008: *uttameṣūtkaṇṭhā*

eS.P, Vol. 1, Ch. 14, p. 842: *dṛṣṭaśrutānumiteṣu prāptiparihārārthamupāyaparibhāvanā manasaścintā*; S.P, Vol. 2, Ch. 18, p. 1009: *dharmārthakāmeṣu cintā*

[f]S.P, Vol. 1, Ch. 14, p. 826: *śāstraparibhāvanayārthanirdhāraṇaṃ matiḥ*; S.P, Vol 2, Ch. 18, p. 1009: *uttarapakṣe matiḥ*

[g]S.P, Vol. 1, Ch. 14, p. 842: *manasyanubhūtārthapratibimbanaṃ smṛtiḥ*; S.P, Vol. 2, Ch. 18, p. 1009: *avyāhatakramā smṛtiḥ*

[h]S.P, Vol. 1, Ch. 14, p. 827: *artheṣu yathāmatyūho vitarkaḥ*; S.P, Vol. 2, Ch. 18, p. 1009: *tattvānupātī vitarkaḥ*

[i]S.P, Vol. 1, Ch. 14, p. 831: *aniṣṭāgamotprekṣaṇaṃ śaṅkā*; S.P, Vol. 2, Ch. 18, p. 1010: *phalānusāriṇī śaṅkā*

[j]S.P, Vol. 1, Ch. 14, p. 838: *indriyāṇāṃ nimīlanaṃ nidrā*; S.P, Vol. 2, Ch. 18, p. 1010: *vivikteṣu nidrā*
[k]*nidrādikāryaṃ suptam* (S.P, Vol. 1, Ch. 14, p. 838); *niśīthe svāpaḥ* (S.P, Vol. 2, Ch. 18, p. 1010)
[l]S.P, Vol. 1, Ch. 14, p. 839: *nidrātyayeṣu cetanālābhaḥ prabodhaḥ*; S.P, Vol. 2, Ch. 18, p. 1010: *apararātre prabodhaḥ*
[m]S.P, Vol. 1, Ch. 14, p. 835: *varṇanānurāgādijanmā manaḥsaṅkoco vrīḍā*; S.P, Vol. 2, Ch. 18, p. 1011: *svaślāghādiṣu vrīḍā*
[n]S.P, Vol. 1, Ch. 14, p. 835: *iṅgitākāragopanamavahittham*; S.P, Vol. 2, Ch. 18, p. 1011: *sarvatraivāvahittham*
[o]S.P, Vol. 1, Ch. 14, p. 822: *abhīṣṭaprāpteradhikā'spṛhā dhṛtiḥ*; S.P, Vol. 2, Ch. 18, p. 1011: *sambhavādhikye dhṛtiḥ*
[p]S.P, Vol. 1, Ch. 14, p. 821: *manorathāvāptau manaḥprasādo harṣaḥ*; S.P, Vol. 2, Ch. 18, p. 1011: *parārthasiddhau harṣaḥ*
[q]S.P, Vol. 1, Ch. 14, p. 822: *ātmano vikṣobhajanmāntaśśarīrocchvāso romāñcaḥ*; S.P, Vol. 2, Ch. 18, p. 1012: *parāpadāne romāñcaḥ*
[r]S.P, Vol. 1, Ch. 14, p. 823: *harṣādibhirmanovikāsaviśeṣo hāsaḥ*; S.P, Vol. 2, Ch. 18, p. 1012: *smitamātrakārī hāsaḥ*
[s]S.P, Vol. 1, Ch. 14, p. 826: *puruṣārtheṣvāparisamāpteḥ prayatnaviśeṣa utsāhaḥ*; S.P, Vol. 2, Ch. 18, p. 1012: *mahārambheṣūtsāhaḥ*
[t]S.P, Vol. 1, Ch. 14, p. 849: *manaḥpīḍādijanmā śarīrāvasādo glāniḥ*; S.P, Vol. 2, Ch. 18, p. 1012: *vratātiśayādiṣu na glāniḥ*
[u]S.P, Vol. 1, Ch. 14, p. 843: *manaḥpratikūleṣu taikṣṇyaprabodhaḥ krodhaḥ*; S.P, Vol. 2, Ch. 18, p. 1013: *svaśaktidharṣaṇeṣu krodhaḥ*
[v]*iṣṭavirahādibhirmanovaidhuryaṃ śokaḥ* (S.P, Vol. 1, Ch. 14, p. 852); *mahāduḥkheṣu śokaḥ* (S.P, Vol. 2, Ch. 18, p. 1013); *śokānutpattiḥ* (S.P, Vol. 2, Ch. 18, p. 1016)
[w]S.P, Vol. 1, Ch. 14, p. 855: *duḥkhakrodhānandajaṃ nayanajalamaśru*; S.P, Vol. 2, Ch. 18, p. 1013: *duḥkhādyatiśayayoge'śrūdgamaḥ*
[x]S.P, Vol. 1, Ch. 14, p. 832: *śramānurāgādyapekṣamanaḥkṣobhajanmā vapurjalodbhedaḥ svedaḥ*; S.P, Vol. 2, Ch. 18, p. 1013: *prārambhā-tireke svedaḥ*
[y]S.P, Vol. 1, Ch. 14, p. 849: *niratiśayaprayāsajanmā manaḥśarīrāyāsaḥ śramaḥ*; S.P, Vol. 2, Ch. 18, p. 1014: *āyāsātiśayayoge śramaḥ*; S.P, Vol. 1, Ch. 18, p. 1017: *śramānutpattiḥ*
[z]S.P, Vol. 1, Ch. 14, p. 856: *sarvendriyāstamayaḥ pralayaḥ*; S.P, Vol. 2, Ch. 18, p. 1015: *mahāprahārādiṣu mūrcchāgamaḥ*
[aa]S.P, Vol. 1, Ch. 14, p. 824: *ātmotkarṣasambhāvanayā parāvajñānaṃ garvaḥ*; S.P, Vol. 2, Ch. 18, p. 1015: *garvānutpattiḥ*
[ab]S.P, Vol. 1, Ch. 14, p. 825: *madirādyupayogānmanassammohānandasambhedo madaḥ*; S.P, Vol. 2, Ch. 14, p. 1015: *madānutpattir-madyapānapratiṣedhenaiva siddha*
[ac]S.P, Vol. 1, Ch. 14, p. 844: *kṛtāparādheṣu krodhāpraśāntiramarṣaḥ*; S.P, Vol. 2, Ch. 18, p. 1014: *amarṣānutpattiḥ*
[ad]S.P, Vol. 1, Ch. 14, p. 844: *pararddhyasahanamasūyā*; S.P, Vol. 2, Ch. 18, p. 1015: *asūyānutpattiḥ*
[ae]S.P, Vol. 1, Ch. 14, p. 845: *samāneṣu dānamānādyamarṣaṇamīrṣyā*; S.P, Vol. 2, Ch. 18, p. 1015: *īrṣyānutpattiḥ*
[af]S.P, Vol. 1, Ch. 14, p. 846: *doṣato vāgdaṇḍapāruṣyamugratā*; S.P, Vol. 2, Ch. 18, p. 1015: *ugratānutpattiḥ*
[ag]S.P, Vol. 1, Ch. 14, p. 858: *padārthanindanaṃ jugupsā*; S.P, Vol. 2, Ch. 18, p. 1015: *jugupsānutpattiḥ*
[ah]S.P, Vol. 1, Ch. 14, p. 828: *atiśayitapadārthadarśanāccitavistāro vismayaḥ*; S.P, Vol. 2, Ch. 18, p. 1015: *vismayānutpattiḥ*
[ai]S.P, Vol. 1, Ch. 14, p. 836: *manaso'sthairyaṃ capalatā*; S.P, Vol. 2, Ch. 18, p. 1015: *capalatānutpattiḥ*
[aj]S.P, Vol. 1, Ch. 14, p. 837: *śṛṅgārādibhirmānasaḥ kriyāvidveṣa ālasyam*; S.P, Vol. 2, Ch. 18, p. 1015: *ālasyānutpattiḥ*

[ak]S.P, Vol. 1, Ch. 14, p. 829: *raudrādidarśanānmanaso vaiklavyaṃ bhayam*; S.P, Vol. 2, Ch. 18, p. 1015: *bhayānutpattiḥ*
[al]S.P, Vol. 1, Ch. 14, p. 830: *raudrādijanmā cittacamatkārastrāsaḥ*; S.P, Vol. 2, Ch. 18, p. 1015: *trāsānutpattiḥ*
[am]S.P, Vol. 1, Ch. 14, p. 834: *bhayānurāgādyapekṣacittavikṣobhajaḥ kampo vepathuḥ*; S.P, Vol. 2, Ch. 18, p. 1016: *kampānutpattiḥ*
[an]S.P, Vol. 1, Ch. 14, p. 832: *bhayānurāgādijanmā ceṣṭāpratighātaḥ stambhaḥ*; S.P, Vol. 2, Ch. 18, p. 1016: *bhayānutpattiḥ*
[ao]S.P, Vol. 1, Ch. 14, p. 847: *kāmādyabhiṣaṅgātiśayādasambaddhapralāpa unmādaḥ*; S.P, Vol. 2, Ch. 18, p. 1016: *unmādānutpattiḥ*
[ap]S.P, Vol. 1, Ch. 14, p. 850: *daivādibhiḥ samrambhavighāto viṣādaḥ*; S.P, Vol. 2, Ch. 18, p. 1016: *viṣādānutpattiḥ*
[aq]S.P, Vol. 1, Ch. 14, p. 854: *viṣādādijanma kānteranyathātvaṃ vaivarṇyam*; S.P, Vol. 2, Ch. 18, p. 1016: *vaivarṇyānutpattiḥ*
[ar]S.P, Vol. 1, Ch. 14, p. 853: *sattvatyāgādanutkarṣo dainyam*; S.P, Vol. 2, Ch. 18, p. 1016: *dainyānutpattiḥ*
[as]S.P, Vol. 1, Ch. 14, p. 833: *harṣādijanmā vāgbhaṅgaḥ svarabhedaḥ*; S.P, Vol. 2, Ch. 18, p. 1016: *svarabhedānutpattiḥ*
[at]S.P, Vol. 1, Ch. 14, p. 854: *pratīkārāśayā manovyākulatāvegaḥ*; S.P, Vol. 2, Ch. 18, p. 1017: *āvegānutpattiḥ*
[au]S.P, Vol. 1, Ch. 14, p. 851: *itikartavyatāyāmanāloko mohaḥ*; S.P, Vol. 2, Ch. 18, p. 1017: *mohānutpattiḥ*
[av]S.P, Vol. 1, Ch. 14, p. 857: *manasaḥ kriyāsvapāṭavaṃ jāḍyam*; S.P, Vol. 2, Ch. 18, p. 1017: *jāḍyānutpattiḥ*
[aw]S.P, Vol. 1, Ch. 14, p. 858: *vairāgyādibhirmanaḥkhedo nirvedaḥ*; S.P, Vol. 2, Ch. 18, p. 1017: *nirvedānutpattiḥ*. The *udātta* character does not experience boredom though he/she must perform certain duties repeatedly and monotonously. Bhoja provides the following verse to illustrate absence of *nirveda* in the *udātta* character:
tāmantikanyastabalipradīpāmanvāsya goptā gṛhiṇīsahāyaḥ|
krameṇa suptāmanu saṃviveśa suptotthitāṃ prātaranūdatiṣṭhat)||
Translation: Placing objects of worship and a lamp by the holy cow's side, the guardian king (Dilipa) and his queen would rest after she had rested, sleep after she had slept, and wake up in the morning after she had
Source of the verse: *Raghuvaṃśa* (2. 24)
The verse illustrates how king Dilipa, an *udātta* character, took routine care of the cow without getting bored of the monotony inherent in his chore

Dhananjaya. I also discuss below how Dhanika's observations can be integrated into Bhoja's Sankhya-based model for analysing narratively communicated mental states by invoking character typology.

As already mentioned, characters are classified into different types based on which of the three guṇas is predominantly functional in their psychophysical apparatus. Here, one must make a distinction between primary (*aṅgibhūta*) and secondary (*aṅgabhūta*) characters. The primary characters, including the hero (*nāyaka*) and the heroine (*nāyikā*), pervade a literary work more completely than secondary characters.[33] It is an important tenet of Sanskrit poetics that primary

[33]S.P, Vol 2, Ch. 21, p. 1086: *kathāśarīravyāpī—nāyakaḥ*.

characters be characterised in a uniform manner throughout a literary work.[34] If the hero is, for example, initially depicted as an *udātta* character, he must not be depicted as an *uddhata* character at a later point in the narrative. Understood from the standpoint of *guṇa* theory, this means that the *sattva-guṇa* predominates in an *udātta* type of hero (or heroine) all along the course of a literary work. This also means that the hero's (or heroine's) life-goals and the nature of his (or her) mental states (that are intimately related to character type and life-goals) remain unaltered in the narrative. Unlike that of primary characters, the characterisation of secondary characters (including those that support the primary character and those that are in opposition to the latter[35]) need not be uniform throughout a work. A single secondary character may, for instance, be depicted as *uddhata* and *udātta* at different points in the same narrative.[36] From the perspective of *guṇa* theory, these secondary characters can be regarded as exhibiting a preponderance of different *guṇas* at different times. The dynamism in a secondary character's personality also indicates that his/her life-goals (and consequently, the nature of his/her mental states) are prone to change from time to time. The preponderance of *sattva*, *rajas*, or *tamas* in a given character at a given point in time must be inferred from the expression of one or more of the eight unique traits differentiating that character type from other character types. The primary characters consistently exhibit one or more of their unique traits at all times. The secondary characters, however, can transiently exhibit traits unique to different character types at different times.[37]

Explication of Bhoja's Model for Analysing Mental States Communicated Through Literature

In order to get a holistic picture of Bhoja's model, it is important to understand how the various components of that model operate in an interrelated manner. Bhoja's model is based on the premise that to understand a mental state communicated through literature, one must understand the personality type to which the character

[34]D.R.AV, Ch. 2, p. 38: *aṅginastu rāmādeḥ—ārambhopāttāvasthāto'vasthāntaropādānamanyāyyam.*

[35]Bhoja classifies secondary characters (male/female) into three groups, *upanāyaka/upanāyikā*, *anunāyaka/anunāyikā*, and *pratināyaka/pratināyikā*. Characters belonging to the first group are those that are respected by the hero/heroine and are equal to or greater than the latter but do not occupy an exalted position (S.P, Vol. 2, Ch. 21, p. 1086: *nāyakābhyarhaṇīyaḥ sama utkṛṣṭo vānavāptapada upanāyakaḥ*). The second group comprises of characters that are slightly inferior to the hero/heroine but play an important role in the narrative (S.P, Vol. 2, Ch. 21, p. 1087: *nāyakātkiñcidūnaḥ kathāśarīre viśeṣopayogavānanunāyakaḥ*). Both these two groups of characters are favourably disposed towards the hero/heroine. Characters included in the third group are opposed to the hero/heroine and are potentially capable of overthrowing the latter (S.P, Vol. 2, Ch. 21, p. 1088: *nāyakapratikūlavṛttistaducchedārhaḥ—pratināyakaḥ*).

[36]D.R.AV, Ch. 2, p. 38: *na cāvasthāntarābhidhānamanucitamaṅgabhūtanāyakānāṃ.*

[37]D.R.AV, Ch. 2, p. 38: *mahasattvāderavyavasthitatvāt.*

that has the said mental state belongs. Any mental state can be communicated to the audience through one of several stimulus-response pairs. Hogan (2003: 83) calls such stimulus-response pairs 'mini-narratives'. Not all mini-narratives of a mental state are functional in the case of all literary characters. That is, not all characters respond to a particular stimulus in the same way. The character's personality type delimits how he/she will respond to a specific stimulus and even if he/she will respond to it at all.[38] Making sense of narratively elaborated mental states using Bhoja's model hinges, therefore, on deciphering character typologies. This demands a close reading of literary works. The analysis of character mental states employing Bhoja's model can be carried out by following the steps mentioned below:

1. Determining if the character is primary or secondary: The primary character pervades a literary work more completely than secondary characters.
2. Once a character has been identified as primary, zeroing in on the personality type that he/she exemplifies: The primary character consistently demonstrates a set of unique traits that demarcate him/her as *uddhata*, *lalita*, *udātta* or *śānta*.
3. Once a character has been identified as secondary, zeroing in on the personality type that he/she exemplifies: The secondary character may not consistently demonstrate traits unique to a single personality type. Rather, he/she may demonstrate traits unique to more than one personality type at different points in the narrative.
4. Further qualification of characters identified as *uddhata* or *lalita*: Within the framework of Bhoja's model, *udātta* and *śānta* characters always belong to the superior subtype. However, *lalita* characters can be superior or intermediate and *uddhata* characters intermediate or inferior. Evaluating how many of the twenty-four identity-determining traits a *lalita* or *uddhata* character possesses can aid us in subtyping them. Lalita characters possessing twenty-four and eighteen traits are considered superior and intermediate respectively. Likewise, *uddhata* characters possessing eighteen and twelve traits are regarded as intermediate and inferior respectively. In the case of *lalita* intermediate, *uddhata* intermediate and *uddhata* inferior subtypes, we must evaluate not just the presence of some traits but also the absence of others, such as wisdom.
5. Incorporating other components of the model into the analysis, these include life-goals, hierarchy among life-goals, source of self-identity, criteria by which pursuable objects (objects whose acquisition is important for attaining life-goals) are evaluated, evaluation of means relevant for attaining life-goals, evaluation of traits that are appropriate for securing goal-relevant means,

[38]Apart from the character's personality type, the spatiotemporal context in which he/she is situated determines what object he/she can regard as worth pursuing (acquiring, protecting, augmenting, and enjoying within an ethical framework), avoiding, or confronting and thereby direct his/her mental states at. The more true an author's depiction of a character's mental state is to the latter's personality and spatiotemporal context, the more authentic will the audience experience its communication as. Being true to character's personality and spatiotemporal context confers on a literary work the quality of propriety (*aucitya*).

evaluation of the match between goal-relevant means and trait strength, and the effect that success or failure in acquiring (and protecting) and enjoying pursued objects has on self-identity. Table 3 summarises the above-listed components in relation to the four character types.

6. Explaining a character's mental states by invoking his/her personality type: Bhoja's model, really speaking, is a model that explains character personality rather than character mental states. A character's personality, however, is not something that is unrelated to his/her mental states. Rather, it forms the basis of the latter. If we must understand literary mental states, we must go beyond mere

Table 3 Comparative summary of important components in Bhoja's model for analysing four literary character types

Component	*Uddhata*	*Lalita*	*Udātta*	*Śānta*
Life-goals	Material prosperity (*artha*)	Pleasure due to material enjoyment (*kāma*)	Ethical conduct (*dharma*)	Eternal freedom from sorrow (*mokṣa*)
Hierarchy among one's life-goals	Disregard for *kāma*, *dharma*, and *mokṣa*,	Subordination of *artha* and *dharma* to *kāma*; disregard for *mokṣa*	Attainment of *artha* and *kāma* within the framework of *dharma*; does not aim for *mokṣa*	Attainment of *artha* and *kāma* within the confines of *dharma* along with the realisation that such attainment cannot provide *mokṣa*
Source of self-identity	A set of traits	A set of traits	Non-transgression of ethical dos and don'ts	Eligibility for the attainment of *mokṣa*
Criteria for evaluating pursuable object	'Can acquiring it enhance my social worth?'	'Can enjoying it give me pleasure?'	'Does ethical authority enjoin its acquisition and enjoyment?'	'Is it permanent or impermanent?' 'Are its acquisition and enjoyment enjoined by ethical authority?'
Evaluation of goal-relevant means	Likely erroneous	Likely erroneous in intermediate subtype	Non-erroneous	Non-erroneous
Evaluation of traits that are appropriate for securing means	Likely erroneous	Likely erroneous in intermediate subtype	Non-erroneous	Non-erroneous

(continued)

Table 3 (continued)

Component	*Uddhata*	*Lalita*	*Udātta*	*Śānta*
Evaluation of means-trait strength match	May not evaluate	May not evaluate	Correctly evaluates	Correctly evaluates
Effect, on self-identity, of success in acquiring and enjoying pursued objects	Reinforcement of trait-based self-identity when pursued object is acquired	Reinforcement of trait-based self-identity when pursued object is enjoyed	No reinforcement of trait-based self-identity because such self-identity does not even exist in the first place; acquisition and enjoyment of pursued objects within the framework of ethics reinforces self-identity that is of the nature of 'I am a person who will never transgress ethics'	No reinforcement of self-identity based on traits or non-transgression of ethics because such self-identity does not even exist in the first place; self-identity of the nature of 'I am eligible to attain eternal freedom from sorrow' becomes more and more reinforced as one gradually withdraws from the material world and the list of pursuable objects becomes smaller and smaller
Effect, on self-identity, of failure in acquiring and enjoying pursued objects	Disruption of trait-based self-identity; attempts to correct this disruption by devaluing the object of one's pursuit	Disruption of trait-based self-identity; the unwise *lalita* intermediate subtype attempts to correct this disruption by devaluing pursued object; the wise *lalita* superior subtype introspects on the reasons for failure	Failure does not disrupt self-identity because one knows that one has not transgressed ethics; introspection on the reasons for failure	Failure does not disrupt self-identity because one knows that one has not transgressed ethics; introspection on the reasons for failure and on the impermanence of the pursued object

identification of the stimulus-response pairs through which they are communicated to us. A narrative analysis that incorporates the steps outlined above can bring clarity to our knowledge about a character's personality. And the clearer we are about a character's personality, the more obvious it is to us why, at some point in the narrative, he/she (1) communicates to us a mental state through one, and not another, stimulus-response pair and (2) communicates to us one, and not another, mental state. The primary characters' mental states can be explained as resulting from a single, stable personality that pervades an entire narrative. The secondary characters' mental states, on the other hand, may have to be explained by invoking a dynamic personality that, at different points in the narrative, approximates different personality types.

Conclusion

In this chapter, I have detailed a model, derived from Sanskrit poetics, for analysing the myriad mental states that authors communicate to us through their characters. According to this model, borrowed from the eleventh-century aesthetician Bhoja, in order to understand narratively depicted mental states, it is imperative to figure out the personality types of characters that are regarded as the substratum of those mental states. Though I have not attempted to analyse the mental states of any particular literary character, I have spelt out the tools with which such an exercise can be carried out.

As a possible implication of this chapter, it is tempting to explore the utility of Bhoja's model for understanding the mental states of real-world individuals. It is important to realise that concepts such as *guṇas* and *puruṣārthas* were primarily formulated to understand real-world entities. Bhoja only borrowed these already existing concepts for the limited purpose of building a model that can be used to analyse literary characters. In other words, by lending themselves to be classified on the basis of *guṇas* and *puruṣārthas*, literary characters are only mirroring real-world individuals. To address the question of whether real-world individuals faithfully represent one or the other character type, I presume that they feel, think and behave like different character types at different times and thus possess a dynamic, continuously evolving personality. Therefore, like secondary characters in a literary work, real-life individuals allow for flexibility in personality.

A second implication of this chapter is that it anticipates a tool for self-reflection and personal growth. The four personality types discussed in this chapter exist along a hierarchical continuum with the *uddhata* (inferior) type, abounding in tamas, occupying the lowest position and the *śānta* (superior) type, distinguished by an excess of *sattva*, the highest. As one approaches the *śānta* end of this continuum, one gets closer and closer to a state that promises eternal freedom from displeasure. Conversely, as one approaches its *uddhata* end, one experiences displeasure that progressively becomes more and more frequent as well as intense. In between the two ends, various degrees of pleasure and displeasure are experienced. The extent of one's displeasure is not just an indicator of which end of the continuum one is close to in a given instant. It also informs one about the life-goal one is preoccupied with at that instant. Reorienting one's life-goal can help reduce the burden of one's displeasure. To put it more specifically, an *uddhata*-like individual must graduate from merely acquiring an object to enjoying it, a *lalita*-like individual from merely enjoying an object to acquiring as well as enjoying it within an ethical framework, and an *udātta*-like individual from merely acquiring and enjoying an object within an ethical framework to realising its impermanence. It is important to reiterate that terms such as '*uddhata*-like' are used here to characterise an individual at specific instants in time and not for categorising him/her in general.

References

Primary Sources

Bhāvaprakāśana of Śāradātanaya. Ed. Yadugiri Yatiraja and K.S. Ramaswami Sastri. 2nd edition. Baroda: Oriental Institute, 1968.

Daśarūpaka of Dhanañjaya, with the *Avaloka* commentary of Dhanika. 5th edition. Bombay: Nirnaya Sagar Press, 1941.

Kāvyānuśāsana of Hemacandra. Ed. Kasinatha Sarma. Bombay: Nirnaya Sagar Press, 1901.

Kāvyaprakāśa of Mammaṭa, with the *Bālabodhinī* commentary of Vamanacharya Ramabhatta Jhalakikar. Ed. Raghunath Damodar Karmarkar. 6th edition. Poona: Bhandarkar Oriental Research Institute, 1950.

Nāṭyaśāstra of Bharata, with the *Abhinavabhāratī* of Abhinavagupta, vol. 1. Ed. K.S. Ramaswami Sastri. 2nd edition. Baroda: Oriental Institute, 1956.

Rasagaṅgādhara of Jagannātha Paṇḍitarāja. Ed. Mathuranath Sastry. 6th edition. Bombay: Nirnaya Sagar Press, 1947.

Sāṅkhyakārikā of Īśvarakṛṣṇa with the *Māṭharavṛtti* commentary of Māṭharācāraya and the *Jayamaṅgala* commentary of Śrī Śaṅkara. Ed. Visnu Prasad Sarma and Satkarisarma Vangiya. Varanasi: Chowkhambha Sanskrit Series Office, 1970.

Sāhityasāra of Acyutarāya. Ed. Vasudeva Sarma. Bombay: Nirnaya Sagar Press, 1906.

Śṛṅgāraprakāśa of Bhoja. Ed. Rewa Prasad Dwivedi and Sadashivakumara Dwivedi. New Delhi: Indira Gandhi National Centre for the Arts, 2007.

Secondary sources

Cattell, R. B. (1950). *Personality: A systematic, theoretical, and factual study*. New York: McGraw-Hill.

Hogan, Patrick Colm. (2003). *The mind and its stories*. New York: Cambridge University Press.

Pollock, Sheldon. (2016). *A rasa reader*. New York: Columbia University Press.

Raghavan, V. (1978). *Bhoja's Śṛṅgāraprakāśa* (3rd ed.). Madras: Punarvasu.

Sperry, L. (2015). *Mental health and mental disorders: An encyclopedia of conditions, treatments, and w-being*. Santa Barbara, California: Greenwood.

Part IV
The Self and Alternative Epistemologies

Brain and Self: A Neurophilosophical Account

Georg Northoff

Background Ia: Conceptual Determination of the Self in Philosophy

You read these lines. You're winning a game of tennis while your girlfriend is watching. You feel pride. Who experiences that pride? You. You are the subject of the experience of boredom. Without you as subject of this experience, you could not experience anything at all, not even boredom. This subject of experience has been described as the 'self'. Your 'self' makes it possible for you to experience things. In other words, it is a necessary condition for experience and thus also for consciousness. It is clear, therefore, that there is much at stake when it comes to the self.

The concept of self has been subject to intense philosophical discussion over the centuries. Different philosophers have suggested different concepts of self. Because of time and space constraints, we will only focus on those that are relevant in the attempt to map the interface between philosophical and neuroscientific accounts of the self.

There are four main different concepts of self discussed in current philosophy. First is 'the mental self', which is based on our thoughts and a specific mental substance. Second, is the 'empirical self'—this concept of the self represents and reflects the biological processes in one's body and brain.

G. Northoff (✉)
Brain Imaging and Neuroethics Research Unit, Institute of Mental Health Research, Royal Ottawa Mental Health Centre, 1145 Carling Avenue, Ottawa, Canada
e-mail: Georg.Northoff@theroyal.ca
URL: http://www.georgnorthoff.com

S. Menon et al. (eds.), *Self, Culture and Consciousness*,
DOI 10.1007/978-981-10-5777-9_16

Third is the 'phenomenal self', from which originates our experience in consciousness. Our consciousness is accompanied by an awareness of our self, referred to as pre-reflective self-awareness or phenomenal self. Finally, and most recently, philosophers speak of a 'minimal self'. This concept of the self is based on our body and its physiological processes. I will discuss each of these different concepts and how they relate to the brain in this chapter.

Before we do this, I have to shed some light on several related concepts. We experience our self in daily life during, for example, the act of perceiving certain objects, persons or events in our environment. For example, while making a list of all the things you have to do today, you experience not only the act of thinking and writing, but an awareness and experience of your own self. Hence, your self as the very subject of experience seems to be part of that experience. In other words, your self is a content of your consciousness. This is described as self-consciousness. However, there is more to the self than the self itself and our experience of it in self-consciousness. You wake up every morning. Every day. Every week, every year. Your body changes. You become older. You get wrinkled and your hair turns white. Despite all these bodily changes, you nevertheless have the feeling that you are the same self. You still experience your self as being the self of 20 years ago.

You are one and the same person. There is thus a temporal dimension to your self that seems to be coherent and persistent over time. You and your self are continuous across time. The temporal dimension of your self has consequently been discussed under the umbrella of what is called 'personal identity' in philosophy. While our discussion will touch upon the temporal dimension of the self and thus upon personal identity, we will not explicitly discuss it.

In a world of over seven billion people, there are many, many selves: you, your friends, your family and so on. Most interestingly, you can relate to them—you can communicate with other selves and sometimes even feel their emotions as in, for instance, the grief someone might feel when they lose a loved one. Or you might experience pain when your boyfriend's arm is broken. How is this possible? In philosophy, this is called 'intersubjectivity'. Finally, your self is not isolated from the rest of the world. You can share others' experiences and feel connected to the world. The world, with its specific objects, persons and events, has meaning to you —you can more or less relate to it and appropriate it for your own self. How is such basic integration of your own self within the world possible? And how is that related to your brain and its neuronal mechanisms? That shall be the focus in the following.

Background Ib: Empirical Investigation of the Self in Neuroscience

How can we investigate the self? In order to experimentally address the self, we need some quantifiable and objective measures that can be observed from third-person perspective. How can we obtain such measures? Psychologists focusing on memory observed that items related to ourselves were better remembered than those unrelated (Northoff et al. 2006). For example, as a resident of Ottawa, I recall the recent thunderstorm that wiped away several houses locally much better than a person who, perhaps living in Germany, just heard about it in the news.

There is thus superiority in the recollections of those items and stimuli that are related to one's self. This is described as the self-reference effect (SRE). The SRE has been well validated in several psychological studies. Most interestingly, it has been shown to operate in different domains. Not only in respect to memory, but also in relation to emotions, sensorimotor functions, faces, words and so on. In all these different domains (see below for details), stimuli related to one's own self, known as self-specific stimuli, are recalled much better than those that are unrelated to one's own self, known as non-self-specific stimuli.

How is the SRE possible? Numerous investigations (see, e.g., Klein 2012; Klein and Gangi 2010 for summaries) show that the SRE is mediated by different psychological functions. These range from personal memories, including autobiographical memories, over memories of facts (semantic memories) to those cognitive capacities that allow for self-reflection and self-representation. Hence, the SRE is by itself not a unitary function, but rather a complex multifaceted psychological composite of functions and processes.

How can we link the SRE to the brain? Before the introduction of functional imaging techniques such as fMRI at the beginning of the 1990s, most studies conducted focused on the effect of dysfunction or lesions in specific brain regions caused by brain tumours or stroke. These revealed that lesions in medial temporal regions that are central in memory recall, such as the hippocampus, change and ultimately abolish the SRE effect.

With the introduction of brain imaging techniques such as fMRI, we could then transfer the experimental paradigms of comparing self- and non-self-specific stimuli to the scanner and investigate the underlying brain regions. The basic premise here is that if self-specific stimuli are recalled better than non-self-specific stimuli, they must be processed by the brain in a different way. This might be, for instance, by higher degrees of neural activity and/or different regions.

This led to the investigation of numerous experimental designs of SRE-like paradigms in the fMRI scanner. For example, subjects were presented trait adjectives that were either related to themselves (such as, for me, my hometown, Ottawa) as opposed to (Sydney, an unrelated city for me). In other tests, subjects were presented with images of the own face and these were compared with faces of other people. Also, autobiographical events from the subject's past were compared with those from other people. One's own movements and actions could also be compared with those of other people, implying what is called ownership (e.g., my movements) and agency ('I myself caused that action').

The stimuli belonged to different domains, such as memory, faces, emotions, verbal, spatial, motor or social. Most of the stimuli were presented either visually or auditorily and the presentation of these stimuli was usually accompanied by an online judgement about whether the stimuli are related and personally meaningful or not to the research subject.

On the whole, we can see that current neuroscience can investigate the self in various experimental ways using mainly functional brain imaging. However, any empirical research relies on certain presuppositions. This also holds true for current neuroscientific research on the self, which aims to reveal the neuronal mechanisms underlying our experience or sense of self. However, before examining the neuroscientific findings, we need to briefly shed some light on the concept of the self and how it has been defined in philosophical discussions.

Philosophical Concepts Ia: Mental Self

What is the self? What must it look like in order to presuppose experience and be the subject of our experience? The self has often been viewed as a specific 'thing'. Stones are things, the table on which your laptop stands is a thing. And in the same way the table makes it possible for the laptop to stand on it, the self may be a thing that makes experience and consciousness possible. In other words, metaphorically speaking, experience and consciousness stand on the shoulders of the self.

However, another question is whether the self is a thing or, as philosophers such as René Descartes suggest, a substance? A substance is a specific entity or material that serves as a basis for something like a self. For instance, the body can be considered a physical substance, while the self is can be associated with a mental substance.

Is our self real and thus does it exist? Or is it just an illusion? Let us compare the situation to perception. When we perceive something in our environment, we sometimes perceive it not a real thing, but an illusion that in reality does not exist. The question of what exists and is real is what philosophers call a metaphysical question. Earlier philosophers, such as René Descartes, assumed that the self is real and exists.

However, Descartes also assumed that the self is different from the body. Hence, self and body exist, but differ in their existence and reality. Thus, from this perspective, the self cannot be a physical substance and is a mental substance instead. It is a feature not of the body, but of the mind.

However, the characterisation of the self as a mental entity has been questioned. For example, the Scottish philosopher David Hume argued that there is no self as a mental entity. There is only a complex set or 'bundle' of perceptions of interrelated events that reflect the world in its entirety. There is no additional self in the world;

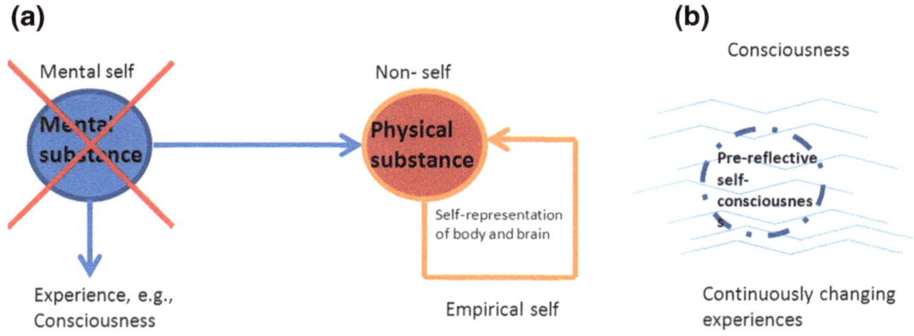

Fig. 1 Concepts of self. **a** Mental self and its replacement by an empirical self. **b** Phenomenal self as pre-reflective self-consciousness. The figure schematically illustrates different concepts of self, the self as mental substance (**a**) and the phenomenal self (**b**). **a** The self is determined as mental substance (*left*) that is distinguished from the body (and brain) as mere physical substance (*right*). Thereby the self as mental self controls and directs the body, following the earlier French philosopher Descartes. This is denied in current empirical approaches to the self (e.g., *vertical red lines*). They reject the notion of the self as mental substance and claim that such mental self does not exist. All there is the body as physical substance with the brain allowing for the representation of both body and brain in the brain's neural activity. Such self-representation may then amount to what can be described as the empirical self. **b** The phenomenal self no longer claims to be outside and prior to any experience. Instead, the phenomenal self is supposed to be 'located' or part of the experience itself in the gestalt of pre-reflective self-consciousness. This is indicated by the *insertion of the circle* within the midst of the experience, e.g., consciousness, itself

instead there is nothing but the events we perceive. Everything else, such as the assumption of a self as mental entity, is an illusion. The self as mental entity and thus as a mental substance does not exist and is therefore not real.

To reject the idea of self as mental substance and to dismiss it as mere illusion is currently popular. One major proponent of this view today is the German philosopher Metzinger (2004). In a nutshell, he argues that through our experience, we develop models of the self, so-called 'self-models'. These self-models are nothing but information processes in our brain. However, since we do not have direct access to these neuronal processes (e.g., all those processes and activities of the cells, neurons, in the brain), we tend to assume the presence of an entity that must underlie our own self-model. This entity is then characterised as the self (Fig. 1).

According to Metzinger, the assumption of the self as a mental entity results from an erroneous inference from our experience. We cannot experience the neuronal processes in our brain as such. Nobody has ever experienced their own brain and its neuronal processes. Therefore, the outcome of our brain's neuronal processes, the self, cannot be traced back to its original basis, the brain, in our experience.

Where then does the self come from? We assume that it must be traced back to a special instance different from the brain. This leads us to assume that the mind and the self are mental entities rather than a physical, neuronal entities originating in the brain itself. Metzinger argues that the self as a mental entity simply does not exist. Therefore, Metzinger (2004) concludes, selves do not really exist. Hence, the title of his book: *Being No One.*

Philosophical Concepts Ib: From the Metaphysical to the Empirical Self

What is the self if not a mental entity? Current authors, such as Metzinger (2004) and Churchland (2002), argue that the self as mental substance or entity does not exist. How do we come up with the idea of a self or the self-model as Metzinger calls it? The model of our own self is based on summarising, integrating and coordinating all the information from our own body and own brain.

What does such integration look like? Take all that information together, coordinate and integrate it and then you have a self-model of your own brain and body and their respective processes. In more technical terms, our own brain and body are represented in the neuronal activity of the brain. Such representation of the own brain and body amounts then to a model of your self. The self-model is therefore nothing but an inner model of the integrated and summarised version of your own brain and body's information-processing. The self is thus a mere model of one's own body's and brain's processes.

The original mental self, the self as mental substance or entity, is in this line of thinking replaced by a self-model. This implies a shift from a metaphysical discussion of the existence and reality of self, to the processes that underlie the representation of body and brain as a self-model. Since this representation is based on the coordination and integration of the various ongoing processes in brain and body, it is associated with specific higher-order cognitive functions such as working memory, attention, executive function and memory.

What does this imply for the characterisation of the self (presupposing a broader concept of self beyond the self as mental substance)? The self is no longer characterised as a mental substance but as a cognitive function. Methodologically, this implies that the self should be investigated empirically rather than metaphysically.

We therefore need to search for the cognitive processes underlying the special self-representation. The self is consequently no longer an issue of philosophy, but rather one of cognitive psychology and ultimately of cognitive neuroscience. According to this model, the self is no longer a metaphysical matter, but a possible subject of empirical investigation.

Philosophical Concepts IIa: Phenomenal Self

One of the problems one encounters is that such substance or meta-representation cannot be experienced as such. Nobody ever experienced a mental substance or a meta-representation in consciousness. We are not conscious of any such substance or meta-representation. Therefore, instead of speculating about something that lies beyond the scope of our experience, why not start with experience itself, and thus with consciousness. Rather than looking at what lies 'outside' our consciousness, like a substance or meta-representation, the self may be found within that very consciousness itself.

However, this localisation is denied in phenomenological philosophy precisely because it focuses on consciousness itself and what lies 'inside' our experience. More specifically, phenomenological philosophy is interested in investigating the structure and organisation of our experience and thus of consciousness. It focuses on how our experience is structured and organised and reveals phenomenal features as we experience them from the first-person perspective.

How does the phenomenal approach determine the self? Currently, it is argued that the self is an integral part of experience itself (Northoff 2012). The self is always present and manifest in the phenomenal features of our experience, such as intentionality (e.g., the directedness of our consciousness towards specific contents) and qualia (e.g., the qualitative character of our experience; what it is like). Without these features the self would remain impossible.

Consequently, phenomenological philosophers, such as Zahavi (2008), consider the self to be an inherent part of consciousness itself. Here, the self is supposed to be always already accompanied by some kind of consciousness of the external world, even if we are not aware of the self being part of that experience. Phenomenological philosophers therefore speak of what they call pre-reflective self-awareness (or pre-reflective self-consciousness).

The concept of pre-reflective self-consciousness contains two main terms, 'pre-reflective' and 'self-consciousness'. 'Pre-reflective' means that the experience of the self does not stem from any reflection or cognitive operation. Instead it is already always there as an unavoidable part of our experience as such. The self is thus pre-reflective. It is simultaneously an inherent part of our experience and thus of our consciousness. The self is consequently no longer outside of our consciousness, but an integral part of it, hence the second term, 'self-consciousness'. Such an approach suggests an intimate and intrinsic link between self and consciousness.

Characterising the self in terms of self-consciousness implies a significant shift. The self is no longer metaphysical as Descartes proposes. Nor is it empirical as advocated by Hume and others such as Metzinger and Churchland. Instead, the self is part of experience and of consciousness itself, and can therefore be characterised as the 'phenomenal self'. Such a phenomenal self is open to systematic investigation of the phenomenal features of our experience, which would complement the metaphysical, empirical and logical approaches to the self.

Philosophical Concepts IIb: Minimal Self

How can we describe the pre-reflective self-consciousness in more detail? It is always already there in every experience so that we cannot avoid it or separate it from the experience. The self is always present in our consciousness and thus in our subjective experience. Even if we do not focus on the self as such, we cannot avoid or remove its presence. Hence, the term pre-reflective self-conscious describes an implicit or tacit experience of our self in our consciousness.

Since the self as pre-reflectively experienced is the basis of all phenomenal features of our experience, it must be considered as essential for any subsequent cognitive activity. Such a basic and fundamental self occurs in our experience before any reflection. For instance, when reading the lines of this book, you experience the contents and, in addition you also experience your self as reading these lines.

Hence, your immediate experience and consciousness comes with both the content and your own self. Since the experience of such self occurs prior to any reflection and recruitment of higher-order cognitive functions. This is why this concept of self is sort of a minimal version of the self. Current phenomenological philosophers such as Gallagher (2000) or Zahavi (2008) speak therefore of a 'minimal self' when referring to the self as implicitly, tacitly and immediately experienced in consciousness.

How can we describe the concept of the 'minimal self'? The minimal self refers to a basic form of self that is part of any experience. As such, it is not extended across time like it is in the experience of the self as a continuity across time in personal identity. Instead the minimal self describes a basic sense of self at any particular given moment in time, but does not yet provide a link between different moments in time and thus continuity across time.

How can such continuity across time be constituted? Cognitive functions such as memories and autobiographical memories in particular may be central. In this model, the self may become more complex. One might speak of a cognitive, extended or autobiographical self, as does, for example, the Portuguese-American neuroscientist, Damasio (see, e.g. Damasio 1999, 2010).

Another important feature of the minimal self is that although we experience it, we may not be aware of it as such. This means that we might not be able to reflect upon it in order to gain knowledge of it. We are, to put it in technical terms, only pre-reflectively aware of the minimal self. In contrast to such pre-reflective awareness, there is no reflective awareness of the minimal self. How can we become reflectively aware of the minimal self? For that to be possible the different moments or points in time need to be integrated and, as philosophers say, represented. For such representation to occur, cognitive functions are needed which make it possible to put and link together the different time points.

Finally, the minimal self may also occur prior to verbalisation and thus linguistic expression. Rather than being tied to specific linguistic concepts, as is the case with more cognitive concepts of the self, the minimal self must be considered pre-linguistic. It is an experience, a sense of self that can barely be put into concepts. We can experience it as self but are not really able to describe these experiences in terms of concepts and thus articulate them in a linguistic way.

Thus, the minimal self is pre-linguistic and pre-conceptual and will therefore, speculatively, not be affected by second language acquisition. It is the kind of experience, an implicit sense of self, which most likely subjects will take with them as more or less stable when moving to a new country where they have to acquire a new language. However, at the same time, the minimal self provides the essential basis upon which more cognitive forms of self are developed. These are then central and instrumental in providing the ability to learn a second language.

Philosophical Concepts IIc: Social Self

How does the self interact with other selves? So far, we described the self in an isolated and purely intra-individual way. However, in daily life, the self is not isolated from others but always related to other selves. This is called inter-individualism rather than intra-individualism. This raises questions about what is described as the 'problem of other minds' or, more generally, questions concerning intersubjectivity. Here we will here give a brief description of the problem of intersubjectivity.

How can we make the assumption of attributing mental states and thus self and mind to other people? Philosophy has long relied on what is called the 'inference by analogy'. What is the 'inference by analogy'? 'Inference by analogy' goes like this. We observe person A to show the behaviour of type X. And we know that in our own case the same behaviour X goes along with the mental state type M. Since our own behaviour and that of the person A are similar, we assume the other person A to show the same mental state type M we experience when exhibiting behaviour X.

What kind of inference do we draw here? There is similar or analogous behaviour between ourselves and the other person. In addition, my own behaviour is associated with a particular mental state. Since now the other person shows the same behaviour, I infer that she also shows the same mental state as it is associated with my own behaviour. Hence, by indirect inference and analogy via our own case, we claim to obtain knowledge of the other person's mental state. How can we make such inference? We may make it on the basis of our own mental states and their associated behaviour. And what we do may also hold true for the other person who in the same way attributes mental states to us by inferring them from the comparison between our behaviour and their own mental states.

Why do we make such inferences? Because it seems to be the easiest and best way for us to explain the other people's behaviour. The assumption of mental states thus seems to be the best explanation for your behaviour. The 'inference by analogy' may thus be considered an inference to the best possible explanation.

The inference by analogy describes intersubjectivity in a very cognitive and ultimately linguistic way when attributing mental states and a self to other persons.

There might, however, be a deeper level of intersubjectivity. We also feel another person's mental states, as when sharing the emotional pain experienced by one's spouse when her father died. Such sharing of feeling is described as empathy and sheds light on a deeper pre-cognitive and pre-verbal dimension of intersubjectivity. This has been emphasised especially in phenomenological philosophy (see, for instance, Zahavi 2008).

However, both empathy and the attribution of mental states to another person are puzzling: despite that the fact that we do not experience the other's mental states and consciousness, we nevertheless either share them (as in empathy) or infer them (as by analogy). We have no direct access to other persons' experience of a self and its mental states in first-person perspective and nevertheless share their mental states and assume that they have a self. How is that possible?

This is where we need to introduce yet another perspective. There is first-person perspective—tied to the self itself and its experience or consciousness of objects, events or persons in the environment. Then there is the third-person perspective—this perspective allows us to observe the objects, events or persons in the environment from the outside, rather than from the inside. The picture is not complete.

What is the second-person perspective? The second-person perspective has initially been associated in philosophy with the introspection of one's own mental states. Rather than actually experiencing one's own mental states in first-person perspective, the second-person perspective makes possible to reflect and introspect about one's own mental states. An example of this is when you ask yourself whether the voice you heard was really the voice of your good friend (see also Schilbach et al. 2013).

The second-person perspective thus allows us to put the contents of our consciousness as experienced in first-person perspective into a wider context, the context of oneself as related to the environment. In other words, the second-person perspective makes it possible to situate and integrate the purely intra-individual self with its first-person perspective into a social context. This transforms the intra-individual self into an inter-individual self. Another way of thinking of second-person perspective is to call this concept of the self, the 'social self'.

How can we define the concept of the social self? The concept of the social self describes the linkage and integration of the self into the social context of other selves. This shifts the focus from experience or consciousness in the first-person perspective to the various kinds of interactions between different selves as associated with the second-person perspective. As we have already indicated, there may be different kinds of social interactions including affective pre-cognitive and more cognitive ones that involve meta-representation, as described above.

Neuroscientific Findings Ia: Spatial Patterns of Neural Activity During Self-specific Stimuli

How can we relate the various philosophical concepts of the self to the neuroscientific findings of self-reference? Above, we discussed that psychology, and later neuroscience, quantified the self in terms of the self-reference effect (SRE). The SRE describes the different impact of self-referential and non-self-referential stimuli on psychological (e.g., reaction time, recall, etc., see above) and neural (e.g., degree of activity, regions, etc., see below) measures. Below we want to briefly highlight some of the main findings of recent imaging studies on the self-reference effect.

What results did the various imaging studies yield in the fMRI? Two different kinds of regions showed up. First, one could see that the regions specific for the respective domains like emotions or faces were recruited. For instance, there is a region in the back of the brain that processes specifically faces (as distinguished from, say, houses), this is called the fusiform face area. This region is obviously active during the presentation of faces, no matter whether it is one's own face or another person's face. Importantly, clear differences between self- and non-self-specific stimuli could not be observed in these domain-specific regions in most studies (Northoff et al. 2006).

What about other regions that are not specific to particular domains (also known as domain-independent regions) involved in the neural processing of the self? Meta-analyses of the various studies demonstrated the involvement of a particular set of regions in the middle of the brain. These regions include the perigenual anterior cingulate cortex (PACC), the ventro- and dorsomedial prefrontal cortex (VMPFC, DMPFC), the supragenual anterior cingulate cortex (SACC), the posterior cingulate cortex (PCC) and the precuneus. Since they are all located in the midline of the brain, they have been coined 'cortical midline structures' (CMS).

The self-specific stimuli—those that were personally relevant for the subjects—induced higher neural activity in these regions than non-self-specific stimuli, or those that remained irrelevant und unrelated to the person. This was observed in the various domains for faces, trait adjectives, movements/actions, memories and social communication. Therefore, the CMS seem to show a special significance to the self and self-reference.

However, there is also some differentiation within the CMS. The self-specific stimuli may be presented in different ways to the subject in the scanner. If subjects have to make judgements requiring cognitive involvement, the dorsal and posterior regions such as the SACC, DMPFC and PCC are recruited to a stronger degree. If, in contrast, stimuli are merely perceived without any judgement, and thus without any cognitive component, the ventral and anterior regions such as the VMPFC and PACC were highly involved (Fig. 2).

This led to the assumption that the different regions mediate different aspects of self-reference. The ventral and anterior regions, such as the PACC and VMPFC, may be more involved in the representation of the degree of self-reference in the

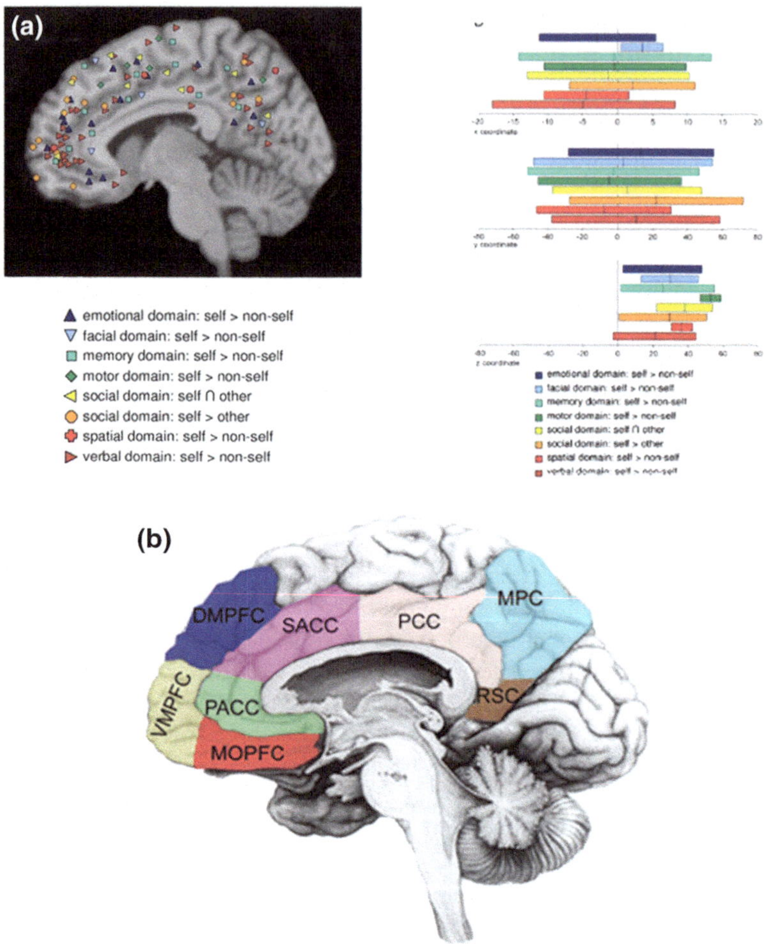

Fig. 2 The figure demonstrates the results of a meta-analysis on imaging studies of self-reference (Northoff et al. 2006) (**a**) and anatomical illustration of the midline regions (**b**). **a** *Top* The figure on the *top left* depicts all the imaging studies on the self as plotted in their obtained location on one brain. This includes self-referential stimuli in various domains or functions like memory, social, spatial, etc., as indicated in the lower text with the colours as shown above and on the right. On the *top right* three different coordinates (*x*, *y*, *z*) are shown that determine the direction (medial-lateral, inferior-superior) of the location in the brain. One can see that all studies locate in the midline regions of the brain (*top left image*) as seen in the *x*-coordinates that describe the medial-lateral location (*top right image*). **b** *Bottom* This figure shows the anatomical regions in the midline of the brain. *MOPFC* Medial orbital prefrontal cortex, *PACC* perigenual anterior cingulate cortex, *VMPFC*, *DMPFC* ventro- and dorsomedial prefrontal cortex, *SACC* supragenual anterior cingulate cortex, *PCC* posterior cingulate cortex, *MPC* medial parietal cortex, *RSC* retrosplenial cortex

stimulus. However, dorsal regions, such as the SACC and the DMPFC, may be related to monitoring and reflection of the stimulus and its self-reference when we become aware of the stimulus as self-specific.

Finally, the posterior regions, such as the PCC, may be implicated in integrating the stimulus and its degree of self-reference into the autobiographical memory of the respective person. These regions seem to be implicated in the recall and retrieval of especially personally relevant and autobiographical information from the past of that person. Thus, it can be concluded that specific regions in the midline of the brain, the cortical midline structures, seem to be involved in the neural processing of self-reference or attributing personal relevance or self-relevance to stimuli.

Neuroscientific Findings Ib: Temporal Patterns of Neural Activity During Self-specific Stimuli

In addition to the spatial patterns of self-reference, its temporal patterns have also been investigated using the EEG. Again self-specific and non-self-specific stimuli have been compared with each other while the subjects undergo EEG measurement. This revealed early changes during self-specific stimuli at around 100–150 m after stimulus onset.

More specifically, self-specific stimuli induced different electrical activity changes already at 130–200 m after their onset when compared to non-self-specific stimuli. This was accompanied by later changes at around 300–500 m. Hence, the temporal pattern between self- and non-self-specific stimuli shows both early and late differences.

In addition, different frequencies of neural activity were investigated. The neural activity oscillates rhythmically in different frequency ranges in the fluctuations of the neuronal activity.

One frequency often induced by stimuli are gamma frequencies in the range of 30–40 Hz. Interestingly, some EEG (and MEG) studies observed higher power in the gamma range in anterior and posterior midline regions during self-specific stimuli than non-self-specific stimuli. The question though is whether such increase in gamma power is specific to self-specific stimuli since it can also be observed in other functions independent of self-reference (see below).

Neuroscientific Findings Ic: Social Patterns of Neural Activity During Self-reference

How can we investigate the earlier described social nature of the self? Various studies have been conducted to investigate different kinds of interaction between different selves. Pfeiffer et al. (2013) and Schilbach et al. (2012) distinguish two different methodological approaches. One investigates social cognition, the cognition of mental states in other people, from third-person perspective. Here, social cognition is investigated in an 'offline' mode. More recently this 'offline' methodological strategy has been complemented by an 'online' mode. In the 'online mode', social interaction is investigated from the 'inside', by taking on the perspective of the interacting selves (rather than the observer's point of view).

Besides conducting several studies, the same group has recently investigated the neural overlap between emotional processing, resting-state activity, and social-cognitive processing (Schilbach et al. 2012). They conducted a meta-analysis including imaging studies from all three kinds of investigations, resting-state, emotional and social-cognitive. In a first step, they analysed the regions implicated in each of the three tasks. This yielded significant recruitment of neural activity in especially the midline regions like the ventro and dorsomedial prefrontal cortex and the posterior cingulate cortex (bordering to the precuneus). In addition, neural activity in the temporo-parietal junction and the middle temporal gyrus was observed.

In a second step they overlaid the three tasks, emotional, social-cognitive, and resting-state, in order to detect commonly underlying areas. This indeed revealed the midline regions, the dorsomedial prefrontal cortex and the posterior cingulate cortex, to be commonly shared among emotional and social-cognitive tasks and resting-state activity. Based on this neural overlap, the authors concluded that there may be an intrinsically social dimension in our neural activity which might be essential for consciousness of both our own self and other selves. If this is true, it will have radical consequences, not only for the concept of the self, but also for consciousness in general.

Neurophilosophical Discussion Ia: Different Forms of Specificity

So far we have covered philosophical approaches to the concept of the self. We also discussed neuroscientific findings about self-reference. Now, the question is how both philosophical concepts and neuroscientific findings are related to each other. *This requires what one may describe as a neurophilosophical discussion. A neurophilosophical discussion directly relates empirical findings in neuroscience to concepts in philosophy.*

How are the neuroscientific findings about self-reference related to the philosophical concepts of self? Are the philosophical concepts of self empirically plausible and thus compatible with the neuroscientific findings of self-reference? In order to address these questions, one should start by investigating the degree of specificity for the self of the neuroscientific findings. One can thus speak of neuronal specificity of the cortical midline structure for the self. The concept of neuronal specificity describes the quest for the exclusive association of a particular

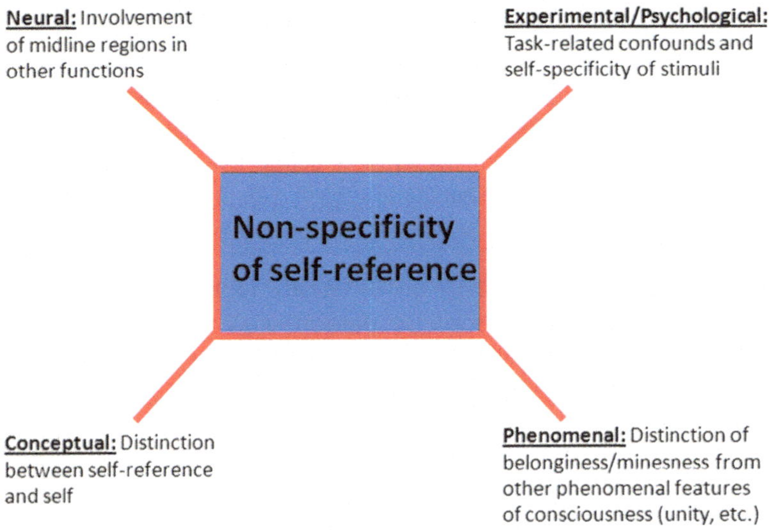

Fig. 3 Non-specificity of self-reference in current imaging studies. The figure shows different domains (neural, experimental/psychological, conceptual, phenomenal) where current imaging studies on self-reference suffer from non-specificity. *Upper left* There is neural non-specificity because the often-observed midline regions are also implicated in functions other than self-reference (as, for instance, in mind-reading, emotion, autobiographical memory, etc.). *Upper right* There is experimental/psychological non-specificity because presentation of self-referential stimuli is often associated with a task like judgement yielding task-related confounds. Moreover, the self-specificity of the stimuli may be confounded by other aspects of the stimuli. *Lower left* There is conceptual non-specificity because the studies do not distinguish between self-reference (of tasks and stimuli) and the self itself in their experimental paradigms. They infer from self-reference to the self which though is an inference between two different concepts that are not identical and do not imply each other. *Lower right* There is phenomenal non-specificity because the experiential, i.e., phenomenal features characterising the self, e.g., mineness/belongingness are not properly distinguished from the ones associated with consciousness in general, e.g., unity, qualia, etc

neuronal measure, like the activity of a certain region or network, with exclusively one specific function.

And they may also be discussed in the context of psychological functions associated with the self and thus psychological specificity. Furthermore, one may question the ability of the experimental designs and measures to really tap into the self. This is called experimental specificity (Fig. 3).

One may also raise the question whether the results really reflect the experiential and thus phenomenal features related to the self. Experience may, for instance, be confounded by features that are not directly related to one's own self. One may thus want to speak of phenomenal specificity. Finally, one may want to discuss how the results relate to the different concepts of the self and whether they correspond exclusively to one specific concept. If they do, this would imply conceptual specificity.

Neurophilosophical Discussion Ib: Neuronal Specificity of Midline Regions

Let's start with neuronal specificity. The concept of neuronal specificity describes whether the spatial and temporal patterns of neural activity observed in studies about self-specificity are really specific to the self. We roughly distinguished two kinds of different regions, the domain-specific regions and the domain-independent regions.

Domain-specific regions are those that are related to the processing of a content in specific modality (e.g., sensory) or domain (e.g., verbal, sensory, motor). Depending on the stimuli and/or the task, domain-specific regions were activated in the above described imaging data.

Are these domain-specific regions specific for the self? No, because the imaging data show that the very same regions are also recruited when applying stimuli that are not related to the self at all. For example, you are shown a house in Brazil. For you, a resident of Canada (and has no connection to Brazil), this image has no degree of self-specificity or self-relatedness to you. It nevertheless activates your fusiform face area.

The self-specialist may now want to argue that at least the degree of neural activity in the fusiform face area or other domain-specific areas may be different between self- and non-self-specific stimuli. However, empirical data are not clear. While some studies report some difference, though small in sensory regions, the majority of studies did apparently not observe differences between self- and non-self-specific stimuli in these domain-specific regions. Hence, it seems as if the domain-specific regions like the sensory and motor cortex remain unspecific for the self. This implies neuronal unspecificity.

What about the domain-independent regions like the cortical midline structures? There has been much discussion whether these regions are specific for self-specific stimuli as distinguished from non-self-specific stimuli. Is the self 'located' in the midline regions? Initial enthusiasm was in support of the theory that the midline regions are specific for the self, recent investigations implicated the same set of regions in a variety of different functions.

Let me be more specific. Tasks requiring the need to understand other people and their mental states—mind-reading as described in theory of mind in psychology—strongly recruit the midline regions. Emotional stimuli and emotional tasks also led to strong activation in the midline regions. In addition, various kinds of social tasks that require social exchange and reciprocity, also recruit these regions. Finally, day dreaming or mind-wandering and other forms of introspection also recruit these regions.

The involvement of the midline regions in various functions other than self-reference sheds some doubt on the neuronal specificity of the midline structure for the self. Hence, even the domain-independent regions like the midline regions do not seem to show any specificity for the self.

The same diagnosis of neuronal unspecificity also is true of the reported gamma synchronisations. Gamma synchronisation is not specific to the self but has been observed in a variety of different functions, including sensorimotor, working memory, attention and episodic memory retrieval. Hence, there is neuronal unspecificity in both a temporal and spatial sense with regard to the self.

Neurophilosophical Discussion IIa: Psychological and Experimental Specificity

Most of the FMRI above compared self- versus non-self-specific stimuli, such as a grand piano for a professional pianist compared to a saw for a carpenter. In addition to the mere perception, subjects were required to make a judgement after each stimulus, to judge whether it was self- or non-self-specific. This raises questions about what exactly the study is measuring—the perception or the judgement of the stimulus? Is it capturing the effect of the stimulus itself, or the task related to that stimulus?

Most likely the results reflect a mixture between stimulus- and task-related effects. This, therefore, casts some doubt on whether the midline regions show psychological specificity for the self. The judgement about self-specificity requires various cognitive functions such as attention, working memory, judgement and autobiographical memory retrieval. What about when research investigates the self in relation to more basic functions such as movements and actions? Even when subjects perform some motor tasks, we face the same confusion of different functions: the self's components, such as ownership (my own movement), as well as agency (whether I am the agent of the movement), may be confounded by the neural mechanisms underlying the execution of the movement/action by the person.

Such psychological unspecificity highlights the need in neuroscience to specify the experimental design and measures. We need measures that are specific to the self as distinguished from the various associated sensorimotor, affective and cognitive functions. We also need experimental designs to segregate stimulus-related effects and task-related effects. For example, we might do this by spacing perception and judgement temporally apart from each other.

Neurophilosophical Discussion IIb: Self-specificity and Other Functions

We also need to discuss the relationship between the self and other functions. Recent imaging studies demonstrated strong neural overlap between the self and reward, the self and emotions, and the self and decision-making. For example, when receiving a reward in relation to a specific stimulus, such as money, regions of the reward system like the ventral striatum (VS) and the ventromedial prefrontal cortex (VMPFC) become active (Northoff 2012). These same regions are also active when the same stimulus is conceived of as self-specific, rather than non-self-specific by the respective subject.

The same effects can be observed in emotions where emotional and self-specific stimuli have been shown to overlap in the anterior midline regions especially. Finally, the same effect can be observed in decision-making: if external cues are provided when making a decision (such as a higher or lower price of the same kind of apples), lateral cortical regions become active. If, in contrast, no such external cues are provided, we need to come up with some internal criterion to guide our decision about which apples to purchase (Nakao et al. 2012). Such internal criterion can only stem from our self. Studies comparing both kinds of decision-making show predominant involvement of the midline regions in internally guided when compared to externally guided decision-making (Nakao et al. 2012).

Together, this neural overlap between the self and other functions, such as reward, emotions and decision-making, raises questions about the relationship between them. Different models could be imagined. Self- and self-specificity could be an independent function, just like attention, working memory, emotion, sensorimotor and so on. However, in that case, one would expect specific regions in the brain and specific psychological functions to subserve specifically, and exclusively, self-specificity. However, at this point in time, this cannot be supported empirically.

Finally, one could also suggest that self and self-specificity are basic functions that underlie and provide the basis for all other functions—sensorimotor, affective, cognitive, social. In this sense, self and self-specificity would occur prior to the recruitment of the other functions. Self-specificity would then always be present, making its involvement and manifestation in the various functions unavoidable. Rather than searching for self-specificity in relation to specific functions, such as language, one would then need to look for more basic functions that must occur prior to sensorimotor, affective or cognitive functions.

One could, for instance, imagine that the strong involvement of the self in language acquisition requires the recruitment of midline regions. Such involvement of midline structures may be implicitly presupposed in many of the tasks or paradigms described above when presenting self- and non-self-relevant words, known as trait adjectives. While the linguistic tasks themselves seem to involve the lateral cortical regions more, their degree of activity may nevertheless be dependent upon the midline regions and their high resting-state activity. Hence, future studies should investigate the relationship between midline regions and the lateral networks implicated in language, which psychologically may correspond to the relationship between self and language.

Neurophilosophical Discussion IIIa: Phenomenal Specificity of Self-reference

The assumption of self-specificity brings us back to the concept of the self as a 'minimal self' (see above). To recap, the minimal self describes a basic sense of self that occurs immediately and is always already part of our experience of the world. The question now is how the concept of the minimal self is related to the neuroscientific results discussed above. To answer this question, we briefly have to shed light on the experience of the minimal self as manifest in pre-reflective self-consciousness.

Various phenomenal features such as qualia and first-person perspective characterise our consciousness. If the minimal self is part of any experience (rather than being outside of it), the self should be manifest in these phenomenal features too. What experiential and thus phenomenal features does the self add? One may assume that the self, first and foremost, makes possible the generation of qualia. Without self, there is no point of view and hence no qualitative features in our experience.

Phenomenological philosophers assume that the special contribution of the self consists in what they describe as 'belonginess' or 'mineness' (Gallagher 2000; Zahavi 2008): the contents of our experience are experienced as belonging to a particular self; they are experienced as mine. For instance, I experience my friend's laptop on which I write for a while as my laptop though I do not own it. This goes along with an experience of a feeling of belong thus being related to my self. However, such experience is not possible for the person sitting beside me who, though looking at the same laptop, does not experience any relation to the self. Instead, s/he may experience mineness or belonginess of the CD lying beside the laptop because s/he is a composer and it is a CD of her/his work.

This relation to the own self is particularly important when one needs to acquire a second language. The foreign language will appear as totally strange, as having no relation to one's own self and thus no self-relevance will be detected in any of the words. Why? Because none of the new words are yet associated with any experiences in specific contexts and situations.

The words thus do not yet elicit any sense of relation to the self. However, once one immerses oneself more and more into a new culture or learning context and gains new experiences, the novel words will become associated with self-relevance, thus inducing a sense of self. In short, the novel language will increasingly become associated with one's own self and become part of it. It is to be supposed that this self-relevance of language may facilitate the acquisition and learning of the new language.

Neurophilosophical Discussion IIIb: Phenomenal Specificity and Phenomenal Limits

In order to account for phenomenal specificity, neuroscience needs to demonstrate which neuronal mechanisms underlie the experience of mineness and belonginess. We also need to distinguish those that underlie other phenomenal features of experience, including intentionality, unity, first-person perspective, qualia and spatiotemporal continuity.

One would therefore require distinct experimental measures and designs for each of these phenomenal features. Only then would we be able to achieve phenomenal specificity and to clearly distinguish the phenomenal or minimal self from phenomenal consciousness. In short, we need to experimentally distinguish between self- and non-self-specific phenomenal measures.

However, the phenomenological philosopher may want to raise the following question: is such phenomenal specificity with the experimental distinction between self and non-self-specific phenomenal measures really possible at all? The minimal self is considered part of the experience and thus of consciousness more generally. Any consciousness of the world goes along with an experience of the self in a pre-reflective way. And the opposite holds true too. Any experience of the self is part of an experience of the world. Both the experience of self and experience of world are thus intrinsically linked.

What does this intrinsic link between the experience of self and the experience of the world imply for the phenomenal specificity of the self? It means that we will remain unable to properly and clearly segregate experimental measures for the minimal self from those of our experience in general. Why? Because these phenomenal features are always already 'infected' by the self—they are encoded and ingrained into the self. Hence, the requirement of maximal experimental and phenomenal specificity may have reached its phenomenal limits. If so, we may be forced to acknowledge that there may be principal limitations to what we can and cannot investigate experimentally when it comes to the minimal self.

Neurophilosophical Discussion IVa: Minimal Self and Body

What about self and body? We experience our own body as our own body. The body plays a central role in the concept of the minimal self in a dual way. First, because of its basic and minimal character, the minimal self is supposed to be subserved by functions of brain and body. The sensorimotor functions of the body that link it with the environment are thus supposed to yield and constitute the minimal self. Phenomenologically-minded neuroscientists therefore consider the minimal self to be embodied and embedded. This leads us to the characteristic feature of the body, namely, that it can be experienced in consciousness. The body is not only an objective body that can be observed from a third-person perspective.

This is the body the neuroscientist and the doctor investigate. It can also be experienced from a first-person perspective. This is the body we consciously experience, also known as 'lived body'.

The lived body is my body as distinguished from others' bodies. Hence, we experience the lived body in relation to our self—in terms of mineness and belongingness. Thus, the experience of the body, the lived body, may be regarded as the first and most fundamental manifestation of the phenomenal or minimal self. Our self in its most basic and minimal form is thus essentially a bodily self.

This relationship to the self is also reflected in what we described earlier as ownership and agency. Ownership describes the fact that I experience my body as my body, rather than some other body. Neuroscientifically, the ownership of the body has been associated with neuronal activity in specific regions of the brain such as the sensory cortex and the parietal cortex. The parietal cortex mediates the spatial position of the body in the world.

Agency is the experience that it is I, rather than some other person, that causes action and movement. I, my self, am the agent of the lines I am currently writing here on my laptop. Neurally, regions such as the premotor cortex and the motor cortex have been associated with agency; these are regions that are implicated in generating movement and action in general.

How is the experience of such bodily self mediated? In determining this, sensorimotor function is considered central, especially the coordination and integration between sensory and motor circuits in the brain. For instance, when generating an action and movement, a copy of such sensorimotor coordination, a so-called efference copy that signals a forward model, is sent to the sensory cortex.

Why? By receiving an efference copy of the intended and to-be-performed action, the sensory cortex can prepare itself to and thus 'anticipate' potentially incoming sensory stimuli. It can thus predict more easily the next sensory state on the basis of what the motor cortex is currently doing. Through this process sensory and motor functions become intrinsically linked together and provide the integration of the body within the environment, known as embeddedness.

Neurophilosophical Discussion IVb: Body and Proto-Self

Sensorimotor functions are not only mediated by cortical regions such as the motor and the sensory cortex. In addition, they are already processed in subcortical regions like the periaqueductal grey, the superior and inferior colliculi, and the basal ganglia like the pallidum, the caudate and the subthalamic nucleus. In addition to the sensorimotor functions, these regions are central in regulating and controlling the vegetative and thus the inner visceral or homeostatic functions of the body. This in turn is central in eliciting emotions.

How are these regions related to the minimal self? Investigations that did not include a strong cognitive or task-related component like a judgement (see above) demonstrated neural activity in these subcortical regions during self-specific stimuli. Because of their involvement in various functions, these regions are definitely not specific for self-referent stimuli. What this shows is that they nevertheless participate in constituting a self, a minimal or phenomenal self.

Current neuroscientists like Jaak Panksepp and Antonio Damasio do, therefore, speak of a bodily self or 'proto-self' that occurs prior to the minimal or phenomenal self. They call this the 'core or mental self'. These subcortical regions seem to coordinate and integrate map the inputs from the body at each moment in time. This allows to represent the body as one's own body in a most basic way. Panksepp (2004) goes so far as even to characterise the term 'self' as 'Simple Ego-type Life Form', to indicate the basic and most fundamental nature and relevance of the self for the body as a biological organism.

Neurophilosophical Discussion Va: Difference Between the Concepts of Self and Self-reference

The concept of conceptual specificity focuses on the question: do the neuronal findings really reflect the self or some other function? First, we must see what exactly is investigated in the imaging studies. Remember that all experimental paradigms are based on the self-reference effect. The self-reference effect assumes the distinct processing of stimuli (e.g., items, objects, persons, or events) that are related to the self when compared to those that are not related to the self.

For instance, a picture of Ottawa has a specific self-reference to me since I live here. While to you as a resident of Australia, it has no self-reference whatsoever. Due to such difference in self-reference, the picture of Ottawa will be processed differently by your brain and my brain.

The self-reference effect presupposes the distinction between the self and a specific content to which the self may refer or not. What is experimentally investigated in the above described experimental paradigms is thus not so much the self itself, but rather the degree of reference of a particular content to the self, known as self-reference. Some neuroscientific authors do, therefore, also speak of self-related or self-referential processing that describes the processes that are assumed to constitute the relation of a particular content to the self.

Why is this important? It means that the experimental paradigms do not target and measure the self as such, but rather the degree of the relation of a specific content to the self. For instance, the degree of neural activity in the midline structures reflects the degree of self-reference, rather than the self itself. The same holds true for the gamma oscillations, which at best correspond to the degree of self-reference, rather than to the self itself.

This means that the various empirical findings remain unspecific with regard to the concept of the self itself. They tell us about self-reference as the relationship of particular contents to the self, but not about the self. This means that the empirical findings are conceptually unspecific with regard to the concept of self. How can we resolve this conceptual unspecificity? In order to close the gap between the concepts of self and self-reference, we may need to shift our focus from the self to self-consciousness.

Neurophilosophical Discussion Vb: Self as Brain-Based Neurosocial Structure and Organisation

What does this imply for the self? Our self may be considered as intrinsically linked to the body. This is called embodied self. Furthermore, since it is based on self-reference, our self may also be intrinsically linked to the environment. This is called the embedded and social self. Our self cannot consequently be regarded as an entity located somewhere in the brain and isolated from both body and environment. Instead our self seems to be intrinsically social, as suggested by the advocates of the concept of a social self (see above).

What does this intrinsically bodily and social nature imply for the conceptual characterisation of the self? Our self may be described as structure and organisation, rather than as an entity—be it mental or physical. Such structure and organisation needs to develop through childhood and adolescence with persistent changes even throughout adulthood. Despite all the changes there may also be persistence and continuity across time, which then accounts for what can be described as identity. Identity may describe the persistence and continuity of self over time which, in an exploratory study, has recently been associated with the midline structures and their high intrinsic activity.

We can also see that this concept of self as structure and organisation is embodied and embedded. Hence, the virtual structure of the self spans across brain, body and environment. At the same that very same virtual structure is dependent upon the respective environmental context. Freud's characterisation of the ego as structure and organisation surfaces here in a more specific way as being integrated in body and environment that is embodied and embedded. Put differently, the ego consists in a relation, the one between brain, body and environment, and can thus be determined in an intrinsically relational way. Future investigation might link the different features Freud attributed to the ego to the self.

What, though, do we mean by the concepts of structure and organisation? The structure must be virtual in that it spans across the physical boundaries of brain, body and environment. Does this mean that we have to revert to a mental structure and organisation as distinct from the physical structure and organisation of the brain? No! The results from neuroscience clearly link the self with neuronal processes related to both intra-individual experiences and inter-individual interaction. There is thus a neuronal basis for the distinct aspects of the self within the context of brain, body and environment. We therefore reject the mental characterisation of the structure and organisation that is supposed to define the self.

How can we define the concepts of structure and organisation in a more positive way? One way is to characterise structure and organisation as social. This distinguishes it from mental or physical features. The social characterisation would then be the underlying basis that links and integrates between the purely physical and the purely mental. The self would then be based on the brain, but would also extend beyond it to the body and the environment. This means that conceptually, we need to characterised the concept of the self as brain-based, rather than brain-reductive (as the proponents of the empirical self tend to do). The brain-based nature of the self also excludes both mind- and consciousness-based approaches to the self.

If the social characterisation of the structure and organisation as related to the self is indeed basic and fundamental, one would assume that our brain's neural activity is intrinsically neurosocial: the brain cannot avoid including the social environmental context in the encoding of stimuli into its own neural activity. The neural activity is thus by default neurosocial rather than merely neuronal. This is supported by the above described neural overlap between resting-state activity and the neural activity changes during emotional and social-cognitive tasks.

Whether the brain encodes its neural activity in an intrinsically neurosocial way remains unclear at this point. What is clear is that the exact characterisation of the brain's neural activity will be essential if we are to develop a truly neurophilosophical, brain-based (rather than brain-reductive) and neurosocial (rather than merely neuronal) concept of the self.

Neurophilosophical Discussion Vc: Self, Belief and Valuation

This chapter has so far focused mainly on the self itself and its nature and structure. In contrast, I have left open the capacities of the self. One such capacity of the self is to belief and value. Valuation is a central component in our life. Normative judgement, moral attitudes and, ultimately, many decisions we make in daily life rely and are based on valuation and belief. Psychologically, valuation is closely related to reward, namely to assign reward value to otherwise neutral or value-less stimuli, events and persons.

Where though does the value come from? It may be closely related to the self that links and relates the seemingly neutral and value-less stimuli to itself and makes possible thereby the attribution of value to the stimuli. If so, one would expect a close relationship between self and reward, including their underlying neural correlates. We did indeed conduct imaging studies that directly compared self and reward. Subjects were presented with stimuli upon which they had either to gamble obtaining reward or to judge their degree of self-relatedness (De Greck et al. 2008).

What are the results? We first analysed those regions related to the gambling and thus the reward. This yielded, as expected, the ventral tegmental area (VTA), the ventral striatum (VS), and the ventromedial prefrontal cortex (VMPFC) all well known to typically mediate reward. In a second step, we then investigated the neural activity changes in exactly these reward-related regions during the presentation of the same stimuli when they had to be judged with regard to their self-relatedness. Most interestingly, high self-related stimuli induced high activity in all three—VTA, VS and VMPFC—whereas those stimuli judged by the subjects as low self-related did not induce any activity change in these regions at all. These data illustrate the close relationship between reward and self-relatedness in that the former's regions, VTA, VS and VMPFC, also mediate self-relatedness.

Are reward and self-relatedness identical? De Greck et al. (2009, 2010) conducted the same paradigm in patients with alcohol and pathological gambling. These results show normal reward-related activity in VTA, VS and VMPFC whereas neither high or low self-related stimuli induced activity changes. This suggests that self and reward though being closely related are not identical since otherwise they could not dissociate from each other in their neural activities in VTA, VS, VMPFC.

What do these data tell us about the relationship between self and valuation? One essential capacity of the self is to give and assign value to otherwise value-less stimuli. Though preliminary, the data suggests that this capacity may be closely related to the reward system and its close relationship with self, the neural overlap between both in typical reward regions like VTA, VS and VMPFC. However, neural overlap is not to be confused with neural identity between reward and self since for that empirical evidence is not given. Hence, the self seems to utilise reward-related regions though apparently in a slightly different yet unclear way when compared to reward. This is well compatible with the fact that self and valuation are not identical but closely related as we experience almost on a daily basis. Our self is more than just belief and valuation to which the neuroscientist may add that neurally our self is more than just the reward system.

References

Churchland, P. S. (2002). Self-representation in nervous systems. *Science, 296*(5566), 308–310.

Damasio, A. R. (1999). How the brain creates the mind. *Scientific American—American Edition, 281*, 112–117.

Damasio, A. R. (2010). *Self comes to mind: Constructing the conscious brain*. New York: Pantheon.

De Greck, M., Enzi, B., Prösch, U., Gantman, A., Tempelmann, C., & Northoff, G. (2010). Decreased neuronal activity in reward circuitry of pathological gamblers during processing of personal relevant stimuli. *Human Brain Mapping, 31*(11), 1802–1812.

De Greck, M., Rotte, M., Paus, R., Moritz, D., Thiemann, R., Proesch, U. … Northoff, G. (2008). Is our self based on reward? Self-relatedness recruits neural activity in the reward system. *Neuroimage*, 39(4), 2066–2075.

De Greck, M., Supady, A., Thiemann, R., Tempelmann, C., Bogerts, B., Forschner, L. … Northoff, G. (2009). Decreased neural activity in reward circuitry during personal reference in abstinent alcoholics—A fMRI study. *Human Brain Mapping*, 30(5), 1691–1704.

Gallagher, S. (2000). Philosophical conceptions of the self: Implications for cognitive science. *Trends in Cognitive Sciences, 4*(1), 14–21.

Klein, S. B. (2012). Self, memory, and the self-reference effect an examination of conceptual and methodological issues. *Personality and Social Psychology Review, 16*(3), 283–300.

Klein, S. B., & Gangi, C. E. (2010). The multiplicity of self: Neuropsychological evidence and its implications for the self as a construct in psychological research. *Annals of the New York Academy of Sciences, 1191*(1), 1–15.

Metzinger, T. (2004). *Being no one: The self-model theory of subjectivity*. Cambridge, MA: MIT Press.

Nakao, T., Ohira, H., & Northoff, G. (2012). Distinction between externally vs. internally guided decision-making: Operational differences, meta-analytical comparisons and their theoretical implications. *Frontiers in Neuroscience, 6,* 31. doi:10.3389/fnins.2012.00031.

Northoff, G. (2012). Immanuel Kant's mind and the brain's resting state. *Trends in Cognitive Sciences, 16*(7), 356–359.

Northoff, G., Heinzel, A., De Greck, M., Bermpohl, F., Dobrowolny, H., & Panksepp, J. (2006). Self-referential processing in our brain—A meta-analysis of imaging studies on the self. *Neuroimage, 31*(1), 440–457.

Panksepp, J. (2004). *Affective neuroscience: The foundations of human and animal emotions*. Oxford University Press.

Pfeiffer, U. J., Timmermans, B., Vogeley, K., Frith, C. D., & Schilbach, L. (2013). Towards a neuroscience of social interaction. *Frontiers in Human Neuroscience, 7,* 22.

Schilbach, L., Bzdok, D., Timmermans, B., Fox, P. T., Laird, A. R., Vogeley, K., et al. (2012). Introspective minds: Using ALE meta-analyses to study commonalities in the neural correlates of emotional processing, social & unconstrained cognition. *PLoS ONE, 7*(2), e30920.

Schilbach, L., Eickhoff, S. B., Schultze, T., Mojzisch, A., & Vogeley, K. (2013). To you I am listening: Perceived competence of advisors influences judgment and decision-making via recruitment of the amygdala. *Social Neuroscience, 8*(3), 189–202.

Zahavi, D. (2008). *Subjectivity and selfhood: Investigating the first-person perspective*. Cambridge, MA: MIT Press.

The Self and Its Good Vary Cross-Culturally: A Dozen Self-variations and Chinese Familial Selves

Owen Flanagan and Wenqing Zhao

How Ought One to Live?

Anglophone ethics asks this question: how ought one to live? It often answers in a way that is parochial in two respects. First, it limits its answers to the philosophy that is the lineage of a 'series of footnotes to Plato' (Whitehead 1979). Second, and relatedly, in so far as it attends to actual moralities, it attends to WEIRD ones, to research on WEIRD people (Western Educated Industrialised Rich Democratic). But there is every reason to think WEIRD people are unrepresentative, possibly the most unrepresentative group imaginable, less representative than our ancestors when the ice melted at the end of the Pleistocene era (Henrich et al. 2010).

Even if questions about how one ought to live are culturally universal, good answers to it are not. One reason is that the self to whom the question and the answer are addressed varies cross-culturally. In this chapter, we review first empirical evidence of what we call 'self-variations', and give examples of a variety of moral ecologies where the differences in self-conception reveal themselves. Second, we provide an extended analysis of a distinctive conception of the self, its nature, boundaries and fate, which both explains and justifies certain characteristic ways of thinking about goodness and wellbeing in Chinese philosophy. To put it in the simplest terms, Chinese philosophy depicts the metaphysical self as intrinsically connected to a family lineage, which explains why continued respectful relation with one's own family and the deceased ancestors are central components to a human life.

O. Flanagan · W. Zhao (✉)
The Center for Comparative Philosophy, Duke University, Durham, NC, USA
e-mail: zhaowenqing711@gmail.com

S. Menon et al. (eds.), *Self, Culture and Consciousness*,
DOI 10.1007/978-981-10-5777-9_17

A Dozen Self-variations

Here are a dozen self-variations that appear in the literature (Flanagan 2017: 228–230):

(1) '*The Philosopher's Self.*' There is disagreement among professionals, philosophers, psychologists, neuroscientists and theologians, as well as among ordinary people that world over, about what it is that is the subject of experience, what I am, where I am, how I came to be, what my fate is, and how I keep track of all I am, do, as well as all that happens to me. The candidate list is vast and ranges from *psyche*, *anima*, *nephesh*, *nafs* and *atman*, to the human organism, the brain, a part of the brain, to the idea that there is no 'Mind's "I"'. The 'I' is a tempting illusion produced by,—again, take your pick—the stream of consciousness, memory, the speed at which the brain keeps track of and updates information about what this individual organism is doing, or, in certain post-apocalyptic circles, each of us is a simulation in the Matrix run by the Artificial Intelligentsia.

(2) *Porous and non-porous selves*. Clifford Geertz, the anthropologist, writes that the standard Western conception of the person is 'a bounded, unique, more or less integrated motivational and cognitive universe, a dynamic center of awareness, emotion, judgment, and action organized into a distinctive whole and set contrastively both against other such wholes and against a social and natural background' (1973: 57). Social and cultural psychologists (Shweder and Bourne 1982; Markus and Kitayama 1991), contrast Geertz's 'independent' self with the 'interdependent' self. This contrast is also described as one between 'individualist' and 'collectivist' or 'idiocentric' and 'allocentric' cultures, typically North Americans and northern Europeans on the one side, and East Asians, South Asians, Amerindians, and Africans on the other (Triandis et al. 1988).

(3) *Self-Constitution*. North Americans experience their traits of character as in them, inside themselves, as traits they have. 'Wherever I go, there I am.' I am shy, adventurous, creative or whatever exactly the Myer-Briggs personality inventory or the Minnesota Multi-Phasic Personality Inventory says I am. East Asians experience themselves, their traits of character, as less matters of individual character or personality, and as more dependent on particular situations and relationships, as co-constituted relationally (Markus and Kitayama 1991, 2010).

(4) *Storied Selves*. Many psychologists and philosophers—Bruner (2002), Ricoeur (1984) and MacIntyre (1981, 1984), for example—say that selves are essentially narrative, that narrative is the essential genre for self-making, self-understanding and self-representing. But Galen Strawson thinks that some selves might be constitutionally episodic, and that, in addition, the demand of narrativity is regimen of responsibility and accountability imposed by certain kinds of cultures, not a discovery of the way the self is necessarily, or the way responsibility must be conceived. A narrative or diachronic self 'naturally

figures oneself, considered as a self, as something that was there in the (further) past and will be there in the (further) future…If one is Episodic, by contrast … one does not figure oneself, considered as a self, as something that was there in the (further) past and will be there in the (further) future' (2004: 430). There is evidence in anthropology that there is variation in whether and how much selves are diachronic, narrative, and understand themselves to be embedded in either or both a personal and cultural history, or, on the other hand, synchronic, present focused, and more ahistorical (personally and culturally). The Achuar of the Ecuadorean Amazon (Descola 1996) and the Piraha of the Brazilian Amazon (Everett 2009) might fit the episodic bill.

(5) *Virtuous Selves*. The quality of a person and their character, the goodness of a self, is normally, across the globe, thought to be a function of virtue. Even if there are universal virtues, they are ranked differently in different traditions. Compassion (*karuna*) is the highest virtue among Buddhists, whereas justice as fairness holds pride of place among North Americans. Confucians rank family loyalty and filial piety (*xiao*) higher than Americans do, and so on. If a good self is a virtuous one by the lights of a moral conception or tradition, then different ecologies call on us to be different selves.

(6) *Ideal Emotional Selves*. Children's books in different cultures and subcultures —for example, among the people of Japan, Hong Kong and Taiwan, on the one side, and the USA and Canada, on the other, and between traditions such as Christianity and Buddhism—model different faces and emotional states as ideal. American books model 'happy-happy-joy-joy' faces and endorse states that are high-arousal positive (HAP) ones—'excited, enthusiastic and elated'; Buddhist books, on the other hand, model low-arousal positive (LAP) faces, a look of calm, and endorse internal serenity and equanimity (Tsai 2007; Tsai et al. 2007a, b, c).

(7) *Authentic and Inauthentic Selves*. North Americans value a kind of self they call 'authentic', where this involves being true to one's personal values, more highly than do East Asians, who think that a good person often stands down from their personal self-preferences when the collective or tradition demands it. There are many ways to describe this difference, every one of which expresses a clear normative preference for one side or the other: wilful versus self-effacing; authentic versus inauthentic; egotistical versus communal.

(8) *Self-Reference*. When 4- to 6-year-old children in America and China are asked to report on daily events, the proportion of self-reference is three times greater among the Americans. American kids focus on what *I* did and what happened to *me*. Overall, American children report on fewer events, usually only the ones that mattered significantly to them personally, and not neutral or third-party actions, such as that the teacher put all the toys away. The American children also make twice as many references as the Chinese children to their own internal states (not just their own doings and deeds), to what I *wanted*, whether *my desires* were met or thwarted, what *my feelings* and *my emotions* were (Nisbett 2004: 87–88).

(9) *Self-Recognition.* North Americans prefer jobs in which personal initiative is encouraged and individual merit is rewarded and publicly acknowledged, Japanese and Singaporeans do not, and Europeans are intermediate (Hampden-Turner and Trompanaars 1993). North Americans believe that individuals can pull themselves up by their own bootstraps more than individuals in other cultures believe this, and are less likely to think that good or bad luck is a major factor in accomplishment, success and desert (Clark 1997).

(10) *Self-Comprehension.* Indians, Chinese and Koreans are less susceptible than Americans to the *fundamental attribution error,* which involves overrating the causal influence of stable character traits and understanding situational or contextual factors. An American is more likely to infer from an individual act of kindness—for example, giving a dollar to a homeless person—that he himself is a generous person, or that whoever did it (a third party) is generous, whereas an Indian is more likely to appeal to particular features of the situation, such as the behaviour of the homeless person (he asked politely) or his appearance (he looked as if he needed it for food, not for drugs or alcohol) (Miller 1984; Flanagan 1991; Markus and Kitayama 1991; Nisbett 2004).

(11) *Self-Serving Bias.* North Americans are more prone that Amerindians, Mexicans, Fijians, South Italians and East Asians to the *self-serving bias,* thinking one is better across various dimensions—looks, intelligence and so on—than others. Keillor Garrison (1985) calls this the 'Lake Wobegon effect', where everyone is, by their own judgement, in the top 5%.

(12) *Positive Self-Illusions.* North Americans are more prone than people of other cultures to believe that they are in control of outcomes, controlling the sex of babies at insemination, winning lotteries, or that bad things, getting cancer, divorced, in car accidents and so on, won't befall them. This is not sensible, given that they are told the relevant base rates. Technically, positive illusions involve mistakes (they are 'illusions'), but they are good mistakes (the 'positive' part) since people who make them are happier than those who do not (Taylor 1989; Taylor and Brown 1994). In North American populations, only moderately depressed people are realists and do not suffer positive illusions.

A Chinese Self, Chinese Selves?

The dozen self-variations provide a template that any scholar might find useful in analysing, understanding and explaining cultural differences in the self, say Confucianism, and what it means for the self to flourish. We discuss one conception of the Chinese self, which presents an interesting example of self-variation. In the origins of the Chinese tradition, there was no language that directly addresses the boundaries or constitutions of the self in the contemporary philosophical sense.

There were pronouns that refer to 'me' or 'I' and notions such as the body, consciousness, memory, mind, spontaneity and freedom that pertain to the 'self'. But it is hard, if not impossible, to locate a word or phrase that means the 'self'. Philosophers such as Ames and Rosemont (2011) think that the Confucian-Chinese self is role-relational and interdependent. They believe that Confucian ethics envision human beings as relational persons constituted by the roles they bear rather than as individual selves (17). This understanding puts the Confucian-Chinese self directly in the 'interdependent self' (versus 'independent self') category. However, the following questions remain: What variation does the Chinese self present? Does the 'interdependent versus dependent' cultural model of the self (see (2) porous and non-porous self) adequately capture the Chinese variation? Although it is hard to locate the 'self' concept in classical Chinese texts, there have been extensive discussions of related concepts, such as the 'soul'. The word for 'soul' in Chinese is *hun-po*. It is constituted by two things: the *hun* and the *po*.

Two Forms of the 'Soul': *Hun* and *Po*

In the Chinese tradition, it is believed that there are two kinds of 'souls': the *hun* and the *po*. In his article on the conception of the soul in pre-Buddhist China, Yü (1987) argues that the notion of *hun-po* has always been an important part of the indigenous Chinese tradition. Moreover, he argues that in the pre-Buddhist Han Dynasty, the view of *hun-po* in Confucian classics such as *The Book of Rites* (*Liji*) bears a strong resemblance to that found in popular Daoist texts, which suggests that there is a common core of beliefs in souls that transcends the boundaries of philosophical schools as well as class lines (363–364). This does not mean that such common beliefs in souls constitute a unified system of religion that is held by all pre-Buddhist Chinese people. Indeed, Yü (2005) argues that the beliefs in *hun* and *po* have separate origins: *Hun* originated in southern China, while the beliefs in *po* were more popular in northern China. The two belief systems converged and formed a stable, dualistic understanding of the soul in 2nd century BC at the latest (136).

So, what are *hun* and *po*? How are they related to the body, the rest of the world, and each other? It is clear that Chinese people, from pre-Buddhism to today, have systematic beliefs in a 'soul' that governs the human vitality, consciousness, intelligence and emotions. We can find similar ideas in the West in ancient Greek, Jewish, Christian and different folk traditions. However, the Chinese understanding of the 'soul', the *hun-po*, takes a distinctive form: The *hun-po* is understood as the essence of the heart-mind (*xin*心). In *The Commentary of Zuo*, it is said, 'I have heard the following: feeling sorrow for something happy, and feeling happy for something sad, are both called losing one's heart-mind. The essence and vigorousness of the heart-mind, is called *hun-po*. After the *hun-po* leaves, how can anybody last?' (*Zuozhuan: Zhaogong 25*, translated by Wenqing Zhao). The *hun* soul is made of an energy-like material that permeates this world—namely, the *qi*.

The *qi* that constitutes *hun* is derived from heaven (*tian*), and thus the *hun* is ethereal and pure. It governs the mind-spirit, in particular, the vitality, consciousness and intelligence of a human being. As is made clear in *The Book of Rites,* 'The bones and the flesh return to the earth, this is the destiny. However, the *qi* of *hun* can go anywhere, can go anywhere' (*Liji* 2B: 196, translation by Wenqing Zhao). The *po* soul is thought to be intimately connected with the human body (*ti*), which includes the bodily senses as well as emotions. It is not clear that there is a unanimous view about whether the *po* soul is in charge of the body, makes up the body or simply *is* the body. Some think that *po* soul is also made of *qi*. However, the kind of *qi* that makes up the *po* is derived not from heaven but from the earth, which makes the *po* soul heavy and impure (see Yü 2005: 139).

The reason why we call *hun* and *po* two forms of the soul, instead of two parts of the soul, is because that they are souls of different nature and final destinations, and can exist in separation from each other. The following passage from *The Book of Rites* categorises the dualistic nature of the *hun-po*: 'The *qi* of *hun* returns to Heaven, the body of the *po* returns to the Earth. Hence, people seek (for the deceased's souls) at both the Yin and Yang world' (*Liji* 9: 36, translated by Wenqing Zhao). As Yü (2005) points out, the understanding of *hun* as the soul of the 'spirit' and the *po* as the soul of the 'body' prevailed in Han dynasty (202–220 BC) and formed the core of the Chinese belief of soul ever since (140). After a person dies, the *hun* and *po* become separated. The person's *hun* soul, which is light and ethereal, ascends to heaven (*tian*), and his *po* soul, which is heavy and impure, is cast down into the earth (*di*). In this context, heaven corresponds with *Yang*, while the earth corresponds with *Yin*. Ritual sacrifices, which aim to find and connect with the ancestors' souls, thus seek them in both the Yang-Heaven and the Yin-Earth. In particular, people find the *hun* soul in the Yang-Heaven, and the *po* soul in the Yin-Earth.

There are different understandings about how long it takes for the *hun-po* to become separated, whether the *po* soul stays with the body until the body decays. Furthermore, it is unclear whether, after people die, their *hun-po* enters a separate realm. In this light, Hu (1937) argues that Chinese people, prior to Buddhism, did not have a clear idea of an afterworld that is separated from this world (see Yü 2005: 142–143). Moreover, although *hun* (and some people think *po* as well) is clearly made of the intangible and invisible material of *qi*, Joseph Needham thought that the pre-Buddhist conception of *hun-po* and afterlife is 'naturalistic'. He argued that unlike in the Buddhist, Hindu, Islamic and Christian traditions, the early Chinese people never had the concept of an afterworld that is separated from this world. Everything is included within nature (*ziran*) (Needham 1974; Yü 2005). Yü (2005) disagrees with Needham on this point. He thinks that early Chinese people believed in a Heaven Chamber (*tian-ting*) that exists in a separate realm. Moreover, the Heaven Chamber monitors the moral activities of humans on earth. This early understanding made it easy for Chinese people to absorb the Buddhist notion of *karma*, afterlife, and reincarnation into the indigenous belief system (144).

We will not rehearse the debate about whether or not early Chinese people believed in a separate realm of afterlife. It is clear that there is a common core of understanding *hun-po* after death: 1. The *hun-po* leaves the body after the person dies; 2. *hun* and *po* eventually become separated, while the *hun* ascends, the *po* descends; 3. The *hun-po* (or sometimes it is believed to be just the *hun*) can exist in the form of a ghost (*gui*) after they leave a person's body (death of the body); 4. The ghost (some believe it is just the *hun*, while others think it is the unity of *hun* and *po*) can stay for a long time without dispersing; 5. The ghosts gain power from ritual sacrifices and offerings, which allow them to stay behind without dispersing; 6. The ghost can only accept offerings from his or her direct male descendants; 7. If such offerings stop, the ghosts will starve and become weaker throughout time. In the end, the starving ghosts will disintegrate and disappear; 8. Before the notion of reincarnation and rebirth were introduced into China, it was believed that the *hun-po* eventually disperses and loses individuality, and that marks the true end of the person (see Yü 1987, 2005).

The most distinctive parts of the *hun-po* narrative are 5, 6 and 7, that tied the fate of one's soul intrinsically to one's family. A warring song from the pre-Qin era indicates that the *hun-po* and the family are deeply connected, 'The family lineage and the temple of worship are destroyed, and the *hun-po* are lost. Even if we return, we have no one to be with' (*Shuo Yuan: Zhi Wu*, translated by Wenqing Zhao). When the family lineage and the family temple (*Zongmiao*) are destroyed (in this case, by another country through war), the *hun-po* of this lineage will be forever lost. In general, the pre-Qin texts point to the following two connections between the *hun-po* and the family:

1. The more good things (food and possessions) that a person has enjoyed in his lifetime, the stronger his ghost will be. Moreover, the more prosperous and powerful his clan is, the stronger 'root' the ghost has on earth. In *The Commentary of Zuo*, there is a story in which people asked Zichan (–522 BC) whether a lord named Bo You, who died of a wrongful death, could turn into a ghost. Zichan thought,

> Sure. In the beginning of life, the initial transformation is called the *po*, after you have the *po*, the vigor part of you is called the *hun*. If you have consumed enough good things, then your *hun* and *po* become strong. That is how you have the essential energy and vigorousness, and then you can reach a vibrant state of mind … This noble man Bo You is the descendant of Duke Mu, the grandson of Zi Liang, and the son of Zi Er. Although he is only the minister of a poor and small country, his family has been in power for three generations. He must have enjoyed a lot of good things when he was alive, and his clan is big, which means that he is well-rooted. He also died of a violent death. It is very appropriate for him to become a (strong) ghost (*Zuozhuan*: *Zhaogong* 7, translated by Wenqing Zhao).

2. After the person passes away and becomes a ghost, they rely on the sacrifices from the family to maintain their power. Without familial sacrifices, the ghost will starve and perish. This is why in the above-mentioned warring song, after the family temple was destroyed, the ghosts were also forever lost. Yü (2005) affirms Creel's (1937) understanding that ritual sacrifices are thought as a way to feed the ancestors' ghosts in ancient China (140). He cites a passage from *The Commentary of Zuo*, in which a man, fearing that his family will be destroyed in

war, cried, 'Even the ghosts need food! The ghosts of my family will starve!' (*Zuozhuan: Xuangong* 4, translated by Wenqing Zhao. See Yü 2005: 141). In another passage in *The Commentary of Zuo* about the order of sacrifice, it is said, 'The new ghosts are bigger, and the older ghosts are smaller. The correct order of worship is to attend the new ghosts first and the older ghosts second' (*Zuozhuan: Wengong* 2, translated by Wenqing Zhao). This seems to suggest that the size of the ghosts become smaller throughout time. The existing members of the family have an extended obligation (first to serve the ancestors who are directly related to them, and then the ancestors who are more distant from them) when it comes to ritual sacrifice.

So far, we have shown how the *hun-po* was generally understood in the sources of the Chinese tradition. In particular, it is tightly connected to family lineage and ritual sacrifice. However, the ancient texts from pre-Qin and Han era did not offer a systematic answer to why the ancestors' ghosts can only accept offerings from their direct offspring. Moreover, how are the soul and ghost related to the family metaphysically? These questions are fundamental for us to understand the Chinese family-oriented self.

The Soul and the Family: Zhu Xi's Ripple Metaphor

One of the distinct features of the Chinese 'soul' (*hun-po*) is that it is deeply connected with one's family lineage. This family-oriented understanding of the 'soul' has its sprouts in classical texts and early beliefs. Later scholars, such as one of the most prominent metaphysicians in Chinese history, Zhu Xi, developed lengthy discussions on how one's *hun-po* is connected to its ancestors through *qi*. Before going into Zhu Xi's understanding of the *hun-po*, we have to first briefly mention his construction of *qi*. In Zhu Xi's metaphysical system, there are two essential concepts, the *li* and *qi*. The *li* refers to the principle that governs the pattern of things, whereas the *qi* is seen as the material of the things. *Li* exists prior to any physical things or social constructions, and also before the existence of *qi*. This means that for any particular *qi* of things, there exists a principle (*li*) by which it is gathering. Everything, from human souls to excrement of an ox, has *li* in them, which explains how they are patterned, and the form they take.

Following the ideas in the early canons, Zhu Xi further expanded on the notion of *qi* and how it is related to the *hun-po* and the family. In *Classified Dialogues of Master Zhu: Book on Ghosts and Spirits,* he talked about the relationship by using a metaphor of the ripples:

> The student asks, 'After a person dies, does the *hun-po* disperse?' The master says, 'Yes, it disperses.' The student follows up, 'When the descendants offer sacrifices to them, there is a synesthetic resonance between them, how is so?' The master replies, 'After all the descendants are the *qi* of the ancestors. Although the dead man's *qi* disperses, his root is still here. If the descendants try their best to be sincere and reverent, then they can call upon

> the dead man's *qi* to unify where the descendants are. It is like the ripples of water: the water coming from behind is not the leading water; the second wave is not the first wave. They are nevertheless the same ripple. The *qi* of the descendants and the *qi* of the ancestors are just like that. The man's soul disperses after he is dead, however, his root is still here. Since the root is here, the descendants can lead and unify the dead man's *qi* to be where the descendants are. This is hard to say. You should be able to see it naturally' (*Zhuzi Yulei: Gui Sheng Pian*, translated by Wenqing Zhao).

Here Zhu Xi talked about five things: 1. The *hun-po* disperses and loses its individual form after the person passes away; 2. When the descendants perform ancestral worship, there is a sympathetic resonance between them and the ancestors; 3. This can happen because the descendants are made by the same *qi* as the ancestors; 4. The descendants who share the same *qi* with the ancestors are the root of the ancestors' *hun-po* on earth; 5. Through maximal sincerity and reverence, the descendants can call upon the ancestors' *hun-po* to gather again.

Exactly how do the ancestors and the descendants share the same *qi*? Zhu Xi used the ripple metaphor to explain the connection: Imagine when one throws a stone into a lake, the first ripples created are like the first-generation ancestors. This wave takes a distinctive form. The first wave extends and creates the second wave of ripples. The second wave, again, takes a different form from the first wave. In this sense, the water coming from behind is not the leading water; the second wave is not the first wave. They are nevertheless the same chain of ripples that extends across the surface of the lake. If we take Zhu Xi's metaphor a little further, it seems that the waves are connected through a chain of effect: all the waves are directly caused by the wave prior, and the continuity of the ripple depends on the waves to extend themselves into creating one wave after another. In this sense, one is connected with his ancestors and descendants in ways that are metaphysically significant, and different from how he is connected with other people, and other people's ancestors and descendants.

Zhu Xi also explained why the *hun-po* of ancestors can only unify for their direct descendants, and why people should only worship their own ancestors:

> Hou-zhi asks, 'So the ancestors are one unified *qi*, and they gather and disperse because of their descendants' ritual offerings?' The master says, 'This is what Master Shang-cai means by 'there is when it is needed; there is not when it is not needed. It all depends on the people.' Ghosts and spirits are things with foundations. The ancestors share the same *qi*, and there is always a place where they unify. The ancestors' *qi* is where the descendants' bodies lie. There is a bloodline that penetrates and connects. This is why 'the god does not love whom that is not his kind, and the people do not worship whom that does not belong to the same tribe.' All because their *qi* are not related to each other. There are sayings, such as: 'The King offers sacrifices to the Heaven and Earth, the dukes and princes offer sacrifices to the mountains and rivers, other nobles offer sacrifices to the five gods of the household. These things are not their ancestors, but the King is the lord of all-under-heaven, the dukes and princes are the lords of mountains and rivers, and the other nobles are the lords of the five gods of the household. When I govern something, his *qi* unifies upon me. This is the way they connect' (*Zhuzi Yulei: Gui Sheng Pian*, translated by Wenqing Zhao).

In this passage, Zhu Xi made clear that the ancestors are one unified *qi* that is rooted in the body of the descendants. As he said, 'The ancestors' *qi* is at where the descendants' bodies lie. There is a bloodline that penetrates and connects.' It seems that Zhu Xi inherited the classical understanding of the *hun-po* as something not entirely separated from the body. In particular, the *po* soul is comprised of the bodily senses and emotions. Even when a particular individual dies, it continues to be rooted in the bodies of the descendants. Zhu Xi's understanding also elevates the importance of the bloodline to a metaphysical level through the *qi*. In this case, the bloodline is one of the two things (the other thing is a role of governing, such that the King governs all-under-heaven) that allows the *qi* of the subject of ritual sacrifice to gather.

Zhu Xi explained that 'the god does not love whom that is not his kind, and the people do not worship whom that does not belong to the same tribe', because their *qi* are not related to each other. Through the continuity of *qi*, Zhu Xi tried to solve the problem of why the ghosts can only take offerings from their descendants.[1] This is also why one should only perform ritual sacrifice and worship to their own ancestors. The rituals of ancestor worship are ways to engage in sympathetic resonance(*Ganying*)with the souls and spirits. The souls of the ancestors will only benefit from, respond to and unify those with whom they share the same bloodline. It is a complex cognitive process of recognising oneself as part of a specific family lineage. The special cases are when one is a king, a duke or a prince, or a noble man with official ranking. In these cases, one serves a special role as the lord of things, such as the rivers and mountains and the five gods of household. These are the only cases when it is appropriate to offer sacrifices to those who are not one's ancestors.

The Family-Oriented Self

In the preceding two sections, we explained some of the threads that make up the Chinese understanding of the *hun-po*. Following the method of philosophical archaeology, we are approaching the Chinese self through identifying the distinctive notion of the 'soul' (*hun-po*) and concepts such as *qi*, heaven and earth that surround it. As mentioned in the beginning of this article, there are at least 12 different categories of self-variations that philosophers, psychologists and anthropologists claim exist. So far, we have been focusing on the philosophy of the soul in Chinese tradition, which falls under '1. The Philosopher's Self.' In closing, we explain how our understanding of *hun-po*, the Chinese soul, its fate, and its inextricable metaphysical connection to a family lineage helps us understand the Chinese self-variation in '2. Porous and Non-porous Selves' and '5. Virtuous Selves.' Again, the 'family-oriented self' is only one important stream of understanding the self within the Chinese tradition. We do not intend to cover all Chinese selves in this chapter.

---

[1]Of course, in Zhu Xi's understanding, it has to be the direct male descendants, because the family name and the bloodline are eventually carried by the men.

First, the Chinese notion of *hun-po* seems to present a new variation to the '2. Porous and Non-porous Selves.' Let us take a look at the existing model that claims to model the differences between the 'porous' (interdependent) and 'non-porous' (independent) self.

[See Flanagan 2017: 243, and Markus and Kitayama (1991)] In the 'Independent View of Self', the self is seen as having firm, non-porous boundaries among the family members, friends and co-workers. In the 'Interdependent View of Self', the self has porous boundaries and overlaps with 'family' and 'friends'.

In the Chinese case, which draws upon the notion of *hun-po*, the family-oriented self would have porous boundaries with other family members, and overlap with them significantly, but it would maintain clear boundaries with non-family. Even close friends do not enter into the familial self.

As we can see from Fig. 2, the family-oriented self shares porous boundaries with other members of the family. However, only the family is considered as the in-group members. The family-in-group has firm and non-porous boundaries with out-group members. This picture is clearly different from the independent understanding of the self as portrayed in Fig. 1a. A typical example for Fig. 1a is the neo-liberal self, which shares no porous boundaries with any other members of the society, including one's family. On the other hand, Fig. 1b—'Interdependent View of Self'—does not capture some of the distinctive features of the Chinese family-oriented self: Fig. 1b allows 'friends' or even 'co-workers' to take up a similar status as the 'mother', 'father' and 'siblings'. In this sense, the family-oriented self is different from the Buddhist interdependent self that has open, porous boundaries with all sentient beings, which is often used as a typical example of interdependent, porous self. The Chinese family-oriented self shares porous boundaries primarily with members of the family-in-group, and seldom with non-family members.

Fig. 1 Independent self vs. Interdependent self

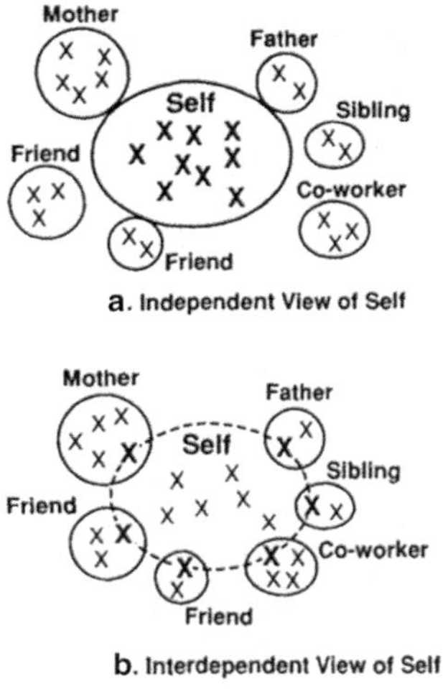

There are the following two ways that the self-variations related to 'porous and non-porous self' are described in the literature of cultural psychology and anthropology: (1) Some people conceive their individual wellbeing in terms of how some unit larger than their individual self is faring; and (2) Some people conceive their individual nature as constituted by others with whom they are in relation with (Flanagan 2017: 242–244). The Chinese family-oriented self is a variation on a variation: For (1), the Chinese family-oriented view conceives individual wellbeing in terms of the wellbeing of the family-in-group as a whole. The Chinese understanding of *hun-po* in ancestral worship embodies an extreme version of this view: Even when a person is dead, the wellbeing of his ghost is still dependent upon the flourishing of his clan. For (2), the Chinese family-oriented view conceives the individual nature as embedded in the family lineage. In Zhu Xi's ripple metaphor, the individual soul is seen as an extension of the generations of ancestors from the past. It is also in the nature of individuals to extend themselves and continue their family lineage.

Of course, Fig. 2 is an over-simplified picture of the 'Family-Oriented Self'. There are questions such as how big the family is, whether the family has to be connected through the bloodline, how to distinguish the immediate family and extended family, and so on. However, the belief that the person shares the same metaphysical material that constitutes his soul (*hun-po*) *only* with his family members has significant implications for the ways that boundaries are drawn. In practice, these differences have implications for ethics. Medical anthropologists and philosophers, such as Fan and Tao (2004), Chan (2004), Zhao (2015), point out that Chinese people operate medically informed consent differently from American people. When Chinese people are giving medical consent, especially in the case where the patient is senile or the illness is terminal, they tend to rely on their family to make the decisions for them. It is seen as a normal and often a virtuous thing for the children to make the difficult decisions for their parents, without the parents knowing about it. When patients are interviewed about this process, many of them say that they think of family consent as equal to their own consent. Some are perplexed to learn that the doctor has to obtain consent from the patient in the West (see Fan and Tao 2004; Chan 2004; Zhao 2015). As one of them said, 'I don't know how to explain it to you. But I just don't see any problems with my parents making health decisions for me. Maybe the Westerners see a problem, but I don't' (Zhao 2015).[2]

[2]An interesting question to ask is how the husband and wife are related to each other. Chinese marriage customs suggest that when the wife is married into the husband's family, she becomes a member of her husband's clan. She adopts the family name of her husband, and addresses the husband's parents as 'mother' and 'father'. When she dies, her husband's clan, and in particular, her sons and unmarried daughters, are responsible for preparing ritual sacrifices and offerings for her. In this sense, the wife is 'adopted' by her husband's family through the special channel of marriage.

Fig. 2 Family-oriented self-schema

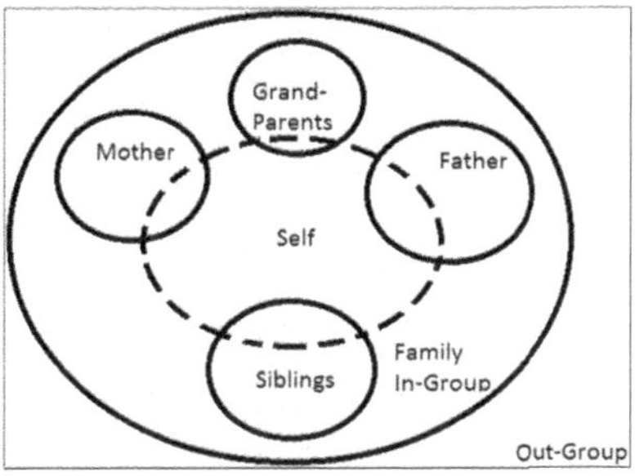

The family-oriented self is interconnected with the core Chinese virtues and their manifestations. The Chinese people rank family loyalty and filial piety (*xiao*) as one of the most prized virtues, and much higher than Americans do (Flanagan 2017: 229). Moreover, the belief that the long lasting of the *hun-po* is contingent upon the continuation of the family lineage gives the Chinese notion of filial piety some distinctive texture. In the *Mencius,* it is famously said, 'Among the three ways of being unfilial to one's parents, the most serious offense is to not have any children' (*Mencius* 4A: 26, translated by Wenqing Zhao). This means that continuing the family lineage is a critical requirement of being filial. To be a filial child, one needs to not only serve one's parents (support them, not disgrace them, and bring them honour), but also to produce children who will continue the family lineage. This goes back to the notion of sharing the same metaphysical material with one's ancestors and descendants, and the importance of the continuation of family lineage and bloodline, which serve as the earthly root of the ancestral *qi*. Some contemporary Chinese scholars, such as Fan (2016), argue that there is a distinctive Confucian-Chinese way of life that can be categorised as 'ethical familism'. Fan writes, 'for Confucians, it is not only intrinsically valuable for us humans to live in this world, it is also intrinsically valuable for us to live in the familist way in the world … one should form stable and harmonious familial relations' (10). He also argues that producing children is of paramount importance for Chinese people: 'By reproducing children, one's family line is enabled to continue, and the unity of one's family is thereby extended over time, including one's ancestors and descendants into a huge, perpetual unity' (9).

Conclusion

We have presented a taxonomy of a dozen self-variations that can be of use to any scholar trying to understand how persons are conceived cross-culturally. We have shown how the taxonomy can be used to illuminate a distinctive way that the self is conceived in China. We propose that the culturally situated notion of *hun-po* can help us dissect some of the important dimensions of the self in the Chinese tradition. In particular, the Chinese 'soul' is deeply embedded in the family lineage, which informs a unique, family-oriented notion of self that help justify ways of thinking about a good life in Chinese philosophy. The Chinese family-oriented self is different from the interdependent self (such as the Buddhist all-sentient-beings-oriented self) and the independent self (such as the neo-liberal self) in ways that are significant. These differences are not straightforwardly captured by the model of 'independent versus interdependent self' in the current literature. To put it in the simplest terms, the Chinese family-oriented self shares porous boundaries with members of the family-in-group. It conceives the individual's nature as relational, as embedded in the family lineage, and it conceives individual wellbeing in terms of the wellbeing of the family. The family-in-group as a whole, on the other hand, has firm and non-porous boundary with other non-family members. This model explains the distinctive texture of the Chinese virtue of filial piety that emphasises not only respect and support of the elderly members of the family, but also producing children who can carry the family lineage. It also helps us understand certain family-oriented ways of decision-making in the Chinese society. Again, it is important to note that the proposed 'family-oriented self' is only one important thread of understanding the self-variations in Chinese tradition.

Acknowledgements Owen Flanagan thanks the Center for Advanced Study in the Behavioral Sciences, where he was a Berggruen Fellow in 2016–17, for its support. Both authors thank the John Templeton Foundation. The opinions expressed in this publication are those of the authors and do not necessarily express the views of the John Templeton Foundation.

References

Ames, R. T., & Rosemont, H., Jr. (2011). Were the early confucians virtuous? In C. Fraser, D. Robins, & T. O'Leary (Eds.), *Ethics in early China: An anthology* (pp. 17–39). Hong Kong: Hong Kong University Press.

Bruner, J. S. (2002). *Making stories: Law, literature, life*. New York: Farrar, Straus and Giroux.

Chan, H.-M. (2004). Informed consent Hong Kong style: An instance of moderate familism. *Journal of Medicine and Philosophy, 29*(2), 195–206.

Clark, C. (1997). *Misery and company: Sympathy in everyday life*. Chicago: University of Chicago Press.

Creel, H. G. (1937). *The birth of China*. New York: Reynal and Hitchcock.

Descola, P. (1996). *The spears of twilight: Life and death in the Amazon Jungle*. New York: New Press.

Everett, D. L. (2009). *Don't sleep, there are snakes: Life and language in the Amazonian Jungle*. London: Profile Books.

Fan, R., & Tao, J. (2004). Consent to medical treatment: The complex interplay of patients, families, and physicians. *Journal of Medicine and Philosophy, 29*(2), 139–148.

Fan, R. (2016). DNA, brain, mind, and soul: A confucian perspective. In: Manuscript Presented at the 5th International Workshop on Bioethics, Multiculturalism and Religion, Houston.

Flanagan, O. (2017). *Geographies of morals: Varieties of moral possibility*. Oxford: Oxford University Press.

Flanagan, O. (1991). *Varieties of moral personality: Ethics and psychological realism*. Cambridge, MA: Harvard University Press.

Geertz, C. (1973). *The interpretation of cultures: Selected essays*. New York: Basic Books.

Hu, S. (1937). The indianization of China: A case study in cultural borrowing. In *Independence, Convergence and Borrowing in Institutions, Thought, and Art* (pp. 224–225). Harvard University Press.

Hampden-Turner, C., & Trompenaars, A. (1993). *The seven cultures of capitalism: Value systems for creating wealth in the United States, Japan, Germany, France, Britain, Sweden, and the Netherlands*. New York: Currency/Doubleday.

Henrich, J., Heine, S. J., & Norenzayan, A. (2010). The weirdest people in the world? *Behavioral and Brain Sciences, 33*(2–3), 61–83.

Keillor, Garrison. (1985) Lake Wobegon Days. Viking.

MacIntyre, A. C. (1981). *After virtue: A study in moral theory*. Notre Dame, IN: University of Notre Dame Press.

MacIntyre, A. C. (1984). *After virtue: A study in moral theory* (2nd ed.). Notre Dame, IN: University of Notre Dame Press.

Markus, H. R., & Kitayama, S. (1991). Culture and the self: Implications for cognition, emotion, and motivation. *Psychological Review, 98*(2), 224–253.

Markus, H. R., & Kitayama, S. (2010). Cultures and selves: A cycle of mutual constitution. *Perspectives on Psychological Science, 5*(4), 420–430.

Miller, J. G. (1984). Culture and the development of everyday social explanation. *Journal of Personality and Social Psychology, 46*(5), 961–978.

Needdam, J. (1974). *Science and civilization in China.* Cambridge: Cambridge University Press.

Nisbett, R. (2004). *The geography of thought: How asians and westerners think differently ... and why*. New York: Free Press.

Ricoeur, P. (1984). *Time and narrative*. 3 vol. Chicago: University of Chicago Press.

Shweder, R. A., & Bourne, E. J. (1982). Does the concept of the person vary cross-culturally? In R. A. Shweder, & R. Alan LeVine. *Culture theory: Essays on mind, self, and emotion* (pp. 158–99). Cambridge: Cambridge University Press.

Strawson, G. (2004). Against narrativity. *Ratio*, n.s., *17*(4), 428–452.

Taylor, S. E. (1989). *Positive illusions: Creative self-deception and the healthy mind*. New York: Basic Books.

Taylor, S. E., & Brown, J. (1994). Positive illusions and well-being revisited: Separating fact from fiction. *Psychology Bulletin, 116*(1), 21–27.

Triandis, H. C., Bontempo, R., Villareal, M. J., Asai, M., & Lucca, N. (1988). Individualism and collectivism: Cross-cultural perspectives on self-ingroup relationships. *Journal of Personality and Social Psychology, 54*(2), 323–338.

Tsai, J. L. (2007). Ideal affect cultural causes and behavioral consequences. *Perspectives on Psychological Science, 2*(3), 242–259.

Tsai, J. L., Louie, J. Y., Chen, E. E., & Uchida, Y. (2007a). Learning what feelings to desire: Socialization of ideal affect through children's storybooks. *Personality and Social Psychology Bulletin, 33*(1), 17–30.

Tsai, J. L., Miao, F. F., & Seppala, E. (2007b). Good feelings in christianity and buddhism: Religious differences in ideal affect. *Personality and Social Psychology Bulletin, 33*(3), 409–421.

Tsai, J. L., Miao, F. F., Seppala, E., Fung, H. H., & Yeung, D. Y. (2007c). Influence and adjustment goals: Sources of cultural differences in ideal affect. *Journal of Personality and Social Psychology, 92*(6), 1102–1117.

Whitehead, A. N. (1979). *Process and reality.* Free Press.

Yü, Y.-S. (1987). O Soul, Come Back! A study in the changing conceptions of the soul and afterlife in pre-buddhist China. *Harvard Journal of Asiatic Studies, 47*(2), 363–395.

Yü, Y.-S. (2005). *Views of life and death in later Han China* (东汉生死观) Shanghai Guji Chubanshe 上海古籍出版社.

Zhao, W. (2015). A confucian worldview and family-based informed consent: A case of concealing illness from the patient in China. In R. Fan (ed.), *Family-oriented informed consent* (pp. 231–43). Berlin: Springer.

The Problem of Qualia: Perspectives on the Buddhist Theories of Experience

Victoria Lysenko

Introduction

The term "qualia" refers to the properties of sensory experience. Being subjective by their very nature, i.e. unique and nonreportable, they constitute the mode we experience the world. It is also a fundamental feature of the human way of being in the world known to us exclusively from the first person perspective. At the same time, Western science always tries to view the first person experience, or any subjective cognition as an objective process that can be measured by scientific tools. This paradox is closely linked with the so called "hard problem of consciousness" (David Charmers): Why is there a subjective component to experience? The achievements of the modern neuroscience allow to map a diversity of human mental functions related to different parts of the brain. Based on these achievements, different forms of artificial intelligence are successfully developed, however further progress meets the seemingly impossible task as to simulate qualia. Qualia are the last fortress of subjectivity difficult to seize by a purely "objective" scientific approach.

Contemporary Western philosophers of consciousness and cognitive scientists somewhat familiar with Buddhism mainly dwell upon the Madhyamaka school with its ideas of emptiness and relativity as well as upon its attitude towards different epistemological problems and studies of consciousness. His Holiness the Dalai

V. Lysenko (✉)
Department for Oriental Philosophies, Institute of Philosophy,
Russian Academy of Sciences, Moscow, Russian Federation
e-mail: vglyssenko@yandex.ru

S. Menon et al. (eds.), *Self, Culture and Consciousness*,
DOI 10.1007/978-981-10-5777-9_18

lama actively cooperates with some leading Western scientists and philosophers within the framework of the annual conferences organized by the Mind and Life Institute founded by Him and Francisco Varela.[1] The fact that Tibetan Buddhists of the Geluk school are modern successors of the Madhyamaka stimulated the interest for this school in the West. Until now, as far as I know, the intellectual heritage of other Buddhist schools such as Abhidharma-Vaibhāṣika and Sautrāntika, as well as the theories of Buddhist logicians and epistemologists are less in demand in this regard.[2] In this paper, I will examine some texts of these two traditions—the seminal work of Vasubandhu (fl. 4th to 5th century AD)[3] and a number of texts of the Buddhist epistemologists Dignāga (c. 480–540 AD)[4] and Dharmakīrti (c. 600–660 AD).[5]

Any comparison between qualia and the Buddhist concepts of experience raises, at least, two crucial questions:

(1) Can the Buddhist system and the contemporary elaboration of qualia meet on a common conceptual basis?
Hence, (2) whether the problems associated with qualia in contemporary philosophical discussions can be formulated in the frame of the Buddhist philosophical systems both Abhidharma and Sautrāntika-Yogācāra?

If we deal with the first issue, we can draw some parallels between the contemporary concept of qualia and the theories of dharmas (units of experience) peculiar to different Abhidharma schools, as well as define the way the Abhidharmikas give account of the difference between first and third person perspectives.[6] If we focus on the second issue, we may suppose that the problem of qualia may be formulated only within those philosophical systems where our cognitive experience is understood in terms of its givenness and accessibility.

I argue that the distinction between "raw sensation" (*nirvikalpaka-pratyakṣa*) and perceptual judgments (*savikalpaka-jñāna*), introduced by Dignāga and developed by Dharmakīrti, as well as their postulation of the constitutive character of *nirvikalpaka-pratyakṣa* for the cognitive experience as such, testifies to the

[1]For the materials of *the Mind and Life Institute* organized by His Holiness the Dalai lama: https://www.mindandlife.org.

[2]For recent comparative analysis of the Abhidharma and the Buddhist epistemological tradition in the context of modern Western philosophical discussions, see: Garfield (2015), Thompson (2014).

[3]*Abhidharmakośabhāṣya* (further on—AKB), Pradhan ed. (1967) and Pruden tr. (1988).

[4]First of all, his *Pramāṇa-samuccaya* ("A Compendium of Valid Cognition") Chapter 1. Steinkellner ed. (2005). Translation from Tibetan into English—Hattori (1968). Translation from Sanskrit into Russian—Lysenko (2011, pp. 115–129); further on—PS and PSV (V-Vārttika, or auto-commentary of Dignāga to PS).

[5]His *Pramāṇavārttika-kārika* ("Commentary on Dignāga's 'Compendium of Valid Cognition'"), Sankrtyayana ed. (1953), further on—PVK; and *Pramāṇaviniścaya* ("Ascertainment of Valid Cognition"), Steinkellner ed. (2007), further on—PVin.

[6]We will not consider here whether the subjectivity problem with regard to experience was formulated in such a way as to bring about the problem of qualia in the Abhidharma texts.

presence of the theoretical conditions under which the problem of qualia could be formulated. Due to the commitment to the soteriological perspective, shared by all the Buddhist thinkers, a non-verbal and non-conceptual awareness (at the last stages of meditation) is accepted by them as a sort of epistemological ideal even for an ordinary cognition. Taking that in account we can understand the reason why the svasaṃvedana, non-conceptual self-awareness, became a hallmark of any experience, including that of conceptualization.

Dharmas and Qualia

One can certainly find a striking similarity between the Abhidharma idea of dharmas as ultimate qualitative units of experience[7] and some concepts of qualia.[8] The very tendency in Buddhism to deal with the world as it is given to us in our experience,[9] the world processed by our somatic and cognitive tools,[10] may in a certain way bring together dharma- and qualia-theories, no matter how dharmas

---

[7]Dharma is defined by Vasubandhu as something that bears/holds its own intrinsic/inherent/ particular characteristics (*svalakṣaṇadhāraṇād dharmaḥ*—AKB I.2). Pradhan (1967, p. 3).

[8]In fact, "dharma" like "qualia" literally means "quality", "property" as distinct from the property holder (dharmin).

[9]Following Husserl, we may call it the "life-world" (*Lebenswelt*). As the Russian scholar Otton Rosenberg pertinently remarked with regard to the Abhidharma analysis: "The epistemological question of the reality of the external world or the reality of the objects in the external world as well as the question of the metaphysical essence of being is still not resolved. The talk is only about the fact that the human being, experiencing a certain penomenon—like seeing the sun—consists of such and such elements that are in such and such relationship, etc. The subject of analysis is a man seeing the sun, but not a man and the sun taken separately.

We must not forget that for the Buddhists all reasoning is aimed at the liberation of living beings, and therefore the question concerning the nature of sun *per se*, is not important. Since we deal 'a man seeing the sun' the Buddhists see in such phenomenon a combination of elements—like mind, "ability" or the "act" of perceiving something luminous, round, etc. All these elements taken together constitute what we call 'a man seeing the sun'. Rosenberg 1991 [first ed. 1918], p. 90.

[10]This question is systematically studied by Peter Harvey in the early Buddhist texts. He writes: "In their discussion of the 'world', the 'early Suttas' focus very much on the world of lived experience. While a purely external, physical world is not denied, this is not the world that people actually live in. They live in the world-as-sensed-and-thought-about, which exists as a constructed experience within a sentient organism. As the Suttas put it: "It is in this fathom-long carcase, (which is) cognitive (*saññimhi*) and endowed with conception (*-mana-*), that, I declare, (lies) the world, and the arising of the world, and the stopping of the world, and the course that goes to the stopping of the world' (S.1.62). This is a statement of the four Holy Truths—on dukkha, its arising, its stopping, and the course going to this—with the world' replacing dukkha, and this 'world' being seen as within the mentally-endowed physical body. The sense of this passage is elucidated by a similar one at S.IV.95: 'Sir, that by which, in the world, one is cognitive (saññi) of the world, holds conceits (-mani) on the world, is called the world in the discipline of the Holy ones'." Harvey (1995, p. 80).

have been interpreted in different Buddhist schools. Thus, we could call the Buddhist approach to reality an experientialist one. To know the world means not only to think about it, but to experience it, to feel and live it through our bodies. Christian Coseru remarks: "however, the notion of a world as experienced does not imply that the elements of existence and/or experience (*dharmas*) are not real in an empirical sense; only that their reality cannot be ascertained independently of any reference to their mode of givenness. From the first-person perspective, the body—as an aggregate of such elements of existence and/or experience—is both the medium of contact with the world and the world with which it comes in contact." Coseru (2015, p. 231).

The somatic or bodily perspective on human cognitive experience, especially in the Abhidharma schools, can be compared with the modern contemporary concept of the embodied or enactive cognition (Varela, Lakoff and others).

In order to better understand the Buddhist approach, we should take into account the fundamental division—common to the overwhelming majority of Indian religious and philosophical traditions—between the ordinary experience within the karman-saṃsāra perspective, and the transcendent or nirvanic perspective overcoming saṃsāra. The first one is the experience of an ordinary, unenlightened person (*pṛthak-jana*), the second—the experience of the Noble ones (*āryas*) or those who achieved enlightenment (*bodhi*) and nirvāṇa (extinction of all affects). Accordingly, there are two approaches: (1) a relativistic, empirical (*vyavahārika*) one which is considered by the Buddhists to be a false, flawed (*mithya-dṛṣṭi*) due to the fact that is springs from ignorance (*avidyā*), and (2) a true (*samyak-dṛṣṭi*), valid (*pramā*) approach giving access to reality as it is (*yathābhūtam*). The refutation of the empirical, erroneous opinions of others and the proof of the validity of their own doctrines constitute the main goal of Buddhist philosophers pertaining to different schools. It explains why they were constantly engaged in disputes with one another and with their Brahmanical opponents.

It is evident from the earliest Buddhist texts that the main preoccupation of the Buddha himself was the emancipation of all living beings from sufferings connected to reincarnation. He repeated time and again: "All my teaching has the sole taste of emancipation". But the Buddha's idea of emancipation has nothing in common with salvation coming from outside (from God or any divine forces). What really matters in early Buddhism is the doctrine of self-emancipation formulated by the Buddha in terms of the Four Truths of the Nobles Ones.

The basic features of experience, according to the Buddha, are three (*trilakṣaṇas*): impermanence (*anityatā*); (2) selflessness,—"non-Self" (*anātman, anattā*), and (3) unsatisfactoriness, misery (*duḥkha*). The principle of impermanence, being applied to the living beings, is at best expressed in the idea of "non-Self" (*anattā, anātman*)—the absence of the permanent and enduring inner core, selfness, or essence (*ātman, svabhāva*) in our personality and in all phenomena we can observe, both inside and outside ourselves. In the contemporary philosophical language, this attitude is coined as "non-essentialism." If, for the Buddha, impermanence simply means that all phenomena (*dharmas*) have their beginning and end, the Abhidharmikas after him developed a more radical view of *kshaṇikavāda* (the doctrine of instantaneity) in which experience is reduced to momentary events each of which has its numerical position within the series of its own category (dharmas are classified in different manners according to a particular kind of experience).

In their soteriological enterprise, the Abhidharmikas elaborated a special practice of mindfulness (*smṛti, sati*) which consisted in discerning and identifying different qualitative mental states (dharmas) in terms of skillful, unskillful or neutral means to reach the final goal of nirvāṇa. In this and other contemplative practices the main obstacles or defilements enslaving the individuals in saṃsāra were supposed to be detected and removed.

Basically, the idea is to "scan" the internal content of individual mental life and to present it as a complex multifactorial process, anonymous by its character.[11] An analysis of the actual content of consciousness was applied to the cognitive activity as a whole, to the experience of emotions and feelings. All this activity led to the elaboration of a comprehensive inventories of observable dharmas (mental events) and to the identification of the status of each dharma according to different classifications developed in the Abhidharma (āyatanas, dhātu, skandhas, etc.). Different schools of Abhidharmikas defended their own lists of dharmas.

The difference between an ordinary experience, on the one hand, and a practice of mindfulness on the other, implied the acceptance of an egological perspective for the former and its elimination for the latter. For an ordinary person (*pṛthak-jana*), all his or her conscious activity, feeling, reactions, etc. are ineluctably linked with his or her "I" or "ego". This ordinary "I" experience, according to the Buddha, is wrongly hypostised in terms of the eternal and unchanging Self (Ātman). A Noble person (*ārya*) is supposed to completely remove the delusion of "I am" or, in contemporary terminology, to abandon the egological perspective.

It goes without saying that the only reality a person deals with is the reality of his or her own mental states; his or her living world, therefore the first person experience is the only one that a person can have. However, the individual soteriological task or, so to say, "nirvanic program" consists in depersonalizing this experience and transforming oneself into "nobody" or, in Buddhist terms, to treat his or her own mental states as impersonal, anonymous dharmas. In doing so, a person overcomes her feeling of "mineness" with regard to her own mental states and their objects,[12] as if she were a detached observer. One's selfhood (*atta-bhāva* Pali) is

---

[11]In this so called anti-essentialist model, the cause and effect relationship was developed into a more complex causal interactive network of cooperating factors involved in every mental or physical event. Nothing is produced by a single cause, but when we analyze a causal complex, we should single out some main (direct) causes (*hetu*), as well as the conditions (*pratyaya*) under which these causes emerge. Their status depends on their contribution to the causal processes. Thus, the occurring of dharmas was placed both in the diachronic and in the synchronic perspective. In the Abhidharma texts we find the very complicated theories of causes, conditions, and their results—4 hetu—6 pratyaya—5 phala).

[12]Indian philosophy develops a special concept for appropriation: *ahaṃkāra*—an "Ego-maker", in Yogācāra this function is attributed to the *kliṣṭa-manas* (afflicted-mind).

recurring in a pure non-afflicted form. In this case, no metaphysical permanent Self or Subject enters the scene. But, in spite of all this, the experience continues to be subjective, though in a quite minimal way.[13]

In the Sarvāstivāda or Vaibhāṣika school, the dharmas occur as the discrete factors of qualitative identity and as the "quanta" of dynamic process. They may obtain their identity only within certain individual series in connection with other dharmas where some dharma, fulfilling its function, disappear conditioning the appearance of the other dharmas belonging to the same series.

The dharmas form a true "texture" of the individual's reality, every moment of which is unique. Some contemporary authors interpret dharmas as tropes—individualized properties, "atoms" of qualitative identity. The phenomenal world of our experience emerges from them in the same way as a picture printed by the dot matrix printer.[14] Thus, according to the theory of dharmas, our experience appears as a totality of qualitatively heterogeneous flows of micro-events or phenomena.

Subjectivity Without a Subject

In the Abhidharma literature, we find the terms *svasantāna* and *parasantāna*: one's own series of experiential events and the other persons' series of such events. These terms clearly indicate the demarcation between the first person and the third-person perspectives. However, the feeling that a certain experience belongs to me and not to someone else, in Buddhist terms, is not making a particular factor separated from the fact that some experience is just taking place. The Buddhists do not deny that experience feels as "mine". To them it is tightly connected with "I" as "ego", a kind of "lower self" (Horner) which, contrary to the immutable upanishadic ātman, is in constant change. The experience of selfhood or personhood (atta-bhāva in Pāli texts) does not necessarily imply the existence of the unchanging Self (ātman). The Buddhists argue that the idea of Self or ātman as an ontological entity is a false mental construction that hypostatizes the linguistic use of the personal pronouns (like in Wittgenstein). Discarding the idea of the permanent "Self" (ātman), the

---

[13]Alexander Piatigorsky (1929–2009), a Russian Buddhist philosopher, made a pertinent remark: "But whether Buddhist psychological theory assumed a position of "outsider", of so to speak, a "third party"?—I am sure that it did not. In European science the removal of researcher from the object of research, his "dismissal", is the highest principle and the basic condition of a correct research. But in the ancient Buddhist science "the matter of study" was primarily the mentality of an observer, a researcher. And so, in the study of almost any "object" in Buddhist psychology, this estrangement (i.e. the presentation himself as "the other", "alien", "second") itself lies within the science, within the knowledge, as a part of their contents." Piatigorsky (1973, p. 277).

[14]The Vaibhāṣika idea that all dharmas exist (*sarva-asti*) that is, the dharmas of the past, present and future are reals (*dravya*), but only the dharmas of the present time are active or have activity (*karitrā*), may suggest to us the comparison of dharmas' activity with neuron spikes.

Buddha continues to use the words "I" (*aham*), "you" (*tvam*), "self" (*atta*) but, as he says, without allowing them to lead him astray (Dīghanikāya I. 202[15]). It means that such designations as "I", "me", "mine" are used in a non-reifying way. In the modern Buddhological literature,[16] the Buddhist interpretation of this selfhood is compared with the phenomenological concepts of "minimal self" or "narrative Self".[17]

The Buddha proposed to replace the "subject-object" centered language by the language of impersonal dharmas and their interdependent origination—*pratītyasamutpāda* [*paṭiccasamuppāda* in Pāli] constituting a valid knowledge. So such questions as "Of whom (S) is old age and dying (P)?" that tacitly imply the subject-object relationship, the Buddha finds "unsound" (*akalya*). Being formulated "correctly," they will have the following shape: "The condition of what (P') is P?" And the "correctly" formulated answer is "P is a condition of P'".

When the Abhidharmikas talk about an individual, personal experience are they well aware of the crucial problem as to how to justify subjectivity and selfhood in the absence of the ātman? They rightly notice that the problem arises as soon as we try to distinguish the internal (*adhyātmika*) and external (*bahya*) elements (*dhātu*) of experience:

> But how can one speak of personal [internal] dhātus or of external dhātus since there is no ātman? The mind [citta] is the object of the idea of self, the mind is what persons falsely grasp as their self. This mind receives, metaphorically, the name of ātman." AKB I, 39b. Pruden (1988, p. 108).

---

[15]For the texts of the Buddhist Pāli Canon(Tipitaka), with translations see: http://metta.lk/tipitaka/index.html.

[16]See specially the fruitful discussion on subjectivity and the "minimal self" in Siderits et al. (2011), between indologists and buddhologists (Joel Krueger, Georges Dreyfus, Jonardan Ganeri, Ram-Prasad Chakravarti, Mattheu MacKenzie, Mark Siderits, and phenomenologist Dan Zahavi).

[17]Dan Zahavi introduced the term "minimal self": "To be consciousof oneself, consequently, is not to capture a pure self that exists in separationfrom the stream of consciousness, but rather entails just being consciousof an experience in its first-personal mode of givenness; it is a questionof having first-personal access to one's own experiential life. Thus, the selfreferred to is not something standing beyond or opposed to the stream ofexperiences but is rather a feature or function of its givenness. In short, theself is conceived neither as an ineffable transcendental precondition, nor asa mere social construct that evolves through time; it is taken to be an integralpart of our conscious life with an immediate experiential reality." Zahavi 2005, p. 106.

Shaun Gallagher states that the narrative self is constituted with the past and the future in the various stories that we and others tell about ourselves. It is a kind of ordinary self that we experience in our daily lives. The minimal self is a self which is composed of minimal factors; all of the unessential features are stripped away. It is the consciousness of oneself as an immediate subject of experience, in time. It is the 'I' who is experiencing 'now-here'. The minimal self has the two aspects; the sense of agency and the sense of ownership. The ownership is the sense that it is my body that is moving, and the agency is the sense that I am the initiator or source of the action. They are indistinguishable in the normal experience of willed action. We have both senses at the same time. It is possible to simplify various notions of the self into two categories; the 'minimal self' and the 'narrative self'." Gallagher (2000, p. 15).

Thus, for the Buddha and the Buddhist thinkers the word ātman is a purely conventional designation of the impermanent mind. The word *adhyātmika* refers, in fact, not to the ātman, as its etymology may suggest, but to certain cognitive phenomena which are "close to mind", like the six *indriyas* (faculties: sight, hearing, taste, smell, touch, mind) and the six *vijñānas* or the six modalities of cognitive discernment (visual, auditive, gustatory, olfactory, tactile, mental).

"Now the organs and the consciousnesses are close to the mind to which one gives the name of *ātman:* they are in fact the point of support of them; then one qualifies them as "internal," or "personal," whereas the visible and the other objects of the consciousness are held to be "external"". (*Ibid.*)

For the Vaibhāṣikas "external" (bahya) objects are color-forms, sounds, smells, tastes, and sensations but not mental phenomena (*dharmas*). Being direct realists in their presentation of the cognitive process, the Vaibhāṣikas believed that color-forms etc. belong to external objects. However, they differentiate between the two modalities of sense-perception: one, at a distance from the organ of sense (for sight and hearing), and the other, through the contact between the organ of sense and its object (for taste, smell and touch).

The objects of sight and hearing, being regarded as "general", are seen by many persons. The apprehension of smell has also its objective dimension:

> It can be the case that smell - the same atomic group of a smell-which produces a consciousness of smell in one person, is also perceived by another. Now this does not hold for the organs. Consequently smells, tastes and tangibles should be compared to visible matter and sounds. AKB I.39b. Pruden (1988, p. 111).

Inasmuch as smell, taste, and tangibility are regarded as some external fields of cognitive discernment, they may have a common nature for different persons, but *indriyas* always keep their individual, personal character and that fact makes the Buddhist interpretation of their function a step forward to our concept of qualia.

In some other cases, the terms "general" (*sādhāraṇa*) and "specific" (*asādhāraṇa*) are used to classify cognitive discernments (5 sense-consciousness and mental consciousness) not in relation to their object but with regard to their appropriate sense-ability/organ (*indriya*): eye-consciousness (*cakṣur-vijñāna*) etc. From the Vaibhāṣika point of view, it is a specificity of the organ which is person's privacy and not of the object which she shares with other persons that constitutes the modality of cognitive discernment.

> An organ of sense [of one person] cannot serve as a support (*āsraya*) for cognitive discernment [of another person] (my translation—V.L.). "Consciousness takes the name of an organ because the organ is its footing. Because the organ is "its own" (specific—*asādhāraṇa*): the organ of a person is the footing of the visual consciousness of this person. Visible matter, on the contrary, is general and like a visible thing it is perceived by the visual and mental consciousness, by one person and by the other. The same holds true for the organs of hearing, smell, taste, and touch and for their objects, sounds, smells, tastes, and tangible things. Let us conclude: the name "consciousness" is given to its organ because the organ is its footing and because the organ is its own thing (specific V.L.—*asādhāraṇa*). But the same is not true for its object." AKB I, 39d. Pruden (1988, p. 126).

The object whether it is *viṣaya,* the object of the five sense-discernments, or *ālambana,*—the object of *mano-vijñāna* (mental discernment), may be general, cognized not only by me, but also by any other person. We can speak about an intentional or positional cognitive activity not only from the first person perspective but also from the third-person perspective.[18]

Unlike the Vaibhāṣika's direct realistic understanding of cognitive process (objects exist outside us) within the perspective of samsaric world peculiar to the ordinary human beings, some concepts of the Sautrāntika, the school of Buddhism often regarded as representative realism, can be compared with "our" qualia on a more solid ground. Consider, for example, the notion of *ākāra* (mental aspect, mental form)[19] through which our cognition makes acquaintance with the external objects. The latter, however, remain inaccessible to us. As these *ākāras* are nothing else but manifestations or reflections in our consciousness of some particular features of the external objects, they may be interpreted in the contemporary terms of "aspectuality" or of "aspectual shape" (Wittgenstein, Searle).

Qualia, Svalakṣaṇas and "Phenomenal Consciousness"

It were the Buddhist epistemologists Dignāga and Dharmakīrti who came closest to our contemporary concept of qualia. Thus, in his theory of *pramāṇas* (instruments of valid cognition), Dignāga systematically elaborated the idea of raw sensation, i.e. perception free from mental constructions or conceptualizations—*nirvikalpaka-pratyakṣa* or *kalpanā-apoddha*. This idea could be considered as a key prerequisite to qualia inasmuch as it is tightly connected with the first person immediate experience. Dignāga distinguishes it from a "cognition with mental constructions" or a conceptualizing cognition—*savikalpaka-jñāna*. According to him, *nirvikalpaka* and *savikalpaka* are the mutually exclusive modes of cognitive activity, and not just different phases of cognition of one and the same object as some other Indian philosophers believed (the realistic schools of the Vaiśeṣikas, Nayāyikas and Mīmāṃsākas). At the first instant of cognition, an object is grasped as something indefinite, and in the next moment its characteristics take on verbal and conceptual form of a perceptive judgment: "this is a cow". According to Dignāga, a raw or immediate perception and a perceptual judgment are different modes of cognition inasmuch as their objects are

---

[18]The causal relatedness between the modality of discernment and its instrument is what the Vaibhāṣikas justify in the following way: "The footing of consciousness is its organ, for consciousness changes according to the modality of the organ. For example: when the organ of sight becomes the object of attention (employment of eye salve, etc.); when it is alert or when it is sluggish and weak, consciousness reproduces modality. It is accompanied by pleasure or pain, it is alert or weak. On the contrary, the object has no influence on the modality of consciousness. Consequently, it is the organ, and not the object, which is the footing of consciousness." AKB 45a-b. Pruden (1988, p. 125).

[19]For more details, cf. Kellner (2014).

different. The object of immediate perception is *svalakṣaṇa*, specific characteristic, while the cognition with mental constructions refers to general characteristics (*samānya-lakṣaṇas*). He particularly emphasizes the fact that an initial perceptual experience refers to something that never goes beyond its own actually present content, and thus may characterize only itself (the literal sense of the word *svalakṣaṇa*).[20] The *svalakṣaṇa* is tantamount to a mere presence, sheer actuality. Mental constructions appearing in the next moment in the form of a perceptual judgment cannot refer to the *svalakṣaṇa* since its moment has already elapsed. Though the perceptual judgment brings nothing new and for this reason does not constitute a separate instrument of valid cognition (*pramāṇa*), it paves the way to *anumāṇa* (logical inference)—another instrument of valid cognition directed at mental objects designated as common features (*sāmānya-lakṣaṇa*). By their very nature perceptual judgments cannot grasp the vanishing, constantly elusive reality of *svalakṣaṇas*, nevertheless Dharmakīrti calls them pragmatically true, for a correct conceptualization allows us to act purposefully (for example, to take a mango and eat it for our pleasure), but at the foundation of this expedient behavior we may always find some private immediate perceptive experience. Actually, we grasp not a mango as a whole but its aspects such as patch of color, shape or smell, always from a specific point of view, always in some particular circumstances and with some particular interests (when we are hungry we see it only as food, but looking aesthetically at this fruit, we may be captivated by how its surface reflects light from the sun etc.).

Supposedly, since Dignāga-Dharmakīrti's distinction between raw and mentally constructed perception we may treat the *svalakṣaṇas* (particulars of color, smell, etc., or particulars of the inner experience—joy, pain, etc.) as "qualia" grasped by the primary, direct, preverbal, and pre-conceptual perception. In the next moment, we mentally construct or conceptualize our past raw experience of something as, for instance, the perception of blue sky or yellow mango and thus articulate our inner feelings regarding it as "pleasure", "suffering" etc.

In a similar manner, qualia are represented in the contemporary literature which describe them as intrinsic, non-representational, non-repeatable, and non-reportable properties. If to ask a Buddhist what he thinks about the case of Mary, a famous thought experiment with a brilliant scientist who is, for whatever reason, forced to investigate the world from a black and white room via a black and white TV monitor, his answer will be that all information Mary gets about colors different from black and white is just in the form of general characteristics (*sāmānya-lakṣaṇas).* She has no direct experience with various colors *svalakṣaṇas*, unique and private yet non-transmittable in whatever form. From the Buddhist point of view, Mary's knowledge of colors cannot be valid and instrumental (*pramāṇa*) until she perceives them immediately as svalakṣaṇas.

[20]We have only a clear reference to the svalakṣaṇa's nature in the PS 1.5 cd: "The object of sense is the form (rūpa) which is to be cognized [simply] as it is and which is inexpressible" (translation —VL). Steinkellner, 2005.

These phases or modes of cognitive activity may also be compared with Ned Block's two forms of consciousness: "phenomenal consciousness" and "access consciousness".[21] Phenomenal consciousness consists of "what it is like to perceive, to experience, to feel, to be," i.e., it is an irreducibly subjective experience of qualia. The "access consciousness" is our ability to report about this experience, to think about it; this includes categorization, conceptualization, and verbalization being also representable. Like *svalakṣaṇas,* qualia are experienced but not reportable as such. Block like the Buddhists, believes that they may be subjected to introspection or meditation (by some advanced practitioners of the mindfulness meditation).

What problems concerning qualia in the contemporary philosophical discussions can be formulated in the framework of the Buddhist philosophical systems?

Self-Awareness, Epistemic Idealism and "Other Minds"

We may suggest that the *svalakṣaṇas* constitute the first person perspective while *sāmānya-lakṣaṇas* refer to the third person, inter-subjective perspective. That doesn't exclude that other persons can also experience *svalakṣaṇas* from their first-person perspective. The question is how can we make sure that the other persons' experience does exist? In Western philosophy we call it "the problem of other minds".

This problem was also formulated in Indian philosophy and debated in the Buddhist texts, for instance, in Vasubandhu's *Vijñaptimātratāsiddhi* (Proof of the Consciousness-only), but especially in a small treatise by Dharmakīrti: *Santānāntarasiddhi* (Proof of the Existence of Other Streams of Consciousness) with Vinītadeva's commentary thereon.[22]

In his defense of the Yogācāra epistemic idealism Dharmakīrti used an argument from constant co-perception (*sahopalambhaniyama*) according to which a property, i.e. color and its perception are always experienced together. This is why, it seems impossible to prove that color exists independently from our perception. The similar rule can be applied to any other object or property—an apple is never found apart from our perception of it, cold is never distinct from our feeling of it, etc. Dharmakīrti tries to prove that this position doesn't lead to a solipsism, because a justification of the existence of other minds, whether advocated by a direct realist (an "externalist" who believes in external objects) or by an idealist ("internalist" for whom only internal mental states exist) use in fact the same kind of inference by analogy (I know that my mind controls my bodily actions when I see the same

---

[21]Block (1995, 2004).

[22]The text was edited in Tibetan and translated into Russian by Th. Stcherbatsky (1916): For English translation see Harish C. Gupta; cf. Stcherbatsky (1969, pp. 84–121).

bodily actions of another person, I can infer that they are also guided by that person's mind).[23]

Further, Dharmakīrti uses the proof which could also contribute to the conjectural Buddhist formulation of the qualia problem. He argues that "my" experience is "mine" because of its immediate and undeniable character of self-awareness. It is based on Dignāga's concept of *svasaṃvedana*—self-awareness as a form of the immediate or raw perception (*nirvikalpaka-pratyakṣa*).

According to Dignāga, our cognitive discernment (*vijñāna*) is given to us in two different manifestations/appearences/forms (*dvyābhāsatā, dvirūpatā*): as an object-appearance (*viṣaya ābhāsatā*) and also as its own appearance (*svābhāsatā*), the latter assumes the presence of a cognizer's aspect (*grahakākāra*), the former the presence of its object, i.e. the the aspect of an object to be cognized (*grāhyākāra*).[24] This thesis is accepted in the epistemology of both Sautrāntika and Yogācāra schools. Sautrāntika understands the latter as a self-awareness of the perceiver, while the former as a *sarūpya*—taking over by consciousness the aspect of its object. For him, self-awareness (*svasaṃviti*) itself points to the awareness as a result (*phala*) of the two appearances mentioned above (Kellner 2010) and thus constitutes a third factor of cognitive experience. However, this factor is not a separate cognitive episode referring to the appearance of consciousness both as an object and a subject. It pertains to the very nature of consciousness as an immediate (*pratyakṣa*) subjective experience.

One may suppose that it is the *svasaṃvedana* which can give access to "how it feels like" to perceive, cognize or to categorize something (according to Dharmakīrti, the awareness of mental construction is also *pratyakṣa* or immediate, while mental construction itself is *parokṣa* or not immediate).

Dignāga proves the necessity of self-awareness with a number of arguments.[25] I will consider here just one well known as the "memory argument".

PSV to PS I.11d. One could not remember the cognition of an object that was not experienced [before]. Just as it is impossible to recall the color [without its sense cognition in the past—V.L].[26]

---

[23]For the detailed analysis of Dharmakīrti's arguments, cf.: Sharma (1985) and Henkel (2012).

[24]Birgit Kellner compares the usage of "appearance" to "what the philosopher Roderick Chisholm referred to as a descriptive phenomenological usage, where the verb "to appear" simply informs how things present themselves to a cognizing subject (Chisholm 1987: 199f.)." (Kellner 2010, p. 209).

[25]For their exposition and a detailed analysis see Kellner (2010). See also Coseru's translation: [PSV ad PS I 9a]: Every cognition comes about with a double appearance, namely that of itself [i.e. as awareness of itself] and that of the object. The awareness of itself as [possessing] this double appearance is the result [of the intentional act]—[PS I 9b]: because the determination of the object [to be cognized] conforms to it". Coseru (2015, p. 239). I do not enter here into the debates on the intentionality or non-intentionality of self-awareness in Williams (1998), Kellner (2010), Arnold (2012), Coseru (2015). I would like to thank Dr. Alex Yianopoulos for sending me a draft of his paper "Is Svasaṃvitti Intentional?".

[26]PSV I.11d.: *na hy ananubhūtārthavedanasmṛtīrūpādismṛtivat*. Steinkellner (2005). For the further analysis of Dignāga's "memory argument" see Kellner (2010, pp. 211–213).

Dharmakīrti uses this "memory argument" to prove the causal continuity and personal identity of experience within the person's particular *svasantāna* or individual series.

PV 3.179: How could one recall [some past episodes of cognition] which were not experienced? [In that case] our knowledge of our own past perceptions would not differ from the knowledge of the other people's past experience.[27]

Highlighting the self-awareness of all cognitive events, Dharmakīrti explains not only how our memory works, but also gives an account of the subjective, first person dimension of our experience, of our enduring feel of "mineness" even without any substantial enduring Self. In his auto-commentary to the PVin I. 14, Dharmakīrti refers to emotions, etc. (*caitta*) in order to demonstrate that their immediate character unequivocally indicates their first person origin.

According to him, if cognition is not immediate pleasures etc. can be neither pleasant, nor unpleasant etc. as if they are experienced by other people". cf. Stein

Among the most self-evident examples of *svasaṃvedana* Dharmakīrti mentions different mental events (*caitta*) such as suffering, pleasure, desires (today they are sometimes regarded as qualia par excellence). Why? Because in the case of internal mental states, their appearance and apprehension are identical with their experience one lives through. We become aware of our desires, joys and suffering at the moment when we experience them, and this experience always occurs in the form of a non-verbal and non-conceptual awareness. It means that we know suffering as an unpleasant experience even if we don't identify it as "I suffer", "it hurts".

This, I think, is the most convincing proof of Dignāga-Dharmakīrti's idea that self-awareness is simultaneous with the appearance of the object and subject aspects of cognition. It is neither the reflection of the past cognitive event (mental state), nor the introspection into the internal cognitive events. To be sure, both reflection and introspection are the consecutive mental events of the highest order in relation to those states that constitute their objects. Here I am talking about simultaneous events of the same order or, better, of simultaneous aspects of one and the same event.

In Indian philosophy there were two kinds of cognitive theories:

- self-illumination (*svaprakāśa*), consciousness illuminates both its object and itself, like a lamp;
- other-illumination (*paraprakāśa*)—the illumination of the other—cognition illuminates its object only; in order to be illuminated itself cognition requires a further cognition.

The similar speculations are rampant in the contemporary Western philosophy.[28]

[27]PV 3.179: *smṛtir bhaved atīte ca sāgṛhīte kathaṃ bhavet| syāc cānyadhīparicchedābhinnarūpā svabuddhidhīḥ||* Sāṇkṛtyāyana (1953).

[28]There were a number of attempts to make a comparison between the Indian and Western concepts of self-awareness. Cf., for example: MacKenzie (2007, pp. 40–62); MacKenzie (2008, pp. 245–266).

(1) first order, self-illumination thesis, reflexivist theories
(2) higher order thesis, reflectivist theories.

In phenomenological theories we find a pre-reflective self-consciousness belonging to the very nature of consciousness distinguished from reflective self-consciousness; Husserl, for example, speaks of consciousness that always involves a self-appearance (*Für-sich-selbst-Erscheinens*). Husserl (1959). One can also cite Merleau-Ponty's words that "<...> the most precise consciousness of which we are capable is always, as it were, affected by itself or given to itself, and that the word consciousness has no meaning independently of this duality." Merleau-Ponty (1962, p. 426). Sartre also developed similar ideas: "This self-consciousness we ought to consider not as a new consciousness, but as the only mode of existence which is possible for a consciousness of something." Sartre (1967, p. 20).

From this point of view, the Buddhist approach is quite close to the phenomenological one and taken together these two approaches may provide us with some refreshing arguments for those cognitive scientists who try to defend qualia against the attacks of the functionalists and hard materialists.

At the same time, as it shows the usage of "qualia", this concept cannot be entirely applied to Buddhism because for the Buddhists-Yogācārins the internal demarcations between the object-subject aspects and self-awareness are not real in the absolute sense, they are nothing but the conventional forms of mental constructions. The same concerns the division between subjectivity and objectivity in a more general perspective: it doesn't entail any ontological implications since both are included within the scope of personal experience which in its soteriological form is supposed to become "impersonal" or, better, "super-personal".

The concept of qualia does not counterbalance any general objectivist strategy in Buddhism as it does in the West. Nevertheless, this concept helps us to inquire into the first person perspective with regard to the Buddhist worldview and thus expand its explanatory power beyond the traditional Western borders.

Original Sources and Translations

Hattori, M. (1968). Ed., tr. Dignāga. *On Perceptin being the Pratyakṣapariccheda of Dignāga's Pramāṇasamuccaya.* From the Sanskrit fragments and Tibetian versions transl. and annot. by M.Hattori. Cambridge, MA: Harvard University Press, Cambridge.

Pradhan, P. ed. (1967). *Abhidharmakośa-bhāṣya by Vasubandhu.* K. P. Jayaswal Research Institute, Patna. (in Sanskrit).

Pruden, L.M. tr. (1988). Vasubandhu. *Abhidharmakośabhāṣyam.* Vol.I [Translated] by Louis de La Vallée Poussin; English translation by Leo M. Pruden. Berkeley: Asian Humanities Press.

Sāṇkṛtyāyana, R. ed. (1953) *Pramāṇavārttikālaṃkāra* (Prajñākaragupta). *Being a commentay on Dharmakīrtit's Pramāṇavārtikam.* Patna. (in Sanskrit).

Stcherbatskoi, F.I. ed. (1916). *Tibetskii perevod sochinenii 'Santānāntarasiddhi' Dharmakīrti Santānāntarasiddhiṭikā Vinitadevy vmeste s tibetskim tolkovaniem, sostavlennym Dandvarkharamboi* [Tibetan Translation of the Dharmakīrti's

Santānāntara-siddhi with Vinītadeva's commentary (Ţikā) thereon with Tibetan Commentary by Dandar Lha-rampa]. Petrograd (Biblioteka Buddhica XIX).

Stcherbatsky Th. tr. (1969). Establishment of the Existence of Other Minds. A free translation of Dharmakīrti's Santānāntara-siddhi and Vinīadeva's Santānāntarasiddhi-ṭīkā. *Papers of Th. Stcherbatsky*. Translated for the first time into English by Harish C. Gupta. (Soviet Indology Series No.2) Calcutta, 73–121.

Steinkellner, E. ed. (2005) *Dignāga's Pramāṇasamuccaya*, Chapter I. A hypothetical reconstruction of the Sanskrit text with the help of the two Tibetan translations on the basis of the hitherto known Sanskrit fragments and the linguistic materials gained from Jinendrabuddhi's Ţīkā by Ernst Steinkellner. [E-Reader Version]. Retrieved from http://ikga.oeaw.ac.at/Mat/dignaga_PS_1.pdf (in Sanskrit).

Steinkellner E. ed. *Dharmakīrti's Pramāṇaviniścaya: Chapters 1 and 2, Critically* edited. Beijing, Vienna: China Tibetology Publishing House and Austrian Academy of Sciences Press, 2007.

Tipitaka. Buddhist Pāli Canon. Texts and Translations into English [E-Reader Version]. Retrieved from: http://metta.lk/tipitaka/index.html.

References

Arnold, D. (2012). *Brains, Buddhas, and Believing. The Problem of Intentionality in Classical Buddhist and Cognitive-Scientific Philosophy of Mind*. Columbia University Press.

Block, N. (1995). On a confusion about a function of consciousness. *Behavioral and Brain Sciences, 18*(2), 227–287.

Block, N. (2004). Consciousness. In R. Gregory (Ed.), *The Oxford companion to the mind* (2nd ed.).

Coseru, C. (2015). Taking the intentionality of perception seriously: Why phenomenology is inescapable. *Philosophy East and West, 65*(1), 227–248.

Gallagher, S. (2000). Philosophical Conceptions of the Self: Implications for Cognitive Science. *Trends in Cognitive Sciences*, *4*(1), 14–21.

Garfield, J. L. (2015). *Engaging Buddhism: Why does Buddhism matter to philosophy?* Oxford: Oxford University Press.

Harvey, P. (1995). *The Selfless Mind: Personality, Consciousness and Nirvāṇa in Early Buddhism*. Routledge.

Henkel, J. E. (2012). How to Avoid Solipsism While Remaining an Idealist: Lessons from Berkeley and Dharmakīrti. *Comparative Philosophy, 3*(1), 58–73.

Hussel, E. (1959). Erste Philosophie II 1923–24. *Husserliana* VIII. Den Haag: Martinus Nijhoff, 189–412.

Kellner, B. (2010). Self-awareness (svasaṃvedana) in Dignāga's Pramāṇasamuccaya and -vṛtti: A close reading. *Journal of Indian Philosophy, 38,* 203–231.

Kellner, B. (2014). Changing Frames in Buddhist Thought: The Concept of Ākāra in Abhidharma and in Buddhist Epistemological Analysis. *Journal of Indian Philosophy, 42,* 275–295.

Lysenko, V. G. (2014b). Buddhist Atomism in the Light of Modern Concepts of 'Emergent Property' and 'Qualia'. In M. T. Stepanyants (Ed.), *Philosophy and Science in Cultures of East and West. Russian philosophical studies*, XIII (pp. 131–143). Washington: The Council for Research of Values and Philosophy.

MacKenzie, M. D. (2007). The Illumination of Consciousness: Approaches to Self-Awareness in the Indian and Western Traditions. *Philosophy East and West, 57*(1), 40–62.
MacKenzie, M. D. (2008). Self-Awareness without a Self: Buddhism and the Reflexivity of Awareness. *Asian Philosophy, 18*(3), 245–266.
Merleau-Ponty, M. (1962). *Phenomenology of Perception*, (trans English) C. Smith. London, Routledge.
Piatigorsky, A. M. (1973). Zamechaniya o strukture teksta Dhammasangani III [Observations on the structure of Dhammasangani. TEXT (III)]. In *Proceedings of the Tartu State University*. Issue VI. Tartu, 276–281 (in Russian).
Rosenberg, O. O. (1991). *Trudy po buddismu* [Writings on Buddhism]. Moscow: Nauka publishers. Chief Editorial Board "Oriental Literature", 1991 (in Russian).
Sartre, J. -P. (1967). Consciousness of Self and Knowledge of Self. In N. Lawrence & D. O'Connor (Eds.). *Readings in Existential Phenomenology* (pp. 113–142). Englewood Cliffs, NJ: Prentice-Hall. [English translation of Conscience de soi et connaissance de soi," Bulletin de la Société Française de Philosophie, 42 (1948): 49–91].
Sharma, R. K. (1985). Dharmakīrti on other minds. *Journal of Indian Philosophy, 13,* 55–71.
Siderits, M., Thompson, E., & Zahavi, D. (Eds.). (2011). *Self, No Self?: Perspectives from Analytical, Phenomenological, and Indian Traditions*. Oxford: Oxford University Press.
Thompson, E. (2014). *Waking, dreaming, being: Self and Consciousness in Neuroscience, Meditation, and Philosophy*. Columbia University Press.
Williams, P. (1998). *The Reflexive Nature of Awareness: A Tibetan Madhyamaka Defence*. London: Curzon Press.
Zahavi, D. (2004). *Subjectivity and Selfhood: Investigating the first-person perspective*. MIT Press.

Getting Stuck on Myself: The Cognitive Processes Underlying Mental Suffering

Marieke van Vugt

Many have demonstrated that we spend considerable time in task-unrelated spontaneous thought (Killingsworth and Gilbert 2010; Smallwood and Schooler 2015). While sometimes beneficial for skills such as creativity (Baird et al. 2012), such spontaneous thought can also impair performance and even lead to accidents. Moreover, when spontaneous thought takes a life of its own, as in the case of rumination, it becomes an important risk factor for depression and other forms of psychopathology. Diagnosed depression has a societal cost roughly equivalent to 1% of the total economy in Europe (Sobocki et al. 2006).

Milder forms of such 'perseverative cognition' may not lead to depression, but still cause impairments in concentration (e.g., van Vugt and Broers 2016). An important question is therefore: what turns innocuous and even beneficial spontaneous thought into impairing rumination? In this chapter, I will shed a new light on this process by looking at it through the lens of cognitive science with suggestions for how to study this process. I will then use that to suggest ideas for new interventions for preventing mind-wandering turning into rumination.

Perseverative Cognition and Rumination Defined

Clinical psychology has defined many terms referring to phenomena similar to rumination, e.g., cognitive reactivity, repetitive negative thinking, associative elaboration and perseverative cognition. Perseverative cognition is a more general term, defined as 'repeated or chronic activation of the cognitive representation of one or more psychological stressors' (Brosschot et al. 2006; Verkuil et al. 2010).

M. van Vugt (✉)
Institute of Artificial Intelligence & Cognitive Engineering,
University of Groningen, Groningen, The Netherlands
e-mail: m.k.van.vugt@rug.nl

S. Menon et al. (eds.), *Self, Culture and Consciousness*,
DOI 10.1007/978-981-10-5777-9_19

Such psychological stressors could, for example, be fears about whether you do well enough at work, or memories of a fight at home. In contrast, 'rumination' refers to the more specific process of self-related negatively-valenced thinking that is repetitive and interfering with ongoing cognitive processes (Marchetti et al. 2016; Nolen-Hoeksema and Morrow 1991; Treynor et al. 2003). While rumination pertains to depressed patients, I refer to perseverative cognition as one that occurs in all people. Since I intend to look at rumination and associated phenomena through the lens of cognitive science, I will frame this chapter around 'perseverative cognition'.

Theories of Perseverative Cognition and Rumination

There are many theories of how perseverative cognition and rumination arise, which I have summarised in three different classes. The first class of theories ascribe rumination to an increased attentional bias to negative-valence information (Dalgleish and Watts 1990). This theory states that attentional focus is narrowed such that people are unable to engage with something else (Whitmer and Gotlib 2013). These theories are based, among others, on experiments that show that rumination is associated with quicker responses to negative information (Bradley et al. 1997) and that negative memories are retrieved more easily in people who suffer from depression (Clark and Teasdale 1982). A second class of theories suggests that rumination arises from a defect in inhibition, in particular inhibition of negative self-focused thought processes (Fox et al. 2015; Whitmer and Banich 2007). This theory is, among others, based on experiments in which participants have to switch tasks or suppress associations, and trait ruminators show impairments. This defect in inhibition then leads to a tendency to perseverate: after attention has been captured, the person is not able to let go of negative information (Altamirano et al. 2010). This theory is supported by experiments that show that ruminators are better at maintaining the current task set, but have difficulty changing or updating working memory. A final version of this class of theory is the impaired disengagement hypothesis (Koster et al. 2011), according to which people are not able to let go of self-referring negative information. This has been supported by experiments in which people had to remove negative information from working memory (Joormann and Gotlib 2008). A third class of theories focuses more on the content of what is rehearsed during rumination: even with the same cognitive capacities, if habits of repetitive negative thinking are stronger, then one can get stuck in this repetitive negative thinking as if it were an attractor (Cramer et al. 2016). Similarly, neural data shows indirectly that ruminators have a more narrow network of associations that is evoked by task stimuli (Harel et al. 2016). While these three approaches all predict impairments in the ability to concentrate and perform a cognitive task (as has also been observed for depressed patients), they make different predictions about the relation between cognitive control capacity and task performance.

Using Mind-Wandering as a Theoretical Framework for Understanding Rumination

To compare these different explanations of perseverative cognition, it is helpful to cast them all in terms of the same computational cognitive theory. The most relevant domain for this is mind-wandering, given that rumination is a thought process that occurs spontaneously and is task-independent, just like mind-wandering (Christoff et al. 2016; Smallwood and Schooler 2015). The model is built on an integration of the most important theories of mind-wandering. One influential theory states that mind-wandering occurs whenever cognitive control fails (McVay and Kane 2010), i.e., a problem in goal management. On the other hand, there is also evidence that people can strategically mind-wander when the task allows them to do so, suggesting that mind-wandering can be a conscious choice (Smallwood et al. 2009). Yet other theories of mind-wandering suggest it is caused by failures of meta-cognition (allowing a process that is not intended to take over). This is based on experiments that show that when people indicate that they were not aware of where their attention was, mind-wandering is more prevalent (Mrazek et al. 2013). This has led to researchers to suggest that we need to distinguish between intentional and unintentional mind-wandering (Seli et al. 2016). Essentially, people can choose to strategically mind-wander when task demands are low. On the other hand, when task demands are high, people with low attentional control may still mind-wander, even though that is not adaptive, and they do not intend to mind-wander. Consequently, mind-wandering can be a person's goal. Mind-wandering often involves a shift of attention from the external world to the internal world—a phenomenon referred to as 'perceptual decoupling' (Smallwood et al. 2011). Such a shift may occur when there is very little stimulation in the external world, but also when internal stimuli are sufficiently powerful that they actively turn attention away from the internal world. For example, Klinger (1987) suggested that mind-wandering arises from a person's preoccupation with non-task goals and concerns—consistent with the idea of a competing mind-wandering goal. In short, there is a multitude of factors that could generate mind-wandering.

A Computational Theory of Mind-Wandering

To organise these different theories of mind-wandering, I have developed a computational theory of this model. Computational theories can help to arrive at a fine-grained understanding of a cognitive process, because they can make quantitative predictions about the consequences of different mental strategies. For example, Marewski and Mehlhorn (2011) used this approach to distinguish between different theories of how heuristics are involved in decision-making. Similarly, my computational model of mind-wandering (to my knowledge the first computational theory of this process), formalises some of the main theories of how mind-wandering arises (Taatgen et al., submitted; van Vugt et al. 2015).

The main assumption of my model is that mind-wandering arises when the competition between the goals of mind-wandering and performing a task is won by the goal of mind-wandering. This assumption implements the idea that mind-wandering is driven by current concerns (which strengthen the mind-wandering goal), perceptual decoupling (which weakens the task goal), and intentional mind-wandering (consistent with the idea that not only the cognitive task but also mind-wandering can be a goal). The other main assumption of my model is that mind-wandering consists predominantly of retrieving items from memory, in line with some experimental findings (Ellamil et al. 2016) that demonstrate that at the moment that someone reports mind-wandering, there is an increase in neural activity in the medial temporal lobe, a system associated with memory retrieval. Mind-wandering in my model can be ended by retrieving a memory that reminds the model of the main task, which returns it to the cognitive task at hand. The memory that is being retrieved at a particular point in time is determined by the activation of the different memories of the model—the most active memory wins the competition. This memory will then spread activation to other memories that are associated, which affects what memory is being recalled next, and thereby shapes the series of memories ('thoughts') that are being retrieved. In other words, an individual's associative memory architecture shapes the trajectory of mind-wandering that occurs.

This simple computational model of mind-wandering is implemented in the adaptive control of thought-rational (ACT-R) cognitive architecture (Anderson 2007). ACT-R is a computational theory of cognition that has to date been used to model many different tasks ranging from free recall to task switching, and driving a car. Its parameters and structure has been validated in all these previous applications, thereby providing constraints on my mind-wandering model.

My model reproduces some hallmark phenomena in the mind-wandering literature satisfactorily: it roughly gets the proportion of off-task thinking correct and it reproduces the increase in response time variability before off-task responses relative to on-task responses (Fig. 1). It also reproduces other features of behaviour, such as the proportion of errors, average response times, and the proportion of

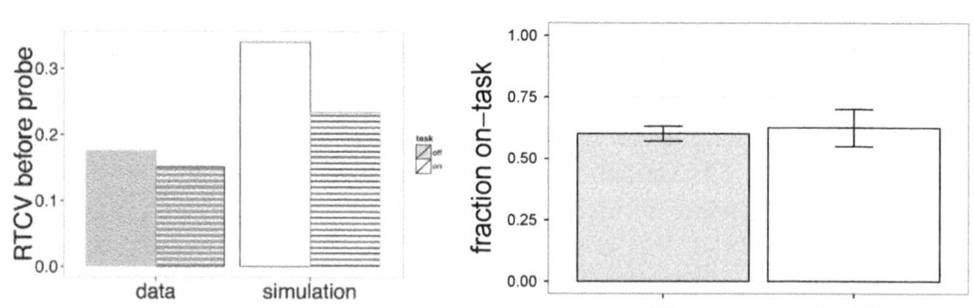

Fig. 1 Computational model of mind-wandering can predict increase in response time variability when off-task (*left*) and fraction of off-task responses (*right*). *White* Empirical data, *grey* model. *Striped* On-task; *open* off-task

thoughts about the past, present and future. In short, on the basis of two main assumptions and leaving most parameters at their defaults, my model can simulate introspection.

Reframing Theories of Perseverative Cognition in a Computational Model

We can now recast the three classes of theories of perseverative cognition in terms of this mind-wandering model. Attentional bias theories posit that rumination is the result of an increased focus on negative information. In my model, that could be implemented as an increase in the activation of negative-valence memories, making them more likely to be retrieved when a person is mind-wandering. Inhibition-based theories posit that rumination arises from an inability to disengage from negative information or conversely, the tendency to perseverate in negative self-referential thinking. In other words, this theory is about a defect in task switching (where the tasks are the cognitive task on the one hand, and ruminating on the other). In general, switching between tasks has previously been modelled in cognitive architectures as requiring the execution of a task-preparation step (Sohn and Anderson 2001), which takes some time, and thereby explains the delay that is generally observed in task performance when participants switch tasks relative to when they stay with the same task. If this task-preparation step is more difficult to perform (in the model this would be implemented as a lower activation for the task-preparation action) then switching from the mind-wandering/rumination process to the main task becomes more difficult. As switching to the cognitive task becomes more difficult, the person will perseverate more in ruminating. The third class of theories posits that rumination arises from particular habits of thought. In my model (which is reported in more detail in van Vugt and van der Velde 2017), the habits of thought are modelled by the activations of memories as well as the pattern of associations between them. Memories that are more often used (and therefore stronger habits) have higher activations, and are therefore more easily retrieved again. Memories can also spread activation to other memories (making the other memories easier to retrieve), thereby

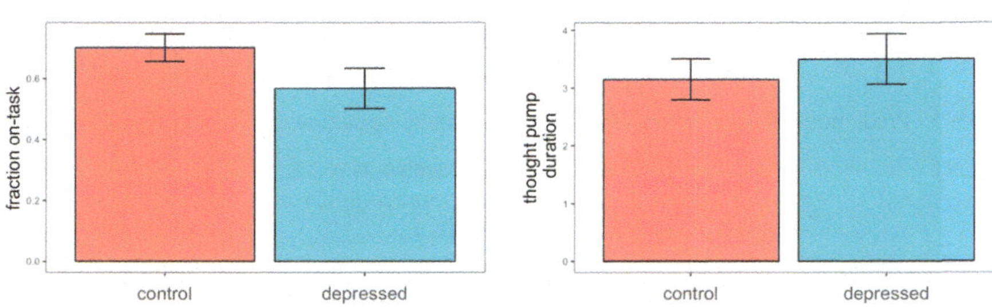

Fig. 2 Model simulations of a computational model of rumination based on manipulating habits of thought. *Left* Ruminating model is more frequently mind-wandering. *Right* Ruminative model has a longer duration of mind-wandering. Ruminative model: *green*; Healthy model: *red*

modelling the associations between memories. The associative network can be configured such that one gets stuck in repeatedly retrieving negative, self-related memories (Bar 2009; Harel et al. 2016). In a preliminary implementation of that mechanism (Fig. 2), I showed that the depressed model (green) is mind-wandering more often than the control model (red), and the duration of mind-wandering is longer. What is remarkable is that the increase in mind-wandering and associated reduction in task performance automatically fell out when we modelled rumination by changing the model's associative network.

Distinguishing Theories of Perseverative Cognition Based on Cognitive Control

While each of the theories of perseverative cognition has been developed to explain a different set of behavioural data, it may be possible to distinguish these theories of perseverative cognition empirically by examining how the duration of perseverative mind-wandering responds to changes in cognitive control (Koster et al. 2011; Marchetti et al. 2016). It is known that cognitive control is an important factor affecting rumination and depression and we will see below that the three classes of theories predict different effects of cognitive control on the duration of mind-wandering for each of these (Table 1). In the modelling framework I am using, cognitive control is implemented as the tendency to check whether the current goal is in line with the intended goal (goal monitoring), and preparing for upcoming tasks when possible (O'Reilly et al. 2010; Taatgen et al. 2009). This means that changing cognitive control would not be associated with any changes in (perseverative) mind-wandering according to theories that focus on an attentional bias towards negative information. The inhibition account of perseverative cognition instead predicts that when there are more cognitive control resources, this helps to strengthen the current task focus, and any spare time is used to prepare upcoming tasks. Together, these control strategies prevent mind-wandering from occurring (Hertel 1997). In this theory, even when mind-wandering occurs, this ensures that the task goals are on general more active, and therefore it is easier to return to the main task, which means that the duration of mind-wandering will be briefer when cognitive control resources increase. In contrast, the habits of thought account suggests that additional cognitive control resources can be used both by the main task and by the mind-wandering [in accord by previous studies of craving: Andrade et al. (2012) and rumination:

Table 1 Predictions of different theories of perseverative cognition

Theory	Cognitive control resources used by MW	Incidence of MW when more cognitive control	Duration of MW when more cognitive control
Attentional bias	No	No change	No change
Problem in inhibition	No	Less	Shorter
Habits of thought	Yes	No change	Longer

MW = Mind-wandering

Altamirano et al. (2010)]. When cognitive control resources are used by the mind-wandering process, this causes the mind-wandering to perseverate even longer.

Table 1 summarises these predictions for how the different classes of theories of perseverative cognition would respond to changes in cognitive control. Note that the three classes of theories make qualitatively different predictions for the duration of mind-wandering episodes (last column). This suggests that we can determine what mechanism is generating perseverative cognition by measuring the duration of mind-wandering. I will next review methods to measure mind-wandering, and discuss how those could be used to measure the duration of this cognitive process.

Methods for Measuring Mind-Wandering in the Laboratory

Since mind-wandering is a spontaneous—rather than task-driven—process, it is a challenge to measure it in the lab, and doing so requires triangulating between different measures. The most common approach is to intersperse self-report questions in a cognitive task, in which participants are asked whether they were mind-wandering at the moment the prompt appeared, and what the mind-wandering was about. Using these methods, it has been demonstrated that participants make more errors and their response time is more variable (Bastian and Sackur 2013) in the periods prior to off-task responses relative to on-task responses.

The problem with thought probe responses is that they effectively create another task within the task, and are only taken at discrete moments, thereby making it impossible to track mind-wandering duration. For that purpose, we would need more continuous neural or physiological measures. In functional magnetic resonance (fMRI) studies, mind-wandering has been associated with activation of the default mode network (Christoff et al. 2009; Mason et al. 2007), a network of brain areas, including the medial prefrontal cortex, posterior cingulate and medial temporal lobe, that are more active during resting periods than during demanding cognitive tasks. EEG studies have most robustly associated mind-wandering with a phenomenon called 'perceptual decoupling'—reduced event-related potentials to external stimuli (Smallwood et al. 2008), but also to increased 9–14 Hz alpha oscillations (Baird et al. 2014; Braboszcz and Delorme 2011).

Although these findings are promising, there is considerable heterogeneity in the observed relations between mind-wandering and brain activity. Such heterogeneity is also evident in physiological findings, in which mind-wandering has sometimes been associated with less phasic pupil responses, and a higher tonic pupil size (Jonathan Smallwood et al. 2011; Unsworth and Robison 2016), but other studies contradict that. Altogether this suggests that we need a more fine-grained understanding of mind-wandering, and the level of perseverativeness of mind-wandering may be a crucial distinction.

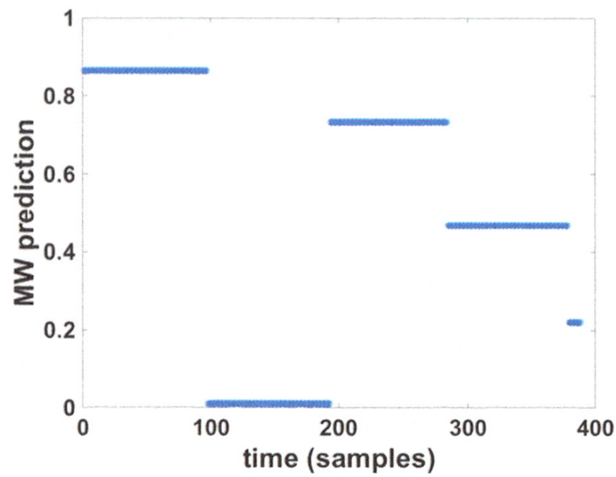

Fig. 3 Elastic net classifier trained on thought-probe-derived samples of intracranial EEG classifies degree of mind-wandering across the whole task for a case study of a single patient

If we have a way of experimentally manipulating perseverativeness, then we can compare neural and physiological activity between perseverative and regular mind-wandering conditions. On the basis of this contrast, we can train machine learning classifiers to distinguish normal and regular mind-wandering on a single-trial level, similar to previous work that has distinguished between on- and off-task thinking on a single-trial level (Mittner et al. 2014). Having trained this classifier, we can then run it on the continuous data and obtain for each moment in time an estimate of whether a person is engaging in mind-wandering or perseverative cognition (Fig. 3). On the basis of these estimates, it will be possible to derive the average duration of mind-wandering.

Methods for Manipulating Perseverative Cognition in the Lab

To create sufficient trials to train a classifier on, we need to be able to manipulate perseverative cognition experimentally. My colleagues and I have recently developed ways to trigger perseverative cognition in the laboratory, in the context of a mind-wandering task and a complex working memory task (Daamen et al. 2016). In this task, participants memorise letters, and the memorisation is interspersed by a task in which they either think about themselves or make other semantic judgments. A robust finding is that self-related thought is associated with a reduction in memory accuracy. A reasonable explanation is that that impairment is caused by a lingering thinking about the self—indeed, perseverative cognition. In line with that idea, participants in this task report thinking more about the words they have to judge when they are in the self-referential condition relative to the control condition.

Another relevant paradigm that has been developed to manipulate perseverative cognition is the current concerns mind-wandering task—a go/no-go paradigm in which a participants' personal concerns (Klinger 1987) are embedded (McVay and Kane 2013; van Vugt and Broers 2016). The hypothesis is that the concerns trigger local increases in off-task thinking, frequently about these concerns. While there is some evidence for that, it is not always consistent (McVay and Kane 2013; van Vugt and Broers 2016). Nevertheless, when we distinguished the thought probes on the basis of self-reported 'difficulty to disengage from the thought' (Verplanken et al. 2007), we observed an increase in the variability of response time and proportion of off-task thinking.

The last task that could be used to trigger perseverative cognition in the lab is free recall with emotionally-valenced words. In this task, participants memorise lists of positive, negative and neutral words. When people are asked to recall these words in any order, the way in which they organize their recall gives information about how their thought patterns are organised (Howard and Kahana 1999). Moreover, searching for items in memory may in itself trigger certain patterns of mind-wandering that are primed by the memory recalls. Indeed, I showed that depressed patients tend to perseverate on recalling negative information, and that that tendency is reduced after Mindfulness-Based Cognitive Therapy (van Vugt et al. 2012). In this task, we can then designate the period in which people persist in recalling negatively-valenced information as perseverative cognition, and compare that to periods in which they recall more freely. In short, at least three different kinds of tasks can be used to create periods of perseverative cognition in the lab.

Is It Likely That Neural and Physiological Measures Distinguish Between Perseverative Cognition and Mind-Wandering?

Even if we are able to manipulate perseverative cognition in the laboratory, how likely is it that neural or physiological activity can distinguish between perseverative cognition and more general mind-wandering? Several lines of research suggest that perseverative cognition has a unique signature. For example, heart rate variability seems to track perseverative cognition (Ottaviani et al. 2016). Recently, Christoff et al. (2016) suggested that perseverative cognition differs from more general mind-wandering in the pattern of connectivity between default mode network areas and frontal cortex, with a reduced frontal influence during perseverative cognition. Similarly, EEG studies have shown differences in oscillatory activity between depressed patients and controls (e.g., Fingelkurts and Fingelkurts 2015; Henriques and Davidson 1991; Pollock and Schneider 1990), although that has not been linked directly to behaviour. Temporally more precise theories of perseverative cognition should we able to allow us to develop more precise neural signatures as well.

How Could Research on Perseverative Cognition Impact the World?

If, through the methods I discussed above, we are able to determine when and how perseverative cognition occurs, and through which mechanism, then what can we do with it? Once we better understand through what mechanism perseverative cognition occurs, this can inform interventions. Currently, there are many interventions for depression, including anti-depressant medication, cognitive behavioural therapy, and Mindfulness-Based Cognitive Therapy (MBCT, Segal et al. 2002). MBCT consists of mindfulness techniques in which people learn to observe their mental patterns in combination with cognitive restructuring exercises derived from cognitive therapy. It is currently the treatment of choice recommended by the United Kingdom's National Health Services. While MBCT has been very helpful in treating depression and preventing relapse, it also has its limitations and does not work for every individual. Could there be other approaches to target perseverative cognition?

Since mindfulness appears to be a promising ingredient of interventions for depression, it is worthwhile investigating other contemplative techniques as well (Dahl et al. 2015). One of the contemplative practices that may be particularly relevant to changing habits of thought is analytical meditation, a practice I have recently started to investigate. This practice may be particularly well-suited to changing habits of thought that are associated with perseverative cognition. Analytical meditation consists of contemplating philosophical ideas (especially those related to the nature of the 'self' and reality) repeatedly, interspersed with resting meditation, in which insights are consolidated, and with testing one's understanding in dyadic debate. Monastic debate is a core activity of life in Tibetan monasteries (Dreyfus 2003; Perdue 1992). The objective of this debate is to reduce excessive self-focus. The debate tackles the perspectives of different philosophers in the Buddhist tradition about what 'self' actually means: is it the body? Is it the mind? When did it start? When did it end? (Epstein 2008; Siderits et al. 2013) Carefully contemplating these (and more) questions slowly causes a realisation to dawn that while the self is usually a focus point of our thinking, and we act as if it is permanent and independent from everything else. In fact, however, it is very difficult to delineate where your 'self' ends and someone else's 'self' begins. This realisation is useful because previous studies have shown that excessive self-focus is associated with rumination, and therefore reducing it should reduce rumination (Ayduk and Kross 2010).

The emphasis on critical reasoning is what analytical meditation differs from mindfulness practices, which focus more on the training of attention and an attitude of non-judgementality. Analytical meditation also differs from (mindfulness-based) cognitive therapy in that it is not directly focused on a person's destructive thought habits, but rather on the underlying assumptions about the nature of the self as something that needs to be protected at all costs against an outside world with which it competes. Potentially, radically changing a person's view of their self can leave maladaptive thought patterns (which depend crucially on the self, since they are self-referential) without basis.

Further Deconstructing the Self in Debate

The individual contemplation of the nature of the self is complemented by dyadic or group debates. In their debate practice, monastics use reasoning accompanied by a highly stylised physical form to deconstruct habitual patterns of self-focus. Our preliminary discussions with the monks suggest that this physical form helps to powerfully turn the intellectual understanding into embodied experience.

Debates centre around the interaction between one or more challengers and one or more defenders. The challenger (or 'clarifier of reasoning' (Lempert 2012)) has the objective of making the defender contradict him- or herself. While the challenger is allowed to say anything, the defender is bound to be consistent with his previous assertions (Liberman 2008). The topic of the debate is decided by the challenger, who will often prompt the defender to start by quoting the relevant part of the text under discussion. After both agree on this, the challenger will derive consequences from this quoted text and earlier statements that the defender can either agree with, state that the reason is not established, or ask the challenger to explain himself. The debate continues back and forth in this way, and both parties are not allowed to spend a long time thinking, otherwise they run the risk being ridiculed (Liberman 2015). If one party needs more time to think, they may rehearse earlier statements to keep up the pace of the debate. In this way, debating helps to solidify the relevant lines of reasoning and deepen conviction in the conclusions of one's reasoning (Dreyfus 2003)—in other words, debate can serve to change habits of thought through extensive reasoning.

The emphasis on developing new habits of thought make debate an interesting target of investigation. Preliminary results from the first studies we have done with the monastics suggest that during the debate, prefrontal 4–9 Hz theta oscillations, frequently associated with concentration (Gevins et al. 1997; Yamamoto and Matsuoka 1990), increase over time, demonstrating that the debate is an attention training (van Vugt et al., in preparation).

Conclusions

In this chapter, I have looked at perseverative cognition, a process of negative self-referential thinking that is associated with psychopathology such as depression, through the lens of computational cognitive science. I have outlined the different theories that exist to explain perseverative cognition: a bias in attending to negative information, a defect in inhibition, or habits of thought that make one get stuck in rehearsing negative patterns of thinking. I have translated these theories into a computational cognitive theory, and demonstrated that they can be distinguished by looking at how cognitive control increases would affect the perseveration of mind-wandering. Interestingly, the perseveration of thinking is what makes it so harmful and interfering with cognition and wellbeing. As a result of this perseveration, new perspectives and external perceptions are discounted or ignored. Instead of planning ahead for the next trial of a task, or adapting to circumstances, people spend their time internally focused, rehearsing the same thought patterns over and over. This means that interventions that reduce perseverative cognition are important. I suggested that analytical meditation practices, inspired by age-old contemplative practices of Tibetan monks, may hold promise as interventions for reducing perseverative cognition. The main reason for this suggestion is that negative self-referential thinking lies at the core of perseverative cognition, and monks are experts at deconstructing self-referential thought, one of the key ingredients of perseverative cognition. Maybe in the future, some of these methods will allow people to be less of a victim of their own self-referential thinking, thereby unlocking the potential of their minds.

References

Altamirano, L. J., Miyake, A., & Whitmer, A. J. (2010). When mental inflexibility facilitates executive control beneficial side effects of ruminative tendencies on goal maintenance. *Psychological Science*. doi:10.1177/0956797610381505.

Anderson, J. R. (2007). *How can the human mind occur in the physical universe?* Oxford University Press.

Andrade, J., May, J., & Kavanagh, D. J. (2012). Sensory imagery in craving: From cognitive psychology to new treatments for addiction. *Journal of Experimental Psychopathology, 3*(2), 45–127. doi:10.1016/jep.024611.

Ayduk, Ö., & Kross, E. (2010). From a distance: Implications of spontaneous self-distancing for adaptive self-reflection. *Journal of Personality and Social Psychology, 98*(5), 29–809. doi:10.1037/a0019205.

Baird, B., Smallwood, J., Lutz, A., & Schooler, J. W. (2014). The decoupled mind: Mind-wandering disrupts cortical phase-locking to perceptual events. *Journal of Cognitive Neuroscience, 26*(11), 607–2596. doi:10.1162/jocn_a_00656.

Baird, B., Smallwood, J., Mrazek, M. D., Kam, J. W. Y., Franklin, M. S., & Schooler, J. W. (2012). Inspired by distraction mind wandering facilitates creative incubation. *Psychological Science*, 956797612446024. https://doi.org/10.1177/0956797612446024.

Bar, M. (2009). A cognitive neuroscience hypothesis of mood and depression. *Trends in Cognitive Science, 13*(11), 63–456.

Bastian, M., & Sackur, J. (2013). Mind wandering at the fingertips: Automatic parsing of Subjective states based on response time Variability. *Frontiers in Psychology, 4,* 573. doi:10.3389/fpsyg.2013.00573.

Braboszcz, C., & Delorme, A. (2011). Lost in thoughts: Neural markers of low alertness during mind wandering. *NeuroImage, 54*(4), 7–3040. doi:10.1016/j.neuroimage.2010.10.008.

Bradley, B. P., Mogg, K., & Lee, S. C. (1997). Attentional biases for negative information in induced and naturally occurring dysphoria. *Behaviour Research and Therapy, 35*(10), 27–911. doi:10.1016/S0005-7967(97)00053-3.

Brosschot, J. F., Gerin, W., & Thayer, J. F. (2006). The perseverative cognition hypothesis: A review of worry, prolonged stress-related physiological activation, and health. *Journal of Psychosomatic Research, 60*(2), 24–113. doi:10.1016/j.jpsychores.2005.06.074.

Christoff, K., Gordon, A. M., Smallwood, J., Smith, R., & Schooler, J. W. (2009). Experience sampling during fMRI reveals default network and executive system contributions to mind wandering. *Proceedings of the National Academy of Sciences USA, 106,* 24–8719.

Christoff, K., Irving, Z. C., Fox, K. C. R., Spreng, R. N., & Andrews-Hanna, J. R. (2016). Mind-wandering as spontaneous thought: A dynamic framework. *Nature Reviews Neuroscience, 17*(11), 31–718. doi:10.1038/nrn.2016.113.

Clark, D. M., & Teasdale, J. D. (1982). Diurnal variation in clinical depression and accessibility of memories of positive and negative experiences. *Journal of Abnormal Psychology, 91*(2), 87–95. doi:10.1037/0021-843X.91.2.87.

Cramer, A. O. J., van Borkulo, C. D., Giltay, E. J., van der Maas, H. L. J., Kendler, K. S., Scheffer, M., et al. (2016). Major depression as a complex dynamic system. *PLoS ONE, 11*(12), e0167490. doi:10.1371/journal.pone.0167490.

Daamen, J., van Vugt, M. K., & Taatgen, N. A. (2016). Measuring and modeling distraction by self-referential processing in a complex working memory span task. In *Proceedings of the 38th Annual Conference of the Cognitive Science Society.* Austin, TX: Cognitive Science Society.

Dahl, C. J., Lutz, A., & Davidson, R. J. (2015). Reconstructing and deconstructing the self: Cognitive mechanisms in meditation practice. *Trends in Cognitive Sciences, 19*(9), 23–515. doi:10.1016/j.tics.2015.07.001.

Dalgleish, T., & Watts, F. N. (1990). Biases of attention and memory in disorders of anxiety and depression. *Clinical Psychology Review, 10*(5), 589–604. doi:10.1016/0272-7358(90)90098-U.

Dreyfus, G. (2003). *The sound of two hands clapping: The education of a Tibetan buddhist monk.* University of California Press.

Ellamil, M., Fox, K. C. R., Dixon, M. L., Pritchard, S., Todd, R. M., Thompson, E., et al. (2016). Dynamics of neural recruitment surrounding the spontaneous arising of thoughts in experienced mindfulness practitioners. *NeuroImage, 136,* 96–186. doi:10.1016/j.neuroimage.2016.04.034.

Epstein, M. (2008). *Psychotherapy without the self: A buddhist perspective.* Yale University Press.

Fingelkurts, A. A., & Fingelkurts, A. A. (2015). Altered structure of dynamic electroencephalogram oscillatory pattern in major depression. *Biological Psychiatry, 77*(12), 60–1050. doi:10.1016/j.biopsych.2014.12.011.

Fox, E., Dutton, K., Yates, A., Georgiou, G. A., & Mouchlianitis, E. (2015). Attentional control and suppressing negative thought intrusions in pathological worry. *Clinical Psychological Science: A Journal of the Association for Psychological Science, 3*(4), 593–606. doi:10.1177/2167702615575878.

Gevins, A., Smith, M. E., McEvoy, L., & Yu, D. (1997). High-resolution EEG mapping of cortical activation related to working memory: Effects of task difficulty, type of processing, and practice. *Cerebral Cortex, 7*(4), 85–374. doi:10.1093/cercor/7.4.374.

Gratz, K. L., & Roemer, L. (2004). Multidimensional assessment of emotion regulation and dysregulation: Development, factor structure, and initial validation of the difficulties in emotion regulation scale. *Journal of Psychopathology and Behavioral Assessment, 26*(1), 41–54. doi:10.1023/B:JOBA.0000007455.08539.94.

Harel, E. V., Tennyson, R. L., Fava, M., & Bar, M. (2016). Linking major depression and the neural substrates of associative processing. *Cognitive, Affective and Behavioral Neuroscience, 16*(6), 26–1017. doi:10.3758/s13415-016-0449-9.

Henriques, J. B., & Davidson, R. J. (1991). Left frontal hypoactivation in depression. *Journal of Abnormal Psychology, 100*(4), 45–535. doi:10.1037/0021-843X.100.4.535.

Hertel, P. T. (1997). On the contributions of deficent cognitive control to memory impairments in depression. *Cognition and Emotion, 11*(5–6), 83–569. doi:10.1080/026999397379890a.

Howard, M. W., & Kahana, M. J. (1999). Contextual variability and serial position effects in free recall. *JEPLMC, 25,* 41–923.

Joormann, J., & Gotlib, I. H. (2008). Updating the Contents of Working Memory in Depression: Interference From Irrelevant Negative Material. *Journal of Abnormal Psychology, 117*(1), 92–182.

Killingsworth, M. A., & Gilbert, D. T. (2010). A wandering mind is an unhappy mind. *Science, 330*(6006), 932.

Klinger, E. (1987). Current concerns and disengagement from incentives. In F. Halisch, & J. Kuhl (Eds.), *Motivation, intention, and volition* (pp. 337–347). Berlin, Heidelberg: Springer. https://doi.org/10.1007/978-3-642-70967-8_23.

Koster, E. H. W., De Lissnyder, E., Derakshan, N., & De Raedt, R. (2011). Understanding depressive rumination from a cognitive science perspective: The impaired disengagement hypothesis. *Clinical Psychology Review, 31*(1), 45–138. doi:10.1016/j.cpr.2010.08.005.

Lempert, M. (2012). Interaction rescaled: How monastic debate became a diasporic pedagogy. *Anthropology & Education Quarterly, 43*(2), 56–138. doi:10.1111/j.1548-1492.2012.01166.x.

Liberman, K. (2008). Sophistry in and as its course. *Argumentation, 22*(1), 59–70. doi:10.1007/s10503-007-9070-y.

Liberman, K. (2015). The logic is made to dance. Rhythm in Tibetan debating. *Etnografia E Ricerca Qualitativa*, (3/2015). https://doi.org/10.3240/81722.

Marchetti, I., Koster, E. H. W., Klinger, E., & Alloy, L. B. (2016). Spontaneous thought and vulnerability to mood disorders: The dark side of the wandering mind. *Clinical Psychological Science*. doi:10.1177/2167702615622383.

Marewski, J. N., & Mehlhorn, K. (2011). Using the ACT-R architecture to specify 39 quantitative process models of decision making. *Judgment and Decision Making, 6*, 439–519.

Mason, M. F., Norton, M. I., Horn, J. D. V., Wegner, D. M., Grafton, S. T., & Macrae, C. N. (2007). Wandering minds: The default network and stimulus-independent thought. *Science, 315*(5810), 5–393. doi:10.1126/science.1131295.

McVay, J. C., & Kane, M. J. (2010). Does mind wandering reflect executive function or executive failure? Comment on Smallwood and Schooler (2006) and Watkins (2008). *Psychological Bulletin, 136*(2), 97–188. doi:10.1037/a0018298.

McVay, J. C., & Kane, M. J. (2013). Dispatching the wandering mind? Toward a laboratory method for cuing 'spontaneous' off-task thought. *Frontiers in Psychology*, 4. https://doi.org/10.3389/fpsyg.2013.00570.

Mittner, M., Boekel, W., Tucker, A. M., Turner, B. M., Heathcote, A., & Forstmann, B. U. (2014). When the brain takes a break: A model-based analysis of mind wandering. *The Journal of Neuroscience, 34*(49), 95–16286. doi:10.1523/JNEUROSCI.2062-14.2014.

Mrazek, M. D., Franklin, M. S., Phillips, D. T., Baird, B., & Schooler, J. W. (2013). Mindfulness training improves working memory capacity and gre performance while reducing mind wandering. *Psychological Science, 24*(5), 81–776. doi:10.1177/0956797612459659.

Nolen-Hoeksema, S., & Morrow, J. (1991). A prospective study of depression and posttraumatic stress symptoms after a natural disaster: the 1989 Loma Prieta earthquake. *Journal of Personality and Social Psychology, 61*(1), 21–115.

O'Reilly, R. C., Herd, S. A., & Pauli, W. M. (2010). Computational models of cognitive control. *Current Opinion in Neurobiology, 20*(2), 61–257. doi:10.1016/j.conb.2010.01.008.

Ottaviani, C., Thayer, J. F., Verkuil, B., Lonigro, A., Medea, B., Couyoumdjian, A., et al. (2016). Physiological concomitants of perseverative cognition: A systematic review and meta-analysis. *Psychological Bulletin, 142*(3), 59–231. doi:10.1037/bul0000036.

Perdue, D. (1992). *Debate in Tibetan buddhism*. Snow Lion Publications.

Pollock, V. E., & Schneider, L. S. (1990). Topographic quantitative EEG in elderly subjects with major depression. *Psychophysiology, 27*(4), 44–438.

Segal, Z. V., Williams, J. M., & Teasdale, J. D. (2002). *Mindfulness-based cognitive therapy for depression: A new approach to preventing relapse*. New York: Guilford.

Seli, P., Risko, E. F., Smilek, D., & Schacter, D. L. (2016). Mind-wandering with and without intention. *Trends in Cognitive Sciences, 20*(8), 17–605. doi:10.1016/j.tics.2016.05.010.

Siderits, M., Thompson, E., & Zahavi, D. (2013). *Self, no self?: Perspectives from analytical, phenomenological, and Indian traditions*. Oxford: OUP.

Smallwood, J., Beach, E., Schooler, J. W., & Handy, T. C. (2008). Going AWOL in the brain: mind wandering reduces cortical analysis of external events. *Journal of Cognitive Neuroscience, 20*(3), 69–458.

Smallwood, J., Brown, K. S., Tipper, C., Giesbrecht, B., Franklin, M. S., Mrazek, M. D., et al. (2011). Pupillometric evidence for the decoupling of attention from perceptual input during offline thought. *PLoS ONE, 6*(3), e18298. doi:10.1371/journal.pone.0018298.

Smallwood, J., Nind, L., & O'Connor, R. C. (2009). When is your head at? An exploration of the factors associated with the temporal focus of the wandering mind. *Consciousness and Cognition, 18*(1), 25–118.

Smallwood, J., & Schooler, J. W. (2015). The science of mind wandering: Empirically navigating the stream of consciousness. *Annual Review of Psychology, 66*(1), 487–518. doi:10.1146/annurev-psych-010814-015331.

Sobocki, P., Jönsson, B., Angst, J., & Rehnberg, C. (2006). Cost of depression in Europe. *The Journal of Mental Health Policy and Economics, 9*(2), 87–98.

Sohn, M.-H., & Anderson, J. R. (2001). Task preparation and task repetition: Two-component model of task switching. *Journal of Experimental Psychology: General, 130*(4), 78–764. doi:10.1037/0096-3445.130.4.764.

Taatgen, N. A., Juvina, I., Schipper, M., Borst, J. P., & Martens, S. (2009). Too much control can hurt: A threaded cognition model of the attentional blink. *Cognitive Psychology, 59*(1), 1–29.

Taatgen, N. A., van Vugt, M. K., Daamen, J., Katidioti, I. and Borst, J. P. (submitted). The resource-availability theory of distraction and mind-wandering.

Treynor, W., Gonzalez, R., & Nolen-Hoeksema, S. (2003). Rumination reconsidered: A psychometric analysis. *Cognitive Therapy and Research, 27*(3), 59–247. doi:10.1023/A:1023910315561.

Unsworth, N., & Robison, M. K. (2016). Pupillary correlates of lapses of sustained attention. *Cognitive, Affective and Behavioral Neuroscience, 16*(4), 15–601. doi:10.3758/s13415-016-0417-4.

van Vugt, M. K., & Broers, N. (2016). Self-reported stickiness of mind-wandering affects task performance. *Frontiers in Psychology, 7,* 732. doi:10.3389/fpsyg.2016.00732.

van Vugt, M. K., Shahar, B., & Britton, W. (2012). The effects of mindfulness-based cognitive therapy on affective memory recall dynamics in depression: a mechanistic model of rumination. *Frontiers in Human Neuroscience, 6,* 257. doi:10.3389/fnhum.2012.00257.

van Vugt, M. K., Taatgen, N. A., Bastian, M., Sackur, J. (2015). Modeling mind-wandering: A tool to better understand distraction. In *Proceedings of the International Conference in Cognitive Modeling*. Groningen.

van Vugt., van der Velde., & ESM-MERGE investigators. (2017). How does rumination impact cognition? A first mechanistic model. In M. K. van Vugt, A. P. Banks & W. Kennedy (Eds.), *Proceedings of the 15th International Conference on Cognitive Modeling* (pp. 25–30). Coventry, United Kingdom: University of Warwick.

Verkuil, B., Brosschot, J. F., & Gebhardt, W. A. et al. (2010). When worries make you sick: A review of perseverative cognition, the default stress response and somatic health. *Journal of Experimental Psychopathology, 1*(1): 87–118.

Verplanken, B., Friborg, O., Wang, C. E., Trafimow, D., & Woolf, K. (2007). Mental habits: Metacognitive reflection on negative self-thinking. *Journal of Personality and Social Psychology, 92,* 41–526.

Whitmer, A. J., & Banich, M. T. (2007). Inhibition versus switching deficits in different forms of rumination. *Psychological Science, 18*(6), 53–546. doi:10.1111/j.1467-9280.2007.01936.x.

Whitmer, A. J., & Gotlib, I. H. (2013). An attentional scope model of rumination. *Psychological Bulletin, 139*(5), 1036. doi:10.1037/a0030923.

Yamamoto, S., & Matsuoka, S. (1990). Topographic EEG study of visual display terminal (VDT) performance with special reference to frontal midline theta waves. *Brain Topography, 2*(4), 67–257. doi:10.1007/BF01129654.

Part V
Consciousness, Experiential Primacy and Knowing

Beyond Panpsychism: The Radicality of Phenomenology

Michel Bitbol

Pure experience is beyond (or below) the level of *being*, and has *no essence*—that is, no property that would distinguish 'it' from anything else. It is lying at the blind spot of what there is because it constitutes the precondition of anything that *is*—namely, of anything that may *appear*. It permeates the show without showing itself. Realising pure experience requires moving backwards from the act of perceiving or taking hold of something, and merging with perceived appearance instead. Realising it requires *not* seeing things from a particular location, but just to dwell, to dwell *here,* in *this* element which has no other location than being the centre. Realising it means cancelling realisation, dissolving intentions, and letting the *one* previsual eye be wide open.

Conversely, sharpening perceptual differences, pushing towards subtler instrumental differentiation, improving the efficiency of handling by technology, increasing the precision of discrimination by scientific theories, perpetuates the blind spot of living and knowing. Something, which being no *thing* is all the easier to forget, remains in the dead angle of objective knowledge; and no move forward can *ever,* in *principle,* account for what has been left behind by this very push. No amount of scientific effort can recover what has been lost by basing science on narrowing down the field of attention, and focusing it on those phenomenal items that can be pointed out and manipulated intersubjectively.

But science also has an important teaching in store about the issue of lived experience. Indeed, scientific advances progressively squeeze the domain of their own silent lived precondition; they contrast this precondition with the network of their objects and concepts, thus stripping it progressively of its false characterisations. The background premise of perceiving, reasoning and knowing is found to be no entelechy, no 'élan vital', no ghost-like soul and no spirit; for the task of those speculative entities has been carefully identified by scientific research and ascribed

M. Bitbol (✉)
Archives Husserl, CNRS/ENS, Paris, France
e-mail: michel.bitbol@ens.fr

S. Menon et al. (eds.), *Self, Culture and Consciousness*,
DOI 10.1007/978-981-10-5777-9_20

to objective 'mechanisms'. Even the words 'mind' and 'consciousness', which were supposed to capture the residual enigma left in knowledge by objectification, have been redefined in terms of blind cognitive functions. Mind is understood as a system of information-processing that allows problem-solving and decision-making. As for consciousness, it is torn between its original meaning which includes lived experience, and a more abstract meaning that only encompasses the functions of meta-cognition and synthesis of representations. In this context, the famous distinction between access consciousness and phenomenal consciousness (Block et al. 1999: 375) aims at preserving some expressive space for what escapes any characterisation as a cognitive function of 'access': the very fact that there are appearances or *phenomena,* the fact that what is accessed *manifests* itself.

To sum up, scientific advances are correlated with a lexical retreat. What escapes scientific characterisation, what cannot be treated as an objective property or entity, becomes increasingly elusive, because any name which may be given to it is confiscated and endowed with a cognitive meaning. With no name and no method of selective handling, 'this' is considered (Cohen and Dennett 2011) as 'outside the scope of science' (which is correct), and thereby negligible (which is highly disputable).

The stakes of the search of a name for the elusive background are then high; not, of course, to elaborate a self-contradicting objective science of the pre-objective, but at least to avoid neglecting the latter. To begin with, the word 'consciousness', even when it is restricted to *phenomenal* consciousness, may have unwanted connotations. Russell (1921) pointed out that consciousness is supposed to be *intentional* (to be consciousness *of* some object), and that it can hardly encompass non-intentional states like pleasure or anxiety, since the latter are usually non-directional tonalities of experience. Moreover, consciousness is usually associated with reflectivity, whereas what we are trying to pick out is what comes prior to any reflection because it is itself the material to be reflected. 'Pure experience', borrowed from James (1976: 117) and his project of a *radical empiricism,* is a reasonable alternative, and the one I have decided to retain throughout this chapter. Indeed, James' experience accommodates non-directional tonalities, and it is said to be 'pure' in view of its anticipating any conceptual structure as well as any self-realisation. Moreover, James' expression 'pure experience' has been found suitable by the Japanese philosopher and founder of the Kyôtô school, Nishida Kitarô (1990: 3), as a valid expression of the deconstructive way of Zen. Yet, the word 'experience' (as any other words that purport to refer to the referring) also has its drawbacks. It irresistibly suggests that there is a single *subject* of experience, be it a 'thin', non-permanent, occasional, subject (Strawson 2006: 191). When the standard terms consciousness and experience are used, the subject-object polarity is still roaming around. An alternative lexicon has then been suggested by authors who definitely lean towards panpsychism but are reluctant to ascribe elementary entities the kind of elaborated synthetic, reflective and self-centred consciousness which we find realised in human beings. Accordingly, their alternative vocabulary relies on mass terms rather than on count terms. Whitehead (1929: 236) offered the word 'feeling', after Peirce. William Clifford also used this term to designate a kind of subject-free experience ('*sentitur* is all that can be said'); but immediately after, he noticed that a feeling is usually a *complex,* and that we still need a name for its 'element' (Clifford 1879: 84). 'Mind stuff' was then retained as a denomination of the simple, plain, subject-free 'feel' of which the universe is said to be made. Much later, even the mentalistic undertones of the panpsychist vocabulary was found misleading, and names for some sort of 'proto-psychic', and 'proto-mental' *continuum* were sought. Feigl (1960) chose the neutral word 'quality', and Wilber (2000) adopted the fascinating word 'depth', situating it below the level of qualification, to capture the elusive proto-*psychè*. Even Nishida Kitarô (1958) progressively abandoned the Jamesian expression '*pure experience*' in favour of a more idiosyncratic *'place of absolute nothingness',* which aims at denoting a process of self-hollowing out that gives perpetually renewed room to manifested beings.

At this point, it is clear that:

(i) Science is leaving a gaping hole behind itself, a 'hole' which is nothing else and nothing less than the most glaring evidence of our existence.
(ii) Science prevents one from conceiving this hole as some occult entity or property that is formally similar to objective entities and properties yet (mysteriously) inaccessible to its methods.

If we wish to obtain a coherent and exhaustive framework in which to accommodate the manifestation as well as the manifest, the tacit certainty as well as the objects of doubt and inquiry, some dramatic initiative must then be taken.

This kind of initiative can adopt two very different directions that were mixed up until now, and that will be separated from this point. The first one is tantamount to moving in the same 'progressive' direction as scientific research, yet speculating in order to identify a missing ingredient construed as a further property of things. The second one consists in adopting the diametrically opposite direction of 'regression' towards what Husserl called the 'realm of the mothers of knowledge' (Husserl 1976: 174)—namely, the experiential background out of which the world of manifest objects is picked out and constituted. I will document both directions in succession, starting with a critical exposition of the first, 'progressive' one, which identifies with panpsychist metaphysics, and then defending the second, 'regressive' one, as the most promising in terms of making global sense of our being-in-a-world.

Panpsychist metaphysics starts from an assumption it shares with the naturalist view tacitly endorsed by scientists, and from two crucial arguments among the many that are available to it (Skrbina 2005: 250). The shared assumption is that there is only one reality which is *physical*, and optionally that physical reality consists of interacting particles of matter (Nagel 1979). The two key arguments are: (1) 'the phenomenon whose existence is more certain than the existence of anything else (is) experience' (Strawson 2006), and (2) there can be no 'emergence' of a feature such as experience, in view of its radical heterogeneity to any standard physical property.

The first argument was expressed as a compelling *deduction* by Descartes ('I think therefore I am', or 'I am thinking therefore I exist' (Descartes 1985: 127). But such expression was deceitful in so far as its syllogistic form is utterly restrictive. Behind the logical pattern, there is an immediate intuitive certainty (Nishida 1990) popping out from the obvious impossibility of *living* its negation (for living its negation would mean living the absence of living). Even Dennett and his followers, the archetypal opponents of the thesis of certainty, have prudently retreated from (or nuanced) their initial assertion that conscious experience is somehow an illusion (Blackmore 2002). In a metaphysical context, Descartes' chief argument turns out to have momentous ontological consequences. Bertrand Russell has expressed it beautifully, in terms that are reminiscent of William Clifford's construal of mental stuff as a basic constituent of the universe: 'Everything that we know of (the world's) intrinsic character is derived from the mental side' (Russell 1992: 402). In other terms, we can access the world as it is in itself only by the direct acquaintance offered by lived experience (Griffin 1997); any other kind of knowledge is derivative and subject to doubt. The same conviction was later conveyed by Feigl (1960), the main proponent of the mind-brain identity theory. He added that, by contrast with this first-hand, direct, intrinsic, access to what there is, physical science is only concerned by indirect access to observables features and structural knowledge. Whether this almost unavoidable downgrading of the ambition of physics is compatible with the physicalist initial assumption of panpsychism will be examined later.

Let's come now to the no-emergence argument. This one is more difficult to shield against criticism, but it is reasonably sound. It has gained some credit from Galen Strawson's stringent description of what emergence of unexpected macroscopic features out of the interaction of known microscopic features would require. According to him, 'There must be something about X and X alone *by virtue of which* Y emerges, and which is *sufficient* for Y' (Strawson 2006). The emergence of (say) liquidity from an interaction of water molecules is relatively easy to account for by way of their Van der Waals interactions; but it looks like there is nothing in the physico-chemical description of atoms and molecules 'by virtue of which' phenomenal consciousness should arise. This obstacle was willy-nilly confirmed by a chief proponent of the emergentist thesis: Jaegwon Kim. In his book entitled *Physicalism or Something Near Enough,* the expression 'near enough' is meant to challenge a standard emergentist view of phenomenal consciousness. Indeed, after a lengthy reflection in which Kim documents several accepted cases of emergent features and categorises them as '*weakly* emergent' (their emergent status being only *apparent,* due to limitations of our conceptual or perceptive abilities), he concludes that phenomenal consciousness is a unique case of *strong* emergence, since it is truly, absolutely, and necessarily unpredictable given the microscopic properties of physical elements (Kim 2005, 2006). *Strong* emergence, here, can be read as a combination of postulated emergence *plus* irreducible ignorance about the 'by virtue of which' this should take place. If one is not dogmatic about the postulate, ignorance is what is left.

In this gloomy situation (for emergentism), the panpsychist strategy looks inviting. It amounts to broadening our concept of the physical by adding to it a new kind of property. This new kind of property purports to be commensurable to what is missing in the physical picture of the world (namely, phenomenal consciousness), yet simple enough to avoid ascribing full-blown conscious minds to every minute physical entity, let alone to every unorganised aggregate of such entities. Since the 'psyche' of panpsychism usually encompasses too many advanced features of human mind, such as reflectivity, self-consciousness, mental processing and complex emotions, this word has been replaced by a variety of more fundamental terms, some of which have already been evoked when the issue of the name of the missing element was addressed. The most uncommitted word is '*proto*psychism', giving rise to the doctrinal denomination 'pan-protopsychism' (Chalmers 2013). Another, more precise, one is 'experience', giving rise to 'panexperientialism' (Nixon 2010).

But this is only the verbal aspect of a systematic rush towards elementarity. The rush started with Leibniz' 'little perceptions' (Leibniz 1993) of simple monads, which are compared with mathematical differentials, and opposed to the integral 'aperception' of high-level monads, which encompasses reflectivity. The same rush is still developing currently, as a contribution to figuring out what should be the additional ingredient needed in order to overcome the emergentist's obstacle. Attempts at characterising the elementary spark of phenomenal consciousness have been made by way of the standard correspondence between mentality and behaviour, provided the concept of behaviour is stretched far enough to encompass the law-like motion of elementary particles of physics (Lewtas 2013). The bottom-level experience is then characterised as follows: devoid of structure and intelligence, non-representational, fleeting, transient, repetitive, first-order (as opposed to the second-order thoughts of reflectivity), patchy, homogeneous, yet *sentient*. A (non-trivial) task to be performed about this raw material is to understand how it can be added up, or 'integrated', into a full consciousness: this is the well-known 'combination problem' (Seager 1995; Goff 2009), which remains to this day a stumbling block of panpsychism.

Although this utterly simple modality of sentience is often conveniently presented in terms of *what it is like to be* the entity endowed with it, several authors have convincingly argued that, in it, any subject-dependence must be denied. Nagel thus considered that pan-protopsychism should posit entities with no standpoint at all (1979), and Chalmers (2013) evoked something like 'experiences without subjects' or 'unsensed sensa'. The limit of this process of peeling any remnant of dualism from the protopsychic qualities can probably be found in the Whiteheadian concept of 'actual occasion', an activity of realisation which is 'always a fusion of subject and object in the unified event of an experience' (Stenner 2008). From that point on, the usual divide between subject and object can be recovered by a series of operations on actual occasions. On the one hand, 'the subject creates itself through its feelings' (Stenner 2008). On the other hand, 'subjectivity perishes and it becomes an *object* for subsequent subjects' (Griffin 1997). This means that the inert arises in the wake of the processual 'occasions', and that the objectified domain is a stabilised residue of past experience.

From the latter characterisations of Whiteheadian process philosophy, one perceives a strong tension which spreads across the network of panpsychist doctrines. On the one hand, the basis and starting point of this family of doctrines is first-person experience, and on the other hand, the stance of this (as of any other) speculative metaphysics is tacitly objectifying, external, spectator-like, even when it purports to describe the arising of an objectified domain from past experience: it surreptitiously objectifies the very process of objectification. Thus, according to Whitehead (1929), the substance of the world is made of a plurality of 'actual occasions' whose content derives from lived experience (first side). But the so-called 'substance' has a recognisable physical model which is the events of special and general relativity theories (Whitehead 2004); its description (Whitehead 1929) thus tacitly implies that it be treated as an object of representation made of 'real things' (second side). Hartshorne (1978) then aptly called Whitehead's philosophy an 'objective idealism', since it tends to identify the sentient constituents of nature with entities spread in front of the mental gaze. Yet, sentience is a concept isolated from the direct acquaintance of our own pre-objective experience, in contradistinction with nature construed by the culturally dominant view as a horizon of description for objective sciences. This general tendency to mixing up two very different sources of knowledge is confirmed by later proponents of process philosophy, who are even more explicit than Whitehead in their attempt to keep the description of feelings and sentience within the sphere of a physicalist doctrine and a realist epistemology (Griffin 1997). A panpsychism of this kind looks like a baroque combination of idealism and physicalism, which turns out not to be conceptually viable.

Let us then make a closer examination of the main dilemmas of panpsychism. To begin with, the choice of starting from first-person experience is not consistently pursued throughout its doctrinal expressions. At first sight, the sought consistency is achieved by the demand of an ontological continuity in nature (Whitehead 1967), similar to the materialist one, yet in terms of elementary experiences instead of physical properties. In the same way as the materialist requires that the manifest universe be made of physical entities and properties up to the highest level of organisation, the panpsychist here claims that it consists in 'feelings' or elementary experiences down to the lowest level of organisation. But actually, this is a pretence. Indeed, unlike in materialism, the ontological continuity of panpsychism does not go together with existential continuity. The ontological continuity (and homogeneity) reached by materialism arises from constant adoption of the existential stance of intentional directedness and search for objectification of phenomena. Here, everything is categorised as an object of knowledge, and nature itself is construed as an all-encompassing object. True, as we know, a massive feature (the experiential *non*-thing) is forgotten or neglected as a consequence of this goal-oriented existential stance of materialism, but this is the price to be paid for methodological unity and efficiency. By contrast, the ontological continuity of panpsychism (experience all the way down the ladder of nature) relies on adopting *two* incompatible stances: (i) the receptive and reflective stance needed to realise that the unavoidable basis of any inquiry is lived experience, and (ii) the intentional directedness needed to pinpoint the elements of nature to which protopsychic/experiential properties are ascribed, and, before that, to isolate such properties in thought.

A host of difficulties derives from this existential discrepancy underpinning panpsychism. The root-difficulty manifests itself in the ambivalent attitude of panpsychists towards physicalism. As mentioned previously, most of them claim to be physicalists, or at least objectivists. But they are bound not to endorse all the consequences of a picture of the objective world derived from physics, or at least not to take it at face value since that would mean embracing materialism. This half-way position was beautifully and synthetically expressed by Galen Strawson: a panpsychist is likely to be a physicalist, but not a *physicS-alist*. In other terms, despite her overt commitment to philosophical physicalism, a panpsychist philosopher cannot restrict her ontology to the observable objective properties of actual physics, since she is precisely trying to complete her representation of nature with 'feelings', 'depths' or elementary experiences. But this dissociation between an overt physicalism and a critique of the limitations of physics proves quite uneasy to handle.

One consequence of this unstable position is that there seems to be a free choice between many available representations of observable nature. Indeed, standard panspychists do not take physics seriously in its day-to-day research and critique of old-fashioned ontologies, but use it as a vague inspiration for various plausible narratives about the objective world, by borrowing from the popular science expositions or from the history of this discipline. Accordingly, a mereologically inclined panpsychist may take advantage of a familiar mereological view of objectified nature to prop up her doctrine. She may, for instance, establish a one-to-one correspondence between her atoms of sentience and the elementary particles of high-energy physics, trying to figure out what it is like to be such a particle (Lewtas 2013). By contrast, a holistically inclined variety of panpsychist is likely to prefer the field-theoretic representation of quantum physics (or even of classical electromagnetism), thus connecting the power of synthesis of consciousness with the continuum of a field. There exist as many physical representations around as there are functions of consciousness. Moreover, both varieties of panpsychist seem to ignore that any such representation is likely not to be taken at face value, as a picture of nature-as-it-is-in-itself, but rather as a picture-*like* tool for orienting predictions and actions (Bitbol 1996a, b; Van Fraassen 2010). In other terms, the majority of panpsychist physicalists adopt alternately a naïve attitude towards physics, by adhering to the somehow dubious ontological claims of its popularisers, and a distant attitude towards the sound epistemological lessons that can be drawn from it.

Another unwelcome consequence of the panpsychist combination of physicalism and distance with respect to physics, is that one must define the 'physical' *in abstracto,* by using terms broad enough to avoid any explicit reference to *physics* as it stands. A widespread proposition of this sort is borrowed from Neurath's well-known characterisation of physicalism (Neurath 1983): this doctrine is concerned by phenomena *located and extended in space-time*. Panpsychism is then defined as the position that holds that (i) 'the universe is spatio-temporal in its fundamental nature' (Strawson, 2006), and (ii) spatio-temporally located elements are endowed with protopsychical qualities. The problem is that space-time here seems to be given the absolute status of a repository of those 'external' entities and qualities whose 'interiority' is proto-psychical. This may be acceptable from the standpoint of an intentional/objectivistic stance (Ruyer 2012), but certainly not from the standpoint of the receptive/reflective stance. Indeed, in the latter stance, even the spatial oppositions of internal and external, right or left, higher or lower, are supposed to arise from the structure and/or activity of experience taken as an ultimate origin. This conflict of two opposite conceptions of space, according to the epistemic stance one adopts, is no novelty. It was already vivid at the beginning of the eighteenth century, when orthodox Newtonians defended an absolutistic version of space, whereas Leibniz considered that space expresses the system of relations of (sentient) monads, which is represented within each single monad. And the same conflict was brought to completion by Kant, who contrasted a transcendental realist and a transcendental idealist conception of space-time as 'a priori forms of sensibility'. Here again, we find that even the minimal, *non*-physicS-alist, expression of physicalism is at odds with the allegedly experiential starting point of panpsychism.

As Schrödinger (1986: 149) cogently pointed out, once lived experience has been left aside in order to elaborate an objective picture of the world, 'If one tries to put it in or on, as a child puts colour on his uncoloured painting copies, it will not fit. For anything that is made to enter this world model willy-nilly takes the form of scientific assertion of facts; and as such it becomes wrong.' Panpsychism is the unambiguous target of this criticism. It represents a clumsy attempt at overcompensating the consequences of adopting the intentional/objectifying stance needed to do science, by adding to it (or by replacing it with) patches of experience very similar to the patches of colour added on the surface of an uncoloured drawing. As soon as this is done, the new picture of the world looks like a scientific picture, apart from the unfortunate circumstance that its additional elements cannot be put to test as it would be the case of a scientific theory. This does not make panpsychism plainly *wrong*, but rather torn apart between its phenomenological origin and its temptation to mimic a theory of the objective world. As a consequence, panpsychism proves unable to define adequate criteria of validity for its own claims.

To recapitulate, panpsychists have so much focused on the project of recovering an ontological continuity in nature without forgetting its most obvious manifest feature, that they have broken their own existential continuity in doing so, and that this existential gap has generated a host of difficulties or conceptual conflicts. Borrowing bits and pieces from physicalism and attempting to bring them to completion by adding a first-person-like ingredient turns out to be a deadlock.

From that point on, the way is open to a completely different strategy that would not only *accommodate* the glaring fact of lived experience, but that would stick to it consistently throughout. This strategy is phenomenological, and it is clearly a descendent of *transcendental idealism*, in line with the initial intention of Kant and Husserl. But precisely for that reason, it is no caricatural or metaphysical version of idealism: it does not consist in claiming that consciousness is all there is, but rather adopting systematically the receptive/reflective stance, thus making sense of everything else (including science) from the standpoint of that stance. As Husserl (2007: 48) pointed out, the appropriate form of 'idealism is no metaphysical substruction, but the only possible and absolute truth of an *ego* recollecting itself on its own doings and its own donation of meaning'. Husserl's phenomenology is no theoretical idealism, but rather an attempt at describing faithfully a practical life developing within the experiential soil of ideas.

If we now take one more step, by dropping the ego-centred vocabulary of Husserl, and renouncing the claim to truth in favour of lived certainty, it becomes even more obvious that the doctrinal 'ism' of idealism is null and void. Moreover, if, in the course of cultivating the receptive/reflective stance, we unfold all the dimensions of lived experience, including the concrete manifestations of embodiment (in line with Maurice Merleau-Ponty and Michel Henry), even the 'ideal' component of idealism is no longer dominant. It turns out that the advocated strategy is no doctrinal assertion about the nature of being (ideal as opposed to spatio-temporal), but the uncompromising adoption of a way of being (phenomenological as opposed to 'natural').

Let us further realise at this point that the epistemological lessons of modern physics (especially quantum physics) are perfectly consonant with the basic stance of phenomenology. Indeed, from the standpoint of a reflective phenomenological stance, it is natural to understand theories in terms of the mental and performative conditions for acting *in* the world, rather than in terms of what they capture of the nature *of* an allegedly independent world. Here, physical theories are not concerned with *representing* the world as it is in itself, but rather expressing the structure of a wide class of adaptative transformations in the world. A large part of Kant's philosophy was devoted to support this kind of interpretation of the physical theories of his time. He thus showed that the structural core of Newtonian mechanics can be derived from the general preconditions of objective knowledge of spatio-temporally extended phenomena called material bodies (Kant 2004; Bitbol et al. 2009). Far from having an empirical origin and a representational status, this structure is provided in advance by the project of objectifying a fraction of what is presented to us in sensible experience. A similar strategy can be adopted to understand reflectively, rather than representationally, the structure of quantum theory (Bitbol 1998), with the welcome consequence that many of the so-called 'paradoxes' of quantum mechanics (such as the cat paradox and the 'mystery' of non-locality) are dissolved from the outset by this move (Bitbol 2011).

Adopting the phenomenological stance then has momentous consequences for the *premises* of the problem of consciousness. Firstly, as mentioned above, the starting point of inquiry in experience is not only confirmed by phenomenology but also stabilised and systematically explored. Secondly, from a phenomenological standpoint, no reason is left for sanctifying the 'physical' of physicalism. If 'physical' only means 'observable in space-time', adopting physicalism is tantamount to turning objects of perceptive attention and intentional directedness into intrinsically existing beings, with no indisputable experiential warrant in favour of this thesis. And, if 'physical' means falling into the domain of validity of physics, or being encompassed by the concepts of a physical theory, it becomes even more obvious that no ontological status can be granted to it, since (from a phenomenological standpoint) those concepts are nothing more than by-products of an enactive (Varela et al. 1992) process of sense making applied to the bulk of what appears.

The basic slogan of panpsychism is thus completely overturned. We have seen that, according to panpsychists, *there is more to the physical (that is, to nature) than what a physics of observable objective properties can tell us;* and only in this 'more' can the nucleus of lived evidence be accommodated. By contrast, along with a phenomenological approach, the said 'more' does not come after, but *before* any observation and any theorising has occurred. The certainty that there is more to the manifest world than *a set of objects* of attention or thought, is latent but unassailable from the outset; it has been obvious ever since the mythical birth of reason. Antecedently to what appears, there is appearance; antecedently to things that can be shown, there is the show; antecedently to present items, there is the presence. Or, rather, more accurately, it is meaningless to separate what appears from appearance, what is shown from its showing, what is present from presence. The problem of fitting lived experience into an empty and silent universe, and the (panpsychist) solution that consists in dispersing atoms of sentience in the spatio-temporal network of this world, only comes after an artificial split has been accepted within the continuum of what is lived. Upstream to this split, there is no problem but only an ultimate mystery. Upstream to this split, there is the mystery of mysteries: *why is there manifest presence rather than none?* This question and perplexity is even broader that Leibniz' famous question as to *why there is something rather than nothing*, since the latter can still be (mis)understood in the narrow sense of 'why are there things rather than no thing'.

The difference between a problem and a mystery is that a problem is a target for future solutions, whereas a true mystery is the unquestioned origin of any problem and any solution, thereby remaining inaccessible to the process of questioning and solving. Phenomenology is the reminder of this origin.

Many difficulties of panpsychism are automatically dispelled by adopting the phenomenological stance. Firstly, phenomenology is not underpinned by a couple of mutually exclusive positions or attitudes (the reflective and the objectifying); phenomenologists rather understands the need and meaning of a plurality of attitudes, including the objectifying one, from the standpoint of their purely reflective attitude. Secondly, phenomenology is not concerned by the 'combination problem' of panpsychism—that is, the problem of building a whole of synthetic experience out of a plurality of elements of experience dispersed in the world. It bypasses this problem not because it has chosen to privilege a holistic physical representation (as some versions of field-theoretical panpsychism do), but because it bases its enquiry on a 'gestalt-psychological' starting point: the present, global, embodied, human, experience of the researcher. Its typical question is then not 'how does the whole of experience arise from experiential or protoexperiential sparks?', but rather, conversely, 'how can one isolate something like the elementary constituents of a presently given integrated experience?' Thirdly, as a consequence of the previous point, the method of phenomenology for accessing elementary experience is neither rational nor observational, but lived. Unlike panpsychism, phenomenology does not rely on behaviour, and even less on a stretched concept of behaviour applied to the law-like motion of physical objects, to figure out indirectly what it might be like to have a truly simple and transient experience. Instead, it uses a direct method of access that digs through our complex and cumulative kind of experience, stripping it layer after layer of its perceptive or intellectual fabrications, and trying to reach beneath it a deeper layer of pure presence. This method is called the *epochè* (suspension of judgement and attentional focusing), and it is assisted by successive 'reductions' of what appears to *strata* of less and less elaborated structures of appearance. It can also be radicalised by practices of meditation of various types, culminating with the mere ability to see any present object or mental activity as a naked and unfathomable presence (this is the realisation of 'suchness', *Tathatā* in Sanskrit, *Nyoze* in Japanese). Fourthly, the ontological issue is not skipped by phenomenology but rather transformed beyond recognition. Instead of looking for a distantiated ontology of observable objects (as in the natural sciences) or an extension of such an ontology that encompasses conceivable atoms of experience (as in panpsychism), phenomenology adumbrates an ontology of immersion, of connivance, of acquaintance; it looks for an 'oblique ontology' of intertwining (de Saint-Aubert 2006), or, as Merleau-Ponty writes, an 'intra-ontology' (Merleau-Ponty 1964: 279). This is an ontology expressed *from* the inmost recesses of the process of being, rather than an ontology *of* the external contemplation of beings.

Let us start with the third point, the issue of method. Accessing layers of experience from within looks like practising archaeology. This was already mentioned by Freud (1921), in his famous *Gradiva,* and even more poignantly by Ferenczi (1938) in his *Thalassa*, which purported to reach the deepest layer of our experiential phylogenesis and ontogenesis (the oceanic feeling) at the far end of an intensive psychoanalytic cure.

The main tool phenomenologists use in order to carry out this kind of archaeological digging, as mentioned previously, is the *epochè*. From a phenomenological standpoint, practicing the *epochè* does not mean triggering some artificial mutation of the state of consciousness. It rather means *undoing* a primordial mutation which (in our ontogenetic past) transformed a universally open, and unfabricated, experience into an operative state of consciousness striving to acquire skills in the use of objects and in one's adaptation to a social network. The original *epochè* was practised by the Greek sceptics; it means suspending judgement, cultivating doubt, and thereby reaching 'ataraxy' (inner peace); and its original acceptation is 'cessation', remarkably similar to the Sanskrit word 'nirodha', which opens Patañjali's *Yogasûtra.*[1] But the phenomenological *epochè* is somehow different from the sceptic's one, and more profound. It does not imply to end up in doubt, but rather to reach a state of neutrality, neither doubt nor faith. And it does not limit itself to suspending *discursive* judgements whose function is to qualify pre-existent objects of perception or thought. Instead, the phenomenological *epochè* digs down to the

[1] '*Yogaś cittavṛittinirodhah*': 'Yoga is the cessation of mental processes.'

pre-verbal spontaneous judgements by which our perceptive intentional directedness posits objects in front of us. It then radically neutralises the tacit belief in a real external world, which is typical of the 'natural' ontological attitude. As soon as this process of suspension is brought to completion, the phenomenologist is ready to redirect her attention towards the lived sources of beliefs and judgements. But this new modality of attention is by no means easy to acquire, since it completely disrupts old habits, and is of an entirely unanticipated nature. Indeed, the new 'pre-natural' domain to explore does not offer anything special to capture because it merges with the act of capturing, it is not categorisable in an unequivocal way because it includes the effort of categorising. Then, according to Husserl, as soon as the *epochè* has been achieved, we are likely to be completely disoriented. 'We are initially in a situation which is akin to that of a person who was born blind, who has just undergone surgery, and who must now literally learn to see' (Husserl 1972: 171). We are bound to grope around, and to make abundant use of half-deceptive analogies. This is probably why the descriptions of what is left after each step of the *epochè*—namely, the experienced material of successive phenomenological 'reductions'—can vary to a large extent from author to author, and sometimes in successive writings of a single author.

Thus, according to classical empiricism, once the naïve belief in external objects has been suspended, what is left is a field of sense-data. Although he was inspired by empiricism, Husserl considered that the *epochè* opens us to the much broader domain of *transcendental consciousness,* in which we, as subjects, recognise not only the sensory contents but also our own acts of intentional directedness and the so-called 'noematic' forms by which the activity of defining and classifying objects is achieved. This new domain is immanent to the experiencer rather than transcendent with respect to her, as a set of natural objects would be. Heidegger later generalised Husserl's approach and attempted to disconnect it with any remnant of the concept of a subject. According to him (Heidegger 1984: 39), the *epochè*, and the subsequent phenomenological reductions, amount to avoiding exclusive fascination by the manifest *entities* and realising the brute fact that they *are*. Becoming able to do so means recognising the 'ontological difference' between beings and being (between '*Seiende*' and '*Sein*' in German). Pushing this de-subjectifying tendency one step further, Jan Patočka even denied the abstract characterisation of the field disclosed by the *epochè* as 'immanent' *qua* opposed to 'transcendent'. He rather suggested that the *epochè* contents itself with freeing one from any thesis whatsoever, be it a sceptical or reflective thesis; that it merely opens the way towards *what there is,* independently of any construction (Patočka 1995: 187).

A (maybe too) simple theoretical tool offered by Husserl to make sense of the regressive path that goes from elaborated consciousness to elementary experience, is the distinctions he makes between: the (sensory) matter and the (pre-conceptual) form of experience; the *hylè* and the *morphè*; the noesis (flux of lived experience and thought) and the noema (structure of the intentional object). One may then oppose: (i) the non-intentional hyletic and noetic components of consciousness that are immanent and represent what is left after experience has been stripped of its interpretative layers, and (ii) the projected noematic intentional structures that have to be suspended, bracketed, neutralised in order to make the former component appear. Then, the 'pure' experience we are looking for by way of the *epochè* looks like it should identify to the hyletic part of the conscious process as opposed to the morphic (or formal) part. But this simple (and immemorial) separation between matter and form is undermined when Husserl points out that, after all, even the interpretative layer made of intentional acts and noematic structures, partakes of what is the case in pure experience: 'What is lived includes in its real *(reell)* composition not only the hyletic moments (colours, sensory sounds), but also the apprehensions which animate them' (Husserl 1950: 339).

Michel Henry took advantage of the latter reflection to disrupt completely Husserl's hylemorphism. According to him (Henry 1990, 2000: 72), any experience, be it the experience of an 'external' object, or the experience of abstract forms, or the experience of one's own mental acts, ultimately turns out to be nothing else than an experience of the *self-affection of one's flesh*. Russell's examples of pleasure and pain are here taken as paradigmatic, and even intentionality, even the assumption of transcendence of natural objects, must be (and is actually) rooted in the immanent impression that the flesh of the living being is making on itself. In other terms, even the *morphè* is experienced as *hylè*; even the perception of a patch of colour on some 'outer' object is underpinned by a self-perception of the perceiver; not even the most abstract conception of a mathematical structure can dispense with being rooted in some concrete self-sensitive modality of the living. Intentional consciousness is borne by non-intentional experience, and therefore the deepest layer of the archaeological stratification of consciousness (the purest kind of experience) is nothing else than a naked self-sensitivity. '*Original* can only mean this: what comes in itself before any intentionality and independently of it, before the space of a gaze, before the 'outside oneself' of which intentionality is only a name; what comes … before the world, out of the world, foreign to any conceivable 'world', *a-cosmic*' (Henry 2000: 82). In other terms, what comes before intentionality, and before the belief in a world, is the non-directional impression of being there: the awareness of being embodied without any notion of the separation between one's own body and anything else, the silent voice of the body whose psychological name is 'cenesthesia'. But once again, unlike the naturalistic programme, which would undertake to account for the latter impression in terms of the interaction of a number of natural objects belonging to the human body, the phenomenological programme adopts the diametrically opposite stance. It starts from the deepest layer of what we are, and then tries to justify the belief in a natural world as a consequence of the multifarious differentiations of such primeval experience.

This phenomenological programme was developed and radicalised in Merleau-Ponty's last and posthumous book, *The Visible and the Invisible*. According to Merleau-Ponty, 'we can accept a world … only after having witnessed its arising from our experience of *raw Being*, which is like the umbilical cord of our knowledge, and our source of meaning' (Merleau-Ponty 1964: 209). This being granted, the first move must be to identify the characteristics of what has been called the 'experience of raw Being'. In order to do so, Merleau-Ponty does not rely so much on discursive thought, as on some very straightforward 'thought-experiments', which are in fact *experiential* experiments. Let me give two examples of those experiential experiments, chosen to be as close as possible to everyday life (Abram 1997).

The first one is to be performed by opening our eyes and weakening progressively the tension towards identifying a variety of objects in the visible environment. If we do so, we cease to see ourselves as a bodily object in front of other objects, and we realise something even more stunning than in Ernst Mach's well-known 'self-portrait'. We perceive distinctly that the seeing emerges from the midst of what is seen, from a sort of hole of transparency which breaks through its dense impenetrability, and that this density is boundless, homogeneous, extending far beyond our skin.

The second experiential experiment is done with our eyes shut. It consists in rubbing a cloth or a smooth stone, progressively concentrating the attention on the sensation that arises from this contact, and suspending judgements about the distinction between objects. Then, what is felt is an undifferentiated mass of sensation in an atmosphere of complete reversibility: one can hardly say if what is felt is the friction of the hand on the cloth, or the caress of the cloth on the hand, and in fact there is no contrast between the two feelings.

The discursive expression of what is experienced in such occurrences could be called radical *situatedness*. Indeed, in view of this kind of experiential experiments, we are not onlookers of a nature given out there, but rather intimately intermingled with the said nature, somewhere in the midst of it (Merleau-Ponty 1964: 152). We are not point-like spectators of what is manifest, but we are a field of experiences which merges with what appears in a certain region of it. From a theoretical standpoint, this immediately gives rise to what has already been called an *intra-ontology*, an ontology of the participant *in* being, rather than an ontology of the observer *of* beings. Here, being is not presented before me as an object of sight, but my vision arises from the middle of being. 'Vision is the tool which allows me to be absent to myself, and to contemplate from within the fission of Being' (Merleau-Ponty 1985: 81).

Then, from an experiential standpoint, the way is open to realising what is usually neglected due to educational prejudice—that is, our true complicity with nature which has too hastily been isolated from us, as if it could be reduced to an object of perception and thought. The preliminary condition for this rebirth of a full-blown vision, is to alleviate consciousness from the symmetrical belief in an *ego* able to posit itself firmly in front of things, after having taken some of these things as an exclusive target of attention (Merleau-Ponty 1964: 77). Indeed, only thus can one gain the patient and unbound sensitivity needed to know what it is like to be in continuity with the world. But as soon as we have settled in this state of coincidence, what do we perceive exactly, or, more exactly, what is perceived through us?

As a precursor of Michel Henry, Merleau-Ponty answers this question with a single word: our *flesh*. We perceive our flesh not as something separate, but as the perceiver itself, a self-perceived perceiver. The flesh is that strange being endowed with complete reversibility, since it is jointly perceived and self-perceiving. The archetypal case of a two-faced kind of perception is the sense of touch, which, unlike distantiated vision, is synchronously appearance of what is touched and self-revelation of the touching in its carnal thickness. Besides, the sense of touch also has a less organic way of revealing the two-fold nature of the flesh, in so far as one can perceive by means of one part of one's own body (say the left hand) the softness and resistance of some other part (say the right hand), while this latter part of the body is also used to touch other things. Here, one is no longer dealing with the fusion of standard perception and intrinsic self-perception, but with a duality of status and functions (*toucher* and *touched)* realised by one and the same body. Now, it turns out that this second, milder, case of reversibility is also realised by vision. As it is clear from the first experiential experiment we described above, the flesh is both *seen* as this (human) visible body, and *seeing* from a certain hollow location of this body in the middle of the head.

This first step of description of our own situation is almost trivial. But dazing triviality is precisely the subject matter of phenomenology. Phenomenology is this unique discipline that sharpens our sensitivity to trivial yet primeval facts of experience which are usually ignored due to our exclusive fascination for the objects of our possible study.

From this point on, Merleau-Ponty turned triviality into a thrilling epic of being, going far beyond what Michel Henry thought acceptable. He invited us to let go any ready-made objectified analysis of the world, especially its division into a plurality of bodies (mine and yours, ours and the inert bodies, etc.), and accordingly not to project it onto our self-analysis of the flesh. He recommended, in other terms, to become fully receptive to experience as it is, with its undifferentiated, unbounded, massive, opaque style of appearing, instead of perceiving it through the glasses of our adult educated system of distinctions. As soon as this is done, the outcome is stunning. Even the identification of the flesh to *this* precise body is bound to melt, since the process of self-definition and positing boundaries between bodies has been voluntarily suspended. The whole surrounding world is then perceived as an extension of 'my' flesh, and it thereby becomes *the flesh of the world, the world as flesh*. 'Where should we locate the boundary between the body and the world, Merleau-Ponty asks, since the world *is* flesh?' (Merleau-Ponty 1964: 182). There are no such things as 'me', 'you', and the world, but a single canvas wherefrom various self-individualising centres of sensitivity emerge, and which leaves patches of half-obscurity between these centres (Barbaras 1993: 304). The role of constituting objectivity, which had been entrusted to the transcendental *ego* or consciousness by Husserl, and which had been extended to the own-body by the first Merleau-Ponty (who wrote *The Phenomenology of Perception*), is further extended to whatever has the status of a flesh by the second Merleau-Ponty (who wrote *The Visible and the Invisible).* But since the flesh is boundless, since the flesh is the whole world, any division between the constituter of objectivity and the constituted objects is meaningless. Just as the flesh is self-perceiving, the world *qua* flesh is self-objectifying.

This conception of a self-objectifying world can be read as a turnabout of Michel Henry's characterisation of the 'original' pure experience. True (as Henry pointed out), the original experience is 'out of the world' in the sense that it comes before any organisation of what appears into a collection of objects ready to be reached and used; but at the same time (according to Merleau-Ponty) this original experience *is* the world-flesh still undifferentiated, capable of objectification yet not objectified.

We can see at this point some similarities and a huge gap between panpsychism and Merleau-Ponty's boundless phenomenology of the world-flesh. In both cases, any difference of nature between mental and mindless, sentient and inert beings is denied. In both cases, experience suffuses what there is. This point of convergence is so deep and so striking that one is surprised to find a recent historian of panpsychism declaring that there is no equivalent of a panpsychist tendency to be seen anywhere in phenomenology (Skrbina 2005). But it is also true that panpsychist approaches and Merleau-Ponty's phenomenology use so deeply antagonistic methodologies that the former similarity can easily be overridden. As noticed previously, panpsychist metaphysics starts its inquiry with the late by-product of an effort to find centres of meaning for an adult activity and objective knowledge; and it then sticks on them sparklets of experience whose simple content is imagined by a kind of behaviourist extrapolation. In other terms, panpsychism takes the ordered *cosmos* of reason and science for granted, and painfully adds to it what looks missing in it. By contrast, Merleau-Ponty's phenomenology is a strategy of extreme deconstruction of the *cosmos* of reason and science, not because it ignores such cosmos but because it purports to reach its fountainhead. Such fountainhead can be exhumed below the strata of perceptive and scientific interpretations, and it turns out to be an unelaborate and unbound embodied experience of self-sensitivity. Then, as soon as this embodied experience is brought to (its own) light, it appears that there is, and can be, nothing else. The usual distinction between experience and its worldly objects relaxes, and what remains is only a world-of-experience. To sum up, the difference between panpsychism and world-flesh phenomenology is between gluing bits of experience onto an allegedly given 'external' world, and (re)discovering the world-as-experience, neither external nor internal. As Merleau-Ponty pointed out, '(In the existentialist branch of phenomenology), writes a critic, one has turned off the mind. It's just the opposite. It has been put everywhere, because we are not mind and body, we are not a consciousness facing the world, but embodied mind, being-in-the-world' (Merleau-Ponty 1996: 92).

This offers a good transition from phenomenology to another 'archaeological' strategy of digging below ordinary consciousness to unearth pure experience: the strategy of Zen.

In Sôtô Zen, as expressed by Dôgen, there exists nothing like our Western *Nature:* a totality of being exposed under the gaze of perception and reason, splitting into a plurality of beings, and having an essence which unfolds by way of a law-like sequence of phenomena. This notion was not used in the Japanese culture, at any rate, and the word *shizen* that was proposed at the end of the nineteenth century as a translation of 'Nature' means 'thus by itself' (Dôgen 2013, p. 233). Instead of the hidden essence and the abstract concept of a whole, this word only evokes the manifest and most spontaneously concrete *thusness*. Another Japanese word, *shô,* also means nature, but in the narrower sense of 'the nature of something'; and its *Kanji* ideogram is connected to that of a sprout born from earth (Dôgen 2013: 252). Here again, the concreteness of presence is retained instead of the abstractness of conceptualisation. The same tendency is even more obvious in the term which is used in Japanese literature, for nature as unspoilt environment: just 'mountains and rivers', or 'mountains and water' (*sansui*). Of course, one could still think at this point that this concreteness is that of ordinary life, with its pre-organisation of the given into a collection of objects: a mountain, a river, some grass, a bunch of trees. If it were the case, one would remain far from the sought pristine vividness of pure experience prior to any level of objectification. But beware: such opposition between the two kinds of experiences (primitive and elaborated) might still be a trace of the kind of duality Zen is trying to overcome. True, *Mountains* and *water* can be understood as names of objects, but they can also be heard as audible evocations of gleams of presence. They are not only designations of forms, but also calling forth of emptiness (of own-being); they are not only denominations of items relative to an activity of categorisation, but also summonings of their absolute present manifestation (Glassman 2002); they carry out both functions at the same time, in good agreement with the celebrated slogan of the Heart Sûtra: 'Emptiness is form and form is emptiness.' But then, the acceptation of the word 'nature' becomes closer and closer to the acceptation of the word 'experience'. In the state of awakening (or enlightenment), nature and experience just merge. Like in Merleau-Ponty, nature arises *qua* present pure experience. This fusion of nature and pristine experience in a state of awakening is confirmed by many texts of Dôgen, such as: 'The mountain and water are the realization as presence of the way of elder enlightened beings' (Dôgen 2012: 102); 'Seeing mountains and rivers is tantamount to seeing the nature of the enlightened one' (Dôgen 2007: 45). Like in Merleau-Ponty, also, the locus of this deep connection between experience and nature is likely to be the body's flesh. Indeed, according to Nishida Kitarô's philosophical reflection on Zen, in the selfless mode of engagement with our environment that is allowed by pre-conceptual experience, 'The world becomes the self's body' (Krueger 2006: 21); in other terms, the self's body is phenomenologically extended to the whole world, and the world is felt as a protraction of the body. True, if, according to the stance of ordinary life, the body is considered as a separate entity, it appears 'transient like dew on the grass'; and the instruction is then to 'cast off body and mind' (Tanahashi 1985: 29). But the body is also the basic instrument we have to realise our thoroughgoing homogeneity with the experienced world, or the world-as-experience, and prescriptions about the position of the body are therefore crucial components of Zen's teachings. These prescriptions pave the way for recognising the world as one's extended experienced body.

Here again, such recognition needs no effort, but rather the cessation of effort. It requires no elaborate sequence of thinking that starts with separating a knowing subject from the objectified nature, later discovers that a basic ingredient is missing in this abstracted concept of nature, and finally tries to compensate for it by reintroducing the appropriate features. It just implies full acceptation of experience as it stands, with its fluid and porous boundaries, with its ability to dissolve and regenerate categories moment after moment, with its flavour and outlines that are pregnant with a concept of world.

As soon as the effort is relaxed, what unfolds is 'pure' experience, as described by Nishida Kitarô. Experience is 'pure' in so far as it is free from discriminations, free from any attempt of looking for the possible objective cause of its qualities, free from any value judgement about them, and even free from questioning about what they might be. Experience is pure and monolithic, because in the absence of discriminations within it, there is no duality between subject and object left. This being granted, pure experience is 'devoid of meaning' (Nishida 1990: 8). It is meaningless because it rests in its own presence, without purporting to establish relations between it and any past or future experience. Meaning is an impulsion of a present object called 'sign' towards what it signifies; judgement is a connection between the present felt quality and other similar qualities experienced in the past; but pure experience does not escape itself towards anything, it just happens to *be*. In particular, pure experience does not *mean* the world, it is not experience *of* the world; it *worlds* so to speak (using the uncommon verb 'to world'), it is melted with the multifarious lived aspects of what can later, after the work of categorisation has crystallised again, be called 'the world'. Pure experience does not turn down the world, nor does it turn down judgement or thought; it just adheres to, or even identifies with, the world and the process of thinking; it is the concrete immediate reality of the world and thinking. For, after all, reality is not a big object, it is just plain existence. The only aspect of thought pure experience does not retain is its basic incentive, which is wanting and craving. 'Thinking … is simply an incomplete state of pure experience' (Nishida 1990: 17). Starting to think is then caused by a feeling of incompleteness, and pursuing an ordered series of thoughts is an attempt at reconstructing the unity of knowledge and will that has been lost due to this very first move.

Panpsychist thought, with its attempt at compensating the inaugural abandonment of the knowing subject in the wake of objectification, is a remarkable instantiation of this process. But as soon as pure experience realises its self-generated state of incompleteness, it immediately self-recovers. It then encompasses every aspect of itself, including the quest for universal forms and the urge towards necessary connections which are crucial components of thought. In other terms, instead of compensating for incompleteness, pure experience comes back to a state of self-completion, not by some regression to an age of innocence but by a comprehensive acceptance of any dimension of itself. Since *worldling* is one of these aspects of experience, any retrospective compensation of the barrenness of an objectified world by panpsychism becomes pointless.

References

Abram, D. (1997). *The spell of the sensuous.* Vintage.
Barbaras, R. (1993). *De l'être du phénomène.* Jérôme Millon.
Bitbol, M. (1996a). *Schrödinger's philosophy of quantum mechanics.* Kluwer.
Bitbol, M. (1996b). *Mécanique quantique: une introduction philosophique.* Flammarion.
Bitbol, M. (1998). Some steps towards a transcendental deduction of quantum mechanics. *Philosophia Naturalis, 35,* 80–253.
Bitbol, M. (2011). Reflective metaphysics: Understanding quantum mechanics from a Kantian standpoint. *Philosophica, 83,* 53–83.
Bitbol, M., Kerszberg, P., & Petitot, J. (Eds). (2009). *Constituting objectivity: Transcendental perspectives on modern physics.* Berlin: Springer.
Blackmore, S. (2002). The grand illusion: Why consciousness exists only when you look for it. *New Scientist,* 26–9, June 22, 2002.
Block, N., Flanagan, O., & Güzeldere, G. (1999). *The NATURE OF CONSCIOUSNESS.* MIT Press.
Chalmers, D. (2013). *Panpsychism and panprotopsychism. Amherst lecture in philosophy.* http://consc.net/papers/panpsychism.pdf
Clifford, W. K. (1879). *Lectures and essays* (Vol. II). London: Macmillan.
Cohen, M. A., & Dennett, D. C. (2011). Consciousness cannot be separated from function. *Trends in Cognitive Science, 15,* 64–358.
de Saint-Aubert, E. (2006). *Vers une ontologie indirecte: sources et enjeux critiques de l'appel à l'ontologie chez Merleau-Ponty.* Vrin.
Descartes, R. (1985). Discourse on the method IV. In *The philosophical writings of René Descartes* (Vol. 1). Cambridge University Press.
Dôgen. (2007). *Shôbôgenzô* (Vol. 3). Sully.
Dôgen. (2012). *Shôbôgenzô* (Vol. 1). Sully.
Dôgen. (2013). *Shôbôgenzô* (Vol. 7). Sully.
Feigl, H. (1960). The mind-body problem: Not a pseudo-problem. In S. Hook (Ed.), *Dimensions of mind* (pp. 24–36). New York University Press.
Ferenczi, S. (1938). *Thalassa, a theory of genitality.* Albany.
Freud, S. (1921). *Delusion and dream, an interpretation in the light of psychoanalysis, of Gradiva, a novel, By Wilhelm Jensen.* Allen & Unwin.
Glassman, B. (2002). *Infinite circle.* Shambala.
Goff, P. (2009). Why panpsychism doesn't help us to explain consciousness. *Dialectica, 63,* 289–311.
Griffin, D. (1997). Panexperientialist physicalism and the mind-body problem. *Journal of Consciousness Studies, 4,* 68–248.
Hartshorne, C. (1978). Physics and psychics: The place of mind in nature. In J. B. Cobb and D. R. Griffin, Jr. (Eds.). *Mind in nature: The interface of science and philosophy.* http://www.religion-online.org/showbook.asp?title=2066
Heidegger, M. (1984). *Les Principes fondamentaux de la phénoménologie.* Gallimard.
Henry, M. (1990). *Phénoménologie matérielle.* Presses Universitaires de France.
Henry, M. (2000). *Incarnation.* Seuil.
Husserl, E. (1950). *Idées directrices pour une phénoménologie.* Gallimard.
Husserl, E. (1972). *Philosophie première* (Vol. II). Presses Universitaires de France.
Husserl, E. (1976). *La Crise des sciences européennes et la phénoménologie transcendantale.* Gallimard.
Husserl, E. (2007). *De la réduction phénoménologique.* Jérôme Millon.
James, W. (1976). *Essays in radical empiricism.* Harvard University Press.
Kant, I. (2004). *Metaphysical foundations of natural science.* Cambridge University Press.
Kim, J. (2005). *Physicalism or something near enough.* Princeton University Press.
Kim, J. (2006). *Philosophy of mind.* Westview Press.

Krueger, J. W. (2006). The varieties of pure experience: William James and Kitaro Nishida on consciousness and embodiment. *William James Studies, 1,* 1–37.
Leibniz, G. W. (1993). *Nouveaux essais sur l'entendement humain.* Flammarion.
Lewtas, P. (2013). What it is like to be a quark. *Journal of Consciousness Studies, 20,* 30–64.
Merleau-Ponty, M. (1964). *Le visible et l'invisible.* Gallimard.
Merleau-Ponty, M. (1985). *L'œil et l'esprit.* Gallimard.
Merleau-Ponty, M. (1996). *Sens et non-sens.* Gallimard.
Nagel, T. (1979). *Mortal Questions.* Cambridge University Press.
Neurath, O. (1983). Physicalism: The philosophy of the Vienna circle. In R.S. Cohen and M. Neurath (Eds.), *Philosophical papers 1913–1946* (pp. 48–51). Reidel.
Nishida, K. (1990). *An inquiry into the good.* Yale University Press.
Nishida, K. (1958). *Intelligibility and the philosophy of nothingness: Three philosophical essays.* East-West Center Press.
Nixon, G. M. (2010). From panexperientialism to conscious experience: The continuum of experience. *Journal of Consciousness Exploration & Research, 1,* 213–401.
Patañjali. (1995). *Yogasûtra.* Motilal Banarsidass.
Patočka, J. (1995). *Papiers phénoménologiques.* Jérôme Millon.
Russell, B. (1921). *The analysis of mind.* George Allen & Unwin.
Russell, B. (1992). *The analysis of matter.* Routledge.
Ruyer, R. (2012). *Néo-finalisme.* P.U.F.
Schrödinger, E. (1986). *What is life & mind and matter.* Cambridge University Press.
Seager, W. E. (1995). Consciousness, information, and panpsychism. *Journal of Consciousness Studies, 2,* 88–272.
Skrbina, D. (2005). *Panpsychism in the west.* MIT Press.
Stenner, P. (2008). A.N. Whitehead and subjectivity. *Subjectivity, 22,* 99–109.
Strawson, G. (2006). *Consciousness and its place in nature.* Imprint Academics.
Tanahashi, K. (Ed.). (1985). *Moon in a Dewdrop, writings of Zen Master Dôgen.* North Point Press.
Van Fraassen, B. (2010). *Scientific representation: Paradoxes of perspective.* Oxford University Press.
Varela, F. J., Thompson, E., & Rosch, E. (1992). *The embodied mind.* MIT Press.
Whitehead, A. N. (1929). *Process and reality.* Cambridge University Press.
Whitehead, A. N. (1967). *Science and the modern world.* The Free Press.
Whitehead, A. N. (2004). *The concept of nature.* Prometheus Books.
Wilber, K. (2000). *Integral psychology.* Shambala.

Is 'Information' Fundamental for a Scientific Theory of Consciousness?

Nithin Nagaraj and Mohit Virmani

Introduction

We live in the Information Age, thanks to rapid developments in communication systems, networking and computer technology. The birth of the Information Age can be traced back to Claude Shannon's paper in 1948, in which he developed a mathematical theory of communication and gave a probabilistic definition to the notion of 'Information' (Shannon 1948). The act of choosing 'one' unique message from 'many' possible messages gives rise to the generation of information which can be precisely quantified by the celebrated Shannon's entropy formula (Shannon 1948). Since then, information theory has been widely applied in diverse fields in science and engineering. Arguably, 'information' may even be a fundamental building block of reality as posited by the theoretical physicist John Wheeler (Wheeler 1989). His 'It from Bit thesis' postulates that, deep down, the foundation of reality is built by the brick of 'information'—and that all things physical are in fact information-theoretic in origin. In this chapter, we are interested in whether 'information' can play a fundamental role in another important aspect of our reality—'Consciousness'.

Consciousness is our 'subjective experience', which is unique and peculiar in its own sense, such as a 'feeling of pain', 'taste of coffee', 'perceived sensation of the colour green', or in a more general sense, the experience felt by any organism, i.e.

N. Nagaraj (✉) · M. Virmani
NIAS Consciousness Studies Programme, National Institute of Advanced Studies, Bengaluru 560012, India
e-mail: nithin.nagaraj@gmail.com

S. Menon et al. (eds.), *Self, Culture and Consciousness*,
DOI 10.1007/978-981-10-5777-9_21

'What's it like to be?' (Nagel 1974). It may be hard to define consciousness, but, the very existence of consciousness is nearly impossible to refute.[1] What is the reality of our experience? This question has troubled philosophers and thinkers in every civilisation.

Having convincingly argued the insufficiency of physicalism or reductionism to explain consciousness, Chalmers (1995), in his path-breaking paper, proposed a new programme for building a non-reductive theory of consciousness. Chalmers argued for treating consciousness as a fundamental entity (like mass, charge, spin) rather than attempting to reduce consciousness to brain function or explained purely in terms of computational or neural mechanisms. Chalmers observed the isomorphism between physically embodied information spaces with phenomenal or experiential information spaces, and thus postulated the notion of 'double-aspect information' (Chalmers 1995, 1996). In his proposed conception of the world, information is truly fundamental and is comprised of dual aspects—corresponding to the physical and the phenomenal features of the world. His notion of information is built on Shannon's idea. Chalmers' speculative proposition of the existence of a fundamental link between 'information' and 'experience' did not receive much attention in the science of mind, until recently, when Tononi (2004) proposed a theory of consciousness borrowing similar (and building on them) ideas from information theory. Tononi's integrated information theory or IIT (Tononi 2004; Oizumi et al. 2014) is an axiomatic non-reductive scientific approach which takes phenomenology of experience and its characteristics as primary and builds a theory to explain consciousness as the capacity of a system of mechanisms (constituted of elements which can either be neurons or logic gates) to integrate intrinsic information (Oizumi et al. 2014). Intrinsic integrated information, though not in the sense of Shannon, plays a central role in this new theory to measure the quantity of experience and to represent its quality as well.

In this chapter, after reviewing Shannon's mathematical formalism of 'information', we trace the fundamental ideas behind IIT and highlight its limitations as well. We then discuss another important clinical measure of consciousness known as Perturbational Complexity Index or PCI (Casali et al. 2013). Both, IIT and PCI belong to the new genre of scientific measures which are rooted in what is called as 'complexity theories of consciousness' (Seth et al. 2008). We explore the link between integrated information, Consciousness and Data compression. We then briefly discuss a novel compression-complexity approach (Virmani and Nagaraj 2016) that we have recently proposed to bridge the gap between IIT and PCI. Thus, the latest trend in scientific theories of consciousness has been on the problem of 'measurement'. To this end, recent scientific literature in the field has witnessed the proposal of a medley of scientific measures which attempts to capture brain

[1]As you are reading this very footnote, can you deny having an intimate and rich experience of seeing, interpreting and understanding?—all these are suffused with consciousness. Even if this were to be a dream or a hallucination, it is impossible to deny the existence of 'experiencing' (something rather than nothing).

complexity and its link to consciousness. We are yet to determine whether information and complexity are fundamental notions necessary to explain and measure consciousness.

'Information'—A Primer

Consider the following list of terms: 'data', 'measurement', 'information', 'knowledge' and 'experience'. These are familiar terms to all of us and we use them in our daily life. However, if asked to provide precise definition to these terms, we may find it difficult especially as we move from 'data' towards 'experience' in this list. Being a subject of experiences, we all have a direct access to our own experience—for example, the visual perception of these words on a white background as we read them. At the other extreme, anything that is presented to me, the subject, prior to any kind of interpretation, can be defined as data (for example, the raw perception of visual input). In between these two extremes, we have the other terms which are not easy to define or understand: 'measurement', 'information' and 'knowledge'. At the centre of this list is 'information' which we intuitively feel to be more than 'data', but somewhat less than 'knowledge' or 'experience'. The idea of information seems to be an important link in this chain and an understanding of information will hopefully connect data with experience.

One might ask, what is the need to define and understand information? Can't one just be content with one's own experience? We already have some idea of what we mean by data—it is the raw sensory input to me, the subject, from the world around me through the five sense organs. For every organism (including humans), it is not possible to survive without making sense of this data and interpreting the content of experience. If I am unable to understand that my visual perception is indicating the presence of a four-legged carnivore running towards me, I will not be able to take evasive action for my survival (like running as fast as I can in the opposite direction or climbing the nearest tree). Hence, it is imperative that data about my experience is interpreted. This is where information comes into the picture. It is this right choice of the 'one' among the 'many' possibilities, that many a times decides life or death. Information, as we shall see, is about quantifying the choice of 'one' among 'many'. This is not the only notion of information and when it comes to understanding and measuring consciousness, we shall need to expand the very notion of information.

What Is Information?

One important definition of 'information' was provided by Claude Shannon (1948) in the context of building a mathematical theory of communication. Shannon, in many ways was a pioneer in giving a quantifiable definition of information.

Though, we currently have several other notions of information, Shannon's original definition of information continues to be widely used today in various disciplines. Thus, it is instructive for us to begin our journey by understanding Shannon Information. This also paves the way for our understanding of other notions of information in the context of characterising conscious experience since these essentially build on top of Shannon's ideas.

Shannon's Mathematical Theory of Communication: The Birth of Information Theory

Shannon not only defined information in precise mathematical terms, but produced an entire theory of information communication (now recognised as Information Theory), establishing fundamental limits for the storage and transmission of information both in the absence and presence of noise. Before we discuss the mathematical details of Shannon's theory, let us understand the context and the motivation for Shannon to propose a definition of information.

The Problem of Communication

In Shannon's own words from his immortal 1948 paper, 'the fundamental problem of communication is that of reproducing at one point either exactly or approximately a message selected at another point'. As shown in Fig. 1.

Shannon makes it clear that semantic aspects of communication are irrelevant to the engineering problem of transmission of a message selected from a set of possible messages. Thus, he is concerned about modelling the purely mechanistic (and probabilistic) aspects of communication and determining the limits of information transmission both in the presence and absence of noise.

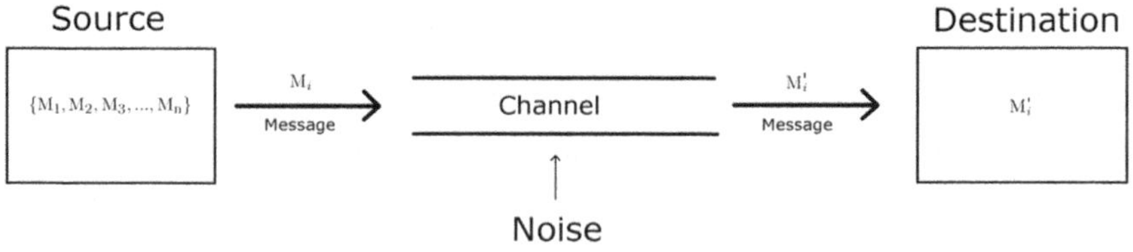

Fig. 1 Shannon's model for communication. A single message M_i is chosen from a finite set of possible messages $\{M_1, M_2, \ldots, M_n\}$ to be transmitted across the noisy channel. It is this choice of 'one' among 'many' that creates an uncertainty or information that can be precisely measured by Shannon's celebrated formula

Shannon 'Entropy': A Measure of Uncertainty, Surprise and Information

One of the key contributions of Shannon's theory is a mathematical definition of information. Shannon was interested in quantifying the uncertainty associated with choosing a single message out of set of all possible messages. Let us denote self-information or the uncertainty associated with an event as a function $i(.)$ where the argument of this function is the probability $P(.)$ of the event whose uncertainty we are quantifying. It is intuitive to demand that the mathematical function $i(.)$ satisfies the following axioms:

1. $i(\text{the certain event}) = 0$ units
2. $i(\text{event } A) \leq i(\text{event } B)$ if and only if $P(\text{event } A) \geq P(\text{event } B)$
3. $i(\text{event})$ is a continuous function of probability of the event
4. $i(\text{event } A \text{ AND event } B) = i(\text{event } A) + i(\text{event } B)$ if and only if events A and B are independent.

Axiom 1 states that there is no uncertainty in the certain event (probability of occurrence = 1) and thus the self-information of the certain event is 0 units. Axiom 2 states that quantity of information conveyed by a less likely event is more than that of a more likely event. Axiom 3 ensures that a slight perturbation in the probability of an event will not lead to drastic changes in the quantity of information. Axiom 4 states the self-information is additive for independent events. As an example, if I throw a die and toss a coin simultaneously, then the self-information of the event of obtaining a 3 on the die and 'Heads' on the coin simultaneously will be the sum of the self-information of the event of obtaining a 3 and self-information of 'Heads' since the two events are independent of each other. This is reasonable since observing 3 on the die gives no extra information about whether the coin will turn out as 'Heads' or 'Tails' and vice versa. There is no information that is 'shared' between the two events. Quantity of self-information of an event can be equally thought of as 'degree of surprise if the event occurred' or 'amount of uncertainty revealed by observing the event'. It is useful to switch synonymously between the notions 'self-information', 'degree of surprise' and 'uncertainty'.

The mathematical function that satisfies all the aforementioned axioms is given by $i(\text{event } A) = -\log_2(P(A))\text{bits}$. Having defined self-information of events, we can define the average uncertainty of information of a random experiment as the expectation of self-information of all the events of the random experiment. In mathematical terms, if X is the random variable corresponding to the random experiment of our interest with N possible outcomes $\{x_1, x_2, \ldots, x_N\}$ with corresponding non-zero probabilities given by $\{p_1, p_2, \ldots, p_N\}$, then the average uncertainty or average information or Shannon entropy of X is defined as:

$$H(X) = E(i(X)) = \sum_{i=1}^{N} p_i(-\log_2(p(x_i)) = -\sum_{i=1}^{N} p_i \log_2(p_i) \text{ bits.} \qquad (1)$$

Note: $E(Y)$ is the expected value of the random variable Y, here $Y = i(X)$. Events with zero probability yield no contribution to $H(X)$, though their self-information would be infinite, and therefore are not considered. Shannon effectively showed that Eq. (1) is the only possible mathematical expression that satisfies all the axioms of information (Shannon 1948). $H(X)$ is measured in bits.[2] An important point to note is that if there are two random experiments X and Y where the number of outcomes of Y are more than that of X and if in both cases all the outcomes are equally likely, then $H(Y) > H(X)$. For example, when an unbiased coin is turned, Shannon entropy is 1 bit whereas when an unbiased die is thrown, $H = 2.59$ bits.

Other Notions of 'Information'

Shannon's notion of information is not the only one in the literature. Stanford encyclopaedia of philosophy (Adriaans 2013) lists other notions of information. These are:

1. Fisher information
2. Kolmogorov complexity
3. Quantum information
4. Information as a state of an agent
5. Semantic information.

The first three (along with Shannon's information) are quantitative whereas the last two are qualitative. Shannon and Fisher's notions of information depend on probability theory, Kolmogorov complexity is based on the notion of computation, quantum information is built from the principles of quantum mechanics, semantic information is based on the core concept of 'true beliefs', and information as a state of an agent refers to the result of the process of being informed. In this chapter, we will not be referring to these concepts of information and will stick to Shannon's definition of information. However, towards the end of this chapter, we will mention the relevance of semantic information to conscious experience. We will briefly visit the interesting link between information and physics.

[2]Since the base of the logarithm is chosen as '2' in Eq. (1), the units of $H(X)$ is bits. Alternatively, if we chose the natural logarithm instead, the units would be 'nats'.

'Information' and 'Physics': 'All Information Is Physical' Versus 'All Things Physical Are Information-Theoretic in Origin'

In his 1996 paper titled 'The physical nature of information', Landaeur proposes that information is always tied to a physical representation (Landauer 1996). Most scientists (especially physicists) would argue that information is just like other physical quantities such as temperature. Information, like temperature, is a secondary quality and can never exist independently of the physical system. The metaphysical creed that is opposed to this view maintains that information is not only independently existing of the physical system, but it is a primary fundamental quality of reality. John Wheeler was a leading proponent of this view, what he called as the 'It from Bit' thesis (Wheeler 1989). In his own words:

> I suggest that we may never understand this strange thing, the quantum, until we understand how information may underlie reality. Information may not be just what we 'learn' about the world. It may be what 'makes' the world. An example of the idea of it from bit: when a photon is absorbed, and thereby 'measured'—until its absorption, it had no true reality—an unsplittable bit of information is added to what we know about the world, 'and', at the same time, that bit of information determines the structure of one small part of the world. It 'creates' the reality of the time and place of that photon's interaction. (Ford and Wheeler 1998: 644)

For Wheeler, deep down, at the very bottom, the physical world has an immaterial source and explanation—'information' which '*may be what "makes" the world*'. Wheeler was clear that '…, *all things physical are information-theoretic in origin and this is a* ***participatory universe***' (emphasis Wheeler's 1989).

For the moment, let us not concern ourselves whether 'information' has a role to play in the fundamental nature of physical reality, as that is a topic worth independent investigation and outside the scope of this chapter.

'Information' and 'Consciousness'

Let us return to the main focus of this chapter—'conscious experience'. We are interested in describing and understanding experience, and we wish to know whether Shannon's notion of information has any role in such an explanation? Chalmers, in his now classic paper (Chalmers 1995), while arguing for a non-reductive theory of consciousness, proposed that 'information' could be the bridge between the physical and the phenomenal.

Chalmer's 'Double-Aspect Information': Bridging Physical and Phenomenal Domains

Chalmers sets out to build a non-reductive theory of consciousness (Chalmers 1996). To this end, he unearths what he calls as 'psychophysical laws'—laws or principles which connects consciousness and physical processes. The first of these is the coherence principle which connects consciousness to awareness, or global availability. The second is the principle of structural coherence that connects the structure of awareness to the structure of consciousness. The third is the law of organisational invariance which states that two systems (for e.g., one made up of neurons and the other made up silicon chips) with exactly the same fine-grained functional organisation will have qualitatively identical experiences. These two psychophysical laws are non-basic components of a theory of consciousness which enable to constrain the final form of the theory. They are not fundamental or basic laws in his theory of consciousness. He proposes a basic fundamental principle in the form of 'double-aspect information'.

The notion of information which Chalmers uses is very much on the lines of Shannon. However, he makes some very interesting observations. Firstly, notice that *information spaces* and *information states* are realised in the physical world. When a die is thrown, an abstract space of possibilities ({1, 2, 3, 4, 5, 6}) is created —this is referred to as an information space, and the individual possibilities are physically realisable states which are referred to as information states (owing to the fact that there exist difference relations between these states). Similarly, a light switch realises an information space consisting of two information states corresponding to the two physically realisable states of the light being ON and OFF respectively. In a physical system, an information space is defined with respect to a *causal pathway* (in the case of the light bulb it is the pathway between the switch and the light) and a space of possible effects at the end of the pathway (light ON or OFF). Two physically distinct/different states which have distinct/different effects in the pathway will correspond to distinct information states. It is the '**difference that makes a difference**' that is defined as information (Bateson 1981). Differences that do not make any difference will not yield distinct information states, and hence carries no information.

Chalmers draws our attention to information that is realised in our phenomenology. The smell of rose is distinctly different from the smell of tea. Thus, the set of all phenomenal experiences of distinct smells creates an information space (just like in the light bulb example) with qualitatively different and distinct smells as information states. Here, it is the intrinsic quality of experience of every smell that creates the difference structure that is necessary for the information space.[3] More complex experiences, such as experiencing an entire visual field, have much richer information spaces with a complex structure.

[3]It feels something like to smell a rose which is intrinsically different from the experience of smelling tea.

Whenever an information space is realised phenomenally, we find that the same information space is realised physically as well. And, when a particular experience realises an information state, the same information state is realised in the physical substrate (neural processes in the brain?) of that experience. In the above example, the smell of tea will realise an information state in the phenomenally instantiated information space and at the same time it will realise an information state in the physically instantiated information space (corresponding to a particular state of the brain areas responsible for smell). The configuration of the brain with respect to its dynamical activity will be different for the two distinct smells.

Chalmers posits that this double life of information spaces may be the key to the fundamental connection between physical processes and conscious experience. A bridge is needed to connect the physical and the phenomenal—and 'information' seems like a good candidate. This leads to the basic principle of 'dual aspect information'—information has two aspects, a physical and a phenomenal aspect. Thus, whenever there is a phenomenal experience, it realises an information state that is also realised in the brain's cognitive system. In a converse fashion, for at least some physically realised information spaces, whenever there is a physical realisation of an information state, it is also realised phenomenally.

Chalmers is very cautious to emphasise that his 'double-aspect information' is largely a 'work-in-progress' and these are mere pointers to building a full-fledged theory (Chalmers 1996). He speculates that information may not be a primitive feature of the physical world (like mass and charge which are primitive), but it could still play a vital role in building fundamental psychophysical laws of a non-reductive scientific theory of consciousness. To our knowledge, there were no serious attempts to flesh these ideas into a full-fledged scientific theory of consciousness with neuroscientific connections, explanations and implications until Tononi.

Tononi's Integrated Information Theory of Consciousness (IIT)

Giulio Tononi (with others) has proposed a theory of consciousness which is based on ideas from information theory (Tononi 2004, 2008; Oizumi et al. 2014). The thesis that Tononi proposes is the following. The brain is arguably the locus of consciousness. Even if you are unwilling to consider the brain as a 'generator' of consciousness, it is hard to argue against the evidence that brain has a very big role to play in our conscious experience, the exact nature of which is still a mystery. Tononi asks some very important questions regarding the brain, which are:

1. Why is our consciousness generated by certain parts of the brain (thalamocortical networks) than others (cerebellum) which may even boast of significantly more number of neuronal connections?

2. Why is our consciousness 'more' in wakefulness than in dream, coma, vegetative states or certain anaesthetic states?
3. What are the conditions which determine whether a system is conscious or not?
4. What contributes to specific qualities of our conscious experience?

To answer these questions and to 'measure' consciousness, Tononi uses tools from information theory while being motivated by phenomenological properties of consciousness. The two of the main phenomenological properties he points to are 'differentiation' and 'integration' which we describe below.

The Property of Differentiation (or Information)

Tononi (2004) describes an interesting thought experiment to illustrate this phenomenological property of differentiation. Assume that you are in a dark room facing a blank screen which alternates between being ON and OFF, and you are required to call out 'light' when the screen is ON, and to call out 'dark' when it is OFF. This task could also be done by a photo-sensitive diode that is placed in front of this blank screen (see Fig. 2). Such a simple light-sensitive diode can easily distinguish between the screen being ON and it being OFF. Though you and the photo-diode are both successful in this task, the key difference in your experience of 'seeing' light consciously is that you are differentiating between a very large number of states as opposed to the photo-diode. You are not only differentiating

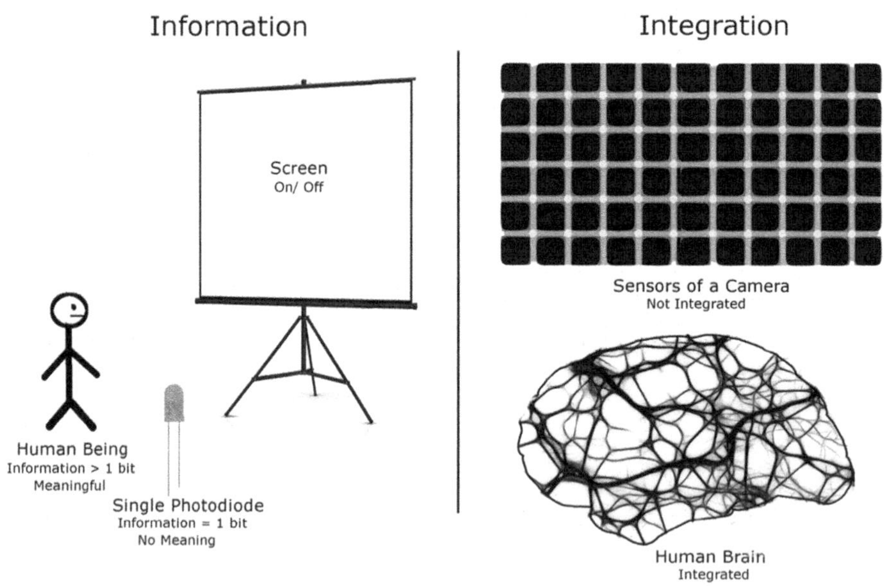

Fig. 2 Information and Integration. **a** *Left* Tononi's thought experiment (2004). **b** *Right* human brain is causally integrated whereas a network of sensors in a camera is loosely integrated

between the experience of seeing light and not seeing light, but you are also ruling out every scene of every movie that you have seen so far in your life. Whereas, the photo-diode is entering one of two states corresponding to the light being ON or OFF, you consciously enter into one out of an enormously larger number states (easily like a million possible states). From the viewpoint of information theory, you are generating a lot more information by this simple act of differentiating whether the screen is ON or OFF whereas the photo-diode is generating only 1 bit of information. If we assume that you have a million equally likely choices out of which you are choosing (ON or OFF), you generate $H = \log_2 10^6$ bits which is a mammoth amount of information.

The Property of Integration

The property of differentiating from a large number of states is one important difference between you and a simplistic photo-diode. However, this property cannot on its own characterise conscious experience. To see why, Tononi (2004) suggests we consider a one-megapixel digital camera. In essence, this is nothing but a conglomeration of one million photo-diodes. Even if we idealise each photo-diode to be binary—capable of distinguishing ON or OFF—the camera itself could differentiate between $2^{1000000}$ states, an enormous number. This corresponds to an information generation of $H = 10^6$ bits which is equal to what you are capable of generating. But, who would believe that the camera is even conscious (let alone that it is as conscious as you are)? The key difference between the camera and you is the property of *information integration*. Each photo-diode in the camera is *causally independent* of other photo-diodes, and '*no information can be integrated among them*' (Tononi 2004). On the other hand, your brain has a high degree of information integration owing to a huge amount of causal interactions and dependencies between different parts of the brain (Fig. 2). When we attempt to disconnect the connections in the brain, it leads to disastrous effect on your consciousness, whereas in the case of the camera, removing element by element leads to gradual degradation in its performance (at no point would we find the effect drastic since the elements are largely independent of each other). The property of information integration has a corresponding phenomenological counterpart in conscious experience—when we consciously 'see' a visual image, it is experienced as an integrated whole and cannot be decomposed into component parts that are independently experienced. For example, we can never experience colours without shape.

IIT: Consciousness as a Capacity to Integrate Information

As per IIT, consciousness is defined as the capacity of a system to integrate information. The inherent ability of a system (for example, a brain network) to differentiate between a large number of available states (property of differentiation or information), and yet indecomposable into a subset of casually independent subsystems (property of integration)—together these two constitute consciousness. The goal of IIT is to identify such an integrated system and to measure its repertoire of available states. A system such as a photo-diode is minimally conscious since it has a very low capacity to integrate information. A system such as the human brain is conscious to greater extent since it can integrate large amounts of information. Thus, as per IIT, consciousness is graded—and the quantity of consciousness is captured by the quantity Φ—which is the measure of 'minimum amount of effective information that can be exchanged across a bipartition of a subset' (Tononi and Sporns 2003). Such a subset of nodes (which could be neurons, or a group of neurons, or even logic gates) which is capable of integrating information, and which is not part of any other subset having higher Φ constitutes a 'complex'. The complexes constitute the boundary within which information can be integrated. Such a complex would be a 'subject' of experience, being the locus where information can be integrated (Tononi 2004). And, since information can only be integrated within a complex and never outside its boundaries, '*consciousness as information integration is necessarily subjective, private and related to a single point of view or perspective*' (Tononi 2004).

We would like to bring to the reader's notice that IIT has undergone several revisions since it was first proposed. The latest version, IIT 3.0 (Oizumi et al. 2014), has 5 fundamental axioms and postulates and theoretically more advanced than its previous versions (IIT 1.0 and 2.0). However, the core idea of defining consciousness as a capacity to integrate information essentially remains unchanged, and only the exact way in which it is computed differs. We will concern ourselves with this core idea, as that suffices for the current discussion.

Unlike the communication system model of Shannon (Fig. 1), the notion of information which IIT uses is quite different. Neurons (or aggregate of neurons) which form the nodes of brain networks could be both senders and receivers, there could be multiple pathways that act as channels and their primary goal is to integrate information (for behavioural control by the subject of experience) and not efficient and reliable transmission of information. Thus, there are significant differences in the notion of information defined by IIT and Shannon's information theory.

Extrinsic Versus Intrinsic Information

Tononi argues that Shannon's notion of information is always with respect to an external observer of the system. Shannon's information theory is always from the extrinsic perspective of an observer of the system's past and current states in order to assess the statistical dependence between inputs and outputs of the system (such as a communication channel of Fig. 1). This is what makes such a notion of information 'extrinsic' as per Tononi (Oizumi et al. 2014: Text S3).

In contrast to Shannon's 'extrinsic' notion of information, IIT claims to introduce a novel notion of information, namely 'intrinsic information'—defined from the intrinsic perspective of the system in terms of the *differences that makes a difference to it*[4] (from within the system onto itself). As per Tononi's theory of consciousness, intrinsic information has the following unique properties:

1. **Causal**: Intrinsic information is causal and not in relation to any observer. Whether there is an observer or not is immaterial since intrinsic information is evaluated by perturbing a set of elements in all possible ways (Tononi 2004; Oizumi et al. 2014).
2. **Compositional**: Intrinsic information is compositional which means that different combinations of elements of a system can simultaneously specify different constraints.
3. **Integrated**: Intrinsic information is maximally integrated. This implies that if partitioning a system does not make any difference to the system, then there is no system in the first place.
4. **Exclusive:** The exclusivity of intrinsic information implies that from the intrinsic perspective there can be one and only one set of causes.
5. **Meaning**: As previously noted, Shannon's extrinsic information is completely devoid of meaning (by its very definition). Shannon's information theory captures the limits of transmission and storage of messages from one point to another point, without worrying about the semantic content of these messages. It is about choosing 'one' unique message from a set of 'many' possible messages by the sender, transmitting the same on a communication channel in the presence of noise, and faithfully reproducing the same at the receiver. The meaning, if any, is only in the 'minds' of the sender and receiver and does not concern the mathematical description of the communication system itself. Clearly, such a definition of information is purely in statistical or probabilistic terms and does not concern itself with semantics of any kind. On the other hand, IIT makes the strong identity claim that *integrated intrinsic information is meaning*. It is not that intrinsic information is interpreted by an external observer, but rather, intrinsic information (which is causal, compositional, integrated and exclusive) is what *defines* meaning to the system.

In short, as per IIT, any network of interconnected nodes (neurons, or group of neurons, or logic gates) which produces intrinsic irreducible integrated information, is defining a subject of conscious experience (complex). Thus, as per IIT, even artificial systems could be conscious. As long as we satisfy the requirements laid out by IIT, we can have a conscious artefact—in fact, we can build one with a non-zero value of Φ.

[4]See also (Bateson 1981).

Limitations of Integrated Information Theory 3.0

Undoubtedly, information is central to integrated information theory, but the theory does face serious challenges in spite of having a strong theoretical grounding. Some of them are enumerated below.

1. **Current-state dependency**: Determining Minimum Information Partition (MIP) is an important step in computing the value of integrated information, $\mathbf{\Phi}^{\mathbf{Max}}$, in the framework of IIT 3.0 (Oizumi et al. 2014). MIP is that partition of a system which makes the least difference to the cause and effect repertoire in a particular state.[5] Therefore, MIP will be different for every current state of the system and hence the value $\mathbf{\Phi}^{\mathbf{Max}}$ will be dependent on the current state of the system. The dependency on the current state makes $\mathbf{\Phi}^{\mathbf{Max}}$ a non-robust measure of consciousness (Virmani and Nagaraj 2016).
2. **Computationally expensive**: Finding out MIP for a system is a rigorous task for a network since one needs to evaluate all possible partitions to find out the one which makes the least difference. As the size of the network increases, the process of finding MIP leads to combinatorial explosion—i.e. the number of combinations under examination grows exponentially, and so rapidly which makes it unmanageable even for the fastest computers to perform the computations in a finite time. Therefore, computing the quantity of conscious experience using IIT 3.0, is currently impossible for large networks (and thus impossible for real brain networks as well).
3. **Inapplicability to continuous data**: Applicability of current measure of conscious experience i.e. $\mathbf{\Phi}^{\mathbf{Max}}$ in the clinic is currently not possible because the measure cannot handle neurophysiological data (for ex. time series data) which is mostly continuous in nature (Barrett and Seth 2011).
4. **Incoherent granularity**: According to IIT 3.0, consciousness arises only when integrated information reaches its maximum within the system. It is very much possible that integrated information reaches maximum at a spatial and temporal grain which does not correspond to the variables recorded in experiments of neurophysiology. This is a potential problem in determining a measure for the quantity of conscious experience accurately by this approach.
5. **Characterisation of qualia and different aspects of experience**: The current framework of IIT 3.0 does not satisfactorily explain the individual qualitative aspects of an experience such as shape, colour or space. In order to characterise qualia, IIT 3.0 refers to the geometry of cause-effect space. In such a geometrical representation, it is not clear how the hard problem of qualia is solved.

[5]Cause and effect repertoire for a system is the probability distribution of its potential past and future states, confined by mechanism in its current state.

Measuring Consciousness in the Clinic: Perturbational Complexity Index (PCI)

The information-theoretic measure—integrated information is not limited to quantifying consciousness in computer simulated networks. But it has also been extended to construct an empirical measure called the Perturbational Complexity Index (PCI), to measure the conscious experience in healthy and non-healthy subjects in the clinic. PCI measures the information content of the response of thalamocortical networks in the brain by a measure of complexity known as Lempel-Ziv complexity (Lempel and Ziv 1976), when the brain is excited by a magnetic pulse (Casali et al. 2013). PCI takes into account the theoretical considerations which indicate that consciousness is contingent on the complex activity resulting in the brain. This complex activity in brain networks is simultaneously information-rich (differentiated in space and time) and integrated (shared by interacting cortical areas). Therefore, it is claimed that PCI is a measure of amount of information encompassed in the integrated response of the thalamocortical system to a direction perturbation by Transcranial Magnetic Simulation (TMS).

PCI is a unique brain complexity measure as it is resistant to noise from muscle activity or those neuronal sources which do not significantly contribute to integration (Casali et al. 2013). PCI, as an objective clinical measure for the computation of consciousness, exhibits promising results when the values of PCI are compared across 3 kinds of subjects: (i) healthy subjects in wakefulness, non-rapid eye movement (NREM) sleep and dreaming states, (ii) subjects induced with sedation by anaesthetic agents: midazolam, xenon and propofol, and (iii) patients emerged from coma (vegetative state, minimally conscious state and locked-in syndrome).

Limitations of PCI

Though PCI claims to be theoretically grounded and shows strong empirical results, but it still faces certain challenges.

- **No direct measurement of integrated information**: PCI only determines the complexity of averaged TMS evoked potentials from a specific target region of the brain, which is not the direct measurement of integrated information. So, it cannot be asserted with certainty what exactly does the PCI value indicates.
- **Weak link between PCI and IIT**: PCI values do not explicitly demonstrate how higher values in wakefulness correspond to higher integration and information-richness (differentiation) in cortical areas and inversely for lower values of PCI. Therefore, the link between PCI and IIT is not very strong, rather it seems at best to be an approximation for measuring integrated information.

Complexity, Information and Consciousness

In PCI, we saw that integrated information of thalamocortical networks can be linked to complexity of brain activity (in response to an appropriate magnetic excitation). PCI made use of Lempel-Ziv complexity, which is related to Kolmogorov complexity (Ming and Vitányi 1997) as well as used to approximate Shannon entropy for ergodic Markov sources. Lempel-Ziv complexity has been used in neuroscience applications for estimation of Shannon entropy of spike trains (Amigó et al. 2004).

Complexity has no unique definition, but one way to characterise it could be as a measure of degree of 'information-richness' in a system. The more information-rich a system is, the higher its complexity. Complexity can also be understood as the amount of variability in time series (electrophysiological measurements from the brain), and this captures the dynamical complexity of brain activity. A periodic signal has well defined structure and low complexity, as such a signal would have low information content. On the other hand, a completely random signal, where the value of the signal at every time instant is independent of values at other time instants, would have a high Shannon entropy but low complexity (in the sense of information integration) because there would be very less correlation between different parts. The highest amount of complexity, in the sense of integrated information, is possessed by neither periodic, nor random, but by signals whose individual parts are highly correlated, yet statistically have high Shannon entropy. Such a mix of paradoxical criteria is met by chaotic signals—these are nonlinear deterministic in origin, but pseudorandom and could have complicated long-range correlations leading to high integrated information and thus high complexity (high amount of Shannon entropy as well).

Interestingly, the brain exhibits chaos at various levels (Korn and Faure 2003). The underlying neural processes in the brain have such a mix of properties since the firing patterns of neurons are known to be chaotic. As per Integrated Information Theory 3.0, consciousness could be a result of integration and differentiation underlying these neural processes. So, it is a good idea to measure consciousness by capturing the correlation and information integration in the activity of neurons through measures of complexity. This leads to the class of 'complexity theories of consciousness' (Seth et al. 2008). Neural complexity (Tononi et al. 1994), causal density (Seth 2005) and IIT are examples of theories in this class.

One of the novel ways to measure the complexity of a system is by using the idea of data compression.

Data Compression, Complexity and Consciousness

Maguire et al. (2014) notes the connection between the unique integration of our experience (consciousness) with our existing memories, and that this integration gives a subjective flavour to our experience. This fact relates to integrated information. Different regions of the human brain are tightly integrated with each other owing to causal interactions between them. For example (Maguire et al. 2014), a video camera which can record several amounts of visual data does not have an 'experience' in the same way as we human beings have. This is because, the memory of the video camera can be selectively deleted (we could erase the past 5 s of memory in the video camera) but the same cannot be done with the human brain (to delete just the past 5 s of our memory is nearly impossible). Thus, the brain acts like a singular unified integrated system.

The idea of data compression goes along well with integrated information (Maguire et al. 2016). In a losslessly compressed file, no single bit is truly independent of the rest, while in an uncompressed text file, every character carries independent information about the text. Maguire observed that, 'the information encoded by the bits of a compressed file is more than the sum of its parts', highlighting connections between data compression and Tononi's concept of integrated information (Maguire et al. 2014). To characterise sophisticated data compression carried out by the brain in order to bind information that we associate with consciousness, the term *compressionism* was coined by Maguire et al. (2016). Thus, it seems that it is possible to capture information integration in brain networks using the concepts of data compression.

The measures of complexity, based on infotheoretic quantities such as entropy or mutual information, have been outperformed by the lossless data compression algorithm-based complexity measures, such as Lempel-Ziv complexity (LZ) (Lempel and Ziv 1976) and Effort-To-Compress (ETC) (Nagaraj et al. 2013; Nagaraj and Balasubramanian 2017). Inspired by this, we have recently proposed a novel measure of brain complexity ($\mathbf{\Phi^C}$) which employs the notion of compression-complexity of time series (Virmani and Nagaraj 2016). Our measure serves as a bridge between IIT and PCI.

Definition of Our Measure

$\mathbf{\Phi^C}$ is defined as the maximally-aggregate differential normalised Lempel-Ziv (LZ) or normalised effort-to-compress (ETC) complexity for the time series data of each node of a network, generated by perturbing each possible atomic partition of an N-node network with a maximal entropy perturbation (MEP) and zero entropy perturbation (ZEP).Intuitively, our measure excites the nodes of the network such that the maximum difference in the input perturbations (in terms of entropy) that makes a maximum difference in the aggregated output response (in terms of compression-complexity) is determined. This gives a measure of dynamical complexity, integrated information as well as the integrated casual power of the network. See Virmani and Nagaraj (2016) for further details. We concisely summarise the algorithm here.

Algorithm for Φ^C

Φ^C for a network[6] is computed by performing the following steps: (i) partition the network into its all atomic bi-partitions, (ii) perturb the atomic node for each bi-partition with maximum entropy and zero entropy time series, (iii) record the output time series from all the other nodes of the network and compute the *compression-complexity* of these individual time series using normalised LZ or ETC for each partition, (iv) compute the aggregate of the differential compression-complexity measures for each bi-partition of network (difference of responses to MEP and ZEP), (v) report the maximum value out of all such computed aggregate differential compression-complexity values obtained in step (iv) as the value of Φ^C.

Strengths of Our Measure

As mentioned previously, IIT and PCI, both have several limitations in spite of being theoretically strong and clinically sound respectively. Φ^C aims to bridge the gap between these two measures and exhibits several benefits over them and other measures of integrated information. Some of the strengths of Φ^C are:

- Compression-complexity approach instead of infotheoretic approach.
- Perturbational approach (inspired from PCI).
- Computationally less expensive.
- Current-state independence.
- Theoretically well-bounded.
- Linearly correlated with entropy of the nodes.
- Captures the co-existence of differentiation, integration, as well as entropy in networks.
- Exhibits a similarity with Φ^{Max} (IIT 3.0) in its behaviour on 3-, 4- and 5-node networks.

[6]We intend to perform extensive experimentation with realistic brain networks in the future. Currently, our results are for Boolean networks with small number of nodes.

Fig. 3 Values of $\mathbf{\Phi^C}$ for two example networks **AC** and **ABC**. **a AC** is a two-node network where node **A** has a logical mechanism **OR** and **C** has **XOR**. **AC** is in the state (1, 1) which means that node **A** and node **B**, both are in ON condition. It has a Compression-Complexity Measure ($\mathbf{\Phi^C}$) value as 0.296, **b ABC** is a three-node network where nodes **A**, **B** and **C** have the logical mechanisms as **OR**, **AND**, **XOR** and states as (1,0,0) respectively. This network has $\mathbf{\Phi^C}$ = 0.548. Note that we have used normalised ETC for computation of $\mathbf{\Phi^C}$

Some Examples

Value of $\mathbf{\Phi^C}$ for two example networks with different number of nodes is depicted in Fig. 3. The first network **'AC'** has two nodes which are fully connected. Each node has a cause-effect relation on the other node as well as with itself via a self-loop. The node **A** follows the mechanism logical **OR** and node **C** follows the mechanism logical **XOR**. Both the nodes happen to be in the **ON** state, so we denote the initial state of the network as **(1, 1)**. On similar lines, $\mathbf{\Phi^C}$ for the 3-node network **'ABC'** (**OR-AND-XOR**) is also calculated. In general, it has been observed that the networks with more number of high entropy gates such **XOR** have higher values of $\mathbf{\Phi^C}$.

These examples are for illustrative purposes only. Our ongoing research is to compute $\mathbf{\Phi^C}$ for larger networks consisting of several thousands of neurons (using spiking neuron models) with prototypical architectures that are encountered in the brain.

Scientific Measures of Consciousness: A Medley

The recent trend and advances in scientific theories of consciousness have focussed on the problem of measurement (Seth et al. 2008). There are several behavioural and brain based measures in the literature. Table 1 lists a few of these measures and compares them qualitatively.

It is evident that there is no single measure that captures the notion of brain complexity and properties that are necessary for consciousness. We need to move towards a theoretically grounded, comprehensive, empirically testable and synergistic confluence of different measures which could be applied in the clinic. A combination of different measures is most likely the way ahead.

Table 1 Measures of consciousness based on scientific theories—a comparison

Sl. No.	Consciousness measure	What does it measure?	Ease of computation	Remarks
1	Chalmer's double-aspect information (Chalmers 1995, 1996)	–	–	Philosophically well grounded, aims to bridge the physical and phenomenal aspects of reality, but not viable for practical measurement
2	Neural complexity (Tononi et al. 1994)	Brain complexity	Medium-low	Mutual information is used
3	Causal density (Seth 2005)	Causal interactions in brain networks as a measure of consciousness	Medium	Granger causality is used
4	IIT 1.0 (Tononi 2004)	Information integration	Low	Effective information defined using mutual information
5	IIT 2.0 (Tononi 2008)	Information integration	Low	Minimum information partition (MIP)
6	IIT 3.0 (Oizumi et al. 2014)	Information integration	Low	MIP, probabilistic model, cause-effect space
7	PCI (Casali et al. 2013)	Information integration in thalamocortical networks	High	Clinically measurable, claims link with IIT, employs Lempel-Ziv complexity
8	Φ^C (Virmani and Nagaraj 2016)	Compression-complexity as a measure of brain complexity	Medium	A bridge between IIT and PCI, inspired by compressionism, uses compression-complexity (LZ, ETC)

Outstanding Questions and Concluding Remarks

As we arrive at the end of this journey, we realise that there are several unchartered territories that have escaped us. We enumerate below a number of these unexplored questions which are nevertheless very important in deciphering the link between 'information' and 'consciousness'.

1. A scientific account of consciousness starts with the brain as the locus of consciousness. If this is an acceptable starting point,[7] then understanding brain complexity is the key to explaining consciousness. Is 'complexity' necessary for

[7]This is challenged by Indian philosophical systems such as Advaita Vedanta—which does not attribute any special role to the brain (or the body) in explaining consciousness.

'consciousness', or the reverse, or are the two concomitant and dependent on a third hitherto unknown aspect of reality?
2. There is overwhelming evidence that causal interactions in the brain (information-processing/integration of some kind) are linked with richness of experience. Has evolution led to such a complexity in trying to optimise for fitness? Is it possible for an organism to be endowed with rich conscious experience (inner mental life), but with a low brain complexity/activity?
3. It appears that the human brain is performing some form of irreversible (lossy) data compression (binding input sensory data with pre-existing memories). Is this the key to explaining 'intelligence', 'understanding' as well as 'consciousness' as *compressionism* claims?
4. Neuroscientific theories of consciousness have ignored the notion of 'semantic information'. IIT equates qualia with a specific geometry of informational relationships between interacting elements that form a complex[8] (Balduzzi and Tononi 2009). But, the notion of information it employs is based on purely probabilistic and structural/dynamical in nature. We are of the opinion that qualia cannot be satisfactorily explained without invoking some version of semantic information. The semantic content of information has a role in our experience, and this is completely ignored in such theories.

Understanding consciousness by a scientific approach (and not limiting it to philosophical discussions) is challenging since scientific methods rely on 'measurements', 'experiments' and 'theories' which consciousness seems to elude. However, such a rigorous approach is useful since it can pave the way to creation of newer technologies and it could be beneficial in clinical applications to treat various kinds of mental disorders.

References

Adriaans, P. (2013). Information. In E. N. Zalta (ed) *The Stanford encyclopedia of philosophy* (Fall 2013 Edition). URL = <https://plato.stanford.edu/archives/fall2013/entries/information/>.

Amigó, J. M., Szczepański, J., Wajnryb, E., & Sanchez-Vives, M. V. (2004). Estimating the entropy rate of spike trains via Lempel-Ziv complexity. *Neural Computation, 16*(4), 36–717.

Balduzzi, D., & Tononi, G. (2009). Qualia: The geometry of integrated information. *PLoS Computional Biology, 5*(8), e1000462.

Barrett, A. B., & Seth, A. K. (2011). Practical measures of integrated information for time-series data. *PLoS Computational Biology, 7*(1), e1001052.

Bateson, G. (1981). Form, substance and difference. In *Essential Readings in Biosemiotics*, (pp 501–518).

Casali, A. G., Gosseries, O., Rosanova, M., Boly, M., Sarasso, S., Casali, K. R., … Massimini, M. (2013). A theoretically based index of consciousness independent of sensory processing and behavior. *Science Translational Medicine, 5*(198), 198ra105.

[8]Remember that, in IIT, a complex is the 'subject of experience' constituted by causally interacting elements that integrate information.

Chalmers, D. J. (1995). Facing up to the problem of consciousness. *Journal of Consciousness Studies, 2*(3), 19–200.

Chalmers, D. J. (1996). *The conscious mind: In search of a fundamental theory*. Oxford: Oxford University Press.

Ford, K. W., & Wheeler, J. A. (1998). *Geons, black holes, and quantum foam: A life in physics*. New York: W.W. Norton.

Korn, H., & Faure, P. (2003). Is there chaos in the brain? II. Experimental evidence and related models. *Comptes Rendus Biologies, 326*(9), 787–840.

Landauer, R. (1996). The physical nature of information. *Physics Letters A, 217*(4–5), 93–188.

Lempel, A., & Ziv, J. (1976). On the complexity of finite sequences. *IEEE Transactions on Information Theory, 22*(1), 75–81.

Maguire, P., Moser, P., & Maguire, R. (2016). Understanding consciousness as data compression. *Journal of Cognitive Science, 17*(1), 63–94.

Maguire, P., Moser, P., Maguire, R., & Griffith, V. (2014). Is consciousness computable? Quantifying integrated information using algorithmic information theory. arXiv preprint arXiv:1405.0126.

Ming, L., & Vitányi, P. (1997). *An introduction to Kolmogorov complexity and its applications*. Heidelberg: Springer.

Nagaraj, N. & Balasubramanian, K. (2017). Dynamical complexity of short and noisy time series. *European Physics Journal Special Topics (Special issue: Aspects of Statistical Mechanics and Dynamical Complexity)*.

Nagaraj, N., Balasubramanian, K., & Dey, S. (2013). A new complexity measure for time series analysis and classification. *The European Physical Journal Special Topics, 222*(3–4), 60–847.

Nagel, T. (1974). What is it like to be a bat? *The Philosophical Review, 83*(4), 50–435.

Oizumi, M., Albantakis, L., & Tononi, G. (2014). From the phenomenology to the mechanisms of consciousness: Integrated information theory 3.0. *PLoS Computational Biology, 10*(5), e1003588.

Seth, A. K. (2005). Causal connectivity of evolved neural networks during behavior. *Network: Computation in Neural Systems, 16*(1), 35–54.

Seth, A. K., Dienes, Z., Cleeremans, A., Overgaard, M., & Pessoa, L. (2008). Measuring consciousness: relating behavioural and neurophysiological approaches. *Trends in Cognitive Sciences, 12*(8), 21–314.

Shannon, C. E. (1948). A mathematical theory of communication. *Bell System Technical Journal*, 27: 379–423 and 623–56.

Tononi, G. (2004). An information integration theory of consciousness. *BMC Neuroscience, 5*(1), 42.

Tononi, G. (2008). Consciousness as integrated information: A provisional manifesto. *The Biological Bulletin, 215*(3), 42–216.

Tononi, G., & Sporns, O. (2003). Measuring information integration. *BMC Neuroscience, 4*(1), 31.

Tononi, G., Sporns, O., & Edelman, G. M. (1994). A measure for brain complexity: Relating functional segregation and integration in the nervous system. *Proceedings of the National Academy of Sciences, 91*(11), 7–5033.

Virmani, M. & Nagaraj, N. (2016). A compression-complexity measure of integrated information. arXiv preprint arXiv:1608.08450v2.

Wheeler, J. A. (1989). Information, physics, quantum: the search for links. In *Proceedings of III International Symposium on Foundations of Quantum Mechanics* (pp. 354–68). Tokyo.

Encircling the Consciousness Conundrum

Ravindra M. Singh

Introduction

Given the grudging official acknowledgement of the beginning of the Anthropocene era with its accompanying devastating consequences and the human hands holding the knife of CRISPR with the potential to change all existing life forms beyond recognition, we are witness to a new-found Faustian exuberance of the human spirit. The pace of change with its concomitant cascading effects is also being celebrated as a kind of flattening of the world and coming into being of the pot-pourri of cultures and civilisations characterised as a global village that transcends formal political boundaries. Philosophy has also not remained unaffected by these developments of such vast proportions as we witness fast but tectonic shifts here as well. Doing speculative philosophy is no longer sneered at nor is calling someone a metaphysician considered a term of philosophical abuse. Invoking Bertrand Russell, Strawson (2008, 2015, 2016, 2017), has provocatively declared that we know next to nothing about the physical. Totally opposed to this is the view of Christof Koch who believes that '[w]e have a detailed and very successful framework for matter and for energy but not for the mind-body problem' (Koch 2009: 16). Many others find any kind of philosophical engagement with individual entities in the name of scientific grounding of philosophy to be a wrongly headed enterprise; claiming to do philosophy, for them, by appealing to a science that was superseded almost 100 years back is passé (Campbell 2015; Campbell and Bickhard 2011).

When we specifically look at the broad area of consciousness studies, we find equally significant transformations. For instance, the failure of reduction has occasioned a rise of interest in emergence and supervenience (Bedau and Humphreys 2008: 213). One very positive consequence of this undermining of

R.M. Singh (✉)
Department of Philosophy, University of Delhi, New Delhi 110007, Delhi, India
e-mail: ravimsingh@gmail.com; rmsingh@philosophy.du.ac.in

S. Menon et al. (eds.), *Self, Culture and Consciousness*,
DOI 10.1007/978-981-10-5777-9_22

reductive enterprise has been that spookiness surrounding the nature of emergence has largely been exorcised. But a little hangover of the ideals of philosophical method left over by logical positivism in the name of science and philosophical rigour remains (Churchland 1981; Van Gulick 2007). The fate of dualism over the decades has been no different. At one point of time its ghost lurked everywhere (Ryle 1949) and extraordinary efforts were made at exorcising it (Dennett 1991). We are not unfamiliar with the relative softening of stance towards dualism in the last two decades (Chalmers 1996; Lycan 2008). The rapid rise and fall of behaviourism, type-identity theory and functionalism and some very recent attempts to rehabilitate them are notable (Skinner 1957; Chomsky 1959, 2000; Smart 1959; Putnam 1960/1975; Clark and Chalmers 1998/2002; Tononi 2004, 2008, 2015; Koch 2009, 2015). The rapid rise, fall and very powerful resurrection of artificial intelligence and computationalism is also equally significant (Turing 1950; Putnam 1967/1975; Pylyshyn 1984; Haugeland 1985; Fodor 1975, 1980, 2008; Searle 1980, 1992, 2013; Dreyfus 1972, 1992; Thelen and Smith 1994; Chemero 2009; Chalmers 1993, 1995; Clark 2014; Kurzweil 2012).

While beginning the task of articulating the issues surrounding the nature of consciousness, I find it appropriate to acknowledge that even after centuries of extraordinary efforts by religious thinkers, mystics, philosophers, psychologists and scientists, no consensus has emerged about not only what consciousness is but even who can be attributed as having it in a systematic and consistent fashion (Rohde 1925; Bremmer 1983; Block 2001; Chalmers 1996; McGinn 1999; Seager 2009; Rees and Frith 2007; Frith and Rees 2007; Koch 2009; Strawson 2015; Tononi and Koch 2015; Tononi et al. 2016). Consequently, views about how consciousness is incarnated within its bodily boundaries or why consciousness is the way it is remain somewhat tentative.[1] But I believe that many proposals are currently on offer that seem to offer some hope of making progress. Of course, there is no shortage of naysayers either, most prominent among them being Levine (1983/2002), Chalmers (1996), McGinn (1999) and Heffernan (2013).[2] I, however, believe that there are good reasons to be optimistic about making progress in understanding the nature of human consciousness; it is not beyond the reach of current human methods of gaining knowledge. Although I do not find Strawson's contention that 'we know exactly what consciousness is' (2016) agreeable, I believe that the study of consciousness is neither beyond human efforts nor is engagement with it any waste of

[1]For details of varieties of philosophical and scientific positions on the matter, see Seager (2009), Kreigel (2007), Frith and Rees (2007), Block (2009) and Edelman and Tononi (2000, Chap. 1 and endnote 4).

[2]One of the best responses that I have come across in the literature that addresses Chalmers's scepticism about the inescapability of the hard problem of consciousness is by Zeki who states that 'Consciousness is often regarded as the most complex problem in the Universe, a belief that I find it difficult to agree with, since we do not know what else exists in the Universe. It is often spoken of as the last frontier in science, implying that we have crossed all the other frontiers, which is not at all evident to me' (Zeki 2007: 580).

resources and human energy (Paller and Suzuki 2014). What follows is a detailing of this sentiment.

The Consciousness Conundrum

As soon as we begin to think about consciousness, the first set of questions that confronts us are about its nature; what is consciousness; where all can it be found; how to situate it in the world. If we situate consciousness in the world like everything else, then how do we deal with the problem that consciousness cannot be accessed in the manner typical of other objects. The presence in the world of a functioning consciousness is itself a precondition for such an accessibility. No scientific method can provide any access to the world without its dependence on the presence in the world of a functioning consciousness. Without consciousness, we would not even know that we or the world exist (Thompson 2015: 79). Such puzzles about the nature of consciousness notwithstanding, the most basic axiom about consciousness that I begin with is the existence of consciousness (Tononi 2012b: 296; Koch et al. 2016: 2). So, at least its existence, accessibility to it and belongingness, even if in a minimal sense, of experiences to the one who is endowed with consciousness are being taken here as beyond any reasonable doubt.[3] A sense of belonging thus appears to be present whenever we are conscious and includes everything that we can call our own in this world—one's body, psyche, family, reputation and so on. So, even if we were to decide to adhere to a physicalist paradigm, mere presence of a life process is necessary but not sufficient for consciousness to happen. Some kind of subjectivity or self has to be added to this process for a living thing to become conscious because it is not existence of mere rudimentary sensuousness that we call consciousness. It is belongingness of experiences of every kind to an owner that consciousness designates.

It is not that there cannot in principle be any life processes in the world without presence of consciousness. But without consciousness we cannot know that such a life process exists. Presence of a subjectivity 'is the only natural means to know them' (Damasio 2012: 12). However, there is no reason to believe that existence of subjectivity or consciousness will not create distortions in its view of the world or of itself. Persistence of perceptual as well as self-illusions demonstrate the truth of this assertion. We are also aware that other experiencing beings exist. But we cannot access their experiences as they themselves do. We can only observe their bodily behaviours and actions.

This brings me to the second underlying assumption that I am making here. Given the necessary boundedness of consciousness with a living body, it cannot be viewed as a nonphysical process but needs to be viewed as 'a key biological

[3]At this juncture, I am not, however, getting into the nature of this belongingness. Nor is it the focus of the present chapter.

adaptation with multiple functions' (Baars 1997: x). As Damasio points out, 'states of mind begin physically, and physical they remain. They can be revealed only when an equally physical construction called self becomes available and does its witnessing job' (Damasio 2012: 322). Accordingly, following Damasio (2012), Edelman (1987, 1988, 1989) and Edelman and Tononi (2000), Tononi (2004, 2012b; Tononi et al. 2016), I assume that viewing of consciousness as discontinuous from the biology that creates it results in placing it outside the laws of physics. And there is too much of overwhelming empirical evidence against such a move. Such efforts have also borne barely any fruit in the past.

Taking the biological grounding of consciousness seriously means that we need to look at human consciousness in evolutionary terms. Making sense of consciousness requires that we also take cognizance of early life forms and identify what having of consciousness does to organisms that have it. When we look at simplest life forms we find that the objective of managing and maintaining of life is discernible even in the case of single cell organisms. This principle is coded in their genome. That is, 'managing and safekeeping life is the fundamental premise of biological value' (Damasio 2012: 25). Display of adaptive behaviour is accordingly not limited to higher animals like humans alone. All living cells display determination to stay alive for as long as their DNA commands them to. They all want to live out their 'prescribed genetic allowance' (Damasio: 35). It is in pursuit of such a biological value, which biologists have termed *homeostasis*, that consciousness 'emerges within the history of life regulation' (Damasio: 25). Homeostasis, however, is not a fixed thing but dynamic in nature and the process continues from one life form to another inaugurating and augmenting new hierarchically organised means of life management. So, consciousness as we encounter at work in humans, results from a stepwise increase in complexity within the biological realm. Though consciousness comes into being because of biological value, it is not creator of this value. The biological value pre-existed consciousness as life maintaining processes can exist without consciousness. It merely provides new ways of managing biological value.

But while many of the mechanisms that are constitutive of different living beings have been investigated and understood to a large extent, even if their complete knowledge is still lacking. But unravelling the Schopenhauerian 'world knot' of consciousness continues to be burdened with unsurmountable difficulties. Although a fairly good knowledge of mechanisms responsible for conscious experiences has been achieved by scholars working in the area of mind/consciousness studies and we are able to correlate changes in the working of bodily mechanisms with corresponding changes in conscious experiences, what still appears doubtful is why mechanisms that are physical in nature produce effects of which we are conscious in and through experiences.[4] However, it is still difficult for many to agree that activities of the neuronal soup in the brain could entirely be responsible for

[4]It is this unbridgeable gulf that has been cited as one of the reason for the proverbial 'explanatory gap' between the neural and the experiential (Levine 1983/2002).

generating conscious experiences even though the overwhelming opinion is moving in the direction of treating 'consciousness as a state of the brain' (Block 2009).

I find that one of the chief reason for inability of sceptics (Levine 1983; Chalmers 1996; McGinn 1999) in appreciating any progress in our understanding of consciousness is that they largely believe that conscious experiences possess qualities that are radically different from everything else. No doubt consciousness is very different because unlike other objects of scientific inquiry, it is not entirley objectively accessible. This ever elusive but unique quality of consciousness is what has come to be designated as the 'subjective feel' of experiences after Thomas Nagel's famed articulation of it (1974).[5] It is also the quality that has come to be designated as the hard problem of consciousness (Chalmers 1996).[6] And there appear good reasons not to overlook the inherently private nature of conscious experiences.

Undermining of this well recognised fact, in my opinion, is also one of the reason for the lack of success of reductionist and eliminativist approaches. Instead of contributing in any way to our knowledge of consciousness, they are inevitably led to dispensing away with the very subject of inquiry.[7] It appears somewhat paradoxical that though reductionists are generally very jubilant about the demise of dualism, yet they are far from successful in escaping dualism in its entirety in their own theorisations.[8] Is dualism, therefore, to be shunned in every possible form and conceptually, is it all that compelling to do so? Does having of human aspirations, needs and desires, fears and anxieties; making decisions to handle them; and eventually succeeding in making human efforts to realise them (all undoubtedly mental in nature) through bodily movements and actions in the real world (something that is uncontroversially physical) entail embracing of dualism of some kind on our part that we cannot escape? Moreover, are problems with dualism owing to our understanding of the mental or do they arise from our current understanding of the nature of the physical as well (Heylighen 2009; Campbell and Bickhard 2011; Campbell 2015; Strawson 2016, 2017)?

When it comes to accounting for the causal efficacy of consciousness, the situation of non-reductive physicalists does not appear to be much better. Prominent among their problems are: (i) accounting for the nature of emergence of conscious

---

[5]This subjective feel is also referred to as the phenomenality that is necessarily associated with every conscious experience, see, Block (1995, 2007, 2014a) for more details.

[6]Strawson (2015), however, considers it incorrect to credit Chalmers as the first to recognise hard problems surrounding the nature of consciousness.

[7]This has been succinctly stated by Searle: "You cannot reduce intentional content (or pains or "qualia") to something else, because if you could they would be something else, and they are not something else". (Searle 1992: 51).

[8]See Lycan (2008), for instance, for articulation of why it is not as easy to walk over dualism as easily as it is often made out to be. Searle (1998) has also accused researchers of still being 'in the grip of a residual dualism' (Searle 1998: 1939). Also, see, Dennett (1991) for a discussion of hangover of the dualist imagery that most self-proclaimed non-dualist accounts suffer in their treatment of the problem.

processes without rendering consciousness to be epiphenomenal in nature; (ii) accounting for downward causation, essentially referring to the influence of the mental on the physical, without accepting drainage of causal powers from one level to the other ad infinitum.[9] Similarly, mere introspection and armchair thinking are not sufficient as an account of conscious states because of the evidential deficit associated with self-delusional and hallucinatory conscious states (Dennett 1991). This means that the subjective proximity of conscious states is no guarantee of their veracity.[10] Similarly, brain studies alone cannot reveal to us what it is to be conscious or undergo experiences (Block 2014a).

In the context of approaches directed at identifying neural correlates of consciousness (NCC) (e.g. Koch 2004; Koch et al. 2016), it is important to be clear about whether 'the neural background required to make a core neural basis [is] sufficient for a phenomenally conscious experience' (Block 2007: 852). That it is not always easy to ascertain this has been widely noted in the literature (Aru et al. 2012a). For instance, Rees et al. (2000, 2002) have reported cases of visuospatial extinction wherein individuals deny having any experiences even when there are robust activations in the identified core areas for the tasks in question. How do we make sense of such cases?

The significant question in this context is whether we can argue for the presence of a phenomenally conscious state independent of its reportability or is it the case that the conscious state is accessible but not reportable? In this regard, Block (2007), has suggested that this may possibly be true, for instance, at least in those cases where mechanisms of cognitive access and/or reportability are damaged or blocked (under anaesthesia, for example). Could it be the case that the brain area that we have identified as the core neural basis is merely necessary for phenomenal consciousness in question but not sufficient? Block has forcefully argued that conscious reporting of experiences fails short of what subjects actually experience (Block 2007, 2009, 2014a). He considers the demand of reportability as an unreasonable legacy of behaviourism.

In this context, while Block (2005) has been persisting with his demand for recognition of two NCCs (one NCC for phenomenality and another NCC for access), Aru et al. (2012a) have emphasised the difficulties involved in aiming to separate NCC for processes directly corresponding to conscious experience from prerequisite processes carrying no actual conscious contents (NCC-pr) and consequent processes following actual conscious experience (NCC-co). The situation is all the more complicated because of presence of consciousness 'even in the absence of reliable behavioural responses' (Koch et al. 2016: 307). However, in the last decade, many novel methods have been developed for isolating NCC, some of the

---

[9]But there have been some attempts to address these issues in the recent literature (Bedau and Humphreys 2008; Campbell and Bickhard 2011; Campbell 2015).

[10]See Dennett (1991) and Dehaene and Naccache (2001) for further discussion of the limitations of first-person approaches and need for a method that not only takes into account the subjective aspect of our experiences but can also objectively anchor them.

prominent ones being no-report paradigm, perceptual awareness scale, attentional blink, online functional imaging studies like event-related potential (ERP), magneto-encephalography (MEG) and magnetic resonance imaging (MRI), different techniques for contrasting non-REM sleep and REM sleep as well as activations associated with different cognitive transitions (Sandberg et al. 2010; Aru et al. Aru et al. 2012b; Tsuchiya et al. 2015; Koch et al. 2016). Equally impressive has been the outcome from the use of perturbational complexity index (PCI) through use of transcranial magnetic stimulation (TMS) and electroencephalogram (EEG) recordings by Casali et al. (2013) 'to quantify the level of consciousness under a variety of conditions and at the level of individual patients' (Koch et al. 2016: 320). Brain regions associated with thought-like experiences, so dear to philosophers, have also been isolated and studied in sufficient details (Koch et al. 2016: 308).

Drawing from their earlier work on global neuronal workspace model (Dehaene and Naccache 2001; Dehaene et al. 2006),[11] Naccache and Dehaene, however, find Block's claim (2007) empirically untestable (2007: 519). Their proposal is that the 'availability of report' in principle be treated as 'the sole criterion for conscious perception' (519). This is irrespective of the fact of these reports being even 'nonverbal or … entirely covert' (519). Reinterpreting studies by Sperling (1960) and Landman et al. (2003) used by Block (2007) to establish his account, Naccache and Dehaene (2007) report lack of 'univocal support in favor of Block's phenomenality without-report hypothesis' (519). In this regard, their alternative global neuronal workspace (GNW) model emphasises the role of consciousness in attentional amplification and global availability of information by 'integrating the supervision, limited-capacity, and re-entry properties' (Dehaene and Changeux 2011: 210; Dehaene and Naccache 2001: 1). Their account also takes cognizance of the fact that considerable cognitive processing is possible without consciousness (Baars 1997, 2007; Dehaene et al. 2006).

Moreover, Dehaene and Changeux's (2011) proposal is in conformity with, still somewhat tentative and counter intuitive, new research findings that attention and consciousness are not only two distinct neurobiological processes but they are dissociable as well (Koch and Tsuchiya 2007; Cohen et al. 2012; Tsuchiya et al. 2012). While acknowledging that many theoretical challenges remain in consciousness research and moving from neuronal correlations to establishing causal links is far from accomplished, they do suggest how it can be undertaken (Dehaene and Changeux 2011: 213, 216). And their suggested way is through 'systematic lesion or interference methods' (216). When applied to humans such methods do have severe restrictions but their predictions are nonetheless testable. For instance, one of their prediction is that 'lesioning or interfering with prefrontal or parietal cortex activity, at sites quite distant from visual areas, should disrupt conscious

[11]For a less jargonised articulation of the issues that do not require mastery of biological brain details, see Baars (1997, 2007).

vision' (216). While this prediction was initially refuted on the basis of clinical observations, subsequent evidence has supported 'the GNW account. In normal subjects, TMS over either parietal or prefrontal cortex can prevent conscious perception and even trigger a sudden subjective disappearance of visual stimuli' (216). Dehaene and Changeux (2011) have suggested that most significant test of their account would be through 'inducing a change in the actual conscious content, rather than a mere elevation of the reportability threshold, by stimulating PFC or other components of the GNW networks' (216). While admitting that no such experiment has yet been attempted, they suggest that 'micro stimulation and optogenetic methods now make it feasible' (216). They also cite evidence in support for their account from clinical situations like 'schizophrenia, anaesthesia, and loss of consciousness due to coma or vegetative state' (216). New paradigms (e.g. Kiani and Shadlen 2009; Terrace and Son 2009) that can behaviourally study metacognition and self-representation in both humans and non-humans offer further hope for identification of neurophysiological mechanisms involved in conscious access (Dehaene and Changeux 2011: 218–19).

I find that one of the reason for consciousness researchers' perpetual gloom and belief in inevitability of failure in the past had been the lack of availability of objective methods to study consciousness. It is this lacuna that the contemporary philosophical and scientific studies of the last two-three decades have attempted to address. The results of these studies are definitely very encouraging as researchers begin to move from establishing behavioral-neuronal correlations 'to address the challenge of establishing a causal link between subjective conscious experience and measurable neuronal activity' (Dehaene and Changeux 2011: 200). But the evolving nature of their work necessitates adoption of some novel methods for the study of 'why of consciousness'. The question that we inevitably face in any study of consciousness is whether the subjective feel of our experiences can be captured through a physiological account. Will we ever succeed in bridging the proverbial gap between the experiential and physiological accounts of the same phenomena?

Encircling the Conundrum

Having noted problems surrounding the nature and theories of consciousness, in the rest of the chapter I will be mainly focusing on the dynamic core hypothesis derived from the evolutionary idea of neuronal group selection. This hypothesis was initially elaborated by Edelman in his trilogy (1987, 1988, 1989) wherein consciousness came to be identified with reciprocal connectivity in the thalamo-cortical system. In the subsequent years, the proposal has, however, been refined and made mathematically more elegant after Tononi joined Edelman as a collaborator. The theory is now largely referred to as the information integration theory

(IIT) (Edelman and Tononi 2000; Tononi 2004, 2008, 2012a, 2015). In subsequent years, jointly articulating their ideas, Tononi, Koch and their co-workers have tended to bring together IIT and NCC approaches (Tononi and Koch 2015; Koch et al. 2016). While the theory initially began with a definite hostility towards cognitive and functionalist approaches (Edelman 1989; Edelman and Tononi 2000: Chap. 1), its more recent accounts have attracted the accusation of being functionalist in nature (Block 2009). Some commentators have also charged the theory of embracing panpsychism (Kouider 2009). These charges have been recently responded to by Tononi and Koch (2015). In this regard, they unequivocally state:

> in sharp contrast to widespread functionalist beliefs, IIT implies that digital computers, even if their behaviour were to be functionally equivalent to ours, and even if they were to run faithful simulations of the human brain, would experience next to nothing (p. 1).

Even with respect to panpsychism, IIT has maintained equally reasonable distance. Tononi and Koch (2015) in this context have clearly and emphatically stated that:

> Panpsychism's beauty has been singularly barren. Besides claiming that matter and mind are one thing, it has little constructive to say and offers no positive laws explaining how the mind is organized and works. IIT was not developed with panpsychism in mind ... Unlike panpsychism, however, IIT clearly implies that not everything is conscious (p. 11).

The updated versions of IIT, contrary to critics' understanding, do not seem to be thus aligned to any version of functionalism or panpsychism.[12]

While putting forward IIT as the most promising theory of consciousness that we currently have, I will however not be confining myself entirely to IIT theorists. I believe that our understanding is likely to be enriched if we can integrate useful insights from other scholars working in the area of consciousness studies. More so when no one yet claims to have the complete picture in hand (Seth 2007). Notable among those whose contributions offer useful cues for understanding the nature of consciousness are Block (1995, 2001, 2007, 2009, 2014a), Baars (1997, 2007), Dehaene, Naccache, Changeux and their co-workers (2001, 2007, 2011), Damasio (2012) and Zeki (2007).[13] In what follows, I not only look at how the theory as proposed by Edelman, Tononi and their collaborators looks in its successive formulations but also aim at assessing how successful IIT has been in encircling the consciousness conundrum.

---

[12]And this change or evolution of the theory has not gone unnoticed. For instance, while Block (2007: 482) aligns his own position on consciousness with Edelman's (2004) 'physicalistic view' as opposed to Searle's (1992) defence of dualism, he subsequently classifies IIT (Edelman and Tononi 1998) among functionalist theories of consciousness (Block 2009: 1111).

[13]I will not, however, be dealing with epiphenomenalists as they tend to undermine consciousness in their accounts and render it causally inefficacious in our interactions of the world (Huxley 1874/1898; Jackson 1982). Such accounts also term all voluntary human actions as illusory in nature (Wegner 2002; Gold 2011).

Information Integration Theory

The first important thing to note about Edelman and Tononi's proposal is that, for them, consciousness, much like James (1910/1997), is not an object but a process. But this immediately confronts us with a problem. As soon as we think of accounting for how the subjective, experiential qualities are generated, we are faced with a unique kind of asymmetry. Instead of just explaining things that are out there in the world, in the case of consciousness we are required to simultaneously explain 'what it is to be us' (Edelman and Tononi 2000: 11). The process that we are trying to explain is 'we ourselves'.

The existence of these difficulties notwithstanding, Edelman and Tononi (2000) view consciousness as a process that is grounded in physical brain states. But not all brain states are conscious. This means that for any theory of consciousness to be successful, it must be able to specify conditions under which a thing in question can possess consciousness. It will also be required of such a theory that it can tell us how the extent of consciousness in a system varies with changes in the underlying brain states. So, the question that we are aiming to answer is why should possession of certain biochemical features lead to in the possessor of that brain with the favor of subjective experience? This fact about brain states is all the more intriguing because it is the activity of only some neurons that results in conscious experience with its characteristic phenomenality and not the activity of others.[14] Moreover, as we shall discuss in what follows, it is not always the case that activation of same set of neurons results in conscious experience. The difference between IIT and other approaches is that IIT focuses on neural processes, rather than brain areas, that support consciousness.[15]

Following James (1910/1997) and Sherrington (1951), Edelman and Tononi (2000) consider consciousness to be also unitary and private in nature. In fact, the private nature of conscious events is closely linked with their unity. Furthermore, the unity of consciousness is closely associated with the coherence of perceptual events. That is, we cannot simultaneously be aware of or entertain two mutually incoherent or contradictory scenes or objects. These two features of unity and coherence of consciousness are tied to its capacity limitation. Also, conscious states are not only coherent and unified, they are stable as well. This accounts for the persistence of the world around us in terms of its stability (Edelman and Tononi 2000: 23–7).

---

[14]For greater detail and elaboration of this aspect, see Tononi (2004, 2012; Tononi and Koch 2015).

[15]There is no denying that some of the early efforts aimed at identification of brain functions tended to entirely preoccupy themselves with the task of identifying brain region(s) responsible for happening of consciousness. That such approaches to consciousness are wrongly headed was one of the reasons responsible for Dennett's strong repudiation of what he called Cartesian theatre and his espousal of consciousness in terms of multiple drafts model (Dennett 1991).

While explicating the nature of consciousness, Edelman and Tononi emphasise the special nature of brain and its interconnections. This is important not only owing to peculiarity of brain's connectivity and representational plasticity but also due to its 'mode of functioning and relation to the body and the world' (2000: 37). That is, our involvement in different tasks, cognitive or otherwise, regularly modifies our brain by modifying its neuronal circuitry and consequent computing abilities. It is not only the case that no two brains are same, but the same brain also undergoes change as a function of its engagement with itself and the world. And this variability 'is not just noise or error' (2000: 47). The variability in fact enables the brain to deal with the unforeseeable conditions of life. It is not only that no existing artefact or computing machine displays as much structural diversity as the brain but the way the world is presented to the brain is ambiguous, context-dependent, and the meaning of the actual signals received is influenced by prior judgement or a kind of 'perspective taking'. Moreover, there is no judge or homunculus in the brain to read the message. So, successful perceptual categorisation is realised by the brain through diverse sensory modalities without the input signals being presented in a coded manner as is the case with computers. The brain is a very unique organ because its interconnectivity results in synchronisation of activity across diverse regions and this happens without there being any command centre in the brain.

According to the proposal put forward by Edelman and Tononi (2000), the brain is distinguished from every other thing by its rich *re-entrant circuitry* which is largely *parallel*. Also, while different regions of the cortex are tied to specific aspects of consciousness, no one region is tied to consciousness. It is the distributed patterns of activity in the brain that underlie conscious experience. One of the suggested ways to identify critical areas involved in conscious experience is thus to contrast activation of regions in conscious and unconscious control of diverse bodily activities. The different dissociative disorders like hysterical blindness and split-brain cases, for example, reveal how life can be affected by loss of integrative functions. For consciousness to be incarnated in the brain, Edelman and Tononi (2000) suggest that the required neural activity has to be highly variable and differentiated both spatially and temporally. This means that functional discrimination of any kind requires lack of homogeneous activation. This explains the phenomena of temporary snow blindness caused by continuous staring at the snowy field of frosty white reported by arctic explorers. Loss of consciousness during epileptic seizures and during deep sleep even when brain is 'highly active' are other examples that support IIT (Tononi 2004: 2).

To account for consciousness in terms of underlying brain processes, Edelman and Tononi appropriate from neuroscience insights about a variety of brain processes at different levels of organisation, such as perceptual categorisation, dynamic processes of corticothalamic organisation, value and non-representational memory. For the purpose of methodological simplicity and clarity they further make a distinction between primary consciousness and higher-order consciousness.

The primary consciousness is considered to occur in animals who have brains much like we do. Animals with such brains are capable of entertaining 'mental scene' but have limited semantic or symbolic capacities but found to be lacking in linguistic ability. It is defined as 'the ability to generate a mental scene in which a large amount of diverse information is integrated for the purpose of directing present or immediate behavior' (Edelman and Tononi 2000: 103).

As regards higher-order consciousness that is uniquely human, it is 'built on the foundations provided by primary consciousness and is accompanied by a sense of self and the ability in the waking state explicitly to construct and connect past and future scenes. In its most developed form, it requires a semantic capability and a linguistic capability' (2000: 103–4). It is only beings with higher-order consciousness who can report conscious states as also talk about consciousness. They alone are also capable of being 'conscious of being conscious' (2000: 104).

Let us now have a brief look at the structures and processes necessary for supporting consciousness as identified by Edelman and Tononi (2000):

1. *Ability of Perceptual Categorisation*: This refers to the ability of the animal to categorise signals received from the world such that it is useful for the individual in question. Besides the ability to control bodily movements, perceptual categorisation is a most important function and can simultaneously occur in various modalities (seeing, hearing, touching, etc.) as well as in a variety of submodalities (within the modality of vision for instance, cognisance of colour, motion, orientation, texture, etc.).
2. *Acquisition of Concepts*: By concepts Edelman and Tononi mean 'the ability to combine different perceptual categorisations related to a scene or an object and to construct a "universal" reflecting the abstraction of some common feature across a variety of such percepts' (2000: 104). For example, when presented with faces of different shapes, colours and sizes, the ability of the brain to abstract the commonality from this variety depends upon acquisition of concepts.
3. *Memory*: Given the inherent ambiguity, novelty and context-dependency of input signals as well as the neural medium of storage which is not capable of high precision, memory is viewed as dynamic and non-representational and primarily serves the organism's adaptive goals. It is realised through synaptic alterations in re-entrant circuits.
4. *Value*: It refers to evolving architectural constraints of an organism that develop as a consequence of experience. This is necessary as nervous systems being selectional in nature, they cannot be pre-programmed nor is there any centralised command centre that can perform online monitoring. Such systems therefore require evolving constraints as part of their architecture.
5. *Re-entry*: It is one of the most important aspect of theory of neuronal group selection that forms the basis of Edelman and Tononi's views about consciousness.[16] Re-entry 'is a process of ongoing parallel and recursive signalling

[16]For more details on re-entry, see Edelman (1987), Edelman and Tononi (2000: 85*f*).

between separate brain maps along massively parallel anatomical connections, most of which are reciprocal. It alters and is altered by the activity of the target areas it interconnects… [It is] the most important integrative mechanism in higher brains… [It] assures integration that is essential to the creation of a scene in primary consciousness' (2000: 105–6).

6. *Integration or binding problem*: It is also most critical for the existence of consciousness as it is responsible for bringing together functionally distinct maps in different higher brain regions and can operate coherently without intervention or further involvement of anyone to guide or supervise the process. The interesting thing about integration is that it happens in the presence of not only very diverse but at times conflicting inputs. It is integration that allows the organism to produce a coherent and contextually appropriate successful behaviour in complex scenarios wherein activations are often not in sync with each other. Integration becomes possible because of re-entry across different levels of neuronal organisation across distanced neuronal groups. It takes place at different levels, i.e., across different levels, modalities and submodalities, and is achieved in real time as the objective of desirable behavioural output is brought within the constraints of real life situations. By achieving unsupervised synchronisation of neuronal activations across widely separated neuronal groups across different brain regions, integration through re-entry can dynamically realise large scale correlations. Integration therefore 'plays a central role in mechanisms leading to consciousness' (2000: 107).

Now that we have a basic working idea of the mechanisms necessary for accounting for consciousness, we can deal with the issue of how primary consciousness arose in the course of evolution. From the above considerations about the brain mechanisms it is clear that cortical systems along with their appendages like basal ganglia and other mechanisms discussed above were in place before primary consciousness came into being. This means that before humans arrived on the world scene, evolutionary transition from reptiles to birds to mammals and consequent emergence of new anatomical connectivity in the form of conceptual memory systems, perceptual categorisation and re-entry between different brain areas were already in place. It is thus re-entrant interactions between these new emergent anatomical structures in the course of evolution that are responsible for the emergence of consciousness.

It accordingly appears reasonable to assume that integration/binding of hierarchically organised but often distant neuronal groups is one distinct feature of conscious experience because it is in conformity with the sense of unity that is accepted as the hallmark of each conscious state since James (1910/1997). Given the theoretical tools made available by Edelman and Tononi (2000), Tononi (2004, 2015), we can not only know what integration is but we can also measure it quantitatively by identifying underlying neuronal processes responsible for integration. They did this by undertaking several simulation studies. These simulation studies do not involve any 'superordinate entity' for coordinating integration between segregated areas. The findings of these studies demonstrate 'how reentry can solve the binding problem by coupling neuronal responses of distributed cortical areas to achieve synchronization and *global* coherence' (Edelman and Tononi 2000: 119).

For further concretising the meaning of neural processes being integrated and then connecting it to the unitary nature of conscious experiences, Edelman and Tononi (2000) propose the concept of 'functional clustering'.[17] In a functional cluster, which has identifiable boundaries, a subset of elements strongly interacts among themselves but weakly with the rest of the system. Also, interacting components cannot be decomposed further into independent or nearly independent components (Tononi 2004). Edelman and Tononi (2000) liken a functional cluster to the way we understand a traditional family whose members interact among themselves much more frequently and intensely than with members of the wider society. Indeed, the existence of synchronous activity among distant neuronal groups over a short time-frame is an indication of existence of functional clustering. The concept of functional clustering can thus be used to characterise processes underlying conscious experience. And the process which achieves this is re-entry.

The other major feature of IIT as proposed by Edelman and Tononi (2000) and Tononi (2004, 2012) is 'differentiation'. It is differentiation that renders conscious states informative. Given the vastness of brain connectivity, billions of conscious states are possible. It is brain's ability to differentiate between them that constitutes information. These proposed concepts can be useful only if we can make use of them to measure degree of differentiation. Measurement of differentiation is critical for IIT because it is only through this measure that the informativeness of conscious states can be measured. In their proposed model, the degree to which a neural system is differentiated is reached by averaging the mutual information between each subset of a neural system and the rest of the system for all possible bipartitions of the system. The system's measure for this neuronal differentiation is called 'neural complexity' (Edelman and Tononi 2000: 126, 130*f.*). Such a measure of neuronal differentiation means that higher the mutual information between each neuronal subset and the rest of the system, greater is the value of complexity. Therefore, functional clustering can be used to measure integration; and complexity can be used to measure the extent of differentiation of the neural processes.

It is appropriate to characterise consciousness/conscious experience in terms of complexity because complexity has been understood as the measure of average mutual information within a neuronal system (Edelman and Tononi 2000: 135). It is complexity of neuronal states that renders them informative and hence conscious. The very possibility and sustenance of any conscious experience therefore requires the presence of a high degree of complexity in the underlying neuronal base. This means that any simultaneous hypersynchronous firing of a large number of neuronal groups would be associated with low complexity and hence absence or very low level of consciousness. This is in fact what we observe in cases of epileptic

[17]For more details see Edelman and Tononi (2000: 120*f.*) and Tononi et al. (1998).

seizures. This would imply that complexity is low or zero in systems that are either totally random and thus disorderly or totally orderly and homogeneous or integrated.

The important thing to note is that Edelman and Tononi's understanding of complexity is quite in sync with wider characterisation of complexity. It thus appears appropriate to apply the concept of complexity to characterise consciousness because biological systems in general are good examples fitting the criterion of a complex system (Edelman and Tononi 2000: 135). The complexity of a system will accordingly have high value in cases where a neuronal system is composed of large parts that are functionally specialised (meaning differentiated) and demonstrate integration. One implication of these considerations is that neuronal groups that are not strongly activated to sustain activation for sufficient duration cannot support conscious experience. The other implication is that the whole brain is not required to be involved in supporting conscious experience because not all activities of the brain are related to what a person is conscious of. So, all regions do not contribute to dynamic coherent states. This means that though every task will require activation and deactivation of certain groups, the underlying neuronal processes need to be both, sufficiently differentiated and integrated, turning the neuronal states informative and thus conscious. So, effective interactions do not involve all regions because only a subset of neuronal groups contributes directly to conscious experience. This is illustrated by automated tasks which do not directly contribute to conscious experience as the neuronal groups involved in such tasks become functionally isolated over time with their regular practice (Tononi 2004: 14, Tononi 2012b: 318).

The other fact that we need to take cognisance of and explain is that our conscious percepts are much more stable than the rapidly changing nature of the incoming stimuli. This means that the visual perception of an object only captures properties that are invariant in the position and orientation of the object in question. So, rapidly varying local details are often ignored by the brain in so far as the overall meaning or gist of the perceptual scene is not affected. The implication of this is that we are now in a position to justify our initial agreement with James' characterisation of consciousness as a process on empirical grounds. Consciousness is neither a thing nor a simple happening in the world because integration and differentiation—properties that are fundamental to conscious experience—can be explained only by taking recourse to a distributed neuronal process and not by local properties of neurons. This entails that consciousness can only be supported by a neuronal group that strongly interacts with itself; has distinct functional borders; and achieves this at the time scale of milliseconds.

Such a neuronal group is thus endowed with a dynamic core. But this dynamic core is a process and not a thing or a place and changes with time. It is defined in terms of neuronal interactions and not in terms a neural location. It is spatially distributed and constantly changing. Even in case of a functional cluster, re-entrant interactions need to be sufficiently differentiated as determined by its complexity. But such a restricted activation scenario of neuronal groups is not confined to any one group as dynamic core does not refer to any *invariant set of brain areas.*

Also, the same neuronal group may be involved in more than one functional cluster. On occasions, it may support conscious experience and on some other occasions it may not. The dynamic core is decided by rapidly shifting functional connectivity rather than anatomical proximity or boundaries. The exact composition of the dynamic core for particular conscious states may vary from person to person.

Having provided an account of Edelman and Tononi's (2000) proposal in the form of dynamic core hypothesis, we are now in a position examine whether it can account for properties of conscious experiences. The most agreed upon features of consciousness are:

i. *Consciousness is an integral process*: As suggested by James, consciousness is not a thing but a process. This is accounted for as discussed in previous sections.
ii. *Consciousness displays unity*: Integration as *sine qua non* of conscious experiences makes them to be not amenable for decomposition into independent constituents (Tononi 2004: 10; Tononi 2012b: 297).
iii. *Privacy:* Consciousness is experienced from a single point of view and not sharable by another person (Tononi 2004, 2012b). The property of functional clusters marks boundaries that make such states effectively private.
iv. *Coherence of conscious states*: A subject cannot simultaneously entertain conflicting conscious states. This feature also follows from integrative feature of conscious states.
v. *Consciousness as a differential process*: Differentiation of conscious experience is related to features like, informativeness, global access and flexibility.
vi. *The informativeness*: Dynamically active groups must have high level of complexity corresponding to high values of average mutual information between them. The loss of consciousness in sleep or generalised epilepsy results in strong integration and loss of functional specialisation and informativeness (Tononi 2004, 2012b).
vii. *Distribution of information, context-dependency, and global access*: Consciousness increases access to other states even if they are not part of the dynamic core. These features are part of the adaptive aspect of consciousness during evolution.
viii. *Flexibility and the ability to respond to and learn unexpected associations*: This must have been evolutionarily absolutely necessary for survival of our own as well as other animals. Without the ability to assess and respond to novel situations, life would not have ever taken the evolutionary leap that we are witness to from the tree of life. The unpredictability of situations in life requires that the system has at its disposal, as far as possible, bodily means to cope with such predictability. The adaptive importance of such a facility can hardly be undermined given the fact that all biological beings continuously 'face an open-ended world full of novelty' (Edelman and Tononi 2000: 150).

ix. *The limited capacity of consciousness*: This refers to 'an upper limit on *how many partially independent sub-processes can be sustained within the core without interfering with its integration and coherence*' (150).
x. *The serial nature of conscious experience:* This refers to the temporal succession that having of all experiences displays (Tononi 2012b: 22).
xi. *Consciousness as a process that is continuous but continually changing*: Edelman and Tononi's view of consciousness as a process and not a thing as well as the hypothesis of the dynamic core that postulates existence of an active and dynamically bounded neuronal core that maintains its unity is in conformity with human experiences.

This brings us to the critical questions and predictions that the theory of consciousness proposed by IIT needs to address for it to be taken seriously (Edelman and Tononi 2000; Tononi 2004, 2012b; Tononi et al. 2016). Some of the critically important ones are:

i. As proposed by Edelman and Tononi there should be evidence for interactions between distinct and distributed neuronal groups as conceived by the concept of a *functional cluster*. It has therefore to be tested whether we can identify such a neuronal group.
ii. Whether activations in a neuronal group spread much more widely when the group is part of the dynamic core supporting conscious experience or not?
iii. We can also test if deviations from the norm in conscious experiences gets reflected in abnormalities in the supporting neuronal processes?
iv. Whether the neuronal processes underlying conscious experiences are really found to be as differentiated and integrated over real time as hypothesised?
v. Whether processes underlying automated sequences display lower complexity?
vi. Whether increase in informational complexity is found to be linked to increase in discriminatory abilities of persons having corresponding conscious experiences? As discriminative abilities are found to be changing with experience (e.g. learning to drive, acquiring musical accomplishments, professional achievement in some game), there should be demonstrable evidence for a corresponding increase in the complexity of the dynamic core as hypothesised (also Tononi 2004: 7).

This brings us to the problem of explaining the nature of qualia or the subjective feel that accompanies conscious experiences. 'Why should the physical, objectively describable fact that a certain neuron is firing correspond to a conscious sensation, to a subjective feel, to a quale?' (Edelman and Tononi 2000: 159). It is indeed because of the desire of philosophers to find a satisfactory answer to the qualia problem that we find them engaged in discussions of sensations, pains, philosophical zombies (things or beings who can have similar experiences but without any accompanying quale/feeling) and so on. Can these be accounted for only in terms of neuronal activations? What about knives, forks, diodes, thermostats, other living things like plants and animals as well as human children? Can't they also be conscious? Such questions are not gratuitous once we take seriously Tononi's proposal that 'consciousness corresponds to the capacity to integrate information' (Tononi 2004: 6).

Here, it is to be noted that though subsequent versions of dynamic core hypothesis in terms of information integration theory have given greater emphasis to developing of quantitative tools to measure informational component, its focus on neuronal processes has not been diluted (Tononi 2004, 2012b). Approaching the issue in terms of quantitative measure of information of a physical system in terms of an objectively measurable quantity called Φ (Tononi 2004: 3; Tononi et al. 2016: 12) in fact safeguards the theory against possible charges of the theory practising any kind of neural chauvinism (Block 1980). Additionally, theory makes many predictions about the quantity and quality of consciousness that can be empirically tested (Tononi 2004; Tononi and Koch 2015; Tononi et al. 2016). The IIT thus principally does not deny ascription of consciousness to anything. Such an orientation is adopted by IIT theorists because their primary concern is with ascertaining 'what experience is and what type of physical systems can have it' (Tononi and Koch 2015: 1). Articulation of consciousness as a measure of differentiated integration of information has, however, led to certain misgivings. As discussed earlier, scholars like Block (2009) and Kouider (2009) consider IIT to be functionalist and panpsychist in nature.

While addressing the charge of functionalism or panpsychism levelled on IIT, it is worthwhile to reiterate that a principled criterion for ascription of consciousness has always been a problem because reliance on behaviour and reportability has proved problematic (Edelman 1989, 2004; Block 2007, 2009, 2011, 2014a, b; Dehaene et al. 2006; Dehaene and Changeux 2011; Tononi 2004, 2012b, 2015; Tononi and Koch 2015; Koch 2009, 2012, 2014; Koch et al. 2016). The distancing of IIT from functionalism and panpsychism and unreasonableness of the charges levelled against IIT can perhaps be gauged from the adoption of a two-pronged strategy by Koch and Tononi and their co-workers in their successive formulations of the theory wherein principled characterisation of consciousness in substrate-neutral terms and the search for physical correlates of consciousness move hand in hand. I thus consider IIT theorists' addressing of the problem of philosophical zombies, taking seriously the challenge from researchers working in the area of artificial intelligence and their resistance against generous ascription of consciousness as demanded by the panpsychists to be not inappropriate indulgences. Nor does responding to such long standing philosophical issues turns IIT functionalist or panpsychist. It is not only that the scientific progress in our understanding of different biological systems and coming into being of sophisticated computational devices has rendered privileging of certain highly evolved human capabilities like language as markers of consciousness unreasonable, but our knowledge has reached a stage wherein 'Consciousness can be dissociated from many other cognitive processes that have traditionally been closely linked to it, including memory, emotions and selective attention' (Tononi and Koch 2015: 15). Such a state of affairs necessitates a nuanced approach to the study of consciousness. It is in pursuance of such a requirement that we witness Tononi, Koch and their co-workers emphasising a distinction between *enabling factors* or *background conditions* of consciousness and *physical substrate of consciousness* (PSC) (Tononi and Koch 2015: 2; Tononi et al. 2016: 4).

This change in vocabulary from NCC to PSC is part of IIT theorists' larger endeavour to move away from the question of how consciousness arises in the brain to articulation of the nature of conscious experience and then identifying requirements that a physical system needs to fulfil for realisation of such an experience. It is a significant theoretical step aimed at addressing long standing questions in consciousness/mind research and needs to be so viewed. Such an engagement does not represent embracing of functionalism or panpsychism on the part of IIT theorists. Given the causal efficacy of conscious states, IIT aims at identifying for each necessary feature of conscious experience its corresponding property of the PSC (Tononi et al. 2016: 1). So, possession of intrinsic states, causal powers, compositionality, differentiation and integration, temporal definiteness, etc., are necessary for a PSC to be able to support consciousness. It is such requirements that enable IIT to provide 'a principled explanation for several seemingly disparate facts about the PSC. For example, IIT can explain why the cerebral cortex is important for consciousness, but the cerebellum is not … [even though it] has four times more neurons than the cerebral cortex' (Ibid. 10). Similarly, counterintuitive as it may appear, IIT predicts that on occasions a neuronal group may contribute to experience even when inactive (Ibid. 10). What matters in such cases is satisfaction of conditions listed earlier. The important thing about IIT is that it makes series of empirically testable predictions.[18]

While on occasions some reservations have been expressed by scholars about the enormity of computational resources required for computing Φ even for lower organisms and more efficient algorithms may be required to calculate Φ (Koch 2009), the theory in recent years has enjoyed unparalleled success. For instance, taking recommendations of IIT seriously that consciousness requires informational differentiation and integration, Casali et al. (2013) have successfully developed an objective measure of consciousness in terms of PCI that is not dependent on human subjects' ability to independently interact with the environment. PCI is an objectively measurable empirical parameter that can discriminate 'the level of consciousness in single individuals during wakefulness, sleep, and anesthesia, as well as in patients who had emerged from coma and recovered a minimal level of consciousness' (Casali et al. 2013: 1). PCI is designed to measure EEG response to TMS stimulation in different conditions characterised by differential loss to consciousness. As predicted by IIT, Casali et al. (2013: 2–4) demonstrate PCI to be successful in:

i. reflecting differentiation and integration in the targeted neuronal groups.
ii. discriminating between consciousness and unconsciousness in healthy individuals.

[18]See Tononi et al. (2016: 11–12) for a list of significant predictions that IIT makes.

iii. sensitive to graded changes in the level of consciousness.
iv. discriminating the level of consciousness in brain-injured patients.

Moreover, PCI has proved to be equally reliable across control subjects in different conditions of wakefulness and sleep as well as subjects who are in coma, vegetative or minimal conscious states. Appreciating the significance of PCI in clinical settings by its success in discriminating conscious states from unconscious ones in the absence of behavioural response from subjects, the editorial in *Science Translational Medicine* titled 'Quantifying the Unquantifiable' has termed it as 'a robust measure that distinguishes conscious from unconscious states well enough to be used on an individual basis'. The theoretical importance of PCI as proposed by Casali et al. (2013) on the basis of predictions made by IIT has been appreciated by other scholars as well (Schiff 2013). Undoubtedly, the challenges facing consciousness research are enormous, but a sound beginning seems to have been made in the form of IIT as evidenced by the baby steps that researchers working in the field have stated to take as made evident by the preceding discussion. So, whatever the future may hold for it, as of now IIT remains one of the most rigorously articulated and empirically testable hypothesis on the topic. After a long and circuitous journey, we are finally in possession of something concrete and elegant enough to be seriously aimed at for refutation. And in that lies both promising science and a big philosophical challenge.

References

Aru, J., Bachmann, T., Singer, W., & Melloni, L. (2012a). Distilling the neural correlates of consciousness. *Neuroscience and Biobehavioral Reviews, 36,* 46–737.

Aru, J., et al. (2012b). Local category-specific gamma band responses in the visual cortex do not reflect conscious perception. *Journal of Neuroscience, 32,* 14–14909.

Baars, B. J. (1997). *In the theatre of consciousness: The workspace of the mind.* New York: Oxford University Press.

Baars, B. J. (2007). The global workspace theory of consciousness. In M. Velmans & S. Schneider (Eds.), *The Blackwell Companion to consciousness* (pp. 46–236). Oxford: Blackwell Publishing.

Bedau, M. A., & Humphreys, P. (2008). *Emergence: Contemporary readings in philosophy and science.* Cambridge, MA: MIT Press.

Block, N. (1980). Troubles with functionalism. *Readings in the philosophy of psychology* (pp. 268–305). Cambridge, MA: Harvard University Press.

Block, N. (1995). How many concepts of consciousness? *Behavioral and Brain Sciences, 18,* 84–272.

Block, N. (2001). Paradox and cross purposes in recent work on Consciousness. *Cognition, 79,* 197–219.

Block, N. (2005). Two neural correlates of consciousness. *Trends in Cognitive Science, 9,* 46–52.

Block, N. (2007). Consciousness, accessibility, and the mesh between psychology and neuroscience. *Behavioral and Brain Sciences, 30,* 481–548.

Block, N. (2009). Comparing the major theories of consciousness. In. M. S. Gazzaniga (ed.), *The cognitive neurosciences IV* (pp. 1111–22). Cambridge, MA: MIT Press.

Block, N. (2011). Perceptual consciousness overflows cognitive access. *Trends in Cognitive Sciences, 15,* 75–567.
Block, N. (2014a). Consciousness, big science and conceptual clarity. In G. Marcus & J. Freeman (eds.), *The Future of the brain: Essays by the world's leading neuroscientists* (pp. 161–76).
Block, N. (2014b). Rich conscious perception outside the focal attention. *Trends in Cognitive Sciences, 30,* 1–3.
Bremmer, J. N. (1983). *The early Greek concept of the soul.* Princeton, NJ: Princeton University Press.
Campbell, R. (2015). *The metaphysics of emergence.* London: Palgrave Macmillan.
Campbell, R., & Bickhard, M. H. (2011). Physicalism, emergence and downward causation. *Axiomathes, 21,* 33–56.
Casali, A. G., et al. (2013). A theoretically based index of consciousness independent of sensory processing and behavior. *Science Translational Medicine, 5*(198), 198ra105. doi:10.1126/scitranslmed.3006294.
Chalmers, D. J. (1993). Why fodor and pylyshyn were wrong: The simplest refutation. *Philosophical Psychology, 63,* 19–305.
Chalmers, D. J. (1995). On implementing a computation. *Minds and Machines, 4,* 391–402.
Chalmers, D. J. (1996). *The conscious mind: In search of a fundamental theory.* New York: Oxford University Press.
Chemero, A. (2009). *Radical embodied cognitive science.* Cambridge, MA: MIT Press.
Chomsky, N. (1959). Review of Skinner's Verbal Behavior. *Language,* 35: 26–58.
Chomsky, N. (2000). New Horizons in the Study of Language and Mind. Cambridge: Cambridge University Press.
Churchland, P. M. (1981). Eliminative Materialism and the Propositional Attitudes. *Journal of Philosophy,* 78: 67–90.
Clark, A. (2014). *Mindware: An introduction to the philosophy of cognitive science.* Oxford: Oxford University Press.
Clark, A. & Chalmers, D. J. (1998/2002). The extended mind. In D. J. Chalmers (ed.), *Philosophy of mind: Classical and contemporary readings.* Oxford: Oxford University Press.
Cohen, M. A., et al. (2012). The attentional requirements of consciousness. *Trends in Cognitive Sciences, 16*(8), 17–411.
Damasio, A. (2012). *Self comes to mind: Constructing the conscious brain.* London: Vintage Books.
Dehaene, S., & Changeux, J.-P. (2011). Experimental and theoretical approaches to conscious processing. *Neuron, 70,* 27–200.
Dehaene, S., Changeux, J. P., Naccache, L., Sackur, J., & Sergent, C. (2006). Conscious, preconscious, and subliminal processing: A testable taxonomy. *Trends in Cognitive Science, 10,* 11–204.
Dehaene, S., & Naccache, L. (2001). Towards a cognitive neuroscience of consciousness: Basic evidence and a workspace perspective. *Cognition, 79,* 1–37.
Dennett, D. C. (1991). *Consciousness explained.* New York: Little, Brown and Company.
Dreyfus, H. (1972). *What computers can't do.* Cambridge, MA: MIT Press.
Dreyfus, H. (1992). *What computers still can't do.* Cambridge, MA: MIT Press.
Edelman, G. M. (1987). *Neural darwinism: The theory of neuronal group selection.* New York: Basic Books.
Edelman, G. M. (1988). *Topobiology.* New York: Basic Books.
Edelman, G. M. (1989). *The remembered present: A biological theory of consciousness.* New York: Basic Books.
Edelman G. M., & Tononi, G. (2000). A universe of consciousness: How matter becomes imagination. New York: Basic Books.
Edelman, G. M. (2004). *Wider than the sky: The phenomenal gift of consciousness.* New Haven, NJ: Yale University Press.
Fodor, J. (1975). *The language of thought.* New York: Thomas Y. Crowell.

Fodor, J. (1980). Methodological solipsism considered as a research strategy in cognitive psychology. *Behavioral and Brain Science, 3,* 63–73.
Fodor, J. (2008). *LOT2*, Oxford: Clarendon Press.
Frith, C., & Rees, G. (2007). A brief history of the scientific approach to the study of consciousness. In M. Velmans & S. Schneider (Eds.), *The Blackwell Companion to consciousness* (pp. 9–22). Oxford: Blackwell.
Gold, H. (2011). Neuroscience on the nature of consciousness. *Neurogenesis, 1*(1), 4–23.
Haugeland, J. (1985). *Artificial intelligence: The very idea.* Cambridge, MA: MIT Press.
Heffernan, V. (2013). *Why I'm a creationist.* Yahoo News July 11. http://news.yahoo.com/why-im-a-creationist-141907217.html.
Heylighen, F. (2009). Complexity and self-organization. In *Encyclopedia of Library and Information Sciences* (3rd ed., 1215–24).
Huxley, T. H. (1874). On the hypothesis that animals are automata, and its history. *The Fortnightly Review, 16*(New Series), 555–80. (Reprinted in Method and Results: Essays by Thomas H. Huxley, New York: D. Appleton and Company, 1898).
Jackson, F. (1982). Epiphenomenal qualia. *The Philosophical Quarterly, 32,* 36–127.
James, W. (1910/1997). The stream of consciousness. *Psychology*, Chapter XI. New York, Henry Holt and Co., 151–75. (Reprinted in N. Block, O. Flanagan and G. Guzeldere (eds). The Nature of Consciousness: Philosophical Debates (pp. 71–82). Cambridge, MA: MIT Press).
Kiani, R., & Shadlen, M. N. (2009). Representation of confidence associated with a decision by neurons in the parietal cortex. *Science, 324,* 64–759.
Koch, C. (2004). *The quest for consciousness: A neurobiological approach.* Englewood, CO: Roberts & Company.
Koch, C. (2009). A theory of consciousness: Is complexity the secret to sentience, to a panpsychic view of consciousness? *Scientific American Mind, 20,* 16–19.
Koch, C. (2012). *Consciousness: Confessions of a romantic reductionist.* Cambridge, MA: MIT Press.
Koch, C. (2014). Is consciousness universal? *Scientific American Mind, 25,* 9–26.
Koch, C., Massimini, M., Boly, M., & Tononi, G. (2016). The neural correlates of consciousness: progress and problems. *Nature Review Neuroscience, 17,* 21–307.
Koch, C., & Tsuchiya, N. (2007). Attention and consciousness: Two distinct brain processes. *Trends in Cognitive Sciences, 11*(1), 16–22.
Kouider, S. (2009). Neurobiological theories of consciousness. In W. P. Banks (Ed.), *Encyclopedia of Consciousness* (pp. 87–100). San Diego (CA): Academic Press.
Kriegel, U. (2007). Intentional inexistence and phenomenal intentionality. *Philosophical Perspectives,* 21: 307–340.
Kurzweil, R. (2012). *How to create a mind.* New York: Viking.
Landman, R., Spekreijse, H. & Lamme, V. A. (2003). Large capacity storage of integrated objects before change blindness. *Vision Research, 43*(2), 149–164.
Levine, J. (1983/2002). Materialism and qualia: The explanatory gap. In D. J. Chalmers (ed.), *Philosophy of mind: Classical and contemporary readings* (pp. 354–361). New York: Oxford University Press.
Lycan, W. G. (2008). Giving dualism its due. *Australasian Journal of Philosophy, 87*(4), 63–551.
McGinn, C. (1999). *The mysterious flame: Conscious minds in a material world.* New York: Basic Books.
Naccachea, L., & Dehaene, S. (2007). Reportability and illusions of phenomenality in the light of the global neuronal workspace model. *Behavioral and Brain Sciences, 30,* 20–518.
Nagel, T. (1974). What is it like to be a bat? *The Philosophical Review, 83*(4), 50–435.
Paller, K. A., & Suzuki, S. (2014). The source of consciousness. *Trends in Cognitive Sciences, 18* (8), 9–387.
Putnam, H. (1960/1975). "Minds and Machines". In Mind, Language, and Reality, Cambridge: Cambridge University Press, 362–385.
Putnam, H. (1967/1975). "The Nature of Mental States". In Mind, Language, and Reality, Cambridge: Cambridge University Press, 429–440.

Pylyshyn, Z. (1984). *Computation and cognition*. Cambridge, MA: MIT Press.

Rees, G., & Frith, C. (2007). Methodologies for identifying the neural correlates of consciousness. In M. Velmans & S. Schneider (Eds.), *The Blackwell companion to consciousness* (pp. 66–553). Oxford: Blackwell.

Rees, G., Kreiman, G., & Koch, C. (2002). Neural correlates of consciousness in humans. *Nature Reviews Neuroscience, 3*(4), 70–261.

Rees, G., Wojciulik, E., Clarke, K., Husain, M., Frith, C. & Driver, J. (2000). Unconscious activation of visual cortex in the damaged right hemisphere of a parietal patient with extinction. *Brain, 123*(8): 1624–33.

Rohde, E. (1925). *Psyche: The cult of souls and beliefs in immortality among the Greeks*. New York: Kegan Paul.

Ryle, G. (1949). The Concept of Mind. London: Hutchinson.

Sandberg, K., et al. (2010). Measuring consciousness: Is one measure better than the other? *Consciousness and Cognition, 19,* 1069–1078.

Schiff, N. D. (2013). Making waves in consciousness research. *Science Translational Medicine, 5* (198): 198fs32:1–3.

Seager, W. (2009). History of philosophical theories of consciousness. In W. P. Banks (Ed.), *Encyclopedia of consciousness* (pp. 50–339). Oxford: Academic Press.

Searle, J. (1980). Minds, brains, and programs. *Behavioral and Brain Sciences, 3,* 57–417.

Searle, J. (1992). *The rediscovery of the mind*. Cambridge, MA: MIT Press.

Searle, J. (1998). How to study consciousness scientifically. *Philosophical Transactions of the Royal Society of London. Series B, 353,* 1935–1942.

Searle, J. (2013). Can information theory explain consciousness? *New York Review of Books, 60,* 30–41.

Seth, A. K. (2007). Models of consciousness. *Scholarpedia, 2*(1): 1328. doi:10.4249/scholarpedia.1328.

Sherrington, C. (1951). *Man on his nature* (2nd ed.). Cambridge: Cambridge University Press.

Skinner, B. F. (1957). *Verbal behavior*. New York: Appleton-Century-Crofts.

Smart, J. J. C. (1959). Sensations and brain processes. *Philosophical Review, 68,* 56–141.

Sperling, G. (1960). The information available in brief visual presentations. *Psychological Monographs: General and Applied, 74*(11), 1–29.

Strawson, G. (2008). *Realistic monism: Why physicalism entails panpsychism. Real materialism and other essays*. Oxford: Clarendon Press.

Strawson, G. (2015). The consciousness myth (revised). *The Times Literary Supplement,* February 27, 2015 (no. 5839). http://www.the-tls.co.uk/articles/public/consciousness-myth/.

Strawson, G. (2016). Consciousness isn't a mystery. It's matter. *The New York Times*, May 16, 2016. http://www.nytimes.com/2016/05/16/opinion/consciousness-isnt-a-mysteryits-matter.html?_r=0.

Strawson, G. (2017). Physicalist panpsychism. In M. Velmans & S. Schneider (eds.), *The Blackwell companion to consciousness*. Oxford: Blackwell.

Terrace, H. S. & Son, L. K. (2009). Comparative metacognition. *Current Opinion in Neurobiology*, *19*, 67–74.

Thelen, E., & Smith, L. (1994). *A Dynamical systems approach to the development of cognition and action*. Cambridge, MA: MIT Press.

Thompson, E. (2015). *Waking, dreaming, being: Self and consciousness in neuroscience, meditation, and philosophy*. New York: Columbia University Press (ebook edition).

Tononi, G. (2004). An information integration theory of consciousness. *BMC Neuroscience, 5,* 42 (1–22). doi:10.1186/1471-2202-5-42.

Tononi, G. (2008). Consciousness as integrated information: A provisional manifesto. *Biological Bulletin, 215,* 42–216. doi:10.2307/25470707.

Tononi, G. (2012a). *Phi: A voyage from the brain to the soul*. New York, NY: Pantheon Books.

Tononi, G. (2012b). The integrated information theory of consciousness: An updated account. *Archives Italiennes de Biologie, 150,* 56–90.

Tononi, G. (2015). Integrated information theory. *Scholarpedia, 10*, 4164. doi:10.4249/scholarpedia.4164.

Tononi, G., Boly, M., Marcello, M. & Koch, C. (2016). Integrated information theory: From consciousness to its physical substrate. *Nature Reviews Neuroscience*. May 2016. doi:10.1038/nrn.2016.44.

Tononi, G., & Edelman, G. M. (1998). Consciousness and complexity. *Science, 282,* 1846–1851.

Tononi, G. & Koch, C. (2015). Consciousness: Here, there and everywhere? *Philosophical Transactions of the Royal Society. B 370*, 20140167. http://dx.doi.org/10.1098/rstb.2014.0167.

Tononi, G., McIntosh, A. R., Russell, D. P., & Edelman, G. M. (1998). Functional clustering: Identifying strongly interactive brain regions in neuroimaging data. *Neuroimage, 7,* 49–133.

Tsuchiya, N., Block, N. &Koch, C. (2012). Top-down attention and consciousness: Comment on Cohen et al. *Trends in Cognitive Sciences*, *16*(11): 527.

Tsuchiya, N., Wilke, M., Frässle, S., & Lamme, V. A. (2015). No-report paradigms: Extracting the true neural correlates of consciousness. *Trends in Cognitive Sciences, 19,* 70–757.

Turing, A. (1950). Computing machinery and intelligence. *Mind, 49,* 60–433.

Van Gulick, R. (2007). Reduction, emergence, and the mind/body problem: A philosophic overview. In N. Murphy & W. R. Stoeger (Eds.), *Emergence: Systems, organisms, persons* (pp. 40–73). New York: OUP.

Wegner, D. M. (2002). *The illusion of conscious will.* Cambridge, MA: MIT Press.

Zeki, S. (2007). A theory of micro-consciousness. In M. Velmans & S. Schneider (Eds.), *The Blackwell companion to consciousness* (pp. 8–580). Oxford: Blackwell.

What Does It Mean for Qualia to be Intrinsic?

S. Siddharth and Sangeetha Menon

Introduction

One of the biggest challenges in understanding the mind is posed by its conscious nature. Every moment of our waking life is animated by experiences. For instance, when one looks at a tree, one has a visual experience of (among other things) green leaves. Or when one reads a book, one has an experience of understanding the words. These experiences are such that they are available only to the subject who is experiencing them. They also feel some way for the experiencing subject, have an associated quality. These *subjective* and *qualitative* properties of experiences, referred to as *qualia* (Tye 2016) pose a philosophical challenge: are they physical properties? If yes, why is it that only certain physical objects, like the human brain, have associated experiential properties, and not others? This is the *Hard Problem of Consciousness* (Chalmers 1995).

One response to this problem has been to posit a second kind of substance different from physical entities, which possess qualia. A prominent proponent of this kind of *dualism* was Descartes (1642/1996). But, why can't physical properties account for qualia? While there have been various arguments for the same, an important strand has been to argue the properties explained by the physical sciences only explain the 'causal-relational properties of matter' and not the intrinsic properties of experience (Lavazza and Robinson 2014: 5).

This insight about the limitation of physical sciences is also the motivation for a different kind of response to the *Hard Problem.* Some argue that there is only one kind of substance, of which qualia are the intrinsic properties and the properties described by the physical sciences their relational properties. Most views of this kind draw motivation from Russell's (1927) insight into the limitation of physical

S. Siddharth (✉) · S. Menon
NIAS Consciousness Studies Programme, National Institute of Advanced Studies, Bengaluru, India
e-mail: Siddharth.nias@gmail.com

S. Menon et al. (eds.), *Self, Culture and Consciousness*,
DOI 10.1007/978-981-10-5777-9_23

sciences, and have come to be known as *Russellian monism* (Alter and Nagasawa 2012).

What does it mean for a property to be intrinsic, though? Why do not physical properties qualify as intrinsic? Broadly, these are the question that will be tackled in this chapter. Independent of the discourse on consciousness, there has been a debate over the precise definition of an intrinsic property. After outlining a few of these definitions, it shall be argued that experiential properties indeed qualify as intrinsic. Further, this chapter shall outline Langton's (1998) interpretation of Immanuel Kant's[1] argument for why we have no knowledge of intrinsic properties and endorse the same, albeit with one exception—qualia. It shall be argued that qualia qualify as intrinsic properties, based on the intuitive and plausible premise that we have direct access to the intrinsic properties of our experiences in simply having them.

What are the implications of the *qualia-as-intrinsic* thesis? This chapter shall argue for the following two:

1a. Experiences cannot combine: it will be argued that combination is a purely relational concept, and thus, experience cannot be characterised as the combination of two or more experiences.
1b. Extrinsic or relational properties, including causal relations, are not reducible to intrinsic properties.

Intrinsic Properties

An intrinsic property (henceforth referred to as an 'IP') can be intuitively understood as the property an object can possess independent of all other objects, by virtue of which an object *is* the object that it is. For example, the shape of a square *S* is its IP, for *S* is indeed a square on account of its shape. On the other hand, being five metres away from a circle is *extrinsic* to *S*, for *S* is not the object it is by virtue of this property. *S* can be the square it is independent of whether it is five metres away from a circle, while it cannot be a square independent of its shape. Thus, an IP is one that an object can have independent of any other object. The question is, how can the notion of independence be captured more precisely. This section shall look at two attempts to capture this notion of independence, by Langton and Lewis (1998) and Vallentyne (1997).

[1]When an argument or view is attributed to Kant, it refers not to the historical Kant, but Rae Langton's interpretation of Kant in her work, unless specified otherwise. Langton's reconstruction of Kant's arguments is based on multiple works of Kant, and cannot be attributed to any one of his works. Hence, while referring to Kant's views, only the relevant section in Langton's work is cited, except when Kant is directly quoted.

Jaegwon Kim defined a property G as *internal* (his term for an IP) if and only if an object x having G could conceivably be the only contingent[2] object in the world —in other words, the object is compatible with *loneliness* (Kim 1982). However, Lewis (1983) points out that loneliness in itself intuitively seems like an extrinsic property—for it does not pertain only to the object in question—but is classified as an intrinsic property by Kim's definition. To address this objection, Langton and Lewis (1998) argue that in addition to being compatible with loneliness, an IP ought to be compatible with *accompaniment* as well—an object x could possess G even if it were accompanied by other contingent objects. To this, they add a further condition that a thing lacking an IP should also be compatible with loneliness and accompaniment. Thus, Langton and Lewis (1998: 334) come up with four conditions for a property G to be intrinsic:

2a. A lonely thing can have G.
2b. An accompanied thing can have G.
2c. A lonely thing can lack G.
2d. An accompanied thing can lack G.

According to these conditions, the loneliness would not qualify as intrinsic, as neither can a lonely thing lack it, nor an accompanied thing possess it.[3]

A similar characterisation of IPs is that of Vallentyne (1997). While Lewis (1983), and later Langton and Lewis (1998) wanted to characterise the independence in terms of logical independence, Vallentyne (1997) seeks to do so in terms of what he calls *contraction* of a given world, 'which is to be understood as a world 'obtainable' from the original on solely by 'removing' objects from it' (p. 211). An $x - t$ *contraction*, as Vallentyne (1997) defines it, is one where all or some objects are removed while still leaving a specified object x at a specified time t untouched. An $x - t$ contraction involves removal of contingent laws as well. Now, Vallentyne defines an IP as,

> *P is intrinsic* $=_{\text{def}}$ for any world w, any time t, and any object x: (a). If Px at t in w, then Px at t in each $x{-}t$ contraction of w, and (b). Likewise for $\sim P$[4] (Vallentyne 1997, p. 212)

A property P is an IP if and only if, when an object x has property P at time t in world w, it shall do so in every $x - t$ contraction of w. Even as we remove every

[2]A contingent truth is one that is not necessarily true—it ought to be conceivable that the proposition is false. For example, that one plus one equals two is a necessary truth, whereas the fact that blue is one's favourite colour is a contingent one.

[3]This definition, as Langton and Lewis themselves note, does not for work for *disjunctive* properties. For example, they point out that the property of being 'either cubical and lonely or non-cubical and accompanied' intuitively seems to be about something other than an object itself, but comes out as intrinsic according to these conditions. Langton and Lewis tweak this definition further to address this limitation. For our purposes, however, understanding IPs as those properties that are compatible with loneliness and accompaniment would suffice.

[4]'$\sim$' indicates logical negation.

object other than x, and every law from the world w, it can conceivably continue to have the property P.

This may seem very similar to the Langton and Lewis' (1998) notion of compatibility with loneliness and accompaniment, but there are some cases where the two differ. For example, consider the property of being the only red object.[5] This property—call it R—can be possessed as well as lacked by both a lonely and accompanied thing and thus classified as an IP according to Langton and Lewis (1998). Intuitively it seems to be an extrinsic property, for an object having it or not depends on the existence of other objects. This indicates that there seems to be something missing in the notion of independence that Langton and Lewis (1998) adopt, which Vallentyne (1997) seeks to address. On Vallentyne's (1997) definition of an IP, R is not classified as intrinsic, for an object may not be the only red object in our world, but may turn out to be one in an $x - t$ contraction (and thus violate condition b of his definition).

To understand the difference between Langton and Lewis' (1998) definition, and Vallentyne's (1997), consider the property R. A lonely red ball in world w_1 where it is the only object will possess R, as will a red ball in world w_2 where it is the only red object amongst other objects. A lonely blue ball in world w_3 will not have it, as neither will a blue ball in world w_4 where there are other objects. Even if there were no other possible world where a lonely object or an accompanied one has the property R or lacks it, it will still be deemed an IP going by Langton and Lewis' definition. On the other hand, per Vallentyne's (1997) definition, if an object has or lacks a property, it must continue to do so in *every* $x - t$ contraction. If there is a possible world w_5 in which a red ball exists with another red cube, this world will not affect the classification of R as an IP per Lewis and Langton's (1998) definition as there is at least one possible world where an accompanied object lacks R, while according to Vallentyne's (1997) definition, it will not be classified as an IP as there is one possible world (w_5) in which the red ball will not continue to lack R.

In outlining these characterisations of intrinsic properties, our purpose here is not to arrive at a precise definition, but to use these definitions to elaborate on our intuitive understanding of IPs. It shall be argued later, in Sect. 4, that qualia indeed qualify as IPs on both these definitions.

Ignorance of Intrinsic Properties

Why don't the physical properties we have knowledge of qualify as intrinsic? Why cannot, for example, spatial properties like location or volume be IPs of physical entities? Why is the property of being five metres long not an IP? An object having the property of being five-metres-long can conceivably continue possessing it on all $x - t$ contractions (or, is compatible with both loneliness and accompaniment).

[5]The example is Vallentyne's.

While such properties may seem to be intrinsic, in this section, it shall be argued that they are not.

Immanuel Kant, the German philosopher has been interpreted as arguing that we have no knowledge of the intrinsic properties of substances. While it is arguable whether this is indeed the view Kant himself had, what is of interest to us is the argument that has been attributed to Kant, prominently by Langton (1998).

Before getting into Kant's argument for our ignorance of the intrinsic properties of objects, it would be pertinent to dwell upon the notion of a *substance*, and its relation to IPs. Descartes (1644/1997) understood a substance as 'a thing which so exists such that it needs to no other thing to exist '(*Principles of Philosophy*, Part 1, Art. 51). This Cartesian definition of a substance is very similar to the notion of an IP. Kant makes a similar observation: 'Substances in general must have some intrinsic nature, which is therefore free from all external relations' (A274, Kant 1781, translated in Langton 1998: 16). A substance is something that cannot be thought of as a predicate of something else, and is a thing that has an *intrinsic nature* free of all relations and dependence on things other than itself. However, nothing can exist without properties—even a substance, to exist, ought to have some property. What kind of properties can these be? If a substance were to be such that it requires no other thing to exist, it ought to possess properties that are compatible with this. In other words, *a substance must have intrinsic properties*. An intrinsic property is thus that property by virtue of which a substance is indeed a substance (Langton 1998).

Kant argues that we have no knowledge of what these intrinsic properties are, a thesis Langton (1998) refers to as *Kantian humility*. This means that when I look at the table in front of me, I do not have knowledge of what the intrinsic nature of whatever substance the table is (or is made of). It is not only our perceptual knowledge that is deficient, but *anything* we can know about the table, including what our latest theories in physics tell us (that it is made up of atoms or quarks or strings, etc.) is not its intrinsic nature, but only about its relational or extrinsic properties. We might, in our daily lives or when engaging in the scientific enterprise assume that we have knowledge of substances, but these according to Kant do not ultimately hold water.

Why are we irredeemably ignorant of the intrinsic properties of substances? Langton (1998) reconstructs Kant's argument as follows:

3a. *Distinction*: There is a real, ontological distinction between intrinsic and relational properties.
3b. *Receptivity*: Any knowledge we have of substances is only through the effect their relational properties have on us.
3c. *Irreducibility*: Relational properties are not reducible to intrinsic properties.

Thus,

3d. *Humility*: We have knowledge only of relational properties, and not intrinsic properties.

A detailed defence of each of the three premises is provided by Langton (1998), which shall be briefly outlined below.

Distinction

The distinction between intrinsic and extrinsic properties is based on the intuition that any relations (extrinsic properties) require *relata* (or substances) of which they are relations. Similarly, for causal powers or dispositions to exist, there need to be things *of which* these they are the causal powers. It may be argued that such intrinsic properties are not required, and that all of concrete reality is purely relational (or structural), a view called *Ontic Structural Realism* (Frigg and Votsis 2011: 260). While the arguments for the necessity of a substance (and intrinsic property) shall not be outlined here (see Frigg and Votsis (2011) for a summary of arguments against ontic structural realism, and Seager (2006) who discusses it in the context of conscious experience), it shall only be noted that Kant posits such a distinction. It shall later be argued in Sect. 4 that the existence of qualia provides a good evidence for the existence of intrinsic properties.

Receptivity

The second premise holds that any knowledge one has of other objects is through what is caused in us (whatever 'we' and 'us' may ultimately be) by these objects. For example, when one looks at a red ball, one has a visual experience of a red, spherical object. We normally assume *redness* and *sphericity*, as they appear to us, to inhere in the ball. Is this assumption warranted, though? One can certainly claim that one has visual experience of a spherical, red thing, but the claim that the ball in itself is spherical and red as it appears to us is a further inference. One's visual experience is the effect the object has on one's cognitive system, which ultimately results in one having the experience of a red ball. One does not access the object—the substance it is and its intrinsic properties—directly, but only through the effect the object's causal powers have on us.

Langton (1998) bases Kant's argument for receptivity on what she refers to as the *principle of succession*. The principle holds that any change in a substance's property can be brought about only by another substance and its properties—a substance by itself, without the influence of another substance cannot undergo change in its properties. We have changing experiences (which includes thoughts), and it is only through these experiences we have knowledge of the world. According to the principle of succession, our experiences can change only under the causal influence of other substances. Thus, our knowledge of other substances is *receptive*. We do not directly know an object, but only know the effect it has on our experiences (Langton 1998: 104–7).

The causal theory of perception outlined above has inspired a family of views in philosophy of science, known as *structural realism*. Proponents of the same include Russell (1912, 1927) and Maxwell (1968), whose views are sometimes also referred to as *epistemic* structural realism (Frigg and Votsis 2011). They reject *naïve realism* —the thesis that properties of objects as they appear to us in perception literally inhere in the object—and argue 'that the properties we ordinarily would say that we observe in external objects are just as much in the mind as those we experience in dreams or hallucinations' (Maxwell 1968: 152). In saying this, Maxwell does not deny the existence of a mind-independent reality. Rather, his claim is that all we can know about this reality are its *structural* properties, where:

> The notion of *form* or *structure* needed here may accurately be said to be logical (and/or mathematical) and, in a sense, abstract; *characterizations*[6] of instances of it will be in terms of logic alone, i.e. the logical connectives, quantifiers and variables—they will contain no descriptive terms. (Maxwell 1968: 153)

For example, science has no predicate that refers to colours and tastes, but only ones that refer to wavelengths and chemical compositions of the object with which we associate those colours and tastes. Often, it is assumed that an object that emits photons of a particular wavelength is literally of the corresponding colour. As noted earlier, such an assumption is not warranted. Rather, the assumption is a case of confusing the cause and the effect—the colour of an object as experienced by the observer is the *effect on* the observer *by* the object. This does not mean the object in itself possess the colour as it appears to us.

Irreducibility

The third premise of Kant's argument is that the relational properties of substances —specifically, their causal properties—are not reducible to their intrinsic properties. Intuitively, it seems that one can separate a substance from the causal powers it has. For example, one can conceive of a different world in which an electron (with its IP, whatever that may be) has the causal power to attract rather than repel another electron. It seems that it does not necessarily follow from the intrinsic property of an electron that it ought to repel another electron. If there were a God creating the world, the electron's dispositional property will have to be a further added by her, over and above its intrinsic property, in the form of contingent laws of nature. Thus, the dispositional properties of substances are not reducible to their intrinsic properties (Langton 1998).

The *receptivity* thesis tells us that we have knowledge only of the effect substances have on us, that is, their relational properties. *Irreducibility* tells us that these relational properties of substances are not reducible to their intrinsic

[6]Emphasis in original.

properties. Thus, it follows for Langton's (1998) Kant that we have no knowledge of the intrinsic properties of substances, or *humility*.

Qualia as *Intrinsic*

Are we, as Kant seems to claim, irredeemably ignorant of the intrinsic properties of substances? This section shall argue that there is one exception to this ignorance—qualia. It shall be argued that in experiencing qualia, we know the intrinsic properties of at least some substances.

First, do qualia fulfil the criteria put forth by Langton and Lewis (1998) and Vallentyne (1997) to qualify as an IP? Consider the sceptical scenario put forth by Descartes (1642/1996), where he—the subject of experience he is, and his experiences—is the only thing that exists, and the external world that he perceives is a hallucination. As improbable as this scenario may seem, it is a perfectly conceivable one. As is the scenario that one usually takes to be the case— 'I' exist along with other contingent objects like other human beings and the rest of the external world. The coherence of both the sceptical and normal scenarios points to the fact that the existence of a particular qualia is compatible with both loneliness and accompaniment (or, an object experiencing that qualia can conceivably continue to do so on all x - t contractions).

Within the discourse on consciousness, that qualia cannot be explained by and characterised in terms of only relations and structures has been put forth as an argument against physicalism. David Chalmers notes 'the hard problem is hard precisely because it is not a problem about the performance of functions. The problem persists even when the performance of all the relevant functions is explained' (Chalmers 1995: 202), where by 'functions' he refers to any causal role a system may perform. Other arguments, like the inverted qualia scenarios (Chalmers 1996: 263–6), where it is at least conceivable that two systems having the same functional or physical organisation result in producing different qualia, highlight that a functional or physical description would not provide a complete description of experiential properties. This is not to say that qualia do not possess dispositional properties, but that these dispositional properties are not *essential* to their existence; neither would a complete description of qualia be exhausted by a description of their dispositional or relational properties. It is on account of this feature of qualia that they seem to be our best candidate for intrinsic properties. Further, the very fact that qualia exist, and a purely structural/relational explanation would not suffice to explain them also serves as an argument against positions (such as ontic structural realism) that hold that all concrete reality is purely structural.

What justifies the claim that we know what qualia *really* are, though? Why does not Kant's argument apply to qualia? Is not our knowledge of our own qualia also *receptive*? The response to such objections is based on the thesis that one has direct access to the qualia of one's conscious mental state simply by being in that state, a thesis known as *real acquaintance* (Goff 2015). The intuition behind this is that by simply being in an experiential state, one knows the *essence* of the experiential properties of that state. One cannot be in a state of pain and not have epistemic access to the essence of what pain is, where 'essential property of *X* is a property *X* has in all possible worlds....one that constitutes the real definition of *X*, that is one that makes the kind of thing it is' (Goff 2015: 7). In other words, the essence of *X* is its intrinsic property. Goff argues for *real acquaintance* from the premise that it best explains the following two intuitively acceptable claims:

4a. A psychologically normal subject can get into a situation where one can be certain that one is in a particular phenomenal state (like being in a state of pain), and not any other state.
4b. A psychologically normal subject can know a priori similarities and difference between various phenomenal properties, like that red is more similar to orange than green (Goff 2015: 10–14).

René Descartes was one of the earliest proponents of a version of the *real acquaintance* thesis—he famously argued that one could not doubt the fact that one was having some experience (and hence, not doubt that one exists), though one could doubt whether the experience itself corresponded to an external reality (Descartes 1642/1996). *Real Acquaintance* extends Descartes' claim, and holds that not only can one be certain that one is having an experience, but also be certain—at least at the moment when one is having the experience—as to the content and intrinsic nature of the experience in simply having it.

In summary, the *real acquaintance* thesis tells us that by being in a given experiential state, we know the intrinsic nature of the experiential properties. Further, these experiential properties that we have real acquaintance of qualify as potential intrinsic properties of substances. Thus, as opposed to Kantian humility, we have knowledge of the intrinsic properties of at least some substance—whatever substance we are when we have the experience that we do.

Kant does consider the possibility that the 'I, as a thinking being (soul)' (A348, Kant, 1781)—as something that thinks, and thus instantiates qualia—is a substance, but rejects it (*Paralogisms of pure reason*, A348-351, Kant, 1781). This would imply that Kant will also have to reject the option of qualia being the IPs of whatever substance souls ultimately are. While his reasons for the rejection of the substantiality of the soul are complicated, and beyond the scope of this chapter, it is worth noting that his rejection is based on his rejection of the proposition that we have access to the intrinsic nature of thinking beings. In other words, Kant rejects *real acquaintance*.

Accepting that qualia are IPs leaves one with two options. One could assume that qualia are the only kind of intrinsic properties of substances, which will lead us to a type of Russellian monism referred to as *Russellian panpsychism* (Chalmers 2013). Or, one could assume that physical substances have non-experiential intrinsic properties, of which we may or may not have knowledge, along with substances with experiential intrinsic properties that we have access to—a dualistic ontology. Cartesian dualism is an example of the second kind, for Descartes (1642/

1996) assumed spatially extended objects to be substances, along with experiential substances.

It is interesting to note that Kant considers the option of panpsychism, saying:

> It could nevertheless be that the something which lies at the ground of external appearances, which affects our sense so that out sense receives representations of space, matter, shape and so on—it could be that this something, when regarded as noumenon (or better, as transcendental object), *is at the same time the subject of thoughts*[7]...Matter is mere external appearance, whose substratum cannot be known through any predicate we can assign to it. So I can well assume that matter in itself is simple ... and that the substance which possesses extension with regard to our outer sense, possesses thought in itself, which can be consciously represented through its own inner sense. (A358, Kant 1781, translated in Langton 1998: 208–9)

After recounting how matter is wholly relational, Kant considers the option that the ground of these relations—the substance—is a 'subject of thought', by which he means a subject of experience experiencing qualia. However, since he otherwise seems to emphasise epistemic humility and ignorance of things in themselves, Langton opines that perhaps this ought to be 'regarded as an unwise and atypical aberration' (Langton 1998: 210). As noted earlier, Kant (1781, A348-351) rejects the substantiality of the soul—and thus that qualia are the IPs of whatever substance the human "soul" ultimately is. Given that he rejects the possibility of qualia being the IP of the human soul, it is understandable that Kant rejects the possibility of qualia being the IPs of all substances.

David Lewis, who also puts forth an argument for epistemic humility—albeit slightly different from Kant's—considers the possibility that fundamental properties that are instantiated in our world are qualia (Lewis 2009). This, he notes, would require one to accept *real acquaintance* (he refers to it as the 'identification thesis'), while Lewis' commitment to materialism requires him to reject it. As a materialist, he contends that qualia are ultimately reducible to physical properties. Given that we are not aware of this reducibility in simply having the experiences that we do, Lewis will have to deny that we not have direct access to the intrinsic nature of qualia. Thus, according to Lewis (2009), *humility* applies as much as to our qualia as to all other properties. Based on this rejection of *real acquaintance*, he also rejects panpsychism.

However, for someone who accepts *real acquaintance*, and is not committed to being a materialist of the Lewisian kind, panpsychism provides an attractive option. On the one hand, it allows one to retain the intuitive notion that qualia are not reducible to physical properties, while on the other it seeks to fill the gap of the missing intrinsic property of matter [for discussions on various issues surrounding panpsychism, see Freeman (2006), Skrbina (2009), Bruntrup and Jaskolla (2016)]. It must be noted that panpsychism (or any *qualia-as-intrinsic* theory) diffuses humility only to the extent that we know the IPs of substances are subjective, qualitative properties—we are still ignorant of which specific qualia these

[7]Emphasis is mine, not in original.

properties are (except in those that are the IPs of whatever substance we ourselves are). For example, while panpsychism tells us that the intrinsic property of a given *quark* is some experiential property, it still does not tell us *what it is like* to be that quark.

Two Implications of the *Qualia-as-Intrinsic* Thesis

Where does accepting the *qualia-as-intrinsic* thesis leave us? It is likely to have profound implications on various debates in philosophy of mind and metaphysics. This section shall look at two possible implications of the same: on the notion of combination of experiences, and the relation between intrinsic and extrinsic properties.

Combination

Can experiences combine? James (1890) contended that the notion of combination was unintelligible in the context of experiences. Given ten different experiences (and subjects of experiences), James (1890) argued that, however one 'arranged' them, one cannot intelligibly think of an eleventh experience that is the composite of the ten, like how we can think of an arrangement of bricks composing a wall. Any such eleventh experience, he argued, would be a totally new fact, irreducible to the ten original experiences. The question of combination of experiences has gained relevance once again in the contemporary debates surrounding Russellian panpsychism. If fundamental, microphysical entities are intrinsically experiential, how do *macroexperiences*—experiences of macrophysical entities like human beings—emerge from *microexperiences*? This has come to be known as the *combination problem* (Chalmers 2016).

There have been some attempts to answer the combination problem, some of which have tried to negate the Jamesian intuition that experiences cannot combine, and posited various kinds of combinatorial mechanisms (Chalmers 2016). However, if one accepts the *qualia-as-intrinsic* thesis, it would seem that the Jamesian intuition holds. If qualia are intrinsic, and cannot be reduced to relations between other entities, neither can they combine. Combination, as James (1890) and Langton's (1998) Kant argue, is a purely relational concept. To understand this, take the example of a hundred bricks combining to form a brick-wall. It is intuitively understandable that there is nothing more to the brick-wall than the hundred bricks and the spatial relations between them. On the other hand, a substance with IPs cannot be simply reduced to or divided into relations between other entities—this is implied by the very definition of an IP. The very notion of division and combination is a relational one, and hence would not apply to substances (and IPs that characterise them). Kant (following Leibniz) says as much: 'Substances in general must have some intrinsic nature which is therefore free from all external relations, and therefore also from composition' (A274, Kant 1781 translated in Langton 1998: 71).

Thus, given the *qualia-as-intrinsic* thesis, the combination of experiences can be ruled out. The notion of combination is a pertinent one to anyone interested in the metaphysics of experience, but is of great relevance to the Russellian panpsychist, who holds that qualia are the IPs of microphysical entities. If Russellian panpsychism were true, it follows that no two microphysical entities can intrinsically combine or compose a further third entity. All physical entities are such that they cannot be divided, or *simples.* This leads one to a position in metaphysics known as *mereological nihilism,* which can be stated as:

> It is impossible for one to bring it about that something is such that the Xs compose it, because (if the Xs are two or more), nothing is such that the Xs compose it.' (van Inwagen 1990: 72)

Mereological nihilism is a very counterintuitive notion, and its implications include that everyday objects like tables and rocks do not exist as unified objects—all that exist are numerous simples that are arranged so as to appear as a unified object (table or rock) to an observer. This is not the place to dwell upon the arguments for and against mereological nihilism—the point is only that if one accepts Russellian panpsychism, there is a strong case to be made for mereological nihilism as well.

Are Extrinsic Properties Reducible to Intrinsic Ones?

The second implication of the *qualia-as-intrinsic* thesis that we would like to consider is the relation between intrinsic and extrinsic properties—specifically, causal powers and dispositions. In Kant's (Langton 1998) argument for *humility*, one of the premises was that the relational properties of a substance are not reducible to its intrinsic property. This premise was justified on the intuition that these causal powers are contingent properties of nature, and it is conceivable that a substance with a given IP could conceivably have dispositions different from the ones they have in our world.

Langton (1998) contrasts Kant's thesis of irreducibility against Leibniz' thesis that all relational properties are reducible to the intrinsic properties of substances. It is his irreducibility thesis that leads Kant to epistemic humility, and Leibniz' reducibility thesis that leads him to the conclusion that one can know the IPs of all substances. In positing that we know the IPs of at least one kind of substance—qualia—we have partially violated the *humility* thesis. What can we say about the reducibility or irreducibility of relational properties to intrinsic properties given the *qualia-as-intrinsic* thesis?

The *real acquaintance* thesis tells us that we have direct access to the intrinsic nature of our experience—we know the IPs of whatever substance our experiences are. Is there something in our experience that also tells us how they relate to other substances, say for example another human, experiential entity? Let us, for the sake of argument, assume that the substance of which a human subject's qualia is the IP manifests as a part of her brain to another human observer. Is there anything in the qualia from which one can a priori arrive at how the qualia-experiencing-subject will appear to another human observer? It seems there is not—nothing in one's experience and its qualia tells us how it appears to other entities.

Perhaps, given the IPs of the other entity, the IPs of both the substances can together help us arrive at their respective dispositions. Consider, for example, E_1, a normal human subject S_1*'s* experience of looking at a plain red wall, and E_2, the experience of an observer S_2 who is currently closing her eyes, but ready to open them and look into S_1's brain (assume for the sake of the example that S_1's skull has been cut open to allow this with no change in her ability to perceive the red wall). Now, when observer S_2 opens her eyes (assume this happens instantaneously) and looks at S_1's brain, she has an experience E_3 of looking at a brain. Now, given E_1 and E_2, there seems to be nothing about these experiences that tell us what experience the substance that is E_1 is disposed to cause in S_2.

Thus, given the *real acquaintance* thesis, the *irreducibility* thesis holds. Perhaps, diluting *real acquaintance* can help one get around irreducibility, by positing that there is something 'extra' to the intrinsic nature of substances over and above what is revealed to us in having an experience. It could be posited that this extra ingredient necessitates the causal powers and dispositions that our experiences in fact have. Adopting such a *partial revelation* stance comes with its own problems, though. Our argument for the *qualia-as-intrinsic* thesis was based on an assumption of full revelation, and any dilution to the same could raise questions of what justifies even partial revelation. One will then be left with the following options:

5a. The IPs of substances is some wholly non-experiential property such that qualia can be intelligibly derived from it. Such kinds of properties have been referred to as *protopsychic* (Chalmers 2013). This leads one back to humility, as we have no conception of what such protopsychic properties might be. Further, it can be argued that experience cannot intelligibly emerge from anything non-experiential (such as protopsychic properties), and any such emergence will have to be brute in nature.

5b. The IPs are both experiential and something 'extra', where this extra component is something wholly non-experiential. Is it possible for a single intrinsic property to be intrinsically both experiential and non-experiential? Such a notion, as Strawson (2006)[8] also notes, appears to be contradictory and incoherent.

[8]Strawson considers the possibility of whether a substance could be intrinsically both experiential and non-experiential, a position he terms *Equal Status Fundamentally Dual* monism. His interest in the same is not whether the non-experiential component could explain the causal powers of our experience, but in evaluating the possibility of a monism in which both experiential and non-experiential reality are fundamental yet part of a single substance. He argues that such a notion is incomprehensible to us on account of violating the law of contradictions.

5c. IPs are wholly experiential, but there is something 'extra' (call it *x*) to these experiences, where *x* is also qualia-like i.e. subjective and qualitative, and such that they necessitate the causal powers and dispositions. Is this an option? If a qualia-like property is instantiated, it necessarily follows that a subject *experiences* it. Can a subject experience qualia-like property, and yet not be acquainted with it (if only at the moment of experiencing)? If yes, it would be very counterintuitive, and one is not even sure if one is talking about experiences, as we know it, anymore. In what sense is *x* still a subjective property? It would be tantamount to saying that *x* is experienced, and yet not available to the subject to be experience. A proponent of this view might retort that such qualia-like properties seem incoherent to us *precisely* because we do not have access to *x*, and this is the point of rejecting full revelation. In such a case, we will never have access to *x*, and this gap in our understanding will always remain. Why do we need to argue for the reducibility of extrinsic properties to IPs at the cost of positing such mysterious aspects? It seems unmotivated and unnecessary.

Thus, if one wants to accept the *qualia-as-intrinsic* thesis, one would be better off adopting *full revelation*, and thus *irreducibility* as well. Any *partial revelation* thesis that denies that we have *real acquaintance* with the intrinsic nature of our experiences seems to face the challenges outlined above, which in turn leads further towards epistemic humility.

Conclusion

At the heart of the *Hard Problem* is the intuition that there is something fundamentally different between qualia and physical properties, because of which the former cannot be reduced to the latter. This chapter argues that this difference can be captured using the intrinsic/extrinsic distinction between properties, and that in having direct access to the qualia we experience, we have knowledge of at least one kind of intrinsic property. Two implications of the *qualia-as-intrinsic* thesis are discussed here—the impossibility of combination, and the irreducibility of extrinsic to intrinsic properties. The arguments for these implications are not meant to be conclusive; rather the purpose of the discussion was primarily to highlight the rich possibilities the *qualia-as-intrinsic* thesis opens up for further investigations in related areas of philosophy of mind, and metaphysics.

References

Alter, T., & Nagasawa, Y. (2012). What is Russellian monism? *Journal of Consciousness Studies, 19*(9–10), 67–95.

Bruntrup, G., & Jaskolla, L. (2016). *Panpsychism: Contemporary perspectives*. Oxford: Oxford University Press.

Chalmers, D. J. (1995). Facing up to the problem of consciousness. *Journal of Consciousness Studies, 2*(3), 19–200.
Chalmers, D. J. (1996). *The conscious mind*. New York: Oxford University Press.
Chalmers, D. J. (2013). Panpsychism and panprotopsychism. In *The Amherst lecture in philosophy 8*, 1–35. Retrieved on March3, 2015 from http://www.amherstlecture.org/chalmers2013/.
Chalmers, D. J. (2016). The combination problem for panpsychism. In L. Jaskolla & G. Bruntrup (Eds.), *panpsychism*. Oxford: Oxford University Press.
Descartes, R. (1996). *Meditations on First Philosophy* (J. Cottingham, Trans.). Cambridge: Cambridge University Press. (Original work published 1642).
Descartes, R. (1997). Principles of philosophy (E. S. Haldane and G. R. T. Ross, Trans.). In E. Chavez-Arvizo (ed.). *Descartes: Key philosophical writings* (pp. 261–337). Hertfordshire: Wordsworth (Original work published 1644).
Freeman, A. (Ed.). (2006). *Consciousness and its place in nature: Does physicalism entail panpsychism*. Exeter: Imprint Academic.
Frigg, R., & Votsis, I. (2011). Everything you always wanted to know about structural realism but were afraid to ask. *European Journal of Philosophy of Science, 1*, 76–227.
Goff, P. (2015). Real acquaintance and physicalism. In P. Coates & S. Coleman (Eds.), *Phenomenal qualities: Sense, perception, and consciousness*. Oxford: Oxford University Press.
James, W. (1890). *The principles of psychology*. New York: Henry Holt & Co.
Kant, I. (1781). *Critique of pure reason*. As translated and quoted in Langton, R. (1998). *Kantian Humility: Our ignorance of things in themselves*. Oxford: Oxford University Press.
Kim, J. (1982). Psychophysical supervenience. *Philosophical Studies, 41*, 51–70.
Langton, R. (1998). *Kantian humility: Our ignorance of things in themselves*. Oxford: Oxford University Press.
Langton, R., & Lewis, D. (1998). Defining 'intrinsic'. *Philosophy and Phenomenological Research, 58*(2), 45–333.
Lavazza, A., & Robinson, H. (2014). Introduction-dualism: what, how and why? In A. Lavazza & H. Robinson (Eds.), *Contemporary dualism: A defence* (pp. 1–15). New York: Routledge.
Lewis, D. (1983). Extrinsic properties. *Philosophical Studies, 44*(2), 197–200.
Lewis, D. (2009). Ramseyan humility. In D. Braddon-Mitchell & R. Nola (Eds.), *Conceptual analysis and philosophical naturalism* (pp. 22–203). Cambridge, MA: MIT Press.
Maxwell, G. (1968). Scientific methodology and the causal theory of perception. In I. Lakatos & A. Musgrave (Eds.), *Problems in the philosophy of science* (pp. 77–148). Amsterdam: North Holland Publishing.
Russell, B. (1912). *The problems of philosophy*. Oxford: Oxford University Press.
Russell, B. (1927/1992). *The Analysis of Matter*. London: Routledge.
Seager, W. (2006). The 'intrinsic nature' argument for panpsychism. In A. Freeman (Ed.), *Consciousness and its place in nature: Does physicalism entail panpsychism* (pp. 45–129). Exeter: Imprint Academic.
Skrbina, D. (Ed.). (2009). *Mind that Abides: Panpsychism in the new Millennium*. Philadelphia, PA: John Benjamins.
Strawson, G. (2006). Panpsychism? Reply to commentators with a celebration of Descartes. In A. Freeman (Ed.), *Consciousness and its place in nature: Does physicalism entail panpsychism* (pp. 184–280). Exeter: Imprint Academic.
Tye, M. Qualia. In E. N. Zalta *The stanford encyclopedia of philosophy* (Winter 2016 Edition). http://plato.stanford.edu/archives/win2016/entries/qualia/.
Vallentyne, P. (1997). Intrinsic properties redefined. *Philosophical Studies, 88*, 19–209.
Van Inwagen, P. (1990). *Material Beings*. Ithaca, NY: Cornell University.

Matter and Consciousness: The Classical Indian Philosophical Approach

V.N. Jha

Introduction

The enquiry into the nature of 'consciousness' has led to a deeper level—the inquiry into the very source of our existence. While some reduce life-forms to physicality, others assume consciousness as a distinct entity to solve this puzzle. The riddle remains unsolved. It is therefore necessary to sit together, again and again, and exchange one another's findings to enrich our understanding regarding the nature of consciousness which is so important a topic that addresses the entire existence of the living and the non-living.

The worldviews of classical Indian philosophers can be formulated as follows:

1. The source reality is One and that One has become Many which we experience. The entire world is real.
2. The source reality is One and the Many that we experience is a superimposition on that One. Thus, One appears as Many. The world is therefore unreal.
3. The source reality is Plural and hence the experiential plurality has emerged from the source plurality. The world is real.
4. The experiential plurality has emerged neither from One of (1) and (2) above nor from Many of (3) above. The world is a flux. Everything is momentary.

All of them, however, agree that the world consists of living and non-living entities. Something is called living because it has *cit* (consciousness) and where there is no *cit* it is called *a-cit* (devoid of 'consciousness').

Those who hold the worldview (1) believe that everything is from consciousness and distinction of living and non-living is done on the basis of the criterion of manifestation and non-manifestation of the same source consciousness. In humans and other beings, consciousness is manifest and in stone and other non-living, it is

V.N. Jha (✉)
Centre of Advanced Study in Sanskrit, University of Pune, Pune, India
e-mail: vnjha46@yahoo.co.in; vnjha1946@gmail.com

S. Menon et al. (eds.), *Self, Culture and Consciousness*,
DOI 10.1007/978-981-10-5777-9_24

non-manifest. Philosophers like Vallabhacarya hold such a view. The philosophers like Sankaracarya are of the opinion that the source reality is indeed consciousness but it cannot undergo any change. Hence the experiential plurality is a superimposition on the singularity of the 'source consciousness'. This is the presupposition behind the worldview (2) above. The worldview (3) above is held by the Indian realists like Nyaya-Vaisesika. The source is plural and the plurality of our experience is derived from that. All that is real is knowable and nameable.

The fourth worldview is that of another type of idealists called Buddhists. Here the basic assumption is that nothing is durable, including consciousness. Our impression that the world is durable is a make-believe kind of impression. The Buddhist too, like the idealists, believe that reality cannot be known and expressed in language. From the above we can easily conclude that in Indian idealism, language cannot present reality and in Indian realism, reality cannot be presented without language.

As regards the models of understanding consciousness, the realists argue that 'consciousness' is the property of a 'conscious' being and a conscious being is one who has knowledge, cognition, desire and volition. Thus, what the personal pronoun like 'I' or 'You' refers to is not 'consciousness' but to 'conscious' being. Consciousness, therefore, should mean 'knowledge', 'cognition', 'desire', and 'volition' and the locus of 'consciousness' is what we call self or soul. For 'consciousness' the Sanskrit term is *cit* or *citi* or *caitanya* and 'conscious' is called *cetana*. While *cetana* is an adjective, *caitanya* is a noun.

Excepting the Nyaya-Vaisesika system of thought all Indian philosophical systems accept 'consciousness' as self or soul whereas the Nyaya-Vaisesika system takes the locus of consciousness i.e. a spiritual substance as the self or soul. Each system of Indian philosophy, thus, developed its own model of understand 'consciousness'. Basically, there will be two models: (a) Samkhya model and (b) Nyaya-Vaisesika model. But when the Samkhya model was adopted by other schools of thought they adjusted that model according to the basic worldviews of their respective systems. For instance, like Samkhya, Sankaracarya also accepted the nature of the self as 'consciousness', but refuted Samkhya because for Samkhya self is not One but Many.

In this way, one needs to analyse the epistemological process of each system of Indian philosophy in order to discover the models of understanding 'consciousness' employed by them. Keeping this in mind one will have to understand the nature of consciousness and self or soul. This will clarify the areas of application of the terms like 'mind', 'spirit', 'self' 'soul' and 'consciousness' which are used so freely nowadays to add to confusion. This will also help in distinguishing 'consciousness' from effects of 'consciousness'. This chapter aims to do this.

Orthodox Systems of Indian Philosophy

Although much has been lost over thousands of years, much ancient literature is still preserved in India and if one peeps into it one can find that right from the Vedic times there has been deep enquiry into the nature and source of this universe of our experience. The enquiry has always been holistic unlike the enquiries of physical sciences. Indian sages have enquired into not only the nature and source of the external physical world but also into one's inner and psychological world. The prominent reason was to discover the nature of the knower and the purpose of his or her coming to this world and his or her destination. In short, they wanted to know the nature and the source of the entire universe, not only physical one but also the psychological. The Upanishadic literature is the testimony of such enquiry. Such deep enquiries of reality, over a period of time, must have paved the way for the development of systems of Indian philosophy. There was absolute freedom of thought and hence multiple systems of thought emerged, some concurring the Vedic thoughts while others challenging some of the presuppositions behind those Vedic thoughts. For the purpose of discussion let us put them into two groups: orthodox and heterodox (Annambhatta 2010).

There are six orthodox systems of Indian philosophy—namely, Sāṅkhya, Yoga, Nyāya, Vaiśeṣika, Pūrvamīmāṁsā and Uttara-mīmāṁsā. Pūrvamīmāṁsā is referred to as simply Mīmāṁsā and Uttara-mīmāṁsā is popularly known as Vedānta. These six are called orthodox because they believe in the authoritativeness of the Vedas.

Sankhya enquired into the origin of the universe. Yoga was engaged in studying the transformation of our inner world. Nyaya kept on developing tools to bring rationality to philosophical discourses. The Vaisesika system offered linguistic modelling of the whole universe of our experience in terms of linguistic categories. Purvamimamsa developed principles of interpretation of language so that philosophers could interpret the Vedic texts cogently and Uttaramimamsa or Vedānta came out with various models of understanding 'reality' through interpreting the Upanishadic texts. Nevertheless, it should be kept in mind that all these six systems discuss the nature of reality, the self and the goal of life.

Heterodox Systems of Indian Philosophy

As against those six orthodox systems, there are three philosophical systems which do not believe in the authoritativeness of the Vedas. These are the Cārvāka, Buddhist and Jaina systems of philosophy. These are referred to as heterodox systems of Indian philosophy. The orthodox systems are called *āstika-darśana* and the heterodox systems are known as *nāstika-darśana*. Here the terms *āstika* and *nāstika* refer to the systems of thought which believes in the authoritativeness of the Vedas and that which does not respectively. The terms *āstika* and *nāstika* do not mean here 'one who believes in God' and 'one who does not believe in God' respectively, because the System of Purvamīmāṁsā is an *āstika-darśana* and yet it does not believe in God. Similarly, there is *nirīśvara-sāṅkhya* and *seśvara-sāṅkhya* (yoga); Vaiśeṣika too did not have God in its sūtra-period.

Brief History of Development

All the six Vedic philosophical systems were initially formulated in the sūtra-style. Kapilamuni wrote Sāṅkhyasūtras; Patañjali composed Yogasūtras; Kaṇāda wrote Vaiśeṣikasūtras; Gautama composed Nyāyasūtras, Jaimini wrote Mīmāṁsa-sūtras and Bādarāyaṇa composed Vedānta sūtras. The thoughts of these systems developed further through a very rich tradition of writing commentaries and sub-commentaries thereon which have recorded the intense dialogues between the heterodox systems on one hand and inter-orthodox systems on the other. These dialogues have added depth to Indian philosophical systems.

Īśvarakṛṣṇa wrote *Sāṅkhyakārikā*; Vyāsa wrote Bhāṣya on the Yogasūtras; Praśastapāda composed Bhāṣya on the Vaiśeṣikasūtras; Vātsyāyana wrote Bhāṣya on the Nyāyasūtras; Śabara wrote Bhāṣya on the Jaiminisūtras; and the Bādarāyaṇasūtras were commented upon by various schools of Vedānta. Śaṅkarācārya wrote Bhāṣya on the Vedāntasūtras; likewise, Nimbārka, Rāmānuja, Mādhva, and Vallabha composed elaborate commentaries from a specific point of view giving rise to various schools of Vedānta.

Thus, Vedānta too is of various types: Advaita of Śaṅkara, Sabhdadvaita of Bhartrhari, Śivādvaita of Kashmir, Śuddhādvaita of Vallabha, Bhedābheda of Nimbārka, Viśiṣṭādvaita of Rāmānuja, Dvaita of Mādhva and Acintyabhedābheda of Caitanyamahāprabhu.

Although popularly one understands by the term Advaita as Śaṅkara's Advaita, the philosophy of Bhartṛhari is called Śabdādvaita, that of Vallabhācārya as Śuddhādvaita and that of Kashmir Śaivism as Śivādvaita since their fundamental presuppositions are different. It is interesting to note that all these Advaita philosophers quote the same Vedic text in support of their respective stand-points and this became possible only because their basic presuppositions were different.

Likewise, two different interpretations of the *Śābarabhāsya* gave rise to two schools of Pūrvamīmāṁsā:

(1) Kumārila and
(2) Prābhākara (Ganganatha 1964)

Murāri created a third school of Pūrvamīmāṁsā by commenting directly on the Mīmāṁsāsūtras of Jaimini (Panse 1990). Indian philosophical systems acquired structural depth because they honoured freedom in thinking and respected others' positions at intellectual levels (Raju 1985). That is why we find that whenever a particular system takes up an issue for discussion it presents first of all what others have to say on that issue. Even Cārvāka's view is not ignored. It is a different matter whether others' views are to be accepted or rejected after a critical examination.

The Vaiśeṣika and the Nyāya systems were distinct in the beginning. God did not find a place in the Vaiśeṣika school and *abhāva* ('absence') as an entity was not discussed in the Nyāya school. But gradually there were changes in these systems and by the tenth century AD we find that these two systems have merged into one system due to their similar worldviews. Thereafter these systems have been referred to by a combined name Nyāya-Vaiśeṣika.

It appears that there has been intense debate between the Naiyāyikas and the Buddhist logicians and philosophers for at least one thousand years, if not more, over a number of philosophical issues. Whether there is something called substance or not; where are qualities located; whether there is a whole independent of its parts or not; whether there is self or not; whether 'absence' should be given a status of an entity or not, whether there is a necessity of accepting philosophical God or not—many such issues engaged the attention of these philosophers.

This dialogical literature of 1000 years is called Prācīnanyāya. It consists of the *Nyāyasūtras* of Gautama, Nyāyasūtrabhāṣya of Vātsyāyana, Nyāyasūtrabhāṣyavārttika *of Udyotakara, Nyāyasūtrabhāṣyavārttikatātparyaṭīkā* of Vācaspatimiśra and *Nyāyasūtrabhāṣyavārttikatātparyaṭīkāpariśuddhi* of Udayanaācārya (tenth century AD) (Vidyabhusana 1920). This bulk of literature is an outcome of an issue-oriented dialogue and hence Prācīna-Nyāya may be characterised as *prameya-pradhāna* (Hiriyanna 1993).

Model of Enquiry

A distinguishing feature of Indian philosophical enquiry into truth is its holistic approach. We live with a physical world outside and a psychological or mental world inside each of us. The universe consists of the sum total of both. Thus, the enquiry should not be limited only to either of the physical world outside as is the case with the enquiry of the physical sciences or only to our mental world. Such an approach cannot give a complete picture of the truth. Therefore, the classical Indian philosophers have all along adopted the holistic approach.

Vatsyayana,[1] the commentator of the Nyayasutra, developed even a universal model of enquiry. He suggested that if one wants to arrive at the truth, one must direct his or her enquiry towards four aspects:

(a) the knower
(b) the knowable
(c) the process of knowing
(d) knowledge.

One can see how big is the canvas of enquiry. Nothing is left out. Nothing can be known in isolation since everything is related to something else. So unless we enquire into totality, the truth cannot be arrived at. It is because of this universality of the model, all thinkers and philosophers adopted this very model to arrive at the truth in their respective areas of investigation. Not only the orthodox systems of philosophy adopted this model but even the heterodox systems of philosophy adopted it.

[1]On Gautama Nyayasutra 1.1.1.

We have said above that the structural depth in Indian epistemology was acquired through dialogue. This universal model of enquiry further paved the way towards that goal. The participants never felt any problem of communication since a common model was adopted.

Worldviews

We have noted that classical Indian philosophers enjoyed complete freedom of thought and that is why there were several philosophical systems and all could co-exist without causing any physical harm to anybody. But the question remains—why did multiple systems of thought emerge? The answer to this lies in the fact that different philosophers cherished different worldviews. While a group of philosophers believed that the plural world of Name and Form has emerged from One source; the others held that the experience of plurality is an illusion and it is created by super-imposing plurality on singularity; while another group believed that the plural world of our experience has come from the plural source and yet another group maintained that the entire universe is a flux and there is nothing which can be treated as durable or permanent.

If we take the Source as One and world of our experience as Many the following four possible worldviews can emerge:

1. +One +Many
2. +One −Many
3. −One +Many
4. −One −Many.

Let us read the above as follows: (1) 'the One has become Many'; (2) plurality is superimposed on One; (3) the world of Many has not come from One but from Many and (4) the world of Many has neither emerged from One of 1 and 2 nor from Many from 3.

A representative example of 1 is the Pure Monism of Vallabhacarya; that of 2 is the Monism of Sankaracarya; that of 3 is the Nyaya-Vaisesika system and other realist systems and that of 4 is the idealism of the Buddhists.

Looked at from another angle, we can say that the worldviews of the idealists are represented in 2 and 4, while 1 and 3 represent the realists' worldviews (Jha 2010a, b).

Language and Reality

In the Upanishadic period, we find statements which support the idealists' views as well as the realists' views. A closer reading reveals that for an idealist the reality cannot be captured by language and for a realist on the other hand language alone can reveal it.

A deeper analysis reveals that in an idealism like that of Sankara, there is no scope of accepting any relation as real since the reality is without a second and in realism a relation must be real because without that no structure can be formed and without a structure language cannot capture it. To illustrate, according to Sankara, Brahman is the only reality and It is devoid of all attributes. Thus, when there are no two entities, how can there be any relation which is real? Language, however, has no capacity to capture anything unless the thought reveals a structure of 'locus—relation—located'. Thus, if someone expresses as Brahman, it has been possible only because the speaker has identified Brahman-hood existing somewhere. Where the Brahman-hood exists alone can be called as Brahman, is the argument of the realists. Thus everything, right from a particle up to God can be captured by language. This is the claim of the realists (Jha 1990).

The Buddhists philosophers too, being idealists, maintained the same position that language cannot capture reality. Thus, for an idealist, reality is beyond mind and language (*a-vanmanasor gocaram*). Dharmakirti, the noted Buddhist logician had deep reflection on this issue and he must have realised that the main reason for the play of language is the role of relation which helps construction and imagination of structures. He therefore wrote an independent treatise on relation called Sambandha-pariksa. In this treatise, he dismisses the reality of all Relations which human beings use to construct the structures of the world of our experience. He made the refutation of realism so simple. For him, if one accepts the reality of a relation then he/she is a realist and if not then he/she is an idealist (Jha 1990).

It is this reason that there is no scope of any relation in Advaita philosophy and the realist systems like Nyaya cannot afford to dismiss any relation. That is why the Nyaya-Vaisesika system of philosophy calls the universe *padartha*—i.e. the referent of language. The universe for them is a grand total of referents of language (Annambhaṭṭa 1992). Bhartrhari explained the universe as a manifestation of the Sabda-brahman.

Outer and Inner World

With the background of what we have discussed so far, let us now look into the whole universe of our experience afresh. The universe of our experience is partly external and partly internal. While a tree is external, happiness is internal. There is awareness of both. Both are expressed in language and this is how one can know the existence of others' inner worlds also.

The Conscious and the Non-conscious

The universe can be classified into cetana (conscious) and a-cetana or jada (non-conscious). The conscious is one where the awareness arises and non-conscious is one where awareness cannot arise. Various models have been developed in classical India for classification of the universe.

Sankhya philosophers divided the universe into two categories called Purusa and *Prakṛti*. Purusa is always cetana and *Prakṛti* is always a-cetana. For all schools of Vedanta this Sankhya model was acceptable although all did not accept that both Purusa and *Prakṛti* are reals as accepted by the Sankhya philosophers (Hiriyanna 1993).

The idealist Vedantin like Sankaracarya did not accept that the world of plurality is real and the only reality for Sankara is the Brahman and the world of plurality is the projection of mind facilitated by a postulated entity called Maya. The Sankara Advaitins classify the universe as *drk* ('knowledge') and *drsya* ('object of knowledge') (Balasubrahmanyam and Revathi 2012).

The Nyaya-Vaisesika philosophers classified the referents of language into bhava 'positive' and abhava 'negative' entities. Positive entities were divided into six categories—substance, quality, action, universal, particular and inherence—and the negative entities were classified into two—relational absence and mutual absence. Relation absence was further divided into pre-absence, destruction and absolute absence. Substance was divided into nine: earth, water, fire, air, ether, time, space, soul and mind. Qualities were put into 24 divisions: colour, taste, smell, touch, number, measure, separateness, conjunction, disjunction, weight, remoteness, nearness, fluidity, moisture, sound, cognition, happiness, unhappiness, desire, hatred, volition, merit and demerit and impression. Actions were conceived as upward movement, downward movement, expanding, shrinking and any movement. Universal is accepted as a class-forming property. Particulars are posited to distinguish one permanent substance from another permanent substance. Inherence posited as a permanent relationship.

Here also, if we analyse closely, we can see that for the Indian logicians too there is an inner world and an outer world. In the inner world, there is a knower which is a substance and in whom a knowledge arises when the knower acquires a knowledge of the world. This knowledge then produces a desire in him or her; subsequently the desire produces volition and as a result he or she takes up an external activity. In him or her occurs pleasure or pain or aversion or merit or demerit or impression. The access to the external world is made possible by the entity called mind which is also a substance in this system through the external senses. The knower has access to the inner world through the mind. In other words, for knowing the external world a set of five external senses are accepted and for knowing the internal world the internal sense-organ called mind is postulated.

Thus, when the knower (self) comes in contact with the mind and mind comes in contact with the senses and the senses come in contact with the external world, the perception of the external world occurs and when the mind comes in contact with the self where the inner world is situated, the knowledge of the inner world emerges.

This self is nether the body, nor the senses, nor the mind, but a distinct entity called *atman*. These logicians accept that as many living beings exist, so many selves exist. Each self is provided with a mind as an inner instrument of knowing. This system also accepts one super self called God. God is supposed to be the creator this universe (Jha 2010a, b).

Udayanacarya wrote two treatises, one called *Atmatattva-viveka*,[2] to establish the reality of a distinct category of entity called self, and the other *Nyayamusumanjali*[3] to establish the reality of God.

Prasastapada,[4] the celebrated author of the commentary on the Vaisesika-sutras, offers the following argument to infer self.

The senses are the instruments of knowing. Someone is using these instruments to acquire knowledge and he or she is the self. As a carpenter uses his instruments to cut a log and he is distinct from the instruments, so the self is distinct from the senses.

The self cannot be the body or the senses or the mind because they are ignorant i.e. they cannot be the locus of knowledge. A body cannot be the locus of consciousness just like a jar cannot be since it is a product of Matter. Moreover, after death, the body remains but here is no consciousness in the body. The senses too cannot be the locus of consciousness because the senses are instruments which are used by the knower. Moreover, even after the damage of a sense-organ we find that the knower remembers the experience caused by that sense-organ. If the sense organs were indeed the knower, then after the damage of a particular sense-organ there would be no knower left and remembrance would not have been possible. Mind too cannot be the locus of the consciousness since it is a mere instrument to be used by the knower to acquire knowledge. If someone holds that there is no need of any other instrument to acquire knowledge then simultaneously there will be many cognitions and remembrances which is against our experience. Moreover, it is an established fact that the mind is an instrument to be used by the knower to acquire knowledge and how can an instrument itself become the knower?

Prasastapada offers some beautiful metaphors to understand the concept of self.

1. Self is like a charioteer and body is like a chariot. As the action in a chariot makes one know the charioteer so the action in the body makes one infer the self.
2. Self is like a blacksmith. As the blacksmith blows the blow-bag to heat the iron-ball, so the action of breathing makes one know the self who is breathing.
3. Self is like a puppet player. As a puppet player holds the string in his hands and makes the eye-lid close and open so also there is self who makes our eyes close and open.
4. Self is like a house-owner. As the owner of the house repairs his house if some part of it gets damaged in the same way if there is some cut in the body the self repairs it and the wound gets healed.
5. Self is like a boy playing with a marble. As the boy pushes the marble to get connected with a desired object so also the self connects mind to a particular sense-organ for acquiring the knowledge of the desired object.

[2]With English Translation by N.S. Dravid, Indian Institute of Advanced Study, Shimla. 1995.

[3]With Bengali Translation by Shrimohan Bhattacharya, 1995.

[4]With Kiranavali of Udayanacarya, Gaekwad Oriental Institue, Baroda, 1971.

6. Self is like an observer peeping through more than one window in a room having many windows. As one and the same person peeps through more than one window in a multi-windowed room, so one and the same self that experiences a tasty food, and later when he remembers it his tongue starts to water.

These are our daily experiences and how can these experiences be explained without postulating an entity called 'self'? Thus, according to Indian logicians, 'self' is different from the body, senses and the mind, and hence it is different from Matter.

The school of Purvamimamsa too holds the same view as that of the Indian logicians since for Purvamimamsa too this world of our experience is real.

Sabarasvamin, the commentator of Jaimini's Purvamimamsa-sutras, establishes self as distinct from Matter as follows (Jha 2001):

1. We can infer self on the basis of breathing, blinking of eyes, etc. We cannot say that the body is breathing since after death the body exists but breathing does not. So one who breathes is different from the body.
2. Self cannot be equated with pleasure, pain, etc. since pleasure or pain cannot be the answer to the question: 'whose is the pleasure or pain?' Therefore, pleasure is not the self but one who has pleasure is the self.
3. Self cannot be even a cognition, since knower cannot be the same as knowledge.
4. Self is known by the fact that one remembers and desires something that the same person had known earlier. This person is neither the cognition nor the desire, but the one who knows and desires. It cannot be the case that one will know x another will remember that x and yet another one will have desire of that x.
5. Therefore, self cannot be explained only by accepting momentary series of cognitions as suggested by the Buddhists.
6. The argument of the Buddhists is that remembrance is similar to momentary cognitions. Hence if it is the same series of remembrances, the remembrance of an earlier experience is possible and as such even if once the past experience has vanished the later remembrance is possible because the series is continued. This is also not tenable. Because, when one remembers, he or she says '"I" had experienced' and the same 'I' am now remembering. This is not possible in 'the series of cognition' theory of the Buddhists since the experiencer's cognition and the rememberer's cognition are not one and the same. Hence self is not momentary cognition or pleasure or pain or any skandha.
7. Moreover, the Brhadaranyakopanisad (6.5) clearly states that 'This self is eternal and not momentary'. Cognition is momentary and hence self is not cognition. It is a distinct category of entity.
8. It is true that it cannot be shown like a pot or a flower. But it can only be experienced. One can experience one's own self only and cannot know directly an other's self. Other's self can only be inferred. This self is eternal.

Uttaramimamsa or Vedanta too classifies the world into conscious and non-conscious.

In this way, all the six systems of Indian philosophy have divided the universe into two sets: *cetana* (conscious) and *a-cetana* or *jada* (non-conscious).

Nature of Self or *Atman*

The astika darsanas accept self as a distinct category of entity called *atman*. Jainism too accepts *atman*. The Carvakas (Sharma 1954) and the Buddhists do not accept *atman* as a distinct category of entity.

For Carvaka, the body of a living being is made of four elements—namely, earth, water, fire, air—and such a body itself is the self. It emerges from Matter and after death it goes back to the elements from which it is made. A section of the Carvakas called Lokayatikas hold the view that the senses, or mind or vital air either individually or a collection of these elements constitutes the self and after death this self goes back to its source elements. Different schools of Buddhist thoughts hold different positions on the nature of self. The Madhyamika school believes that a self is nothing but sunya or 'void'. The Vijnanavadins hold that self consists of a series of momentary cognitions. The Jaina philosophers do accept *atman* as a distinct entity but they say that this *atman* or self is of the size of one's body. The *atman* of an insect is of the size of the body of the insect and that of an elephant is of the size of that elephant. *Atman* is, therefore elastic in nature. It can be small or it can be big depending on the body of the living being.

Among the astika darsanas, the Sankhya system accepts self as purusa or cetana i.e. conscious. There are as many selves as there are living beings. Each self is distinct. The Yoga system too accept self as cetana and plurality like the Sankhya system. The Nyaya school believes that self is a substance in which alone cognition or consciousness can emerge and reside. It is of two types: individual self and the creator self called God. Both are conceived as locus of knowledge or cognition. While the cognition residing in an individual self is a product and hence non-eternal, that residing in God is eternal since it is not a product. This self is, however, different from the body, senses and the mind. In other words, the self is not cetana intrinsically but it alone can be *cetana*. God is always *cetana* since His knowledge is permanent and the substance called God is never without a cognition or knowledge.

The Vaisesikas hold the same view as that of the Nyaya. Vaisesikas did not have God at the beginning of their history although later they too accepted God when gradually both the systems merged into one called Nyaya-Vaisesika system of philosophy by the end of the ninth century AD.

The system of Purvamimamsa accepted self as a distinct category of entity. It is defined as 'a locus of consciousness'. This definition is similar to that of the Nyaya-Vaisesika system. The Mimamsakas also accept a plurality of selves like the Indian logicians (Sampurnananda 1987).

There are several systems of Uttaramimamsa or Vedanta and all of them accept self as intrinsically cetana or caitanya 'consciousness' unlike the Naiyayikas and the Mimamsakas. The Vedantins like Sankaracarya accept self to be only one while Madhvas and others accept plurality of selves. While for Sankara, self is one without a second, the Madhvas accept self to be of the size of an atom and hence a plurality of them exist. Same is the case with the Achintya-bhedaabheda school of Caitanya Mahaprabhu.

We all know that the schools of Vedanta emerged because of the interpretations of the Upanishads. The Upanishads have statements which can be interpreted as identity statements of individual self and the supreme self called the Brahman. There are also statements in the Upanishads which can be interpreted as statements of difference between an individual self and the supreme self. The Vedantins took a particular position and accordingly interpreted the Upanishads to build their own school of Vedanta. And this is how Monism of Sankara, qualified Monism of Ramanuja, Pure Monism of Vallabhacarya, Dvaita of Madhva, Bhedabheda of Nimbarka, Acintya Bhedabheda of Caitanya, etc. came into existence (Hiriyanna 1993).

Models of Understanding Consciousness

If we look into the data given above one can find that while some philosophical systems take consciousness itself to be the self, some others take the locus of consciousness to be the self. Thus, all the Vedanta systems of philosophy equate self with consciousness, the realist system like Nyaya and Mimamsa consider the locus of the consciousness to be the self. Thus, when we say self is cetana or 'conscious', it means that the self is the locus of consciousness (caitanya). One who has cognition, desire, volition, etc. is the self, and cognition, desire, volition, etc. are the various forms or manifestations of consciousness.

On the basis of the treatments of various philosophical schools of this phenomenon we may postulate the following models of understanding consciousness:

Model I: The Carvaka Model

According to this model, consciousness emerges out of Matter and at the time of death, it goes back to Matter, very much like the modern physical sciences. In support of this theory, the Carvakas quote the following sentence of the Upanishad:

'The consciousness emerge from these Matters and merges back in the same Matters, it does not go beyond Matter.' (quoted in the Nyayamanjari of Jayantabhatta).

Model II: The Sankhya Model (Hiriyanna 1993)

According to Sankhya, only the Purusa is *cetana* ('conscious') and the rest of the universe is *jada* ('non-conscious'). Purusa is the self. There are as many selves as there are living beings.

When a person sees an object, the *antahkarana* goes out through the sense-organ to reach the object and take the shape of that object called vrtti and the reflection of the self in that is what is called as consciousness. All the Vedanta systems borrowed this Model of Sankhya to provide epistemological explanation in their own systems of thought.

Model III: The Nyaya-Vaisesika Model (Jha 2012a, b)

The emphasis in this model is on language. Language makes a distinction between a noun and an adjective. Everybody says 'I am conscious' and no one says 'I am consciousness'. If we analyse the sentence 'I am conscious' it means 'I am the locus of consciousness' and never means 'I am consciousness'. Thus, the self has to be the locus of consciousness and not consciousness itself.

It implies that the self by nature is not conscious but becomes conscious when consciousness emerges in him or her. The consciousness is the quality of the self. The self is a distinct category of substance where alone consciousness like cognition, desire and volition can arise. Thus, although self is a substance, it is different from other substances like earth, water, light, air, ether, time and space. Although time is a substance, smell occurs only in the earth and not in any other substance. In the same way consciousness emerges only in the spiritual substance called the self.

Model IV: The Advaita Vedanata Model of Sankaracarya

In this model, there is only one self without a second and that is the only reality that exists. But we act and interact with this plural world. The plurality of this universe is nothing but the superimposition of plurality on a singular self. Thus plurality is an appearance and hence unreal. With this worldview, when one analyses an experience, one finds that 'consciousness' is like water in a pond and the senses are like the outlets of that pond. Thus, when we perceive something with our eyes, our self is delimited by mind, the mind goes out and takes the shape of the object of perception called *vrtti* and in that *vrtti*, when there is reflection of the self, we call it cognition or consciousness. Thus, the one and same self gets various forms due to the delimiters.

The distinctness of this model is this that here the self is consciousness itself. There is no locus-located relationship since the self is one and only one. At the level of behaviour, however, there is simply assumed differentiation.

Model V: The Bhakti Vedanta Model (Manohar 2012)

In this model, an individual self is said to be of atomic size and is also considered to be eternal and of the nature of knowledge. This size of the self is to cultivate humility. The God is the greatest of all and a devotee is just a particle of that infinite, all-pervading and omnipotent God. Here too the self is by nature *cetana*.

Model VI: The Buddhist Model

The Buddhists' postulated self as an aggregate of five skandhas. There is no durable self since everything is momentary in this system of thought.

Model VII: The Jaina Model

Here, the self is elastic in nature and hence it is of the size of a body. In an ant, the self is of the size of the body of the ant and in an elephant it is as big as the body of the elephant.

Nevertheless, the self in Jainism is of the nature of consciousness (Jha 2012a, b).

Conclusions

From all that we have discussed so far, it is clear that Indian thinkers right from the Vedic times have been engaged in knowing the truth of the universe. Their prime interest was to discover the true nature of the self and its relationship with the rest of the universe. It was soon realised that the self cannot be known in isolation and a holistic inquiry was an absolute necessity. They also agreed to respect diverse worldviews and enjoyed complete freedom of thought. The result was the development of diverse models of understanding consciousness by different systems of Indian philosophy. The results of these exercises, I am sure, will help in deriving deeper insights in understanding the true nature of the universe and thereby that of **Matter and consciousness**.

References

Annambhaṭṭa. (1992). *Tarka-sangraha: with English translation by Jivananda Vidyasagara Bhattacharya*. Chowkhamba Sanskrit Series Office.

Annambhatta. (2010). Tarkasangraha of Annambhatta—English Translation (pp. xiii–xxi), Chinmaya International Foundation Shodha Sansthan, Veliyanad, Ernakulam, Kerala.

Balasubrahmanyam, & Revathi. (2012). Vedantasangraha of Ramaraya Kavi, English Translation by Balasubrahmanyam and Revathi, Chinmaya International Foundation, Veliyanad, Kerala.

Ganganatha, J. (1964). Purva-mimamsa: In its sources.

Hiriyanna, M. (1993). *Outlines of Indian philosophy*. Motilal Banarsidass Publisher.

Jha, V. N. (Ed.). (1990). *Philosophy of relations* (No. 66). Sri Satguru Publications.

Jha, V. N. (2010a). *Language, grammar, and linguistics in Indian tradition*.

Jha, V. N. (2010b). *Tarkasangraha of Annambhatta (English Translation with Notes)*. Ernakulam, Kerala: Chinmaya International Foundation.

Jha, V. N. (2012a). Nyaya-Vaisesika Model of Understanding Consciousness. In U. Jha (Ed.), *Proceedings of Models of Understanding Consciousness*. Pune: Centre of Advanced Study in Sanskrit.

Jha, U. (2001). Marathi Translation of the Sabarabhasya: Vol. I Tarkapada, Centre of Advanced Study in Sanskrit, University of Pune.

Jha, U. (Ed.). (2012b). *Proceedings of Models of Understanding Consciousness*. Pune: Centre of Advanced Study in Sanskrit.

Manohar, S. (2012). Vallabh Vedant men cetana-bodh ke vividh pratiman ya star (Hindi). In U. Jha (Ed.), *Proceedings of Models of Understanding Consciousness.* Pune: Centre of Advanced Study in Sanskrit.

Panse, U. (1990). A reconstruction of the third school of Pūrvamīmāṃsā.

Raju, P. T. (1985). Structural depths of Indian thought. SUNY Press.

Sampurnananda. (1987). Mimamsakutuhalam of Kamalakarabhatta. Sanskrit University, Varanasi, pp. 46–55.

Sharma, U. S. (1954). Sarvadarsanasangraha, Hindi Translation, Chowkhamba Vidyabhavan, Varanasi.

Vidyabhusana, S. C. (1920). *A history of Indian logic: Ancient, mediaeval and modern schools.* Motilal Banarsidass Publisher.

Printed by Printforce, the Netherlands